MOLTEN SALTS AND IONIC LIQUIDS

MOLTEN SALTS AND IONIC LIQUIDS

Never the Twain?

Edited by

Marcelle Gaune-Escard
École Polytechnique
Marseille, France

Kenneth R. Seddon
QUILL, The Queen's University of Belfast
Belfast, United Kingdom

WILEY

A JOHN WILEY & SONS, INC., PUBLICATION

Copyright © 2010 by John Wiley & Sons, Inc. All rights reserved.

No part of this publication may be reproduced, stored in a retrieval system, or transmitted in any form or by any means, electronic, mechanical, photocopying, recording, scanning, or otherwise, except as permitted under Section 107 or 108 of the 1976 United States Copyright Act, without either the prior written permission of the Publisher, or authorization through payment of the appropriate per-copy fee to the Copyright Clearance Center, Inc., 222 Rosewood Drive, Danvers, MA 01923, 978-750-8400, fax 978-750-4470, or on the web at www.copyright.com. Requests to the Publisher for permission should be addressed to the Permissions Department, John Wiley & Sons, Inc., 111 River Street, Hoboken, NJ 07030, 201-748-6011, fax 201-748-6008, or online at http://www.wiley.com/go/permission.

Limit of Liability/Disclaimer of Warranty: While the publisher and author have used their best efforts in preparing this book, they make no representations or warranties with respect to the accuracy or completeness of the contents of this book and specifically disclaim any implied warranties of merchantability or fitness for a particular purpose. No warranty may be created or extended by sales representatives or written sales materials. The advice and strategies contained herein may not be suitable for your situation. You should consult with a professional where appropriate. Neither the publisher nor author shall be liable for any loss of profit or any other commercial damages, including but not limited to special, incidental, consequential, or other damages.

For general information on our other products and services or for technical support, please contact our Customer Care Department within the United States at 877-762-2974, outside the United States at 317-572-3993 or fax 317- 572-4002.

Wiley also publishes its books in a variety of electronic formats. Some content that appears in print may not be available in electronic formats. For more information about Wiley products, visit our web site at www.wiley.com.

Library of Congress Cataloging-in-Publication Data:

Molten salts and ionic liquids : never the Twain? / edited by Marcelle Gaune-Escard, Kenneth R. Seddon.
   p. cm.
  Includes bibliographical references and index.
  ISBN 978-0-471-77392-4 (cloth)
 1. Fused salts. 2. Ionic solutions. 3. Coulomb potential. I. Gaune-Escard, Marcelle. II. Seddon, Kenneth R., 1950-

  QD189.M596 2009
  546′ .34--dc22

                                                              2008047048

Printed in the United States of America
10 9 8 7 6 5 4 3 2 1

CONTENTS

Acknowledgements		ix
Preface		xi
Editorial		xiii
Contributors		xvii
1	**Ionic Liquids in the Temperature Range 150–1500 K: Patterns and Problems** C. Austen Angell	1
2	**Conductivities of Ionic Liquid Mixtures with Organic Electrolyte Solutions** D. Bansal, F. Croce, J. Swank, and M. Salomon	25
3	**How Hydrophilic Ionic Liquids Behave in Aqueous Solutions** Marijana Blesic, Kenneth R. Seddon, Natalia V. Plechkova, Nimal Gunaratne, António Lopes, and Luís Paulo N. Rebelo	37
4	**Mass Spectrometric Studies on Ionic Liquid Aggregates** R. Lopes da Silva, I. M. Marrucho, J. A. P. Coutinho, and Ana M. Fernandes	49
5	**Study of Sm–Al Alloy Formation in the Molten LiCl–KCl Eutectic** G. De Córdoba and C. Caravaca	63
6	**Alumina Solubility and Electrical Conductivity in Potassium Cryolites with Low Cryolite Ratio** Alexander Dedyukhin, Alexei Apisarov, Olga Tkatcheva, Yurii Zaikov, and Alexander Redkin	75
7	**Ionic Liquids as Solvents for the Variable Temperature Electrodeposition of Metals and Semiconductors: A Short Introduction** S. Zein El Abedin and F. Endres	85

CONTENTS

8 Predicting the Thermodynamic Behaviour of Water + Ionic Liquids Systems Using COSMO-RS — 101
   M. G. Freire, L. M. N. B. F. Santos, I. M. Marrucho, and J. A. P. Coutinho

9 Metallic Inert Anodes for Aluminium Electrolysis — 123
   I. Galasiu, R. Galasiu, and C. Nicolescu

10 The Behaviour of Phosphorus and Sulfur in Cryolite–Alumina Melts: Thermodynamic Considerations — 133
   I. Galasiu, R. Galasiu, C. Nicolescu, J. Thonstad, and G. M. Haarberg

11 Ionic Liquid–Ionic Liquid Biphasic Systems — 143
   Dirk Gerhard, Friedrich Fick, and Peter Wasserscheid

12 Recent Developments in the Reprocessing of Spent Fuel by Catalyst Enhanced Molten Salt Oxidation (CEMSO) — 151
   Trevor R. Griffiths, Vladimir A. Volkovich, and W. Robert Carper

13 Plasma-Induced Molten Salt Electrolysis to Form Functional Fine Particles — 169
   Yasuhiko Ito, Tokujiro Nishikiori, and Takuya Goto

14 Liquid Electrolytes: Their Characterisation, Investigation, and Diverse Applications — 181
   Keith E. Johnson

15 Protection of a Microstructured Molybdenum Reactor from High Temperature Oxidation by Electrochemical Deposition Coatings in Molten Salts — 193
   S. A. Kuznetsov, A. R. Dubrovskiy, S. V. Kuznetsova, E. V. Rebrov, M. J. M. Mies, M. H. J. M. de Croon, and J. C. Schouten

16 Molten Salt Synthesis of $LaAlO_3$ Powder at Low Temperatures — 219
   Zushu Li, Shaowei Zhang, and William Edward Lee

17 Accurate Measurement of Physicochemical Properties on Ionic Liquids and Molten Salts — 229
   V. M. B. Nunes, M. J. V. Lourenço, F. J. V. Santos, M. L. S. M. Lopes, and C. A. Nieto de Castro

18 Molten Salt Physics and Chemistry in the Current Development of Spent Nuclear Fuel Management — 265
   Toru Ogawa, Kazuo Minato, and Yasuo Arai

19	**An Organic Chemist's Perspective on High Temperature Molten Salts and Room Temperature Ionic Liquids** *Richard M. Pagni*	279
20	**Raman Spectroscopy of High Temperature Melts** *G. N. Papatheodorou, A. G. Kalampounias, and S. N. Yannopoulos*	301
21	**Thermodynamic Properties of LnI$_3$–MI Binary Systems (Ln = La or Nd; M = K, Rb, or Cs)** *Leszek Rycerz, M. F. Butman, and Marcelle Gaune-Escard*	341
22	**Materials Informatics for Molten Salts Chemistry** *Changwon Suh, Slobodan Gadzuric, Marcelle Gaune-Escard, and Krishna Rajan*	355
23	**A Novel Ionic Liquid-Polymer Electrolyte for the Advanced Lithium Ion Polymer Battery** *Daisuke Teramoto, Ryo Yokoyama, Hiroshi Kagawa, Tsutomu Sada, and Naoya Ogata*	367
24	**Solubility of Al$_2$O$_3$ in NaCl–KCl Based Molten Salt System** *Y. Xiao, C. R. Mambote, G. A. Wierink, and A. van Sandwijk*	389
25	**Molten Salt Synthesis of Ceramic Materials** *Shaowei Zhang, D. D. Jayaseelan, Zushu Li, and William Edward Lee*	397
26	**Fuel Cell and Electrolysis Studies with Dual Phase Proton and Oxide Ion Conduction** *Bin Zhu, S. Li, X. L. Sun, and J. C. Sun*	407
Index		419

ACKNOWLEDGEMENTS

Although our names appear on the title page as editors, the production of this volume would not have been possible without the superlative support we received. In particular, Joyce Bartolini was the conference secretary for the Tunisia meeting, and Deborah Poland was the editorial assistant for this volume. The energy, and meticulous attention to detail, of Joyce and Deborah, and their willingness to go that extra mile, have been invaluable, and are fully appreciated. Also without the financial support of the conference sponsors, the original meeting would not have happened, and so we gratefully acknowledge the generosity of: The European Association for Chemical and Molecular Sciences; The Electrochemical Society; US Army—Communications–Electronics Research Development and Engineering Center; US Army—International Technology Center Atlantic; The European Office of Aerospace Research and Development, Air Force Office of Scientific Research, United States Air Force Research Laboratory; US Navy—Office of Naval Research Global; and the International Council for Science: Committee on Data for Science and Technology.

MARCELLE GAUNE-ESCARD

KENNETH R. SEDDON

EUCHEM CONFERENCES ON MOLTEN SALTS

1966	Norway, Trondheim	Tormod Førland
1968	Italy, Pavia	Mario Rolla
1970	France, Trois Epis	Jean Brenet
1972	UK (England), Cirencester	Douglas Inman
1974	Germany, Freising	Alfred Klemm
1976	The Netherlands, Leiden	Jan A. A. Ketelaar
1978	Sweden, Lysekill	Arnold Lunden
1980	Israel, Dead Sea	Yizhak Markus
1982	France, La Gaillarde	Marcelle Gaune-Escard
1984	Denmark, Helsingør	Niels J. Bjerrum
1986	Norway, Geiranger	Harald Øye
1988	UK (Scotland), St. Andrews	John Duffy
1990	Greece, Patras	George Papatheodorou
1992	Belgium, De Haan	Bernard Gilbert
1994	Germany, Bad Herrenalb	Werner Freyland
1996	Slovakia, Smolenice	Vladimir Danek
1998	France, Porquerolles	Marcelle Gaune-Escard
2000	Denmark, Karrebæksminde	Niels J. Bjerrum
2002	UK (England), Oxford	Paul Madden
2004	Poland, Piechowice	Adolf Kisza

EUCHEM CONFERENCES ON MOLTEN SALTS AND IONIC LIQUIDS

2006	Tunisia, Hammamet	Marcelle Gaune-Escard
		Kenneth R. Seddon
2008	Denmark, Copenhagen	Rasmus Fehrmann
2010	Germany, Bamberg	Peter Wasserscheid
2012	UK (Wales), Newport	Kenneth R. Seddon
2014	Greece	Soghomon Boghosian

PREFACE

Between Sunday 17th September 2006 and Friday 22nd September 2006, a meeting of major significance occurred in Hammamet, Tunisia: the 21st meeting of a series of EuChem Conferences on Molten Salts (for a comprehensive list, see preceding page). Not only did the 21st meeting have a metaphorical significance—a coming of age—but it marked a change in focus and title for the meetings. No longer the EuChem Conferences on Molten Salts, they were relaunched, phoenix-like, as the EuChem Conferences on Molten Salts and Ionic Liquids. This was a result of a long series of private conversations with many key figures, which culminated in an editorial (reproduced on the following pages) in *Molten Salts Bulletin* (**2005,** *85*, 1–3). A gauntlet had been thrown down. To condense the arguments: Molten salts and ionic liquids are both liquid salts containing only ions—they are thus the same subject. All that is different is the temperature! Both fields involve the study of Coulombic fluids for academic and industrial purposes, both employ the same principles, both require skilled practitioners (but high temperature systems require greater skills), both speak the same language; all that is different is image, and how superficial is that? There is so much knowledge, both empirical and theoretical, which can be passed from the molten salt community to the ionic liquid community, and *vice versa*. The gauntlet was picked up, and the result was the Hammamet meeting, and this book. We believe this meeting has catalysed a *rapprochement* between the two communities: East has met West, the peaches have touched, the brooklets have mingled, and the Montagues and the Capulets are reunited. Both fields are now stronger for this, and the future meetings are assured to be the prime focus for renewing these exchanges.

Whilst representing some of the key talks at the meeting, this book also provides discussion on the transfer of experimental methods and techniques between the differing temperature régimes. While it is conventionally invidious to highlight specific chapters, the reviews from Austen Angell, Keith Johnson, Richard Pagni, and George Papatheodorou are remarkable for their depth of insight and breadth of pedagogic experience, while the chapter from Tsutomu Sada shows the way forward to the industrialisation of products based on this new technology. We believe this book will also act as a line in the sand—the contents are a snapshot of a strengthened field. The tears of the phoenix are said to contain the healing abilities of purity, and its cry is said to be a beautiful song. May it long continue to sing, and may we be aware enough to listen.

MARCELLE GAUNE-ESCARD

KENNETH R. SEDDON

EDITORIAL

NEVER THE TWAIN ?

Oh, East is East, and West is West, and never the twain shall meet,
Till Earth and Sky stand presently at God's great Judgment Seat;
But there is neither East nor West, Border, nor Breed, nor Birth;
When two strong men stand face to face,
tho' they come from the ends of the earth!
—Rudyard Kipling, *The Ballad of East and West* (1889)

Marcelle has asked me to write a personal editorial for *Molten Salts Bulletin,* and so I am taking her at her word. It will be personal, and hence it will be polemical. And I take as my subject the unfortunate, nay potentially disastrous, split which has emerged between the (high temperature) molten salt community (East) and the (low temperature) ionic liquid community (West). I will argue that this is destructive, cliquish, petty, and (if continuing) will result in the loss of key skills to the whole community. I do not want to wait until Judgement Day for the strong men (and women) of both camps to stand face to face!

I had the honour and privilege of being introduced to low-temperature molten salts, henceforth referred to as ionic liquids, by the fathers of the field–Chuck Hussey and John Wilkes. I was able to learn at the feet of the experts, in depth, the practical and theoretical aspects of ionic liquids, including much unpublished information. In other words, they taught me the tricks of the trade, with enthusiasm and generosity. It led to years of fruitful collaboration with Chuck, and my introduction into the molten salts community. This was dominated by the grandfathers of the field, Bob Osteryoung and Gleb Mamantov. Now, from my experience in the field of second-harmonic generation, whose practitioners were (by and large, but not universally) mean-minded, self-obsessed *prima donnas,* who treated newcomers as potential threats. I had steeled myself for a similar response from the molten salt community. Never had I been more wrong. I was immediately drawn to the core of the subject, and Bob and Gleb (intellectual giants) could not have been more welcoming to a tyro, and immensely generous with both their time and advice. And then there were the conferences: the Gordon Conferences, the Molten Salt Sessions at the Electrochemical Society meetings, the EUCHEM meetings, the Molten Salt series, and the Molten Salts Discussion Group (MSDG). And, as a zenith, the NATO ASI meeting in Camerino,

---

*From *Molten Salts Bulletin,* **85** 1–3 (2005).

two weeks in which all the key people in the field came together to teach the students and post-doctoral fellows. But it is now 2005, and I feel I could be writing an obituary for molten salts: the Gordon meetings were cancelled for lack of support, the Molten Salts meetings (the latest being MS7) will be lucky to have enough participants to be viable, the subject appears to be controlled by cliques, and the MSDG looks like a talking shop for the formerly active. And attending the lectures was intense *déjà vu* — lectures which contained little new content given by the same tired faces. And, most worrying, most of the highly skilled and imaginative practitioners (e.g., Øye, Angell, Freyland, Papatheodorou, Johnson, Gaune-Escard. Bjerrum, etc.) are approaching, or have passed, retirement age.

In contrast, the cheeky interloper, ionic liquids, which was lucky to get one session in ten in a molten salts meeting, which had very few practitioners, and which had no industrial track record, has gone from strength to strength. In June 2005, the first International Congress on Ionic Liquids (COIL) was held in Salzburg, with 410 participants; there have been three ten-session symposia at the ACS National meetings (San Diego, 2001; Boston, 2002; New York, 2003), each being the most attended sessions at their respective meeting; and a NATO ARW in Crete in 2000. The number of published papers on ionic liquids has risen from 20 in 1994 to over 1000 in 2004, and according to the independent ISI web site (http://www.esi-topics.com/ionic-liquids/index.html) there have been papers from 3342 authors in 57 countries, published in 290 journals, from 745 institutions. Tracking the open literature publications, patents in the area are also increasing exponentially, and industrial interest is burgeoning (BASF, DeGussa, Eastman Chemicals, and Iolitec already have commercial processes operating) and there are many large-scale suppliers (including Merck, Cytec, C-TRI, SACHEM, and BASF). Ionic liquids have captured the imaginations of the whole chemical (and, indeed, biochemical and physics) community, in a way that molten salts never did. But it is the same subject!! All that is different is the temperature! Both fields involve the study of Coulombie fluids for academic and industrial purposes; both employ the same principles; both require skilled practitioners (but high temperature systems require greater skills); both speak the same language (unlike the Americans and the British!!); all that is different is image, and how superficial is that?

So what has gone wrong – it is not clear when the twain stopped meeting, and why they are still separated. Maybe it is psychology, maybe personalities, maybe nomenclature. Marcelle—why is this the *Molten Salts Bulletin,* and not the *Ionic Liquids and Molten Salts Bulletin?* Why is MS7 not MS1L7? Why is MSDG not ILDG? It is not through lack of excellent and prominent people straddling the divide (viz. Johnson, Fehrmann, Boghosian. Angell. and Freyland (who gave a fascinating talk at COIL)). There is so much knowledge, both empirical and theoretical, which can be passed from the molten salt community to the ionic liquid community, and *vice versa.* WHY IS IT NOT HAPPENING? Any unbiased observer would note two immiscible groups, with a slightly fuzzy interface. I will lay my cards on the table—I want a homogeneous single phase. Am I a voice in the wilderness? I don't think so! I believe that the members of both communities for whom I have the greatest respect all feel the same way. Are we condemned to be Montagues and Capulets? Do we want to say

"Good night, good night! Parting is such sweet sorrow"? Is nomenclature going to keep us apart (*"What's in a name? That which we call a rose by any other name would smell as sweet"*)? Do we accept that "*What must be shall be*"? I would prefer to think that "*Such young unfurrowed souls roll to meet each other like two velvet peaches that touch softly and are at rest; they mingle as easily as two brooklets that ask for nothing but to entwine themselves and ripple with ever-interlacing curves in the leafiest hiding-places.*" (George Eliot, *Adam Bede*. 1859). So, I implore you all, reverse history, and enjoy the fruits of remingling. I suggest we seek funding from EUCHEM to bring about a joint meeting to map the way forward: Marcelle – that is one with your name on it!

Kenneth R. Seddon

QUILL
The Queen's University of Belfast,
BT9 SAG Belfast, Northern Ireland, UK

k.seddon@qub.ac.uk

Hoc erat in votis. Let us attempt to merge East and West together Ken. And let it be in 2006.

Marcelle

CONTRIBUTORS

C. Austen Angell, Department of Chemistry and Biochemistry, Arizona State University, Tempe, Arizona 85287, USA

Alexei Apisarov, Institute of High Temperature Electrochemistry, S. Kovalevskaya, 22, Yekaterinburg, 620219 Russia

Yasuo Arai, Japan Atomic Energy Agency, Tokai-mura, Naka-gun, Ibaraki-ken, 319-1195 Japan

D. Bansal, MaxPower, Inc., 141 Christopher Lane, Harleysville, Pennsylvania 19438, USA

Marijana Blesic, Instituto de Tecnologia Química e Biológica, ITQB 2, Universidade Nova de Lisboa, Apartado 127, 2780-901 Oeiras, Portugal *and* The QUILL Centre, The Queen's University of Belfast, Stranmillis Road, Belfast BT9 5AG, United Kingdom

M. F. Butman, Ivanovo State University of Chemical Technology, Ivanovo, Russia, *and* École Polytechnique, Mecanique Energetique, Technopôle de Chateau-Gombert, 5 rue Enrico Fermi, 13453 Marseille Cedex 13, France

W. Robert Carper, Department of Chemistry, Wichita State University, Wichita, Kansas 67260-0051, USA

C. Caravaca, CIEMAT, Nuclear Fission Division/Radioactive Waste Unit, Avda. Complutense 22, 28040 Madrid, Spain

J. A. P. Coutinho, CICECO, Departamento de Química, Universidade de Aveiro, 3810-193 Aveiro, Portugal

F. Croce, Dipartamento di Scienze del Farmaco, Universital Degli Studi 'G D' Annunzio, Via Dei Vestini 31, I-66013 Chieti, Italy

R. Lopes da Silva, Departamento de Química, Universidade de Aveiro, Aveiro, Portugal

C. A. Nieto de Castro, Centro de Ciências Moleculares e Materiais, Faculdade de Ciências, Universidade de Lisboa, 1749-016 Lisboa, Portugal, *and* Departamento de Química e Bioquímica, Faculdade de Ciências, Universidade de Lisboa, 1749-016 Lisboa, Portugal

G. De Córdoba, CIEMAT, Nuclear Fission Division/Radioactive Waste Unit, Avda. Complutense 22, 28040 Madrid, Spain

M. H. J. M. DE CROON, Laboratory of Chemical Reactor Engineering, Eindhoven University of Technology, P.O. Box 513, 5600 MB Eindhoven, The Netherlands

ALEXANDER DEDYUKHIN, Institute of High Temperature Electrochemistry, S. Kovalevskaya, 22, Yekaterinburg, 620219 Russia

A. R. DUBROVSKIY, Institute of Chemistry, Kola Science Centre RAS, Fersman St. 14, 184209, Apatity, Murmansk Region, Russia

F. ENDRES, Institute of Particle Technology, Clausthal University of Technology, Robert-Koch-St. 42, 38678 Clausthal-Zellerfeld, Germany

ANA M. FERNANDES, QOPNA, Departamento de Química, Universidade de Aveiro, 3810–193 Aveiro, Portugal

FRIEDRICH FICK, Lehrstuhl für Chemische Reaktionstechnik, Universität Erlangen-Nürnberg, Egerlandstr. 3, D-91058 Erlangen, Germany

M. G. FREIRE, CICECO, Departamento de Química, Universidade de Aveiro, 3810-193 Aveiro, Portugal

SLOBODAN GADZURIC, École Polytechnique, IUSTI CNRS 6595, Technopôle de Château-Gombert, 5 rue Enrico Fermi, 13453 Marseille Cedex 13, France *and* Faculty of Science, Department of Chemistry, University of Novi Sad, Trg. D. Obradovica 3, 21000 Novi Sad, Serbia

I. GALASIU, Institute of Physical Chemistry, "I.G. Murgulescu" of The Romanian Academy, Bucharest, Romania

R. GALASIU, Institute of Physical Chemistry, "I.G. Murgulescu" of The Romanian Academy, Bucharest, Romania

MARCELLE GAUNE-ESCARD, École Polytechnique, IUSTI CNRS 6595, Technopôle de Château-Gombert, 5 rue Enrico Fermi, 13453 Marseille Cedex 13, France

DIRK GERHARD, Lehrstuhl für Chemische Reaktionstechnik, Universität Erlangen-Nürnberg, Egerlandstr. 3, D-91058 Erlangen, Germany

TAKUYA GOTO, Department of Fundamental Energy Science, Graduate School of Energy Science, Kyoto University, Sakyo-ku, Kyoto 606-8501, Japan

TREVOR R. GRIFFITHS, Redston Trevor Consulting Limited, Leeds, LS17 8RF, United Kingdom

NIMAL GUNARATNE, The QUILL Centre, The Queen's University of Belfast, Stranmillis Road, Belfast BT9 5AG, United Kingdom

G. M. HAARBERG, Department of Material Sciences Engineering, Norwegian University of Science and Technology, 7491 Trondheim, Norway

YASUHIKO ITO, Department of Environmental Systems Science, Faculty of Science and Engineering, Doshisha University, Kyotanabe, Kyoto 610-0321, Japan

D. D. JAYASEELAN, Department of Materials, Imperial College, London SW7 2AZ, United Kingdom

KEITH E. JOHNSON, Department of Chemistry and Centre for Studies in Energy and Environment, University of Regina, Regina, Saskatchewan, Canada S4S 7J7

HIROSHI KAGAWA, Pionics Co., Ltd., RITS BKC-TC-9, 1-1-1 Noji-Higashi, Kusatsu, Shiga 525-8577, Japan

A. G. KALAMPOUNIAS, Foundation for Research and Technology Hellas – Institute of Chemical Engineering and High Temperature Chemical Processes, FORTH/ICE-HT, P.O. Box 1414, GR-26504 Rio, Greece, *and* Department of Chemical Engineering, University of Patras, GR-26504, Patras, Greece

S. A. KUZNETSOV, Institute of Chemistry, Kola Science Centre RAS, Fersman St. 14, 184209 Apatity, Murmansk Region, Russia

S. V. KUZNETSOVA, Institute of Chemistry, Kola Science Centre RAS, Fersman St. 14, 184209 Apatity, Murmansk Region, Russia

WILLIAM EDWARD LEE, Department of Materials, Imperial College London, London SW7 2AZ, United Kingdom

S. LI, Institute of Materials and Technology, Dalian Maritime University, Dalian 11606, People's Republic of China

ZUSHU LI, Department of Materials, Imperial College London, London SW7 2AZ, United Kingdom

ANTÓNIO LOPES, Instituto de Tecnologia Química e Biológica, ITQB 2, Universidade Nova de Lisboa, Apartado 127, 2780-901 Oeiras, Portugal

M. L. S. M. LOPES, Centro de Ciências Moleculares e Materiais, Faculdade de Ciências, Universidade de Lisboa, 1749-016 Lisboa, Portugal *and* Departamento de Química e Bioquímica, Faculdade de Ciências, Universidade de Lisboa, 1749-016 Lisboa, Portugal

M. J. V. LOURENÇO, Centro de Ciências Moleculares e Materiais, Faculdade de Ciências, Universidade de Lisboa, 1749-016 Lisboa, Portugal *and* Departamento de Química e Bioquímica, Faculdade de Ciências, Universidade de Lisboa, 1749-016 Lisboa, Portugal

C. R. MAMBOTE, Metals Production, Refining and Recycling, Department of Materials Science and Engineering, Delft University of Technology, Mekelweg 2, 2628 CD, Delft, The Netherlands

I. M. MARRUCHO, CICECO, Departamento de Química, Universidade de Aveiro, 3810-193 Aveiro, Portugal

M. J. M. MIES, Laboratory of Chemical Reactor Engineering, Eindhoven University of Technology, P.O. Box 513, 5600 MB Eindhoven, The Netherlands

KAZUO MINATO, Japan Atomic Energy Agency, Tokai-mura, Naka-gun, Ibaraki-ken, 319-1195 Japan

C. NICOLESCU, Faculty of Industrial Chemistry, University Polytechnica of Bucharest, Bucharest, Romania

TOKUJIRO NISHIKIORI, I'MSEP Co. Ltd., 134 Chudoji Minamimachi, Shimogyo, Kyoto 600-8813, Japan

V. M. B. NUNES, Escola Superior de Tecnologia, Instituto Politécnico de Tomar, Campus da Quinta do Contador, 2300 - 313 Tomar, Portugal *and* Centro de Ciências Moleculares e Materiais, Faculdade de Ciências, Universidade de Lisboa, 1749-016 Lisboa, Portugal

NAOYA OGATA, Tokyo and Chitose Institute of Science and Technology, Chitose, Japan

TORU OGAWA, Japan Atomic Energy Agency, Tokai-mura, Naka-gun, Ibaraki-ken, 319-1195 Japan

NATALIA V. PLECHKOVA, The QUILL Centre, The Queen's University of Belfast, Stranmillis Road, Belfast BT9 5AG, United Kingdom

RICHARD M. PAGNI, Department of Chemistry, University of Tennessee, Knoxville, Tennessee 37996, USA

G. N. PAPATHEODOROU, Foundation for Research and Technology Hellas – Institute of Chemical Engineering and High Temperature Chemical Processes, FORTH/ICE-HT, P.O. Box 1414, GR-26504 Rio, Greece *and* Department of Chemical Engineering, University of Patras, GR-26504, Patras, Greece

KRISHNA RAJAN, Combinatorial Sciences and Materials Informatics Collaboratory (CoSMIC), NSF International Materials Institute, Department of Materials Science and Engineering, Iowa State University, Ames, IA 50011, USA

LUÍS PAULO N. REBELO, Instituto de Tecnologia Química e Biológica, ITQB 2, Universidade Nova de Lisboa, Apartado 127, 2780-901 Oeiras, Portugal

E. V. REBROV, Laboratory of Chemical Reactor Engineering, Eindhoven University of Technology, P.O. Box 513, 5600 MB Eindhoven, The Netherlands

ALEXANDER REDKIN, Institute of High Temperature Electrochemistry, S. Kovalevskaya, 22, Yekaterinburg, 620219 Russia

LESZEK RYCERZ, Chemical Metallurgy Group, Faculty of Chemistry, Wroclaw University of Technology, Wybrzeze Wyspianskiego 27, 50-370 Wroclaw, Poland, *and* École Polytechnique, Mecanique Energetique, Technopôle de Chateau-Gombert, 5 rue Enrico Fermi, 13453 Marseille Cedex 13, France

TSUTOMU SADA, Pionics Co., Ltd., RITS BKC-TC-9, 1-1-1 Noji-Higashi, Kusatsu, Shiga 525-8577, Japan *and* Trekion Co., Ltd., 5-2-1 Aoyama, Otsu, Shiga 520-2101, Japan

M. SALOMON, MaxPower, Inc., 141 Christopher Lane, Harleysville, Pennsylvania 19438, USA

F. J. V. SANTOS, Centro de Ciências Moleculares e Materiais, Faculdade de Ciências, Universidade de Lisboa, 1749-016 Lisboa, Portugal, *and* Departamento de Química e Bioquímica, Faculdade de Ciências, Universidade de Lisboa, 1749-016 Lisboa, Portugal

L. M. N. B. F. SANTOS, CIQ, Departamento de Química, Faculdade de Ciências da Universidade do Porto, R. Campo Alegre 687, 4169-007, Porto, Portugal

J. C. SCHOUTEN, Laboratory of Chemical Reactor Engineering, Eindhoven University of Technology, P.O. Box 513, 5600 MB Eindhoven, The Netherlands

KENNETH R. SEDDON, Instituto de Tecnologia Química e Biológica, ITQB 2, Universidade Nova de Lisboa, Apartado 127, 2780-901 Oeiras, Portugal *and* The QUILL Centre, The Queen's University of Belfast, Stranmillis Road, Belfast BT9 5AG, United Kingdom

CHANGWON SUH, Combinatorial Sciences and Materials Informatics Collaboratory (CoSMIC), NSF International Materials Institute, Department of Materials Science and Engineering, Iowa State University Ames, IA 50011, USA

J. C. SUN, Institute of Materials and Technology, Dalian Maritime University, Dalian 11606, People's Republic of China

X. L. SUN, Institute of Materials and Technology, Dalian Maritime University, Dalian 11606 , People's Republic of China

J. SWANK, U.S. Army Research Laboratory, AMSRD-ARL-SE-DC, 2800 Powder Mill Road, Adelphi, Maryland 20783-1197, USA

DAISUKE TERAMOTO, Pionics Co., Ltd., RITS BKC-TC-9, 1-1-1 Noji-Higashi, Kusatsu, Shiga 525-8577, Japan

J. THONSTAD, Department of Material Sciences Engineering, Norwegian University of Science and Technology, 7491 Trondheim, Norway

OLGA TKATCHEVA, Institute of High Temperature Electrochemistry, S. Kovalevskaya, 22, Yekaterinburg, 620219, Russia

A. VAN SANDWIJK, Metals Production, Refining and Recycling, Department of Materials Science and Engineering, Delft University of Technology, Mekelweg 2, 2628 CD, Delft, The Netherlands

VLADIMIR A. VOLKOVICH, Department of Rare Metals, Ural State Technical University–UPI, Ekaterinburg 620002, Russia

PETER WASSERSCHEID, Lehrstuhl für Chemische Reaktionstechnik, Universität Erlangen-Nürnberg, Egerlandstr. 3, D-91058 Erlangen, Germany

G. A. WIERINK, Metals Production, Refining and Recycling, Department of Materials Science and Engineering, Delft University of Technology, Mekelweg 2, 2628 CD, Delft, The Netherlands

Y. XIAO, Metals Production, Refining and Recycling, Department of Materials Science and Engineering, Delft University of Technology, Mekelweg 2, 2628 CD, Delft, The Netherlands

S. N. YANNOPOULOS, Foundation for Research and Technology Hellas – Institute of Chemical Engineering and High Temperature Chemical Processes, FORTH/ICE-HT, P.O. Box 1414, GR-26504 Rio, Greece

RYO YOKOYAMA, Trekion Co., Ltd., 5-2-1 Aoyama, Otsu, Shiga 520-2101, Japan

YURII ZAIKOV, Institute of High Temperature Electrochemistry, S. Kovalevskaya, 22, Yekaterinburg, 620219 Russia

S. ZEIN EL ABEDIN, Electrochemistry and Corrosion Laboratory, National Research Centre, Dokki, Cairo, Egypt

SHAOWEI ZHANG, Department of Engineering Materials, University of Sheffield, Sheffield S1 3JD, United Kingdom

BIN ZHU, Department of Energy Technology, Royal Institute of Technology (KTH), S-100 44 Stockholm, Sweden *and* Institute of Materials and Technology, Dalian Maritime University, Dalian 11606, People's Republic of China

1 Ionic Liquids in the Temperature Range 150–1500 K: Patterns and Problems

C. AUSTEN ANGELL

Department of Chemistry and Biochemistry, Arizona State University, Tempe, Arizona

Abstract

In this chapter we trace some of the developments in ionic liquids physical chemistry from the crimson metal–molten salts of Humphrey Davy's original melt electrolysis discoveries of elements, to the newest ambient-to-200 °C fuel cell electrolytes. We discuss the origin of a common pattern of behaviour for ionic liquids as they are cooled down, often reaching the glassy state—a pattern that extends from the liquid orthosilicates of the earth's mantle to the subambient temperature ionic liquids of the present era. However, where glasses are rare among orthosilicates and are generally limited to the complex anion "polymerised" silicate liquids (Bockris's "island anions"), they are the general result of cooling the ambient ionic liquids of current interest, even though the anions are always simple. How to maintain high fluidity along with low vapour pressure is the challenge. The highest ambient temperature fluidities are obtained with ionic liquids of the proton transfer variety, and they rival aqueous solutions in conductivity—a consequence of optimising Coulomb cohesion and charge carrier concentration in the absence of solvent. These protic ionic liquids (PILs) show much promise as fuel cell electrolytes and also as biopreservatives. We illustrate the great range of electrolyte character that can be obtained with PILs, using an energy level diagram for proton transfers, and then compare the fuel cell performances of electrolytes of organic cations with that of their inorganic cation cousins.

Molten Salts and Ionic Liquids: Never the Twain? Edited by Marcelle Gaune-Escard and Kenneth R. Seddon
Copyright © 2010 John Wiley & Sons, Inc.

1.1 INTRODUCTION

It is now a little over 200 years since Sir Humphrey Davy, in England, first saw the deep crimson colour of sodium metal dissolved in molten sodium chloride [1]. He was in the process of using electrochemically driven decomposition to produce, for the first time, a free alkali metal element and was observing the consequence of some dissolution of the liberated metal in the liquid salt. This followed Volta's invention of the voltaic "pile" that provided the driving force for the decomposition, by only a few years (which this author finds remarkable).

We select three particular themes to highlight as we follow developments in understanding of ionic liquids from Davy's time to the present era. The themes are (1) ionic liquids that are coloured as a result of particular light-absorbing structural features, (2) ionic liquids providing interconversion of chemical and electrical forms of energy, and (3) ionic liquids transforming to glassy solids in consequence of failure to crystallise. Presently, the low temperature versions of Davy's liquids are being found useful in fuel cells that permit the reverse of Davy's process—namely, the release of electrical energy by conversion of elements (hydrogen and oxygen) to the chemically combined state (water). The narrative will be brief and sweeping and will be strongly and transparently stamped by the author's personal experience over a fifty year period of research in ionic liquids. It is by no means intended to be an account of the field of ionic liquids as a whole.

When the writer was a graduate student in chemical metallurgy at Melbourne University in the 1950s, the understanding of silicate liquids of metallurgical relevance was in a primitive state. Although molten calcium magnesium silicates had long been used in the production of iron by a chemical reduction process (using carbon as the source of electrons to liberate the metal from its combined state), the literature of the time mostly referred to the high temperature liquids in terms of *molecules* of the chemical compounds from which they were formed. Thus the furnace slag contained MgO molecules, CaO molecules, and SiO_2 molecules. In a radical departure from the thinking of the time, John Bockris at Imperial College had recently [2, 3] launched the idea that these highly fluid materials actually consisted of free cations of the alkaline earth elements moving in a sea of silicate anions of varying complexity. The writer's M.Sc. supervisor, thermodynamicist Mervyn Willis [4, 5], liked the idea and encouraged him to accept a fellowship in Bockris's new laboratory in the United States at the University of Pennsylvania. At that time, a subfield of Bockris's "ionic liquids"—one that was designated "molten salts" (because they were more simply constituted than the molten silicates)—was under intense investigation. This was because of the molten-salt-cooled nuclear reactor project that was then under development at Oak Ridge National Laboratory. The reactor was very successfully using molten alkali fluorides and fluoroberyllates as the medium for heat transfer between reactor and turbine. (After highly successful demonstrations, this project was moth-balled in the 1960s as doubts about the future of nuclear energy arose, but it may well be resurrected in the present climate of crisis consequent on the threat posed by carbon energy based global warming.)

Harry Bloom, from the University of Hobart in Tasmania, Australia, was a visiting scientist in Bockris's laboratory. As a student at Melbourne University, Bloom had studied complex anions in molten salts [6] under displaced German physical chemist Erich Heymann, who had come from the early German school of molten salt chemists. The problem of identifying complex anions (like $[CdCl_4]^{2-}$ that, in aqueous solutions, moved toward the anode in an electric field) in a medium containing no solvent was one of the many intellectual basketballs constantly in the air in Bockris's lab. My assignment was to study diffusion of the individual ions in the complex-containing ($CdCl_2$ + KCl) system [7].

In contrast to the field of metallurgical silicate liquids, molten salt studies had been relatively advanced since the early part of the 20th century. The German school had studied a great many lower-melting systems. They had already shown that the Walden rule, Eq. (1.1) [8], that reliably related the equivalent conductivity Λ of aqueous solutions to their viscosities, η,

$$\Lambda \eta = \text{constant} \qquad (1.1)$$

broke down seriously for molten silver halides and needed to be replaced by the "fractional" Walden rule:

$$\Lambda \eta^\gamma = \text{constant} \qquad (1.2)$$

where γ is a constant $0 < \gamma < 1$.

The fractional Walden rule, which implied that the Arrhenius activation energy for conductivity was lower than that for viscosity, found a natural interpretation in terms of one of the ionic species ions being smaller than the other, hence capable of squeezing through smaller gaps in the condensed phase structure. This decoupling of the ionic motions reached an extreme in the case of silver salts of large polarisable anions. These were later to become important components of the "superionic" family of ionic glasses, and thus the stage was already set for the understanding of these materials, which would get much research attention in the second half of the century as electrolytes for solid state batteries. Success in this respect required a breakdown of Eq. (1.1) by some 12 orders of magnitude [9]!

Interestingly enough, the original Walden rule finds a new lease on life in current studies of ionic liquids of the ambient temperature variety, where the Walden plot, in the form log (Λ) versus log (η^{-1}) is being used as a classification diagram to distinguish "normal" ionic liquids from "poor" ionic liquids, on the one hand, and "superionic" liquids on the other hand [10, 11]. An example will be seen at the end of this chapter. The search for "superprotonic" electrolytes for fuel cell applications is in full swing at this time. Superprotonics are the holy grail of the fuel cell discipline and will briefly be considered later.

The study of complex anions was very pertinent to the Oak Ridge National Laboratory nuclear reactor program, which had also benefited from the talents of prewar displaced German physical chemists. At Oak Ridge, people were studying the "weak field" analogues of the "basic" $MgO-SiO_2$ metallurgical slags, represented by solutions of alkali fluorides and the Lewis acid BeF_2 (all ion charges halved relative to the $MgO-SiO_2$ melts). Complexation of the fluoride ions [12], to yield tetrahedral

$[BeF_4]^{2-}$ and $[Be_2F_7]^{3-}$ anions, reduced the Coulomb energy of the liquid, lowering liquidus temperatures and also lowering viscosity [13] to give fluids with the excellent heat transfer characteristics needed for a successful nuclear reactor program.

The high degree of analogy to silicate systems presented by the $LiF-BeF_2$ system was not of great interest to Oak Ridge scientists, but became the centre of attention in the East German laboratory of Vogel. Vogel et al. [14] not only determined the phase diagram but showed that, in solutions rich in BeF_2, the glass-forming solutions underwent a liquid–liquid phase separation [15] during cooling, similar to that which had been identified in silica-rich Li_2O-SiO_2 glasses.

A further analogy was waiting to be made, using molten chlorides closer in nature to those with which the field is currently so involved. With $ZnCl_2$ as the network glass former and pyridinium chloride as the network breaker (or "modifier"), Easteal and Angell [16] demonstrated that the glass-forming properties of the binary chloride system were highly analogous to those of the Na_2O-SiO_2 system.

While complex anions that could be enlarged into glass-forming networks were clearly of interest, complex cations were not to be neglected. The rare earth analogues of Humphrey Davy's crimson alkali metal–alkali halide solutions were then under study at Imperial College in the group of John Bockris's former student, John W. Tomlinson. The writer had the privilege of working under Tomlinson as the Imperial College Stanley Armstrong fellow and studied the deep scarlet solution formed by the solution of metallic cadmium in $CdCl_2$ [17]. Rather than forming the liquid equivalent of blue-purple alkali chloride crystal f-centres (lattice vacancies containing an excess electron), which was Gruen and co-workers' [18] interpretation of Humphrey Davy's crimson alkali halide solutions, the excess cadmium in $CdCl_2$ appeared to form metal dimers, $[Cd_2]^{2+}$.

1.2 VISCOSITY AND GLASS-FORMER PHENOMENOLOGY IN "SIMPLE" IONIC LIQUIDS

While it is now common knowledge that ambient temperature ionic liquids of current interest frequently supercool and vitrify, there was a time when most inorganic ionic liquid researchers would consider the formation of ionic glasses, in the absence of covalently bound networks, to be impossible.

The first evidence to the contrary was provided by Russian researchers engaged in the study of binary phase diagrams [19], who reported that in some simple salt systems there were liquid compositions that failed to crystallise during normal cooling. A useful survey of (alkaline earth + alkali metal nitrate) solutions by the German group of Thilo [20] showed this was systematically related to the relative sizes of monovalent and divalent cations and was connected to the existence in favourable systems, of very low eutectic temperatures.

The observations by Thilo et al. [20] were preceded by Dietzel and Poegel [21], who carried out a detailed study of crystallisation rates and glass formation in the now classic $(Ca(NO_3)_2 + KNO_3)$ system. They found that the composition 38% $Ca(NO_3)_2$ was particularly resistant to crystallisation, although it was well on the $Ca((NO_3)_2$-rich side of the eutectic composition. The 40:60 mole % composition in this system (now affectionately known as CKN) has since become one of the most thoroughly investigated of all ionic liquid systems and has been the test case for sophisticated new experimental techniques. For instance, the neutron spin-echo technique [22], by means of which the intermediate scattering function needed for testing the Götze mode-coupling theory of glass-forming liquids [23] was first determined, employed CKN as the test liquid.

The finding that glasses can be formed in systems as simply constituted as CKN (two cations each with the argon electronic structure, and a small triangular anion) provided the motivation for most of this writer's researches in the physical chemistry of liquids [24].

The especially simple behaviour of the CKN system is illustrated [25, 26] by the way that a spread of precise conductivity data covering several orders of magnitude, obtained when composition and temperature are both varied, can be collapsed onto a single straight line when the data are treated using the relation now known outside the ionic liquid field as the Vogel–Fulcher–Tammann (VFT) [27–29] equation (after the independent (1921–1926) works of early authors who used it to describe the temperature dependence of viscosity for many single-component [29], as well as complex, silicate mixture compositions [28]). For the most typical glass-former transport property, the viscosity, this is expressed by:

$$\eta = \eta_0 \exp(B/[T-T_0]) \qquad (1.3)$$

where η_0, B, and T_0 are constants, and T_0 varies linearly with composition. (In the present author's opinion, Eq. (1.3) should be called the Tammann–Fulcher equation to recognise the weight of Tammann's contribution. After all, Vogel never studied a supercooled liquid, and the equation used by him to describe some liquid viscosities was not of the form of Eq. (1.2), but instead a much less suggestive mathematical equivalent. The idiosyncratic [30] designation, VFT equation, that the writer introduced in 1972 [31]—and which is now widespread in use—was a protest against the chronological convention, and an attempt to credit the superior understanding demonstrated in Tammann's writings of the equation's implications. T_0, of course, represents the temperature at which the viscosity would diverge on extended supercooling. It distinguishes the equation from the Arrhenius law obtained when $T_0 \to 0\,\text{K}$, and implies the existence of some sort of transition that terminates the liquid state. The nature of this transition, and even its existence, has been the focus of unresolved argument among glass theorists for many decades.

The simple relation of the parameter T_0 to the experimental glass transition temperature T_g is shown in Figure 1.2. When the linear increase of T_0 is combined with Eq. (1.3), a simple relation for the conductivity at any temperature and composition is obtained [25]. This permits the data of Figure 1.1a to be collapsed to the single straight line shown in Figure 1.1b.

If B in Eq. (1.3) is replaced by the product DT_0, or T_0/F, as in Eq. (1.4a) and (1.4b),

$$\eta = \eta_0 \exp(DT_0/[T-T_0]) \qquad (1.4a)$$

$$= \eta_0 \exp(T_0/F[T-T_0]) \qquad (1.4b)$$

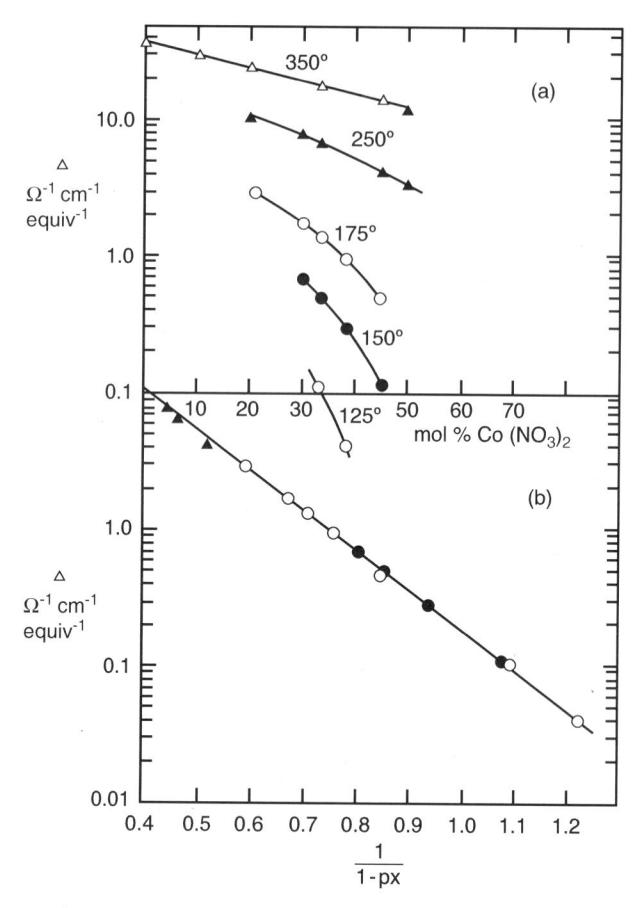

6   IONIC LIQUIDS IN THE TEMPERATURE RANGE 150–1500 K

FIGURE 1.1 Simple behaviour of the physical properties of solutions in the binary system (KNO$_3$ + Ca(NO$_3$)$_2$), which is glass forming in the range 33–45% Ca(NO$_3$)$_2$. (a) Isotherms of conductivity versus mole % Ca(NO$_3$)$_2$. (b) Collapse of the data for part (a) to a single line using an equation derived from Eq. (1.3) on the assumption that T_0 is linear in composition (as demonstrated in Fig. 1.2) (From Angell [25, 26] with permission.)

then the values of the parameters D and F quantify the deviation from Arrhenius behaviour that the data exhibit. This deviation has become known as the fragility of the liquid [32] and can be represented numerically by the parameter F, although a number of other metrics are more commonly used [33]. The calcium–potassium nitrate system exhibits one of the largest values of F that has been observed (though the increase of the Arrhenius slope with decreasing temperature appears to saturate above T_g [34]). Since the dramatic breakdown of the familiar Arrhenius law is one of the most striking features of glass-forming liquids, CKN has attracted a lot of attention from investigators of the physics of glass-forming liquids. When F is large, the value of T_0 lies close to the experimental glass transition temperature, as can be seen in

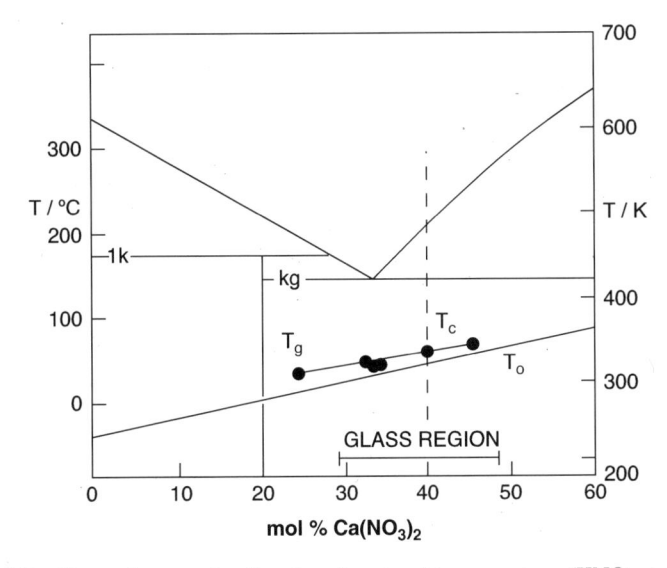

FIGURE 1.2 Phase diagram for the glass-forming binary system (KNO$_3$ + Ca(NO$_3$)$_2$), showing the relationship of the parameters T_0 of Eq. (1.3) to experimental glass temperatures, T_g, and composition. (From Angell [25] with permission.)

Figure 1.2, which shows the relation of each of T_g and T_0 to the phase diagram for this system [25].

1.3 STRONG AND FRAGILE IONIC LIQUIDS

By using the experimentally determined glass transition temperature T_g as a parameter to scale the temperature in an Arrhenius plot, the data from many different types of liquids can be placed on the same diagram. This allows demonstration of the pattern of behaviour that is characteristic of glass-forming liquids [35, 36]. For ionic liquids, a selection of data for cases where only simple ions are present is shown in Figure 1.3 [37], and they are seen to all exhibit rather extreme deviations from Arrhenius behaviour. The lowest temperature cases are data in the low viscosity range for low melting liquids of the 1-butyl-3-methylimidazolium cations neutralised by different anions. Glass-forming derivatives [9] of the silver halides studied by the early German workers are contained in Figure 1.3a and are seen to exhibit the lowest viscosities that have been measured among this group. The manner in which the viscosities of this family of glass formers group together when plotted in T_g-reduced form is shown in Figure 1.3b.

Note how the data for the 1-butyl-3-methylimidazolium salt melts [10] fit smoothly into the overall ionic liquid pattern [35]. It is a matter of interest that the currently much-researched ambient temperature ionic liquids provide a great many examples of this otherwise rather uncommon phenomenon—the single-component glass former.

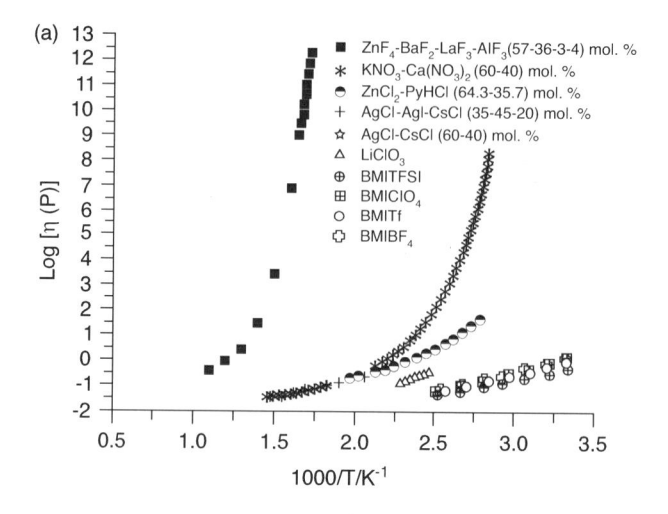

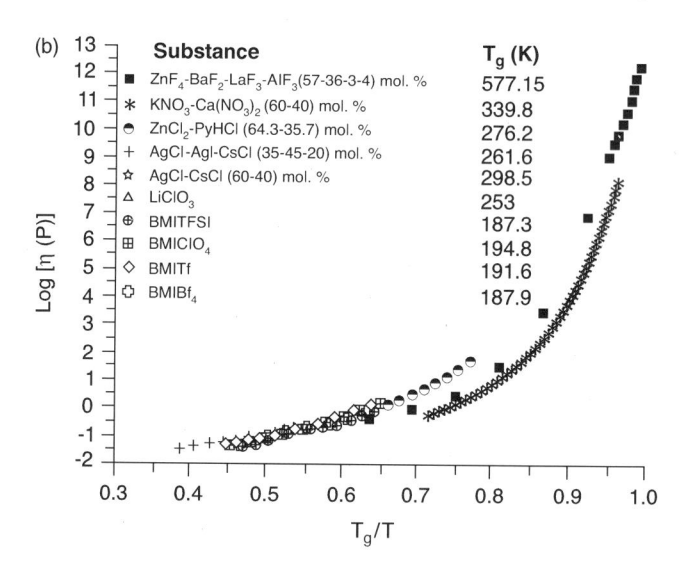

FIGURE 1.3 (a) Arrhenius plots of the viscosities of ionic liquids ranging from molten fluorides to silver halide–alkali halide mixtures, down to [C₄mim][BF₄]. Glass temperatures, at which the viscosities reach values of about 10^{12} Pa·s are given in the legend. (b) The same data, plotted in T_g-reduced Arrhenius form, in order to reveal the strong deviations from the Arrhenius law, known as high fragility, of these liquids [37].

The reason that glass formation among ambient temperature liquids is common is the same as the reason that their melting points are low despite the strong Coulombic forces acting between the ions. When melting points are low relative to intermolecular forces, failure to crystallise during cooling is a common consequence. It is due to the high viscosity at the melting point. Empirically, it is found that when viscosity at the melting point rises above about 0.1 Pa·s (water at ambient, 0.001 Pa·s), the probability of nucleation of the stable phase during normal cooling becomes low [38]. Thus the low melting point is the phenomenon to be understood. It is a consequence of the difficulty that compounds with irregularly shaped charged particles encounter in packing their particles efficiently into three-dimensionally ordered lattices, while maintaining a symmetrical charge distribution. Then the entropy of the liquid state tends to dominate the thermodynamics, stabilising the liquid, and consequently pushing down the melting points. This common glass-forming tendency exhibited by low-melting ionic liquids has been discussed in more detail elsewhere [11] and will not be further dealt with here.

Low melting points relative to forces between particles is a characteristic of the "network" liquids that make up the "glass formers" of classical glass science. In these cases, however, the extended network structure prevents the temperature from having its usual strong effect on the volume of the liquid, and the deviation from Arrhenius behavior in their transport properties is then much smaller. We see this when these liquids, and their partly disrupted binary solutions, are included in Figure 1.3b. Then we get the complete "strong and fragile liquids" pattern [32, 35, 36] for ionic liquids seen in Figure 1.4.

The presence of the single-component glass-forming chloride [39, 40], $ZnCl_2$ in the centre of the diagram, is to be noted. $ZnCl_2$ is a dense tetrahedral network [41, 42] with Zn–Cl–Zn angles averaging only 110° [43], to be compared with SiO_2, Si–O–Si angles averaging 140° with a wide distribution [44]. The intermediate range order

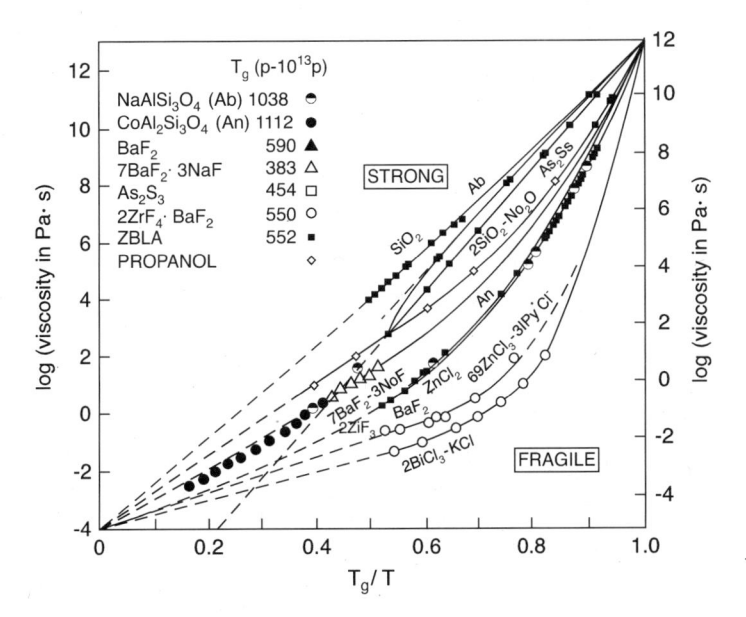

FIGURE 1.4 T_g-scaled Arrhenius plot of ionic liquids with simple halide anions. (From Angell [35] with permission.)

measured by the sharpness of the second and third neighbour peaks in the radial distribution function [43] is much smaller in the case of $ZnCl_2$. This leads to much more fragile behaviour.

The fragility is further increased when the weak network is disrupted by "chemical scissoring" of the Zn–Cl bonds, using a second salt with basic chloride ions. To study this effect, it proved necessary to use a lower melting second component than any of the inorganic chlorides. A suitable choice [45] proved to be the organic salt pyridinium chloride, which had earlier been used by Dieter Gruen in studies of ligand field spectroscopy of nickel and cobalt ions [18]. Data for solutions of the pyridinium chloride–$ZnCl_2$ system [16] are included in Figure 1.4 and are seen to approach the fragile limit of this data collection.

1.4 ROLE OF COMPLEX ANIONS IN DEVELOPMENT OF AMBIENT TEMPERATURE IONIC LIQUIDS

Halides of organic cations have proved very useful for the study of complex anions and their effects on molten salt properties. The dramatically different effects of complexation on the liquid viscosity are well illustrated by the 1976 study of Hodge [46, 47], who used the lower melting 2-methylpyridinium chloride as the source of the Lewis base chloride.

Data seen in Figure 1.5 show that the cohesive energy of an ionic liquid is lowered most strongly when chloride ions are complexed by iron(III) chloride to produce the large singly charged $[FeCl_4]^-$ anion. This observation led to the prediction that, among ionic liquids of a given cation, the tetrachloroferrate(III) salt should be the most fluid and also the most conductive, even more so than the well-studied ionic liquids containing $[AlCl_4]^-$ anions [48, 49]. The veracity of this expectation is

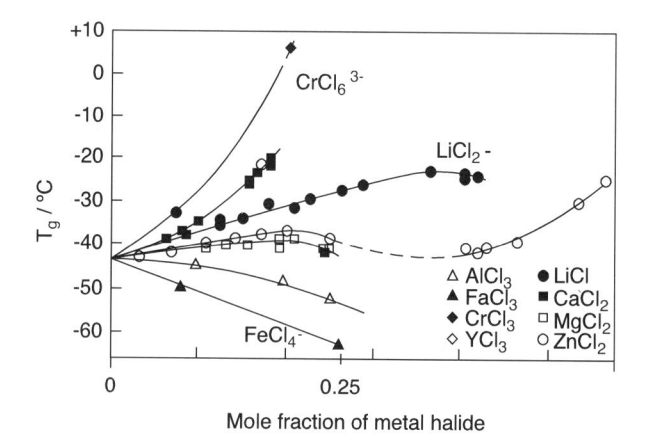

FIGURE 1.5 Variation of T_g of the simple chloride glass former, 2-methylpyridinium chloride, on chloride complexation with various Lewis acids. (Adapted from Angell et al. [46].)

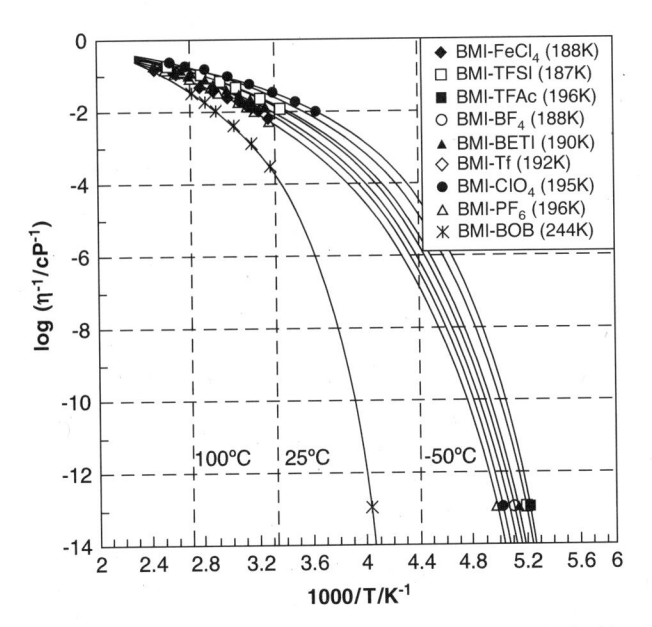

FIGURE 1.6 Fluidity data for a number of salts of the 1-butyl-3-methylimidazolium cation, showing that the tetrachloroferrate(III) is more fluid than any of the fluorinated anion salts, in accord with its lowest T_g seen here (log $\eta^{-1} = 10^{-13}$). (From Xu et al. [10] with permission.)

demonstrated in Figure 1.6 for 1-butyl-3-methylimidazolium salts. For the 25°C conductivity, the tetrachloroferrate(III) is half an order of magnitude above any others in the group. The increase of ionic mobility by complexation seen here is no different in principle from the use of BeF_2 additions to alkali fluoride melts utilised in the molten salt reactor technology discussed in the introduction. There, the addition of 1 mole of BeF_2 replaced two high charge intensity fluoride anions with a doubly charged anion of much larger dimensions, for a net decrease in cohesion.

The fluorinated anion ($[BF_4]^-$ and $[PF_6]^-$) salts that are frequently cited in the literature of the ambient temperature ionic liquids [50–52] are to be understood in the same terms. Unfortunately, like these two fluorinated anions, the tetrachloroferrate anion is slowly hydrolysed by water, and applications of ambient temperature ionic liquids usually require the presence of water-stable anions. The fluorinated species, however, are unpolarisable, and so minimise the effect of the van der Waals contributions to the liquid cohesion. This characteristic of fluorinated anions was first noted in a 1983 study of binary solutions of lithium halides with organic salt halides by Cooper, where the replacement of iodide anions by the equivalent amount of $[BF_4]^-$ anions was predicted to lead to an increase of ambient temperature conductivity, and then shown to do so (in a note added in proof) by some two orders of magnitude [53]. Cooper's further extensive syntheses and measurements on viscosity and conductivity of a family of tetraalkylammonium tetrafluoroborate salts, unfortunately, remained unpublished (for over two decades) until 2003 [10].

In the meantime, Cooper and O'Sullivan [54] reported, in 1992, the first systematic study of ambient temperature water-stable ionic liquids, using triflate and related anions combined with dialkylimidazolium-type cations. This announcement, presented at one of the Electrochemical Society's International Molten Salt Conferences, and eventually published in the Proceedings [54] was immediately followed by a communication by Wilkes and Zaworotko [55], who had been working on similar salts, and the field of high fluidity, water-stable, ionic liquids was launched.

The only significant improvements in fluidity of water-stable, noncorrosive, ionic liquids since that time have come from MacFarlane's group (dicyanamides [56–58]) and the study of protic (as opposed to aprotic) cation salts [59], with which we will concern ourselves after discussing a further important aspect of the organic cation halides.

1.5 SPECTROSCOPIC ASPECTS OF ORGANIC CATION HALIDES

An aspect of organic cation halide salts which has escaped emphasis so far but which is important in understanding the high fluidities of the tetrachloroferrates(III), and so is deserving of further exploitation, is the high Lewis basicity of the halide anions. This is due to the relatively weak effective positive charge intensity of the larger organic cations, which leaves the electron density natural to the chloride anion almost unperturbed—and hence available for donation to stronger Lewis acidic cations. This effect is manifested in the sharp and intense electronic spectra of transition metal cations observed when their halides are dissolved in organic cation halides.

Figure 1.7, from the work of Gruen and McBeth [45], compares the spectra of Ni^{2+} dissolved in molten pyridinium chloride (Fig. 1.7a) with that found in the LiCl–KCl eutectic and LiCl itself (Fig. 1.7b). The sharp spectral lines are consistent with an almost undistorted tetrahedral structure that gives rise to the brilliant blue colour of the solution. Nickel(II) can also serve as a probe for structural changes in binary solutions, for instance, in the binary system (2-methylpyridinium chloride + $ZnCl_2$), sharp and highly temperature-dependent changes can be seen in the vicinity of stoichiometric complexes characteristic of the solvent system. These are reminiscent of the unique and informative spectra of such species as the previously unknown $NiCl_3^-$ anion, which was detected by Brynestad and Smith [60] in a narrow composition region of the (CsCl + $AlCl_3$) system, in which the chloride anion activity changes precipitously across the $[AlCl_4]^-$ stoicheiometry [61]. In systems with organic cations, because of the high chloride ion basicity in the first component, comparable chloride ion activity changes can occur in the crossover stoicheiometry of less dominant anion species than $[AlCl_4]^-$. The consequent possibility of unusual complex anion stoichiometries and symmetries remains to be exploited.

An even clearer use of ionic liquids in transition metal coordination chemistry was that of Abkemeier and Angell, who needed a room temperature molten chloride for the first application of the newly invented diamond anvil high pressure cell in the study of pressure effects on coordination numbers. They used ethanolaminium

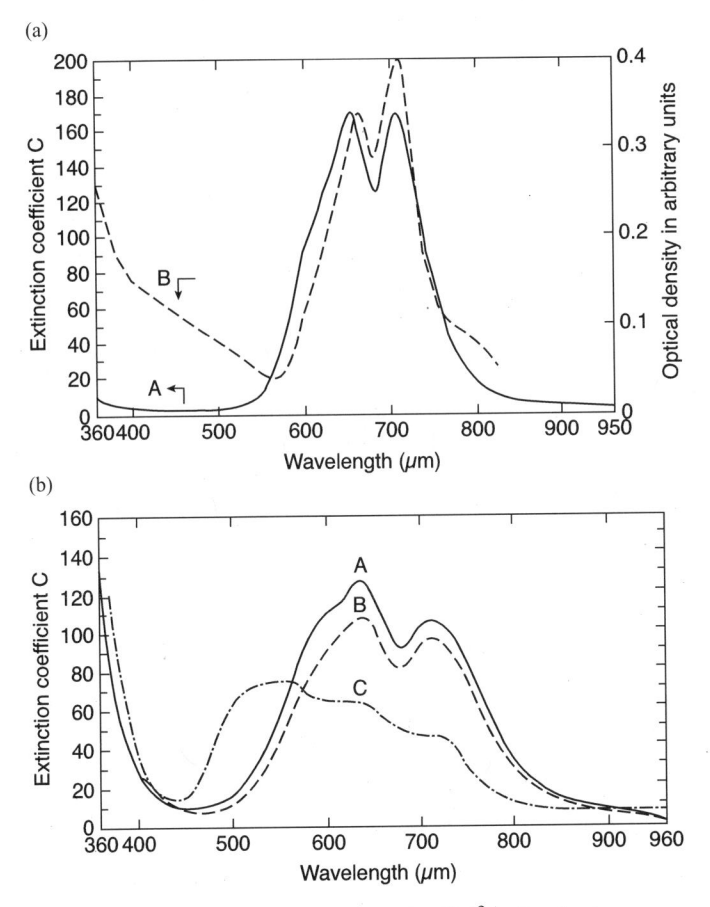

FIGURE 1.7 (a) The electronic (visible) spectrum of Ni^{2+} dissolved at a concentration of 1×10^{-4} M in pyridinium chloride. (b) The equivalent spectrum in LiCl–KCl solutions at two different LiCl contents. The tetrahedral character is almost lost in the LiCl-rich solution, in which the second nearest neighbor to the nickel(II) is most probably Li^+. (From Gruen and McBeth [45] with permission.)

chloride as the ambient temperature noncrystallising component of a mixture with ethylammonium chloride to study nickel(II) spectra during pressure increases up to 1 GPa [62], and observed the conversion from $[NiCl_4]^{2-}$ at ambient pressure to $[NiCl_6]^{4-}$ at higher pressures, the equilibrium constant change yielding a ΔV within 10% of the volume of 2 moles of chloride ion.

Advantage has recently been taken of this sharpness and intensity of complex species electronic spectra to study the time dependence of specific structural changes in ionic liquids, and thereby to permit comparison of chemical order relaxation times with the relaxation times that determine the glass transition in the same solution [63]. In solutions containing large components of both an aprotic ionic liquid chloride and a source of –OH ligands, such as sorbitol, or different sugars, cations like Co^{2+}

distribute themselves between octahedral {Co(OH)$_6$} sites and tetrahedral {CoCl$_4$} sites, and the distribution is temperature dependent. A sudden change in temperature can cause a shift in the populations of the two types of sites, and this is registered very sensitively by the change in intensity of the [CoCl$_4$]$^{2-}$ peaks (by contrast, the octahedral absorption bands are very weak [63], because the d-d transitions are strongly forbidden in octahedral symmetry). The change in the tetrahedral band intensities, as a function of time, monitors the change in chemical order of the cobalt(II) species, and it is a question of interest whether or not this chemical order relaxation is slaved to the overall structural relaxation of the glass-forming system. Previous studies of related spectroscopically monitored chemical orders have suggested that it is, but none have been as sensitively detected as that manifested in an organic cation-based ionic liquid system.

Some data for the latter case are shown in Figure 1.8, both for up-jumps and down-jumps to the same final temperature. When similar studies are made for the volume of a glass former, the approach to equilibrium is found to be strongly asymmetrical, with the approach from a lower temperature being much slower than the approach from a higher temperature. (This is known as the "nonlinearity" of relaxation, one of the three "non"s of glass former phenomenology.) Figure 1.8 shows that in the case of chemical order relaxation, the equilibration kinetics can be linear. The observations, of course, are being made on a very minor component of the solution, but so are many

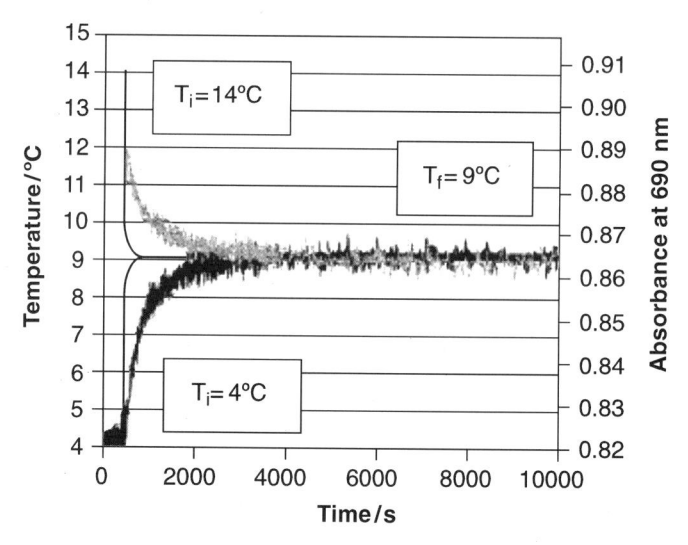

FIGURE 1.8 Time dependence of the intensity of the cobalt(II) tetrahedral absorption in the system described in text, following 4 °C temperature up-jumps and down-jumps to the same final temperature. (The much faster equilibration of the temperature is shown by the less noisy curves near the left axis of the figure.) In contrast to the effect of the same jumps on the density of a liquid near its glass temperature, the relaxations are exponential and also symmetrical, consistent with the longer time scale required for the coordination change at 9 °C relative to the enthalpy relaxation time monitored by calorimetry. (From Martinez and Angell [63] with permission.)

probe studies of glass former kinetics [64, 65]. The relation between chemical order relaxation and physical order relaxation (monitored by volume or enthalpy changes) is one of the important questions in glass science that has not yet been properly investigated.

1.6 OTHER SPECTROSCOPIES

There are many spectroscopic tools that are applicable to the study of ionic liquids. One of particular value for the case of protic ionic liquids is proton NMR spectroscopy. The reason is that the ^{1}H NMR spectrum is sensitive to how completely the proton is transferred from the donor species (protic acid) to the basic species. Indeed, the proton chemical shift on the N-H proton can be used in the same way as an indicator ion is used, to probe the change in proton activity as the 1:1 stoicheiometry is approached and then passed.

An example of such a titration curve was given long ago by Angell and Shuppert [66], who studied the effect on the proton transferred to pyridine by HCl (giving pyridinium chloride) when the acid strength of HCl was increased by complexing with a Lewis acid, $AlCl_3$. This is an analogue of the formation of superacids by adding a strong Lewis acid (like SbF_5) to a strong protic acid (like HSO_3F). Indeed, these authors showed that with excess $AlCl_3$, the ^{1}H chemical shift approached the free gaseous pyridinium cation value.

The ^{1}H chemical shift is currently being used as a way of characterising the effective acidity of protic ionic liquids [67], and thereby using them to control the stability of folded proteins in aquated versions of these media that are being harnessed for protein folding studies [68] and for biopreservation (see last section).

1.7 PHYSICAL PROPERTIES OF PROTIC IONIC LIQUIDS

In contrast to the systems described earlier, the protic ionic liquids, of which ammonium chloride is the elementary inorganic example, are formed very simply by proton transfer from an inorganic acid to an organic base. The base is usually a primary, secondary, or tertiary amine, or a nitrogen atom in a heterocyclic ring (which may or may not be resonance stabilised) [69]. The properties of the ionic liquid formed in this manner depend quite strongly on the relative strengths of the acids and bases between which the proton is transferred [70].

In order to simplify the prediction of protic ionic liquid properties, Angell and co-workers [37, 70] have generalised the Gurney proton energy level diagram [71] not just to deal with proton transfers between acid and base species in an aqueous environment, but rather to describe the energy changes when a proton moves between any proton donor and proton acceptor pair, in the absence of any solvent at all. So far, the spacings in the diagram have been based on the known aqueous pK_a values, and hence are only semiquantitative, but these will in time be replaced by energy data obtained from electrochemical studies and other direct measures such as vapour pressure changes [70].

An example of the energy level diagram [37] is provided in Figure 1.9. Acid protic electrolytes result when the proton falls from an occupied level on a superacid to a vacant level on a weak base, for example, from triflic acid to fluoroaniline, or to water, while neutral electrolytes arise when the proton transfers from a moderate acid like nitric or trifluoroethanoic acid to ethylamine or cyclopentylamine. Provided the proton falls across a gap of 0.7 eV or more, the electrolyte is a "good" ionic liquid, meaning the proton resides on the base for at least 99% of the time.

Protic ionic liquids, in which the proton energy change is not greater than 1 eV, seem to have low glass temperatures and also high fluidities and ionic conductances.

	Occupied	Vacant	pK_a	E/eV
Acid Electrolytes	$HSbF_s$	SbF_6^-		
	HBOB	BOB^-		
	HTFSI	$TFSI^-$		
	HTf	Tf^-	-14	0.83
	HSO_3F	SO_3F^-		
	$HClO_4$	ClO_4^-	-10	0.59
	H_2SO_4	HSO_4^-	-9	0.53
	HPO_2F_2	$PO_2F_2^-$		
	HNO_3	NO_3^-	-1.30	0.08
	CH_3SO_3H	$CH_3SO_3^-$	-0.60	0.04
	2-fluoropyridine H^+	2-fluoropyridine	-0.43	0.03
	CF_3COOH	CF_3COO^-	-0.25	0.01
Neutral Electrolytes	H_2O^+	H_2O	0	0
	H_3PO_4	$H_3PO_4^-$	2.12	-0.13
	1,2,4-1H-triazole H^+	1,2,4-1H-triazole	3.00	-0.18
	HF	F^-	3.20	-0.19
	HCOOH	$HCOO^-$	3.75	-0.22
	CH_3COOH	CH_3COO^-	4.75	-0.28
	ImH^+	Im	6.99	-0.41
	hydrazine H^+	hydrazine	7.96	-0.47
	NH_4^+	NH_3	9.23	-0.55
	$EtNH_3^-$	$EtNH_3$	10.63	-0.63
	$Bu_2NH_2^+$	Bu_2NH	11.25	-0.67
Basic Electrolytes	H_2O	OH^-	14	-0.83
	NH_3	NH_2^-		
	OH^-	O^{2-} (Na^+)	28	-1.66

FIGURE 1.9 The proton energy level diagram for predicting the ionicities of ionic liquids resulting from the falling of a proton from an occupied level (on an acid molecule) to a vacant level (on a basic molecule or on a divalent anion; in the latter case the countercation(s) must remain to balance the new anion charge). (From Belieres and Angell [81] with permission.)

PROTIC IONIC LIQUIDS AS FUEL CELL ELECTROLYTES   17

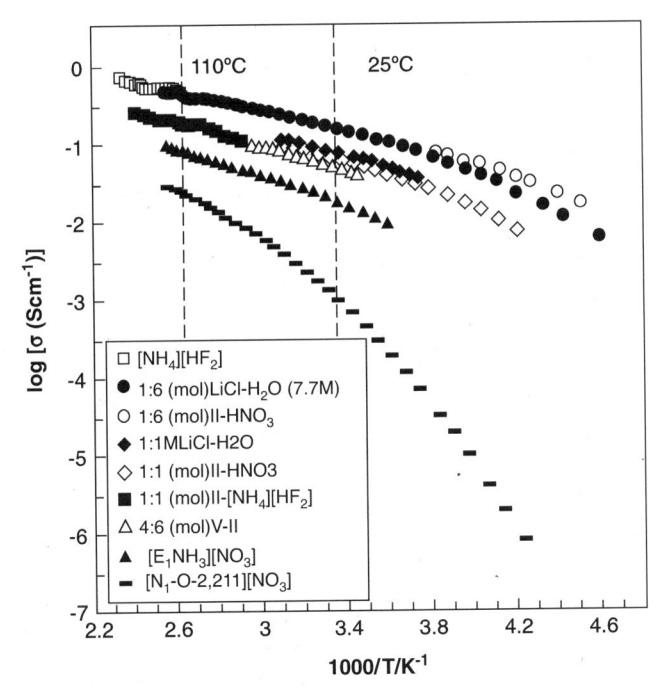

FIGURE 1.10 Conductivities of various protic ionic liquids (and one aprotic case) compared with those of aqueous LiCl solutions to show how the aqueous solutions are not unique with respect to high ionic conductivity. (From Xu and Angell [59] with permission.)

A typical case is ethylammonium nitrate, known since 1914 [72]. The ionic conductances of some protic ionic liquids [59, 73] are shown in Figure 1.10, along with those of aqueous LiCl to show that, for some salts like dimethylammonium nitrate [59] and hydrazinium methanoate [74], the conductivities at ambient temperature can be as high as those found in aqueous solutions [59, 73].

1.8 PROTIC IONIC LIQUIDS AS FUEL CELL ELECTROLYTES

It is astonishing that protic ionic liquids had been known for nearly a century before it was realised [75, 76] that they could serve as the proton carriers in fuel cell electrolytes. In fact, they can serve very well in this role and can provide electrolytes of a type that are simply not available in systems in which water is a major contributor to the acid–base equilibrium.

When the protic salt serving as the fuel cell electrolyte is formed by a transfer in which the protons decrease their energy by about an electron volt, it seems that the fuel cell is able to realise its theoretical voltage, at least for low current densities. The observation implies that, under these conditions, the energy barrier for oxygen reduction to water vanishes, though the reason for this simple behaviour is not yet clear. Figure 1.11 shows the relation between the proton gap and the open circuit cell

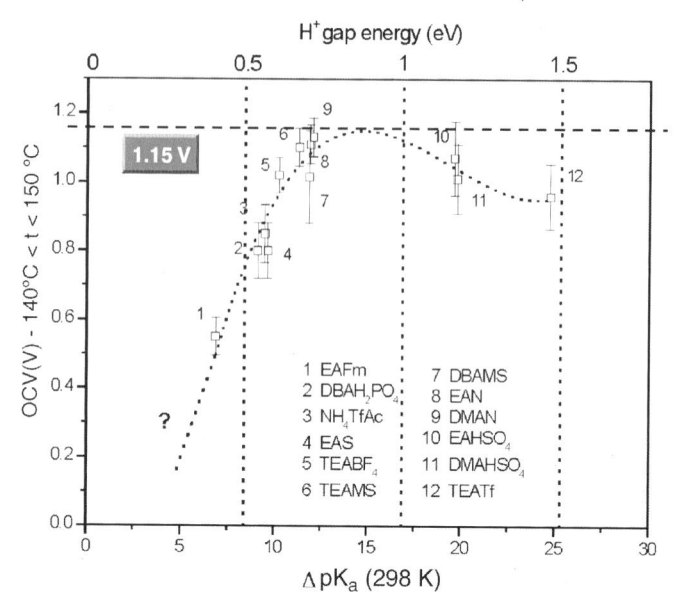

FIGURE 1.11 Open circuit voltages of $H_2(g)$–$O_2(g)$ fuel cells in relation to the proton gap characteristic of fuel cell electrolyte. (From Belieres et al. [82] with permission.)

voltage [77]. When the gap is too small, the proton transfer is incomplete, the conductivity is low (though the fluidity may be exceptionally high), and the fuel cell voltage output is also small.

1.9 INORGANIC FUEL CELL ELECTROLYTES

It is even more astonishing that, until 2006 [77], it had not been realised that mixtures of ammonium salts can be used as protic electrolytes for fuel cells running above 80°C (hydrazinium salts could be used to much lower temperatures [73]). Applied as fuel cell electrolytes, the ammonium salts give more stable performance than electrolytes containing organic cations. The IR-corrected polarisation curves for some inorganic electrolyte fuel cells are shown in Figure 1.12, using the logarithmic ("Tafel plot") form. The plateau current at the theoretical voltage can now be extended out beyond $50\,mA\cdot cm^{-2}$ in inorganic systems to be described in future articles. The occurrence of such high cell voltage–current combinations shows that it is now possible to increase the efficiency of chemical-to-electrical energy conversion considerably above the levels that are currently being obtained in, for instance, the phosphoric acid fuel cell.

An advantage of the phosphoric acid electrolyte, at the moment, is that the conductivity of the molecular liquid is higher, due to the presence of a considerable "dry" proton mobility. This superprotonic component, illustrated in Figure 1.13, is unfortunately not present in the protic ionic liquids, which mostly lie on or below the "ideal" Walden line [10]. It may be necessary to compromise on power output for the

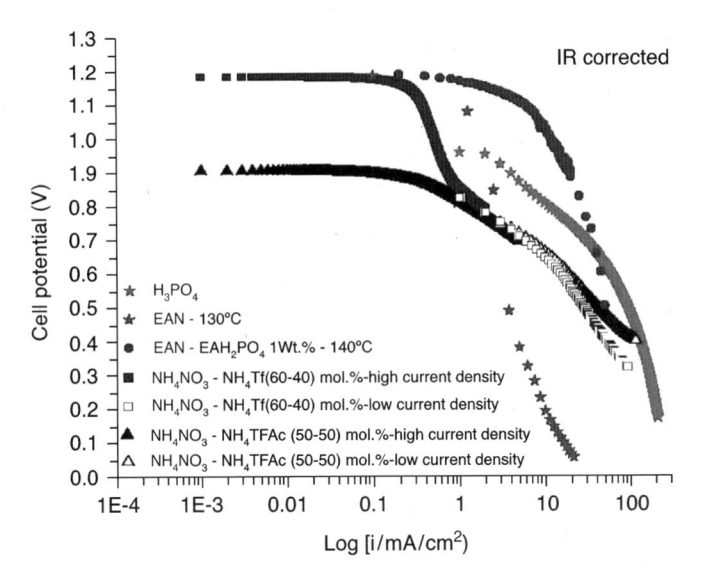

FIGURE 1.12 IR-corrected polarisation curves for hydrogen/oxygen fuel cells utilising inorganic and organic cation ionic liquid electrolytes. (From Belieres et al. [82] with permission.)

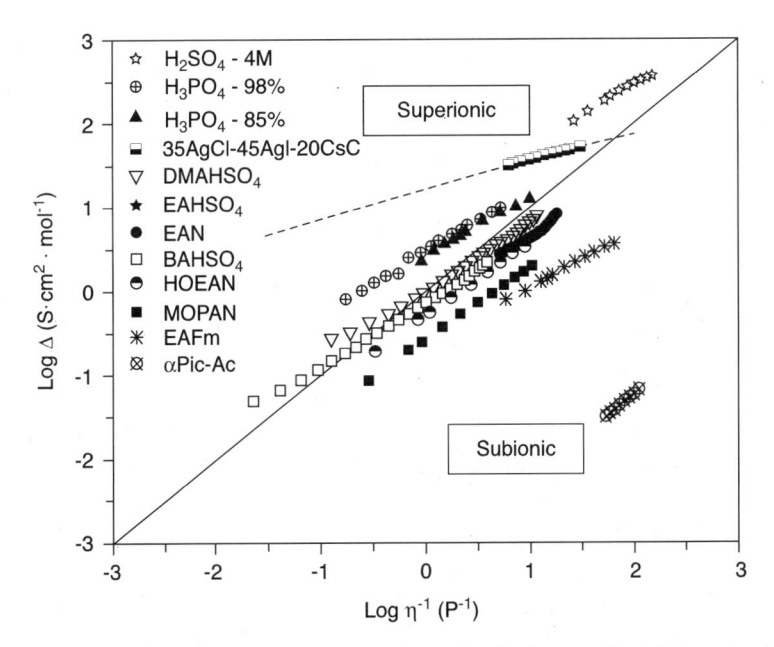

FIGURE 1.13 Walden plot showing the behaviour of equivalent conductivity expected from precise correlation with liquid viscosity solid line, with examples of subionic behaviour in some protic salt cases, and superprotonic behaviour in some (acid + water) systems, and a molten silver caesium halide mixture. (From Belieres et al. [83] with permission.)

sake of energy efficiency, if some breakthrough in conduction mechanisms is not forthcoming.

What are the prospects for development of really fast proton conducting materials, preferably solids? It is known, for instance, that nature has developed nanochannel proteins [78] that promote proton mobilities exceeding, by factors of 20, the normal mobility of protons in dilute acid solutions. Thus it is certainly not impossible that nanochannel structured polymers, or ionic liquid crystals with similar characteristics, could be discovered. The required number density of such channels seems improbable of achievement, however.

1.10 BIOPRESERVATION

An exciting new role for ionic liquids may lie in the storage and manipulation of sensitive biomolecules. Fujita et al. [79] have recently reported that the protein cytochrome c is remarkably stabilised in solution in a nontoxic dihydrogenphosphate ionic liquid, and we have found that lysozyme in 200 mg·cm^{-3} concentration is still folded after three years in ambient solutions that have large ethylammonium nitrate concentrations [80]. Certainly the ability of proteins to recover their native states after repeated unfoldings is enhanced by solution in protic ionic liquids of optimised proton activity [68]. Many developments in this general area are to be expected in the future (for instance, see the recent "Crucible" column by Phillip Ball; http://www.rsc.org/chemistryworld/Issues/2009/April/ColumnThecrucible.asp).

1.11 CONCLUSION

While the 200 years since Humphrey Davy began electrochemistry have seen wonderful advances in many areas, the progress in the specific area of fuel cells since the field began 150 years ago has been less impressive. It is to be hoped that the long-awaited breakthroughs needed to help this source of electrical energy realise its theoretical promise will become one of the new achievements of this very young branch of the very old "ionic liquids" discipline.

ACKNOWLEDGEMENTS

Most of the work described here has been supported by the NSF, under grants from the Solid State Chemistry division, for which the writer is very grateful. Recent studies on fuel cells have been carried out under the auspices of the DOD Army Research Office, grant no. W911NF-04-1-0060.

REFERENCES

1. Davy, H., *Researches, Chemical and Philosophical*, Biggs and Cottle, Bristol, 1800.
2. Bockris, J. O., Tomlinson, J. W., and White, J. L., The structure of the liquid silicates—partial molar volumes and expansivities, *Trans. Faraday Soc.* **52**, 299–310 (1956).

3. Tomlinson, J. W., Heynes, M. S. R., and Bockris, J. O., The structure of liquid silicates. 2. Molar volumes and expansivities, *Trans. Faraday Soc.* **54**, 1822–1833 (1958).
4. Willis, G. M., Physical-chemistry of lead extraction, *J. Metals* **31**, 28 (1979).
5. Davey, T. R. A., and Willis, G. M., Review of developments in extractive metallurgy of lead, zinc, and tin in 1978, *J. Metals* **31**, 17 (1979).
6. Bloom, H., and Heymann, E., *Proc. R. Soc.* **A188**, 392, (1947).
7. Angell, C. A., and Bockris, J. O., A technique for the measurement of diffusion coefficients in molten salts, *J. Sci. Instrum.* **35**, 458–461 (1958).
8. Walden, P., *Z. Phy. Chem.* **55**, 207, 246 (1906).
9. Angell, C. A., Fast ion motion in glassy and amorphous materials, *Solid State Ionics* **9–10** (Dec), 3–16 (1983).
10. Xu, W., Cooper, E. I., and Angell, C. A., Ionic liquids: ion mobilities, glass temperatures, and fragilities, *J. Phys. Chem. B* **107**, 6170–6178 (2003).
11. Angell, C. A., Xu, W., Yoshizawa, M., Hayashi, A., Belieres, J.-P., Lucas, P., and Videa, M., Physical chemistry of ionic liquids, inorganic and organic, protic and aprotic, in *Electrochemical Aspects of Ionic Liquids*, H. Ohno (Ed.), Wiley, Hoboken, NJ, 2005.
12. Hitch, B. F., and Baes, C. F., An electromotive force study of molten lithium fluoride–beryllium fluoride solutions, *Inorg. Chem.* **8**, 201 (1969).
13. Robbins, G. D., and Braunstein, J., Electrical conductivity and mobility in molten LiF–BeF$_2$ mixtures and a comparison with molten silicates, *Phys. Abstr. Papers Am. Chem. Soc.* **[Sept]** 86 (1970).
14. Vogel, W., Reiss, H., and Seifert, J., On the structure and properties of beryllium fluoride glasses containing CoF$_2$, *Glass Technol.* **24**, 133–138 (1983).
15. Vogel, W., Phase separation in glass, *J. Non-Crystalline Solids* **25**, 170–214 (1977).
16. Easteal, A. J., and Angell, C. A., Phase equilibria, electrical conductance, and density in glass-forming system zinc chloride + pyridinium chloride—a detailed low-temperature analog of silicon dioxide + Sodium oxide system, *J. Phys. Chem.* **74**, 3987 (1970).
17. Angell, C. A., and Tomlinson, J. W., Diffusion and elecrical conductance measurements in solutions of cadmium dissolved in molten cadmium chloride, *Discuss Faraday Soc.* **32** (1962).
18. Gruen, D. M., Spectrophotometry of fused salts, *Q. Rev.* **19**, 349, (1965).
19. Protsenko, I., *Zh. Obshch. Khim.* **20**, 1365 (1950).
20. Thilo, E., Wieker, C., and Wieker, W., *Silic. Tech.* **15**, 109 (1964).
21. Dietzel, A. and Poegel, H. J., in *Proceedings III International Congress on Glass*, A. Garzanti (Ed.), Nello Stabilimento Grafico Di Roma Della, Rome, Venezia, 1953, 1954, p. 219.
22. Mezei, F., Knaak, W., Farago, B., Neutron spin-echo study of dynamic correlations near the liquid-glass transition, *Phys. Rev. Lett.* **58**, 571–574 (1987).
23. Bengtzelius, U., Gotze, W., and Sjolander, A., Dynamics of supercooled liquids and the glass-transition, *J. Phys. C Solid State Phys.* **17**, 5915–5934 (1984).
24. Angell, C. A., Success of free volume model for transport in fused salts, *J. Phys. Chem.* **68**, 218 (1964).
25. Angell, C. A., Free volume model for transport in fused salts—electrical conductance in glass-forming nitrate melts, *J. Phys. Chem.* **68**, 1917 (1964).

26. Angell, C. A., Viscous flow and electrical conductance, *J. Chem. Phys.* **46**, 4673–4679 (1967).
27. Vogel, H., *J. Phys. Z.* **22**, 645 (1921).
28. Fulcher, G. S., *J. Am. Ceram. Soc.* **8**, 339 (1925).
29. Tammann, G., and Hesse, W. Z., *Z. Anorg. Allgem. Chem.* **156**, 245 (1926).
30. Scherer, G. W., Editorial comments on a paper by Fulcher, *J. Am. Ceram. Soc.* **75**, 1060–1062 (1992).
31. Angell, C. A., and Bressel, R. D., Fluidity and conductance in aqueous-electrolyte solutions—approach from glassy state and high-concentration limit. 1. $Ca(NO_3)_2$ solutions, *J. Phys. Chem.* **76** (22), 3244 (1972).
32. Angell, C. A., Relaxation in liquids, polymers and plastic crystals—strong fragile patterns and problems, *J. Non-Cryst. Solids* **131**, 13–31 (1991).
33. Richert, R., and Angell, C. A., Dynamics of glass-forming liquids. V. On the link between molecular dynamics and configurational entropy, *J. Chem. Phys.* **108**, 9016–9026 (1998).
34. Weiler, R., Bose, R., and Macedo, P. B., *J. Chem. Phys.* **53**, 1258 (1970).
35. Angell, C. A., Spectroscopy, simulation, and the medium range order problem in glass, *J. Non-Cryst. Solids* **73**, 1 (1985).
36. Angell, C. A., Formation of glasses from liquids and biopolymers, *Science* **267**, 1924–1935 (1995).
37. Belieres, J.-P., and Angell, C. A., Ionic liquids: preparation and characterization of new protic ionic liquids, and the proton transfer energy level diagram, *J. Phys. Chem. B* **111**, 4926–4937 (2007).
38. Uhlmann, D. R., Kinetics of crystallization, *J. Non-Cryst. Solids* **7**, 337 (1972).
39. Goldstein, M., and Nakonecz, M., Volume relaxation in zinc chloride glass, *Phys. Chem. Glasses* **6**, 126 (1965).
40. Easteal, A. J., and Angell, C. A., Viscosity of molten $ZnCl_2$ and supercritical behavior in its binary-solutions, *J. Chem. Phys.* **56**, 4231 (1972).
41. Angell, C. A., and Wong, J., Structure and glass transition thermodynamics of liquid zinc chloride from far-infrared, Raman, and probe ion electronic and vibrational spectra, *J. Chem. Phys.* **53**, 2053 (1970).
42. Wilson, M., and Madden, P. A., Voids, layers, and the first sharp diffraction peak in $ZnCl_2$, *Phys. Rev. Lett.* **80**, 532–535 (1998).
43. Wilson, M., and Madden, P. A., Short-range and intermediate-range order in Mcl2 melts—the importance of anionic polarization, *J. Phys. Condensed Matter* **5**, 6833–6844 (1993).
44. Dupree, E., and Pettifer, R. F., Determination of the Si—O—Si bond angle distribution in vitreous silica by magic angle spinning NMR, *Nature* **308**, 523–525 (1984).
45. Gruen, D. M., and McBeth, R. L., Tetrahedral $NiCl_4^{2-}$ ion in crystals and in fused salts—spectrophotometric study of chloro complexes of nickel(II) in fused salts, *J. Phys. Chem.* **63**, 393–395 (1959).
46. Angell, C. A., and Hodge, I. M., and Cheeseman, P. A., Molten salts, in *International Conference on Molten Salts*, J. P. Pemsler (Ed.), The Electrochemical Society, 1976, p. 138.
47. Sun, X. G., Xu, W., Zhang, S. S., and Angell, C. A., Polyanionic electrolytes with high alkali ion conductivity, *J. Phys. Condensed Matter* **13**, 8235–8243 (2001).

48. Boon, J. A., Carlin, R. T., Elias, A. M., and Wilkes, J. S., Dialkylimidazolium–sodium chloroaluminate ternary salt system—Phase-diagram and crystal-structure, *J. Chem. Crystallogr.* **25**, 57–62 (1995).
49. Carlin, R. T., and Wilkes, J. S., Complexation of Cp2Mcl2 in a chloroaluminate molten-salt—Relevance to homogeneous Ziegler–Natta catalysis, *J. Mol. Catal.* **63**, 125–129 (1990).
50. Rogers, R. D., and Seddon, K. R., Ionic liquids—solvents of the future? *Science* **302**, 792–793 (2003).
51. Wasserschied, P., and Keim, W., *Angew. Chem. Int. Ed.* **39**, 3773–3789 (2000).
52. Rogers, R. D. and Seddon, K. R. (Eds.), *Ionic Liquids: Industrial Applications to Green Chemistry*, ACS Symposium Series Vol. 818, American Chemical Society, 2002.
53. Angell, C. A., Origin and control of low-melting behavior in salts, polysalts, salt solvates, and glassformers in *Molten Salts: From Fundamentals to Application: NATO-ASI*, M. Gaune-Escarde (Ed.), Kluwer Academic Publishers, Kas, Turkey, 2002, pp. 305–322.
54. Cooper, E. I., and O'Sullivan, E. J. M., New, stable, ambient-temperature molten salts, in *Proceedings of the 8th International Symposium on Molten Salts*, Electrochemical Society, 1992, pp. 386–396.
55. Wilkes, J. S., and Zaworotko, M. J., Air and water stable 1-ethyl-3-methylimidazolium based ionic liquids, *J. Chem. Soc. Chem. Commun.* 965–967 (1992).
56. MacFarlane, D. R., Golding, J., Forsyth, S., Forsyth, M., and Deacon, G. B., Low viscosity ionic liquids based on organic salts of the dicyanamide anion, *Chem. Commun.* 1430–1431 (2001).
57. MacFarlane, D. R., Forsyth, S. A., Golding, J., and Deacon, G. B., Ionic liquids based on imidazolium, ammonium and pyrrolidinium salts of the dicyanamide anion, *Green Chem.* **4**, 444–448 (2002).
58. Forsyth, S. A., Batten, S. R., Dai, Q., and MacFarlane, D. R., Ionic liquids based on imidazolium and pyrrolidinium salts of the tricyanomethanide anion, *Aust. J. Chem.* **57**, 121–124 (2004).
59. Xu, W., and Angell, C. A., Solvent-free electrolytes with aqueous solution - like conductivities, *Science* **302** (5644), 422–425 (2003).
60. Brynestad, J., and Smith, G. P., Evidence for a 3-coordinate complex of nickel(II), *J. Am. Chem. Soc.* **92**, 3198 (1970).
61. Smith, G. P., Dworkin, A. S., Pagni, R. M., and Zingg, S. P., Bronsted superacidity of HCl in a liquid chloroaluminate—AlCl$_3$-1-ethyl-3-methyl-^{1}H-imidazolium chloride, *J. Am. Chem. Soc.* **111**, 525–530 (1989).
62. Angell, C. A., and Abkemeier, M. L., Diamond cell study of pressure-induced coordination change for Ni(II) in liquid chloride solvents, *Inorg. Chem.* **12**, 1462–1464 (1973).
63. Martinez, L.-M., and Angell, C. A., *Physica* **314**, 548–559 (2002).
64. Cicerone, M. T., and Ediger, M. D., Relaxation of spatially heterogeneous dynamic domains in supercooled ortho-terphenyl, *J. Chem. Phys.* **103**, 5684 (1995).
65. Blackburn, F. R., Wang, C. Y., and Ediger, M. D., Translational and rotational motion of probes in supercooled 1,3,5-tris(naphthyl)benzene, *J. Phys. Chem.* **100**, 18249 (1996).
66. Angell, C. A., and Shuppert, J. W., NMR studies of the Lewis acid–base reactions between pyridinium chloride and the acids, ZnCl$_2$ and AlCl$_3$, *J. Phys. Chem.* **84**, 538–542 (1980).

67. Belieres, J.-P., Byrne, N., Cherry, B. R., Holland, G. P., Yarger, J. L., and Angell, C. A., On the use of proton NMR to classify the strength of acids and superacids, *J. Am. Chem. Soc.*, submitted for publication.
68. Byrne, N., and Angell, C. A., Protein unfolding, and the "tuning in" of reversible intermediate states, in protic ionic liquid media, *J. Mol. Biol.* **378**, 707–714 (2008).
69. Yoshizawa, M., Ogihara, W., and Ohno, H., Design of new ionic liquids by neutralization of imidazole derivatives with imide-type acids, *Electrochem. Solid State Lett.* **4**, E25–E27 (2001).
70. Yoshizawa, M., Xu, W., and Angell, C. A., Ionic liquids by proton transfer: vapor pressure, conductivity, and the relevance of ΔpK_a from aqueous solutions, *J. Am. Chem. Soc.* **125**, 15411–15419 (2003).
71. Gurney, R. W., *Ionic Processes in Solution*, Dover Publications, New York, 1962.
72. Walden, P., *Bull. Acad. Imper. Sci. (St. Petersburg)*, 405–422 (1914).
73. Sutter, E. J., *Hydrogen Bonding and Proton Transfer Interactions in Hydrazine-Based Binary Liquids and Related Systems*, Ph.D. thesis, Purdue University, 1970.
74. Sutter, E. J., and Angell, C. A., Glass transitions in molecular liquids. I. Influence of proton transfer processes in hydrazine-based solutions, *J. Phys. Chem.* **75**, 1826–1831 (1971).
75. Susan, M. A. B. H., Noda, A., Mitsushima, S., and Watanabe, M., Bronsted acid–base ionic liquids and their use as new materials for anhydrous proton conductors, *Chem. Commun.*, 938–939 (2003).
76. Angell, C.A., Xu, W., Belieres, J.-P., and Yoshizawa, M., International Patent WO 2004114445 (2004).
77. Belieres, J.-P., Gervasio, D., and Angell, C. A., Binary inorganic salt mixtures as high conductivity electrolytes for >100°C fuel cells, *Chem. Commun. (Cambridge)*, 4799–4801 (2006).
78. Wu, Y., and Voth, G. A., A computer simulation study of the hydrated proton in a synthetic proton channel, *Biophys J.* **85**, 864–875 (2003).
79. Fujita, K., MacFarlane, D. R., and Forsyth, M., *Chem. Commun.* **4034**, (2005).
80. Byrne, N., Belieres, J.-P., and Angell, C. A., Reversible folding–unfolding, aggregation protection, multi-year stabilization in high concentration protein solutions, *Chem. Commun.*, 2714–2716 (2007).
81. Belieres, J.-P., and Angell, C. A., Protic ionic liquids: preparation, characterization, and proton free energy level representation, *J. Phys. Chem. B*, **111**, 4926–4937 (2007).
82. Belieres, J.-P., Gervasio, D., and Angell, C. A., Binary inorganic salt mixtures as high conductivity electrolytes for >100 °C fuel cells, *Chem. Commun.* 4799–4801 (2006).
83. Belieres, J.-P., Xu, W., Markusson, H., Gervasio, D., and Angell, C. A., Protic ionic liquids: a new class of fuel cell electrolytes addressing oxygen over-voltage and water electro-osmosis problems, to be published.

2 Conductivities of Ionic Liquid Mixtures with Organic Electrolyte Solutions

D. BANSAL
MaxPower, Inc., Harleysville, Pennsylvania

F. CROCE
Dipartmento di Scienze del Farmaco, Universitat Degli Studi 'G D' Annunzio, Chieti, Italy

J. SWANK
U.S. Army Research Laboratory, Adelphi, Maryland

M. SALOMON
MaxPower, Inc., Harleysville, Pennsylvania

Abstract

Conductivities of ionic liquid-based systems have been studied in several configurations. First we report the conductivities of six liquid-based electrolyte solutions, and second we review the conductivity and transport behaviour of three gelled electrolytes of varying ionic liquid content. For the gelled electrolytes, a PVdF block copolymer was used as the host combined with organic solvents and the ionic liquid 1-ethyl-3-methylimidazolium bis(perfluoroethylsulfonyl)amide ($[C_2mim][N\{SO_2(C_2F_5)\}_2]$). For liquid-based systems, the ionic liquid 1-butyl-1-methylpyrrolidinium bis(trifluoromethylsulfonyl)amide ($[C_4mpyrr][NTf_2]$) was mixed with organic solvents. For both gelled and liquid systems, the organic solvents employed in our studies are those that have application to primary and rechargeable lithium batteries. Conductivities were determined over the temperature range of −60 to 60 °C, and for both gelled and liquid electrolyte systems, it is found that specific conductivities increase as the ionic liquid is added to the organic-based systems. For the liquid electrolyte solution, the conductivity reaches a maximum in the composition range of ~25–35 mass % of the ionic liquid. Both gelled and liquid

---

Molten Salts and Ionic Liquids: Never the Twain? Edited by Marcelle Gaune-Escard and Kenneth R. Seddon
Copyright © 2010 John Wiley & Sons, Inc.

electrolyte systems were found to follow VTF behaviour, which, based on VTF-derived activation energies, are interpreted in terms of free volume theory and viscosity.

For a selected gel system, the transport number for Li$^+$ (t_{Li+}) was determined over the temperature range of -20 to 50 °C, and t_{Li+} was found to decrease as temperature decreased. Although there are no theoretical models to treat and interpret the temperature dependence of transport numbers, we found that a modified (empirical) VTF equation resulted in an excellent fit to the temperature dependence of the transport number, which is another confirmation of a free volume model for transport in these electrolyte systems.

2.1 INTRODUCTION

The use of room-temperature ionic liquids for lithium and Li-ion batteries is an expanding area of research [1]. A major tenant of our work involves the development of suitable electrolyte solutions tailored to specific anodes and cathodes, which are stable over long periods of time (up to 10 years) and possess practical conductivities over the temperature range of -40 to 70 °C. Depending on these specific requirements, we have explored two types of systems containing ionic liquids, namely, Li-based batteries that employ liquid electrolyte solutions and gelled polymer electrolytes. In both approaches, mixtures of organic electrolyte solutions with selected ionic liquids have been studied, and below we summarise the pros and cons we considered in selecting electrolytes with ionic liquids.

Pros: They are nonflammable and environmentally benign, have low vapour pressure due to very high boiling points, have high thermal stability, and have high electrochemical stability at positive potentials (5 + V vs. Li/Li$^+$).

Cons: Ionic liquids generally have moderate to high viscosities, particularly at low temperatures, which reduces ionic conductivity, and unsaturated ionic liquids (see below) react with Li-based anodes at potentials from $\sim$0.5 to 1.2 V versus Li/Li$^+$. These undesirable properties can be mitigated by several methods, which are summarised below.

1. For liquid-based electrolyte solutions, the blending of ionic liquids with a low-viscosity organic solvent will lower viscosity and improve conductivities down to at least -45 °C.
2. For liquid-based electrolyte solutions, the reactivity of the ionic liquid toward reduction at metallic Li and LiC$_6$ anodes can be eliminated by use of "saturated" ionic liquids such as 1-butyl-1-methylpyrrolidinium bis(trifluoromethylsulfonyl)amide ([C$_4$mpyrr][NTf$_2$]) or tetraalkylammonium salts, which are stable at the Li/Li$^+$ reversible potential (e.g., see Ref. 2).
3. For Li systems utilising gelled electrolytes, "unsaturated" ionic liquids such as 1-ethyl-3-methylimidazolium bis(perfluoroethylsulfonyl)amide ([C$_2$mim][N-{SO$_2$(C$_2$F$_5$)}$_2$]) and low vapour pressure organic solvents such as propylene

carbonate (PC) and γ-butyrolactone (γ-BL) can be used even though they are reactive at Li-based anodes. Resistance toward reduction at Li and LiC_6 anodes can be solved by use of a very stable SEI forming additive such as vinylene carbonate (VC) and ethylene sulfide (ES).

Applications of these concepts to both liquid and gelled polymer electrolytes are discussed below.

2.2 EXPERIMENTAL

2.2.1 Liquid-Based Electrolyte Solutions

These studies involved mixtures of the ionic liquid 1-butyl-1-methylpyrrolidinium bis(trifluoromethylsulfonyl)amide ($[C_4mpyrr][NTf_2]$) with a "baseline" quaternary organic electrolyte solution of the following composition:

1 mol·dm^{-3} Li[BF$_4$] in a 1 : 1 : 1 : 3 (by volume) mixture of EC : DMC : DEC : EMC

where EC is ethylene carbonate, DMC is dimethyl carbonate, DEC is diethyl carbonate, and EMC is ethylmethyl carbonate. This quaternary solution was selected as a baseline due to its high anode and cathode stability, and high conductivity over the temperature range of −60 to 30 °C [3]. The organic solvents and Li[BF$_4$] were the highest purity (99%) available from Aldrich and Ferro Corp., and the purity of [C$_4$mpyrr][NTf$_2$] was ≥99.5% (Merck). All materials were used as received. The electrolyte solutions were prepared gravimetrically by adding [C$_4$mpyrr][NTf$_2$] to the baseline solution, and conductivities for the following six solutions were determined:

1. Baseline
2. Baseline + 10 mass% [C$_4$mpyrr][NTf$_2$]
3. Baseline + 25 mass% [C$_4$mpyrr][NTf$_2$]
4. Baseline + 50 mass% [C$_4$mpyrr][NTf$_2$]
5. Baseline + 90 mass% [C$_4$mpyrr][NTf$_2$]
6. 100% [C$_4$mpyrr][NTf$_2$]

The ionic conductivities of these solutions were determined by audio frequency (1 kHz) measurements using a YSI 3200 conductivity metre and calibrated 0.1 cm^{-1} conductivity cells. Measurements were made from high to low temperature in a forced air Tenny Environmental Chamber controlled to ±1–2 °C. The temperatures studied ranged from −60 to 60 °C. The temperature dependence of the conductivities was analysed in terms of Arrhenius behaviour,

$$\sigma = A \exp\left(\frac{-E_a}{RT}\right) \qquad (2.1)$$

and in terms of the VTF equation,

$$\sigma T^{1/2} = A'\exp\left(\frac{-E_a}{R(T-T_0)}\right) \quad (2.2)$$

In Eqs. (2.1) and (2.2), σ is the specific conductivity, A and A' are frequency factors (A' is related to the number of charge carriers as discussed below), E_a is an activation energy, and T_0 is a reference temperature closely related to the glass transition temperature T_g.

2.2.2 Gelled Polymer Electrolyte Systems

The gels studied are based on the block copolymer PVdF-HFP and the plasticising organic solvents ethylene carbonate (EC), propylene carbonate (PC), γ-butyrolactone (γ-BL), and vinylene carbonate (VC) and the ionic liquid [C$_2$mim][N{SO$_2$(C$_2$F$_5$)}$_2$]. Selection of EC and PC is based on the low vapour pressure PVdF-HFP gel developed by Sony [4], and improvement in performance of this base system is achieved by first adding γ-BL to increase conductivities and then the ionic liquid [C$_2$mim][N{SO$_2$(C$_2$F$_5$)}$_2$] to improve cross-linking and safety. Since PC, γ-BL, and [C$_2$mim][N{SO$_2$(C$_2$F$_5$)}$_2$] are known "aggressive" solvents toward graphite anodes in Li-ion cells, small amounts of VC were added to this new formulation to stabilise the passive SEI on graphite [5]. Gelled polymer electrolytes involved dissolving the lithium salt (Li[N{SO$_2$(C$_2$F$_5$)}$_2$]) in the appropriate solvent mixtures heated to ∼45 °C followed by addition of PVdF-HFP and further heating to 90 °C to assure complete dissolution of the polymer. The hot solutions were then quenched to room temperature by casting onto Al foil forming films around 250 μm thick. The following summarises the sources and purities for each material: [C$_2$mim][N{SO$_2$(C$_2$F$_5$)}$_2$] (99% pure, from Strem Chemicals), Li[N{SO$_2$(C$_2$F$_5$)}$_2$] (battery grade, 3M), PVdF-HFP (Kynar PVdF 2801), γ-BL (99%, EM Science), PC (99%, Aldrich), EC (99%, Ferro Corp.), and VC (98%, H.C. Starck). All solids were dried in vacuum at 90–100 °C overnight, and all liquids were dried with 3 Å molecular sieves overnight. Fabrication of the gels was carried out in a dry room of 1% relative humidity. Table 2.1 summarises the compositions of the gels.

TABLE 2.1 Composition of Gels Containing Li[N{SO$_2$(C$_2$F$_5$)}$_2$]

Gel Number	Ionic Liquid[a]/ mass %	EC/ mass %	PC/ mass %	VC/ mass %	GBL/ mass %	Li Salt[b]/ mass %	PVdF 2801/ mass %
1	—	15.61	31.15	2.34	31.15	13.09	6.53
2	4.77	14.86	29.72	2.23	29.72	12.48	6.21
3	62.5	—	—	—	—	18.75	18.75

[a] [C$_2$mim][N{SO$_2$(C$_2$F$_5$)}$_2$].
[b] Li[N{SO$_2$(C$_2$F$_5$)}$_2$].

The ionic conductivities of the gels were determined by AC impedance spectroscopy using cells with stainless steel blocking electrodes and Teflon spacers to fix the thickness. Measurements were made from high to low temperature in the Tenny Environmental Chamber. The impedance tests were carried out from 100 kHz to 0.1 Hz using a Gamry model PC13/300 potentiostat. The specific ionic conductivities, σ, of the gels were determined by using

$$\sigma = \frac{t}{(d/2)^2 \pi} \frac{1}{R_i} \tag{2.3}$$

where t, d, and R_i represent the thickness, the diameter, and the ionic resistance (bulk resistance plus grain boundary resistance), respectively. The temperatures studied ranged from −60 to 60 °C. The temperature dependence of the conductivities was analysed in terms of Arrhenius behaviour, that is, Eqs. (2.1) and (2.2).

In addition, transport numbers for gels 1 and 2 were determined from −20 to 50 °C using the method of Bruce et al. [6–8], which is a direct current (DC) polarisation method. The cell used for these measurements consisted of the gelled electrolyte sandwiched between two lithium metal disks. A small constant DC bias potential ≤0.03 V was applied to the electrodes of the cell, and the current is measured, which falls from an initial value i_o to a steady-state value i_s that was reached after 2–6 h. With the passage of time, anions accumulate at the anode and are depleted at the cathode and a concentration gradient is formed. At the steady state, the net anion flux falls to zero and only cations carry the current: therefore the cation transference number can be evaluated from the ratio i_s/i_o. Since the thickness of the passivating film on the lithium electrodes will vary over the time required to reach a steady-state current, the values of the intrinsic resistance must be measured shortly before the application of the DC bias potential and immediately after the attainment of steady state in order to determine the correct cationic transference numbers t_+, by using the equation

$$t_+ = \frac{i_s(\Delta V - i_o R'_o)}{i_o(\Delta V - i_s R'_s)} \tag{2.4}$$

In Eq. (2.4), the subscripts o and s indicate initial values and steady-state values, respectively, R' is the passivating film resistance R_{film}, ΔV is the applied voltage, and i is the current. The measurement of R'_s and R'_o was achieved by recording two impedance spectra on the cell in the frequency range between 0.1 Hz and 100 kHz before the application of the bias potential, and after steady state has been reached and the DC bias potential has been removed. The resulting spectrum is typically a semicircle and asymmetric in shape due to convolution of two different contributions, that is, the passivation film on the lithium electrodes and the charge-transfer resistance of the lithium oxidation–reduction (deposition–stripping) reaction. The deconvolution of the spectra to obtain R_o and R_{film} was made using the Non-Linear Least-Square (NLLSQ) fit software developed by Boukamp [9, 10] as discussed in detail in Reference 11.

30 CONDUCTIVITIES OF IONIC LIQUID MIXTURES

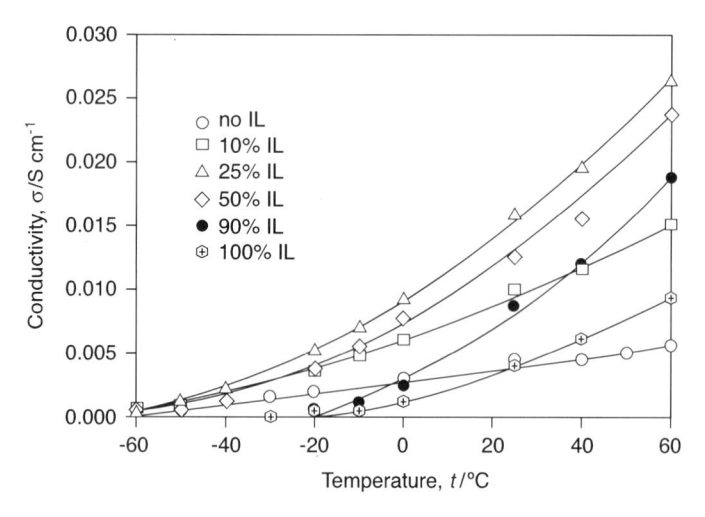

FIGURE 2.1 Linear plot of conductivities of [C₄mpyrr][NTf₂]-based solutions versus temperature over the ionic liquid composition range of 0–100 mass %.

2.3 RESULTS AND DISCUSSION

2.3.1 Liquid-Based Electrolyte Solutions

Figure 2.1 is a linear plot of conductivity for the baseline solution + [C₄mpyrr][NTf₂] as a function of temperature, and Figure 2.2 is the Arrhenius plot for all the experimental data. One of the interesting features of these results is that the

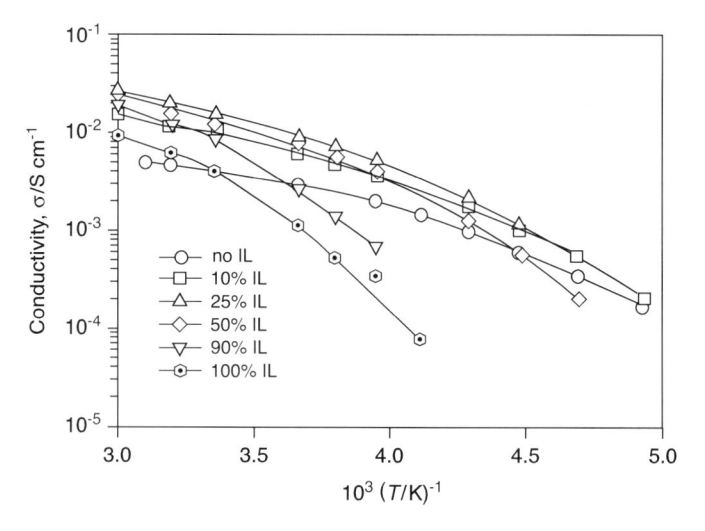

FIGURE 2.2 Semilog (Arrhenius) plots of conductivities of [C₄mpyrr][NTf₂]-based solutions versus the reciprocal of temperature over the ionic liquid composition range of 0–100 mass %.

conductivities increase as [C$_4$mpyrr][NTf$_2$] is added to the baseline solution and reach a maximum at approximately 25 mass% [C$_4$mpyrr][NTf$_2$]. A similar observation was made by Jarosik et al. [12] in mixtures of [C$_4$mpyrr]Br with PC. The reason for this behaviour is complex and certainly depends on multiple factors such as viscosity and free volume as discussed later.

While it appears to us that viscosity is the basic reason for the low conductivities at high concentrations of the ionic liquid, it also appears that free volume effects are involved as originally proposed [11]. The parameters of the semi-empirical VTF equation (2.2) can be interpreted in terms of free volume concepts. In electrolyte solutions based on a salt dissolved in a liquid, cations and anions are solvated (coordinated to) solvent molecules; for example, the solvation number of Li$^+$ in PC and other liquid carbonates is probably 3–4, and for the anion, the solvation number is probably 1 (e.g., see Ref. 13). In studying the conductivities for these solutions, it is the mobility of the whole complex (ion + the solvent coordination sphere) that moves under the applied electric field. When the solvent molecules are removed, one is left with a pure salt which, in the liquid state (the "ionic liquid"), is often more conductive and in which conductivity is based solely on the mobility of ions moving in a viscous environment where there is no solvation sphere. When a molten ionic salt melts, its volume generally increases by 10–25%, but X-ray diffraction (XRD) studies show that the ions in the fused salt are slightly closer to each other than in the solid state [14]. To account for this apparent contradiction of an increase in volume when the interionic distances decrease, it follows that empty volume or "free space" must be introduced into the ionic liquid, and the mobility of ions in the ionic liquid is due to jumps of ions into vacant sites, that is, into the free volume. Our results suggest important contributions to the overall conductivity by this free volume effect, and the conductivity data were therefore fitted to the VTF equation (2.2). The results of this approach are shown graphically in Figure 2.3, and the derived VTF parameters are given in Table 2.2. The VTF A' parameters are related to the number of charge carriers, and, as shown in Table 2.2, the A' parameters generally increase as [C$_4$mpyrr][NTf$_2$] is added to the baseline electrolyte solution, which is in agreement with the free volume model of ion conductance. Similar behaviour is observed for our gelled electrolyte as discussed later. While small additions of 10–90 mass% of the ionic liquid to the baseline electrolyte solution improves the overall conductivities and increases the calculated T_o values, the conductivities of the pure ionic liquid are significantly smaller below room temperature due to the high viscosity of [C$_4$mpyrr][NTf$_2$].

2.3.2 Gelled Polymer Electrolyte Systems

The behaviour of the gels listed in Table 2.1 follows closely that for the liquid-based systems discussed earlier. Again, the conductivities appear to increase as small amounts of the ionic liquid [C$_2$mim][N{SO$_2$(C$_2$F$_5$)}$_2$] are added to the baseline gel (gel 1), but conductivities for a PVdF-[C$_2$mim][N{SO$_2$(C$_2$F$_5$)}$_2$] gel (gel 3) are significantly lower, which we attribute to significantly increased viscosity. The Arrhenius conductivity plots for these gels all show curvature, but the VTF plots

32 CONDUCTIVITIES OF IONIC LIQUID MIXTURES

TABLE 2.2 Derived VTF Parameters for Solutions With and Without [C$_4$mpyrr][NTf$_2$]

% IL	A'/S cm^{-1} $(T/K)^{1/2}$	T_0/K	t_0/°C	E_a/kcal mol^{-1}	E_a/kJ mol^{-1}
0	0.473	152.2	−121.0	0.549	2.299
10	2.637	136.1	−137.1	0.886	3.707
25	5.356	141.5	−131.7	0.927	3.880
50	4.410	156.3	−116.9	0.826	3.456
90	10.949	174.9	−98.3	1.077	4.507
100	7.739	176.2	−97.0	1.167	4.881

shown in Figure 2.4 are linear and again are interpreted in terms of the free volume model of ion conductance.

The derived VTF A' parameters are given in Table 2.3 and are seen to increase significantly as the concentration of [C$_2$mim][N{SO$_2$(C$_2$F$_5$)}$_2$] is increased (Table 2.2) reaching a value of 2.732 S cm^{-1}$(T/K)^{1/2}$ for the gelled electrolyte 3. While small additions of 1–5 mass% of the ionic liquid to the baseline gel number 1 improves the overall conductivities, the conductivities of gel 3 are significantly smaller due to the high viscosity of [C$_2$mim][N{SO$_2$(C$_2$F$_5$)}$_2$].

In a separate study, we determined the Li$^+$ transference numbers for the baseline gel electrolyte at 25 °C and for the 5 mass % gel (gel number 2) over the temperature range of −20 to 50 °C, and the results are given in Table 2.4.

The interpretation of transport numbers as a function of temperature is an interesting problem. If ionic mobilities λ_i^o followed Stokes's law,

$$\lambda_i^o \eta = \frac{zeF}{6\pi r_i} \quad (2.5)$$

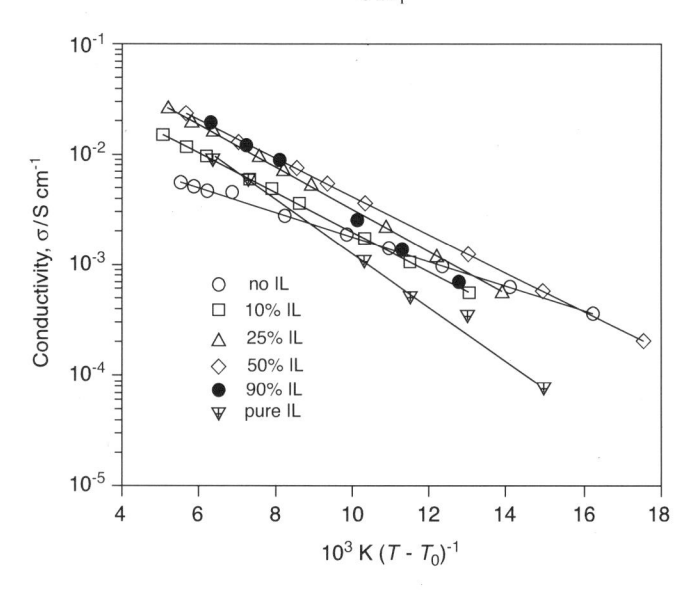

FIGURE 2.3 VTF plot of conductivities of [C$_4$mpyrr][NTf$_2$]-based solutions over the ionic liquid composition range of 0–100 mass %.

TABLE 2.3 Ionic Liquid ([C$_2$mim][N{SO$_2$(C$_2$F$_5$)}$_2$]) Content of Gels and Derived VTF Parameters

Gel Number	$A'/\text{S cm}^{-1} (T/K)^{1/2}$	$E_a/\text{kJ mol}^{-1}$	T_0/K	$t_0/\text{°C}$
1 (no IL)	0.109	0.676	215.7	−57.5
2 (5% IL)	0.437	2.293	169.4	−103.8
3 (63% IL)	2.732	5.233	195.1	−78.1

the Walden product $\lambda\eta$ would be independent of temperature if the ionic radius r_i is constant. It is clear that, at least for the gel we studied, the transport number for Li$^+$ rapidly decreases as the temperature is decreased. Of course, the temperature dependence of mobilities and transport numbers could also arise from changes in ion solvation; that is, the ionic radius r_i is not constant, but how this affects the sign of the temperature coefficient of $\delta t_+/\delta T$ as a function of temperature is not clear. According to Kay [15], no quantitative theory exists to relate the temperature dependence of transport numbers. Since transport numbers are related to mobilities, we sought to determine if behaviour of the gels studied in the present work can be modeled similar to that used for conductivities, that is, to the empirically modified VTF equation (2.6):

$$t_+ T^{1/2} = A^* \exp\left(\frac{-E_a}{R(T-T_0)}\right) \qquad (2.6)$$

Figure 2.5 shows the plot of our transport data as a function of temperature based on the VTF model. Based on the excellent fit of the data to Eq. (2.6), as shown in

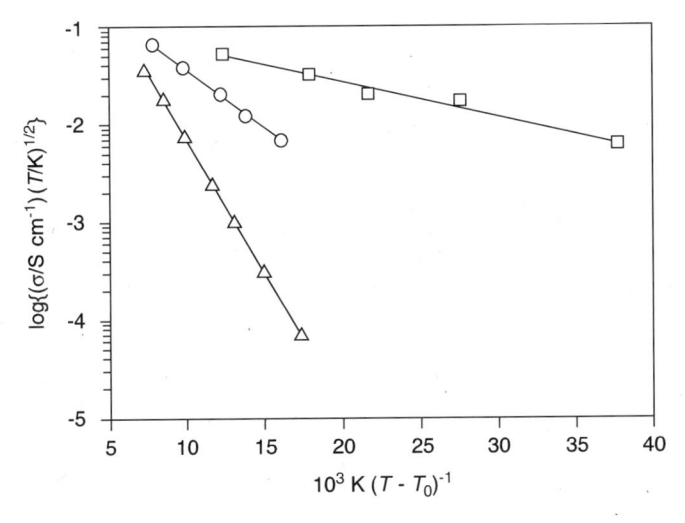

FIGURE 2.4 VTF plots for the conductivities of gels listed in Table 2.1: □, gel 1; ○, gel 2; △, gel 3.

TABLE 2.4 Transport Numbers for Gels 1 and 2

Gel Number	t/°C	T/K	t_+
1	25	298.2	0.41
2	50	323.2	0.66
2	25	298.2	0.47
2	0	273.2	0.38
2	−10	263.2	0.27

Figure 2.5, we conclude that transport behaviour in our polymer electrolyte can be explained in terms of a free volume model as we concluded based on our conductivity measurements. Fitting the t_+ (Li$^+$) data for gel number 2 to temperatures from −20 to 50 °C to the modified VTF equation (2.6) resulted in the following values for the adjustable parameters:

$$A^* = 18.095 \, (T/K)^{1/2}, \quad E_a = 384 \, \text{J} \cdot \text{mol}^{-1}, \quad \text{and} \quad T_0 = 230.9 \, \text{K} \, (-42.3\,^\circ\text{C})$$

While the theoretical significance of these adjusted VTF parameters is not resolved at this time, our analysis again confirms that the transport properties of gel 2, and by inference to all the studies reported here, are heavily influenced by free volume behaviour. Direct comparison of the VTF parameters for conductivity and for transport numbers in our studies is not possible at this time since the former reflects contributions by the two cations Li$^+$ and [C$_2$mim]$^+$, whereas the latter VTF parameters for t_+ reflect contributions only by the Li$^+$ ion. Qualitatively, one can relate the temperature dependence to the mechanism of transport and to solvation effects depending on the system under study, for example, polymer versus aqueous.

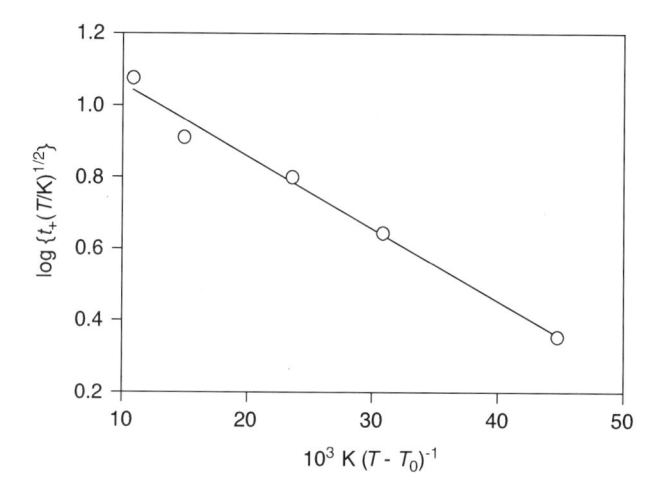

FIGURE 2.5 VTF plot for t_+ in gel 2 (see Table 2.1).

For ionic liquids and many polymer electrolytes, conduction can be related to available free volume [16], and as temperature is decreased, the available free volume also decreases, resulting in a decrease in conductivity and transport number as we observe for the gels studied here.

2.4 CONCLUSION

The fact that we observe an increase in electrolyte solution conductivity when an ionic liquid is added to a liquid organic solvent based electrolyte solution has important significance for future research on these systems for Li and Li-ion batteries. For both liquid electrolyte solutions and polymer electrolytes, conductivities increase as the ionic liquid concentration is increased and, after reaching a maximum, decrease as the ionic liquid concentration further increases.

The conductivities for liquid electrolyte solutions and transport number measurements for one gel were found to follow a VTF model and are interpreted in terms of a free volume model for mobility. For our studies on the gelled electrolyte, the result indicating a significant decrease in transport number as temperature decreases has important implications for the use of gelled polymer electrolytes in Li and Li-ion batteries. It is well known that the conductivities of gelled polymer electrolytes are generally smaller than their liquid-based counterparts and rarely are of practical use for Li-ion cells at temperatures below $-20\,°C$ (e.g., see Ref. 4). The fact that we observe $t_+(Li^+)$ decreases from 0.47 at $25\,°C$ to 0.14 at $-20\,°C$ implies that besides conductivity limitations for gelled electrolytes, Li^+ transport is another significant limitation for use below $-20\,°C$ in Li and Li-ion batteries.

REFERENCES

1. Webber, A. and Blomgren, G. E., Ionic liquids for lithium-ion and related batteries, in *Advances in Lithium-Ion Batteries*, W.A. van Schalkwijk and B. Scrosati (Eds), Kluwer Academic Plenum, New York, 2002, Chap. 6.
2. Howlett, P. C., MacFarlane D.R., and Hollenkamp A. F., *Electrochem. Solid-State Lett.* **7**, A97 (2004).
3. Smart, M. C., Ratnakumar, B. V., Whitcanack, L. D., Chin, K. B., and Surampudi, S., Improved low temperature performance of lithium ion cells with quaternary carbonate-based electrolytes. Paper presented at *11th International Meeting on Lithium Batteries (IMLB11)*, Monterey, California, 23–28 June 2002.
4. Nishi, Y., Li-ion secondary batteries with gelled polymer electrolytes, in *Advances in Lithium-Ion Batteries*, W. van Schalkwijk and B. Scrosati (Eds.), Kluwer Academic/Plenum, New York, 2002, Chap. 7.
5. Biesan, Ph., Bodet, J. M., Perton, F., Broussely, M., Jehoulet, C., Barusseau, S., Herreye, S., and Simon, B., Paper presented at the *10th International Meeting on Lithium Batteries.* Como, Italy, 28 May to 2 June, 2000. See also French Patent Fr 94 04889, April 1994 and U.S. Patent 5,626,981, May 1997.

6. Bruce, P. G., and Vincent, C.A., *J. Electroanal. Chem.* **225**, 1 (1987).
7. Bruce, P. G., Evans, J., and Vincent, C. A., *Solid State Ionics* **28–30**, 918 (1988).
8. Evans, J., Vincent, C.A., and Bruce, P.G., *Polymer* **28**, 2324 (1987).
9. Boukamp, B. A., *Solid State Ionics* **18**, 136 (1986).
10. Boukamp, B. A., *Solid State Ionics* **20**, 31 (1986).
11. Bansal, D., Cassel, F., Croce, F., Hendrickson, M., Plicta, E., and Salomon, M., *J. Phys. Chem. B* **109**, 4492 (2005).
12. Jarosik, A., Krajewski, S. R., Lewandowski, A., and Radzimski, P., *J. Mol. Liquids* **123**, 43 (2006).
13. Salomon, M., Lin, H.-P., Hendrickson, M., and Plichta, E., Temperature effects on Li-ion cell performance, in *Advances in Lithium-Ion Batteries*, W. van Schalkwijk and B. Scrosati (Eds), Kluwer Academic/Plenum, 2002, Chap. 11.
14. Bockris, J. O'M. and Reddy, A. K. N., *Modern Electrochemistry 1: Ionics*, 2nd ed. Plenum Press, New York, 1998.
15. R. L. Kay, Transference number measurements, in *Techniques in Electrochemistry*, Vol. II, E. Yeager and A. Salkind (Eds), Wiley-Interscience, Hoboken, NJ 1986.
16. Ratner, M. A. and Shriver, D. F., *Chem. Rev.* **88**, 109 (1988).

3 How Hydrophilic Ionic Liquids Behave in Aqueous Solutions

MARIJANA BLESIC and KENNETH R. SEDDON

Instituto de Tecnologia Química e Biológica, Universidade Nova de Lisboa, Oeiras, Portugal and The QUILL Centre, The Queen's University of Belfast, Belfast, United Kingdom

NATALIA V. PLECHKOVA and NIMAL GUNARATNE

The QUILL Centre, The Queen's University of Belfast, Belfast, United Kingdom

ANTÓNIO LOPES and LUÍS PAULO N. REBELO

Instituto de Tecnologia Química e Biológica, Universidade Nova de Lisboa, Oeiras, Portugal

Abstract

Self-aggregation behaviour of a number of ionic liquids has been investigated using two techniques: interfacial tension measurement and fluorescence spectroscopy. The list of ionic liquids includes [$C_n$mim]Cl (1-alkyl-3-methylimidazolium chlorides) with different linear alkyl chain lengths ($n = 4$–14, only even numbers), [C_{10}mim][PF_6] (1-decyl-3-methylimidazolium hexafluorophosphate), [C_{10}mim][NTf_2] (1-decyl-3-methylimidazolium bis{(trifluoromethyl)sulfonyl}amide (bistriflamide)), [C_{12}mpip]Br (1-dodecyl-1-methylpiperidinium bromide), [C_{12}mpy]Br (1-dodecyl-3-methylpyridinium bromide), [C_{12}mpyrr]Br (1-dodecyl-1-methylpyrrolidinium bromide), and [$C_{12}C_{12}$pip]Br (1,1-didodecylpiperidinium bromide). Four factors were varied: the length and number of alkyl chains on the ring, the anion, and the structure of the cation (ring). Only [$C_n$mim]Cl ionic liquids with n greater than 8 unambiguously form aggregates in solution. The transitional ionic liquid, [C_6mim]Cl, is able to develop a monolayer at the aqueous solution–air interface but shows no noticeable self-aggregation in the bulk fluid. Ionic liquids used in this study based on more hydrophobic and large anions, namely, [PF_6]$^-$ and [NTf_2]$^-$, do not form micellar aggregates in aqueous solutions. Moreover, the type of the ring structure of the cation and the number of alkyl chains affect the dynamic balance of aggregate formation.

Molten Salts and Ionic Liquids: Never the Twain? Edited by Marcelle Gaune-Escard and Kenneth R. Seddon
Copyright © 2010 John Wiley & Sons, Inc.

The nature of this self-aggregation is discussed in terms of the electrostatic versus hydrophobic contributions.

3.1 INTRODUCTION

The innate amphiphilic nature of some cations, for example, $[C_n\text{mim}]^+$ ($[C_n\text{mim}]^+$ = 1-alkyl-3-methylimidazolium), suggests the possibility of interfacial and aggregation phenomena in aqueous solutions, and their ultimate organisation into micelles displaying surfactant behaviour [1–3]. Ionic liquids are very similar in structure to traditional ionic surfactants (most of the latter are solid salts). Amphiphilic ionic liquids have low melting points, usually a consequence of the presence of bulky and asymmetric cations, compared to considerably smaller size anions, while ionic surfactants typically have higher melting points—for example, sodium dodecyl sulphate (SDS), $\text{Na}[C_{12}H_{25}SO_4]$, has a melting point close to 200 °C. Longer chain ionic liquids self-assemble to form thermotropic liquid crystalline mesophase ionic liquids [4–6] and lyotropic mesophases in concentrated aqueous solutions of 1-decyl-3-methylimidazolium bromide [7].

With the possibility of fine-tuning of the ionic liquids' hydrophobicity by changing the alkyl chain length, the type of head (cation), and/or the nature and size of the counterion (anion), one can affect both the structure and the delicate dynamics of these micellar aggregates. This can lead to modification of their major characteristics such as the critical micelle concentration (CMC) and the aggregation number (micellar size). It is interesting that even some short chain ionic liquids, for example, 1-butyl-3-methylimidazolium tetrafluoroborate, $[C_4\text{mim}][BF_4]$, can form very small aggregates in aqueous solution [8].

On the other hand, the addition of ionic liquids to aqueous solutions of surfactants affects the CMC of the surfactant and offers the possibility of their usage as cosurfactants [2, 9, 10].

Furthermore, it was shown that solvatophobic interactions are present between ionic liquids and the hydrocarbon portion of the surfactant, thus leading to the formation of surfactant micelles in ionic liquids and enhancing the solvation characteristics of the (ionic liquids + surfactant) system [11, 12].

However, previous studies [1, 9] report contradictory conclusions and do not confirm an obvious differentiation between electrostatic and hydrophobic contributions for the self-aggregation of the $[C_n\text{mim}]\text{Cl}$ family. Fortunately, the formation of aggregates, particularly micelles, can be detected using several instrumental methods: equivalent conductivity, turbidity, surface tension, self-diffusion, magnetic resonance, and many others [13]. Hence it is generally possible to obtain comparative results.

The aim of the current contribution is to offer a more systematic study and establish the role of the alkyl chain length, concentration and the nature of the anion, and the type of head group on the aggregation behaviour of the compounds belonging to the $[C_n\text{mim}][X]$ (X = Cl, PF_6, or NTf_2) and $[C_{12}Y]\text{Br}$ (Y = mpyrr, mpy, and pip) families in aqueous solutions, as well as to elucidate those hydrophobic and electrostatic contributions for the building up of the micellar aggregates.

3.2 EXPERIMENTAL

3.2.1 Chemicals

The 1-alkyl-3-methylimidazolium chlorides, [C$_n$mim]Cl ($n=4$–14, even numbers), [C$_{10}$mim][PF$_6$], [C$_{10}$mim][NTf$_2$], and [C$_{12}$Y]Br (Y = mpyrr, mpy, and pip), were all synthesised and initially purified at QUILL (The Queen's University Ionic Liquid Laboratories, Belfast) according to methods described elsewhere [4, 14, 15]. The structural formulae of the cations used in this study are shown in Figure 3.1.

1,1-Didodecylpiperidinium bromide, [C$_{12}$C$_{12}$pip]Br, was synthesised according to the following procedure: Piperidine (8.515 g, 0.100 mol) and 1-bromododecane (27.42 g, 0.110 mol) were placed in a round-bottomed flask (500 cm^3) and dissolved in ethanenitrile (100 cm^3). Potassium carbonate (27.64 g, 0.200 mol) was added to the reaction mixture. The reaction mixture was left to stir under reflux at 70 °C for 3 days. Afterwards, the solid residues were separated by filtration. The filtrate was distilled to yield 1-dodecylpiperidine as a product (22.56 g, 0.089 mol, 89%). The obtained 1-dodecylpiperidine (22.56 g, 0.089 mol) was treated with 1-bromododecane (24.42 g, 0.098 mol) and the reaction mixture was dissolved in ethyl ethanoate (100 cm^3). The mixture was heated in a microwave at 90 °C for 3 h. The halide salt separated as a second phase from the ethyl ethanoate at room temperature. The excess of ethyl ethanoate was removed by decantation. The halide salt was then recrystallised

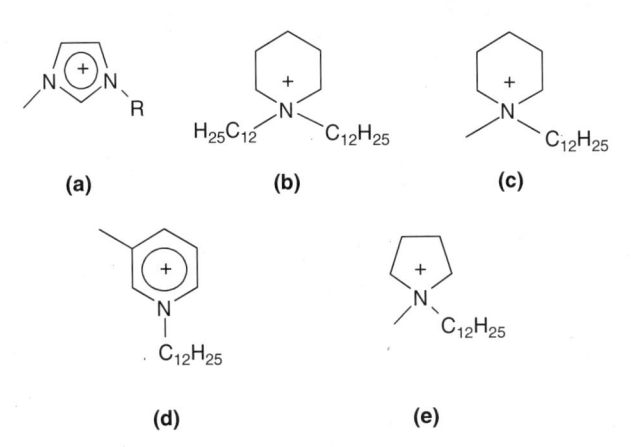

FIGURE 3.1 Structural formulae of the cations used in this study: (a) [C$_n$mim]$^+$ (1-alkyl-3-methylimidazolium), (b) [C$_{12}$C$_{12}$pip]$^+$ (1,1-didodecylpiperidinium), (c) [C$_{12}$mpip]$^+$ (1-dodecyl-1-methylpiperidinium), (d) [C$_{12}$mpy]$^+$ (1-dodecyl-3-methylpyridinium), and (e) [C$_{12}$mpyrr]$^+$ (1-dodecyl-1-methylpyrrolidinium).

from ethyl ethanoate. The ethyl ethanoate was decanted followed by the addition of fresh ethyl ethanoate and this step was repeated six times. After the sixth cycle the remaining ethyl ethanoate (bp: 77 °C) was removed under reduced pressure at 70 °C and the bromide salt was finally dried in vacuo at 70 °C. The product was obtained as a white solid (30.87 g, 0.061 mol, 69% yield).

^{1}H NMR (300 MHz, d-trichloromethane): δ/ppm = 0.73 (t, 6H, J = 7.2 Hz, 3H, N–(CH$_2$)$_{11}$–CH_3), 1.11–1.22 (m, 36H, N–(CH$_2$)$_2$–$(CH_2)_9$–CH$_3$), 1.50 (m, 4H, N–CH$_2$–CH_2–(CH$_2$)$_9$–CH$_3$, 1.75 (m, 6H, N–CH$_2$–CH$_2$–CH_2 (cyclic), N–CH$_2$–CH_2–CH$_2$ (cyclic)), 3.29–3.24 (m, 4H, N–CH_2–(CH$_2$)$_{10}$–CH$_3$), 3.62 (N–CH$_2$ (cyclic)).

^{13}C NMR (75 MHz, d-trichloromethane): δ/ppm = 13.92 (N–(CH$_2$)$_{11}$–CH_3), 19.86, 20.47, 21.53, 22.48, 26.26, 29.02, 29.13, 29.20, 29.23: (N–CH$_2$–(CH$_2$)$_9$–CH_2–CH$_3$), (N–CH$_2$–(CH$_2$)$_8$–CH_2–CH$_2$–CH$_3$), (N–CH$_2$–(CH$_2$)$_7$–CH_2–(CH$_2$)$_2$–CH$_3$), (N–CH$_2$–(CH$_2$)$_6$–CH_2–(CH$_2$)$_3$–CH$_3$), (N–CH$_2$–(CH$_2$)$_5$–CH_2–(CH$_2$)$_4$–CH$_3$), (N–CH$_2$–(CH$_2$)$_4$–CH_2–(CH$_2$)$_5$–CH$_3$), (N–CH$_2$–(CH$_2$)$_3$–CH_2–(CH$_2$)$_6$–CH$_3$), (N–CH$_2$–(CH$_2$)$_2$–CH_2–(CH$_2$)$_7$–CH$_3$), (N–CH$_2$–CH$_2$–CH_2–(CH$_2$)$_8$–CH$_3$), (N–CH$_2$–CH_2–(CH$_2$)$_9$–CH$_3$), 29.40 (N–CH$_2$–CH$_2$–CH_2 (cyclic)), 31.70 (N–CH$_2$–CH_2–CH$_2$ (cyclic)), 58.11 (N–CH_2–(CH$_2$)$_{10}$–CH$_3$), 58.85 (N–CH$_2$ (cyclic)).

^{1}H and ^{13}C NMR analyses showed no major impurities in the untreated, original samples, except for the presence of adventitious water (Karl-Fischer analysis). All samples were then further thoroughly degassed and dried to remove any small traces of volatile compounds by applying vacuum (0.1 Pa) at moderate temperatures (60–80 °C) for typically 72 h.

Doubly distilled deionised water was obtained from a Millipore Milli-Q water purification system (Millipore, USA). [C$_n$mim][X] stock solutions were prepared in a 1.74×10^{-6} M pyrene aqueous solution and all work solutions were prepared from the stock solutions diluting with the same pyrene aqueous solution. Pyrene (Fluka, Germany, 99%) was recrystallised from benzene. For NMR experiments, D$_2$O (Cambridge Isotope Lab., USA; D, 99.9%) was used and, apart from the use of pyrene, all other procedures were the same as for solution preparation.

3.2.2 Interfacial Tension Measurements

Interfacial tension (IFT) was measured using a Drop Shape Analysis Tensiometer Kruss DSA1 v 1.80 working in the pendant drop mode at a constant temperature of 23 ± 1 °C. The needle diameter was chosen as a function of the maximisation of the scale of measurement. IFT is derived from the fit of the pendant drop profile, and care was taken to ensure that the apparatus was calibrated with several solvents of known IFT in the range of interest. The drops were left to equilibrate close to the rupture point and at least three consistent measurements per solution were recorded.

3.2.3 Fluorescence Spectroscopy

Steady-state fluorescence spectra of the pyrene-containing solutions were recorded with a Cary Varian Eclipse Fluorescence Spectrometer and collected from a 1 cm^2 quartz cuvette at 25 °C. Excitation was set to a wavelength of 337 nm and all emission spectra measured were corrected for emission monochromator response and were background subtracted using appropriate blanks. Intensities of first (I1) and third (I3) vibronic bands in the pyrene emission spectra located close to 373 and 384 nm, respectively, were measured and used to determine the ratio I3/I1, a well-known "solvent" polarity scale [16]. Band I1 corresponds to a S0 ($\nu = 0$) ← S1 ($\nu = 0$) transition and band I3 is a S0 ($\nu = 1$) ← S1 ($\nu = 0$) transition.

3.2.4 ^{1}H NMR Spectroscopy

Proton NMR spectra were recorded on a BRUKER AMX300 spectrometer, operating at 300.13 MHz and equipped with 5 mm diameter inverse detection broadband probe head, with the following acquisition parameters: spectral width, 6 kHz; pulse width, 4 µs (70° flip angle); data points, 16 K; repetition delay, 5 s; number of transients, 48; temperature, 298 K.

3.3 RESULTS AND DISCUSSION

As 1-alkyl-3-methylimidazolium chlorides, [C$_n$mim]Cl, are completely miscible with water at all compositions of interest, the first group of experiments examined the behaviour of the aqueous solution–air interface for this series of compounds ($n = 4$–14, even numbers). The results of the IFT for the aqueous solutions as a function of the total concentration of [C$_n$mim]Cl are presented in Figure 3.2a. It is well known that a sharp decrease of the IFT followed by a plateau reveals the aforementioned amphiphilic characteristic of the cation, which is the cause for the possible micelle formation [17]. The break point of each concentration–surface tension curve determines the saturation of the outer monolayer of the surface of the drop, and the consequent onset of aggregation in the bulk, the CMC, when the dissolution permits.

Figure 3.2a clearly shows that the plateaux, which indicate micelle formation, are present only in the case of alkyl chains with a hydrocarbon "tail" equal or greater than $n = 8$. [C$_4$mim]Cl does not achieve any plateau, even in the limit of tending to the pure compound (higher concentration point). For pure and highly concentrated solutions of [C$_6$mim]Cl, it was not possible to determine IFT due to the high viscosity of the solutions. The longer hydrocarbon chain members for the higher concentration show the lower values of the IFT of the plateau region and higher compactness of the surface monolayer. The experimental values for the break points in the IFT measurements are presented in Table 3.1 (along with other CMC results from different techniques [2]). An interesting comparison with the measured CMC values of the [C$_n$mim]Br family, reported by Vanyúr et al. [18], is shown in Figure 3.3. The slightly lower values

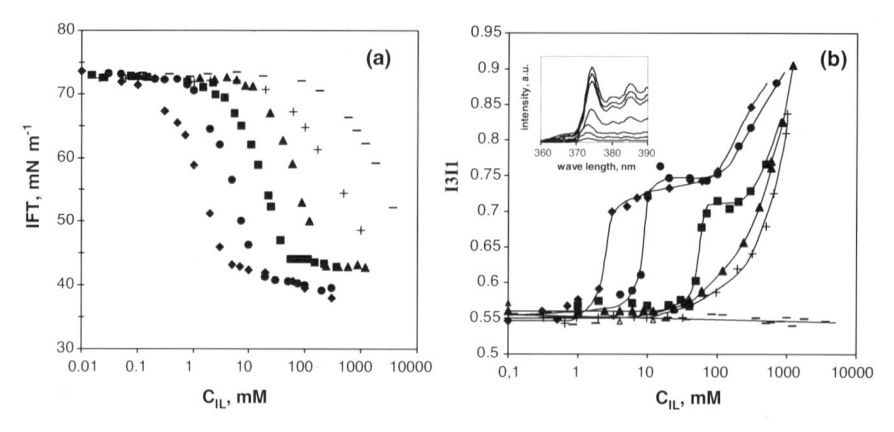

FIGURE 3.2 Monitoring the self-aggregation of [C$_n$mim]Cl using different techniques: (a) IFT and (b) fluorescence for different chain lengths: $n = (-)4, (+)6, (\blacktriangle)8, (\blacksquare)10, (\bullet)12$, and $(\blacklozenge)14$.

obtained for [C$_n$mim]Br are expected since it is well known that the binding of anionic counterions to cationic micelles increases in the order F$^-$ < Cl$^-$ < Br$^-$ < I$^-$ [19]. The binding of counterions not only influences the CMC value but increases the binding of counterions, decreases the head repulsion, and enables denser packing of monomers in monolayers. This leads to lower values for the plateau region at concentrations greater than the CMC. In this regard, we note that Bowers et al. [1] reported the surface tension value of 35 mN· m^{-1}, in the plateau region for [C$_8$mim]I, which is much lower than the value achieved for [C$_8$mim]Cl (43 mN· m^{-1}).

In order to corroborate the results mentioned above, fluorescence measurements that involve a pyrene solvatochromic probe were also used. Intensities of the first (I1) and third (I3) vibronic bands in the pyrene emission spectra were measured and used to determine the ratio I3/I1. The ratio I3/I1 is a function of polarity of the pyrene environment and it increases with decreasing solvent polarity (see Fig. 3.2b inset).

TABLE 3.1 Critical Micelle Concentration, CMC (mM), of [C$_n$mim]Cl ($n = 8$–14), [C$_{12}$Y]Br (Y = mpyrr, mpy, and pip), and [C$_{12}$C$_{12}$pip]Br Measured by Interfacial Tension (IFT), Fluorescence (Fluor), and ^{1}H NMR Spectroscopy[a]

Family	IFT	Fluor	NMR
[C$_8$mim]Cl	220	—	200
[C$_{10}$mim]Cl	55	45	55
[C$_{12}$mim]Cl	15	7	13
[C$_{14}$mim]Cl	4	3	4
[C$_{12}$mpy]Br	10		
[C$_{12}$mpip]Br	11		
[C$_{12}$mpyrr]Br	20		
[C$_{12}$C$_{12}$pip]Br	0.12		

[a]See Reference [2].

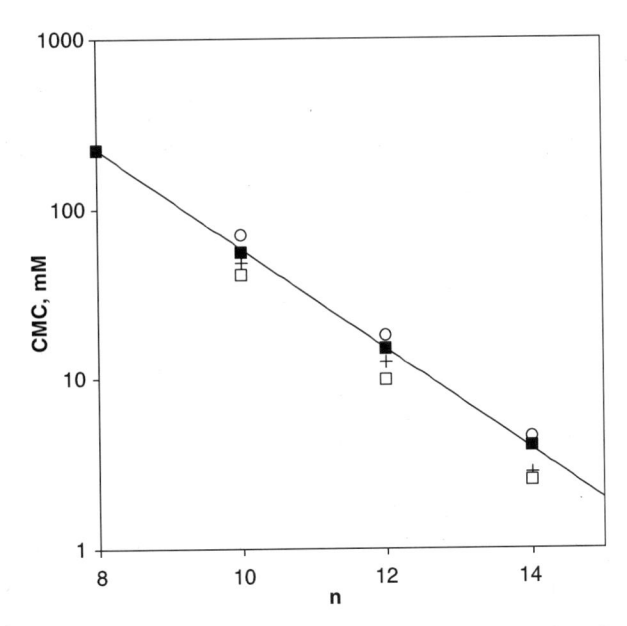

FIGURE 3.3 CMC values for [C$_n$mim]Cl (■) as a function of the number of carbon atoms in the alkyl chain. Data for [C$_n$mim]Br (□) [18], [N$_{111n}$]Cl (○), and ([H$_3$N$_n$]Cl (+) are included for comparison purposes.

The I3/I1 ratio as a function of the total surfactant (in our case the ionic liquid) concentration shows a specific sigmoidal type curve, which increases due to the formation of micelles with a well-defined hydrophobic core into which pyrene partitions preferentially [20]. Figure 3.2b shows the pyrene response for aqueous [C$_n$mim]Cl solutions as a function of [C$_n$mim]Cl concentration. In agreement with the surface tension measurements above, Figure 3.2b provides evidence that in solutions for $n<6$ either no hydrophobic environment is formed or the volume of formed aggregates is too small to "protect" pyrene from the polar environment. Although the I3/I1 ratio shows a typical increase, indicating the presence of hydrophobic environment for tails equal to or greater than 6 carbons, it is only certain that a plateau is reached for $n = 10$, 12, and 14. Consequently, the intersection point between that plateau and the descending part is used for the CMC determination only for these three [C$_n$mim]$^+$ salts. The reason for the absence of a plateau at higher concentrations in the case of [C$_6$mim]Cl or [C$_8$mim]Cl may be attributed to the progressive dense packing of the individual cations that constitute a micelle-like structure. This produces a steady increase in the hydrophobicity sensed by pyrene that reflects the continuous increase in the probe response. However, these observations are speculative, as tentative measurements with the light scattering technique did not clarify this point. Small angle neutron scattering, applied for [C$_8$mim]Cl solution by Bowers et al. [1], did not clarify the size and shape of the micelles either.

The CMC values obtained by the IFT, fluorescence, and ^{1}H NMR spectroscopy [2] are given in Table 3.1. Figure 3.3 summarises the CMC values obtained with the IFT methodology plotted as a function of the number of carbons for [C$_n$mim]Cl, $n = 8$–14. For comparison purposes, data obtained for cationic surfactants of the alkyltrimethylammonium and alkylammonium chloride families, [N$_{111n}$]Cl, [H$_3$N$_n$]Cl, with similar hydrocarbon chain lengths, are depicted. Published data for the [C$_n$mim]Br family measured by conductivity are also included [18]. As one can perceive, there is a good free energy correlation for all the [C$_n$mim]Cl family and the results show the same dependence for micellisation as a function of n as observed for the other two families. It is even possible to note that the CMC values for [C$_n$mim]Cl are located between those of [H$_3$N$_n$]Cl and [N$_{111n}$]Cl. This probably reflects the effect of solute polarity and consequent head-group repulsion. Namely, [N$_{111n}$]Cl is a very stable ammonium salt and more polar than both [C$_n$mim]Cl (where the positive charge is delocalised) and [H$_3$N$_n$]Cl (that can even dissociate to some extent).

Figure 3.4 shows the effect of the anion of the ionic liquid on the surface tension of aqueous solutions of [C$_{10}$mim]X (where X is Cl$^-$, [NTf$_2$]$^-$, or [PF$_6$]$^-$). The concentrations of ionic liquid at which the surface tension starts to lower decrease as the hydrophobicity and bulkiness of the anion increase. This is believed to result from decreased electrostatic repulsion between head groups in the monolayer due to both the relatively strong binding of the hydrophobic counterion to the ionic liquids' cations and the hydrophobicity of the anions themselves. Generally, this may lead to lower CMC values, thus promoting micelle formation [17]. This statement can be confirmed if we compare our IFT versus concentration curve for [C$_4$mim]Cl and a similar curve for [C$_4$mim][BF$_4$] measured by Bowers et al. [1]. The more hydrophobic [BF$_4$]$^-$ anion is more strongly bound to the surface of micelles and consequently is

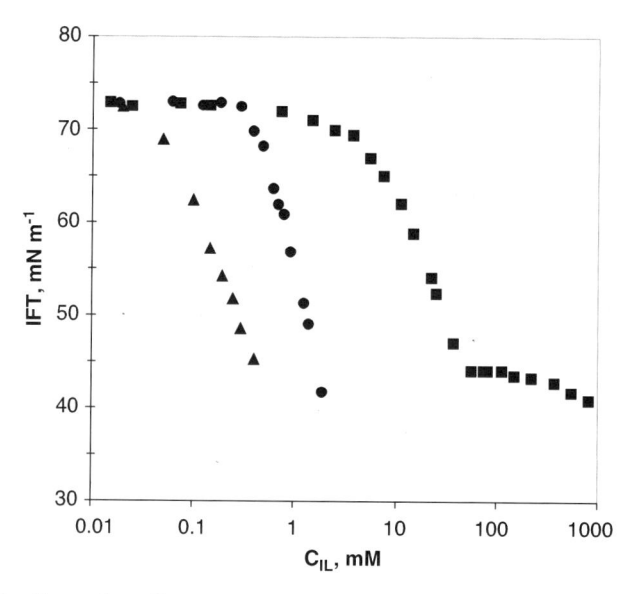

FIGURE 3.4 Counterion effect on [C$_{10}$mim]X aggregation from IFT measurements: X = (▲) [NTf$_2$]$^-$, (●) [PF$_6$]$^-$, (■) Cl$^-$.

more efficient in decreasing the electrostatic repulsion. As a result, [C$_4$mim][BF$_4$] has the CMC at 0.8 ± 0.1 mol· dm^{-3} [1] concentration while [C$_4$mim]Cl did not reach a plateau even at a concentration higher than 3 mol· dm^{-3}. However, in the cases of the [NTf$_2$]$^-$ and [PF$_6$]$^-$ anions (Fig. 3.4), no micelle formation was detected, as the low solubility of the corresponding ionic liquids induced phase separation before any bulk aggregation occurred.

Furthermore, the influence of head type on the aggregation behaviour of ionic liquids was investigated. Three ionic liquids with the same anion and the same alkyl chain length, but with different types of hydrocarbon rings, have been synthesised. Figure 3.5 shows the surface tension curves for [C$_{12}$Y]Br (Y = mpyrr, mpy, and pip) and [C$_{12}$C$_{12}$pip]Br. Surprisingly, two very similar curves with almost the same CMC values are obtained for [C$_{12}$mpy]Br and [C$_{12}$mpip]Br. Since these two molecules have different geometries and volumes, it can only be speculated that the more hydrophobic and "space demanding" [C$_{12}$mpip]$^+$ cation (comparing the boat or chair conformation of an alicyclic six-membered ring with a planar aromatic molecule) has on the other side a stronger bound anion, Br$^-$, than the [C$_{12}$mpyrr]$^+$ cation with the delocalised positive charge. Although the CH$_3$ group on the nitrogen atom could create steric hindrance (it has a volume similar to the Br$^-$ anion) for the approach of a Br$^-$ anion, in order to confirm this statement, a molecule with a bulkier group than CH$_3$ (e.g., an isopropyl group) is needed. A slightly higher value of the CMC obtained for the five-membered heterocyclic ring of [C$_{12}$mpyrr]Br in comparison with the six-membered ring of [C$_{12}$mpip]Br can, *a priori*, be expected, but in general the influence of the head

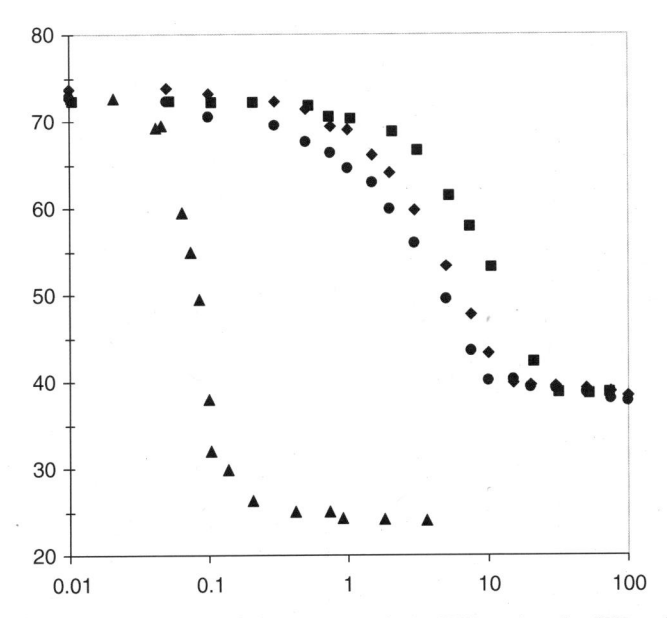

FIGURE 3.5 Monitoring the self-aggregation of [C$_{12}$Y]Br using the IFT technique: (♦) [C$_{12}$mpip]Br, (●) [C$_{12}$mpy]Br, (■) [C$_{12}$mpyrr]Br, (▲) [C$_{12}$C$_{12}$pip]Br.

structure on the CMC is very complex and, without further studies, one cannot draw any clear conclusion about the effects of geometry, volumes, and aromaticity on the head packing in micelles and monolayers.

Finally, the behaviour of a cation with a double chain, [C$_{12}$C$_{12}$pip]Br, was investigated (Fig. 3.5). For this case, the surface tension measurement had to be performed at a temperature higher than room temperature (35°C) since the Kraft point of [C$_{12}$C$_{12}$pip]Br is close to room temperature (the Kraft temperature, T_{Kr}, is the point at which surfactant solubility equals the critical micelle concentration). Since the CMC changes only slightly with temperature, one can compare this result with the others measured at room temperature. The lower value of both the CMC and surface tension in the plateau region and steeper descending part of IFT–concentration curve could be expected, as the addition of each monomer unit gives a more hydrophobic contribution, so the balance between hydrocarbon chain attraction and ionic repulsion is changed.

3.4 CONCLUSION

Here, we suggest general methodologies that can be used to find out whether ionic liquids are capable of forming aggregates in aqueous solutions. More classes of ionic liquids can be checked for their micellar behaviour using the techniques described in this chapter.

Aqueous solutions of ionic liquids of the 1-alkyl-3-methylimidazolium chloride family, [C$_n$mim]Cl, with n ranging from 8 to 14, when scrutinised using interfacial tension, fluorescence, and ^{1}H NMR techniques [2], present evidence for the self-aggregation of the [C$_n$mim]$^+$ cations into micellar aggregates. The exchange of the chloride anion for a more hydrophobic and larger counterion, such as [NTf$_2$]$^-$ or [PF$_6$]$^-$ (besides surface tension diminution), brings the system to a nonmicellar state: the system separates into two phases.

Besides the length of the alkyl chain and the nature of the anion, the type of head group can have a significant effect on the aggregation behaviour. The issue regarding the influence of the head on the aggregation behaviour requires further clarification.

It seems that aggregation is an inherent characteristic of ionic liquids; they aggregate in neat, bulk state [21], and in aqueous [1–3,18] and nonaqueous solutions [8]. Their ability to act as surfactants (longer alkyl chain) or hydrotropes (shorter alkyl chain) can be combined with other unique highly advantageous characteristics of these materials.

ACKNOWLEDGEMENTS

This work was supported by the Fundação para a Ciência e Tecnologia (FCT), Portugal (Projects POCTI/QUI/35413/2000 and POCI/QUI/57716/2004). M.B. thanks FCT for a Ph.D. grant SFRH/BD/13763/2003, and K.R.S. thanks the EPSRC (Portfolio Partnership Scheme, grant no. EP/D029538/1).

REFERENCES

1. Bowers, J., Butts, C. P., Martin, P. J., Vergara-Gutierrez, M. C., and Heenan, R. K., Aggregation behaviour of aqueous solutions of ionic liquids, *Langmuir* **20**, 2191–2198 (2004).
2. Blesic, M., Marques, M. H., Plechkova, N. V., Seddon, K. R., Rebelo, L. P. N., and Lopes, A., Self-aggregation of ionic liquids: micelle formation in aqueous solution, *Green Chem.* **9**, 481–490 (2007).
3. Sirieix-Plénet, J., Gaillon, L., and Letellier, P., Behaviour of a binary solvent mixture constituted by an amphiphilic ionic liquid, 1-decyl-3-methylimidazolium bromide and water. Potentiometric and conductimetric studies, *Talanta* **63**, 979–986 (2004).
4. Bowlas, C. J., Bruce, D. W., and Seddon, K. R., Liquid-crystalline ionic liquids, *Chem. Commun.*, 1625–1626 (1996).
5. Holbrey, J. D. and Seddon, K. R., The phase behavior of 1-alkyl-3-methylimidazolium tetrafluoroborates; ionic liquids and ionic liquid crystals, *J. Chem. Soc. Dalton Trans.*, 2133–2139 (1999).
6. Bradley, A. E., Hardacre, C., Holbrey, J. D., Johnston, S., McMath, S. E. J., and Nieuwenhuyzen, M., Small-angle X-ray scattering studies of liquid crystalline 1-alkyl-3-methylimidazolium salts, *Chem. Mater.* **14**, 629–635 (2002).
7. Firestone, M. A., Dzielawa, J. A., Zapol, P., Curtiss, L. A., Seifert, S., and Dietz, M. L., Lyotropic liquid-crystalline gel formation in a room-temperature ionic liquid, *Langmuir* **18**, 7258–7260 (2002).
8. Dorbritz, S., Ruth, W., and Kragl, U., Investigation on aggregate formation of ionic liquids, *Adv. Synth. Catal.* **347**, 1273–1279 (2005).
9. Miskolczy, Z., Sebők-Nagy, K., Biczók, L., and Göktütk, S., Aggregation and micelle formation of ionic liquids in aqueous solution, *Chem. Phys. Lett.* **400**, 296–300 (2004).
10. Beyaz, A., Oh, W. S., and Reddy, V. P., Synthesis and CMC studies of 1-methyl-3-(pentafluorophenyl)imidazolium quaternary salts, *Colloids Surfaces B:Biointerfaces* **35**, 119–124 (2004); **36**, 71–74 (2004).
11. Anderson, J. L., Pino, V., Hagberg, E. C., Sheares, V. V., and Armstrong, D. W., Surfactant solvation effects and micelle formation in ionic liquids, *Chem. Commun.*, 2444–2445 (2003).
12. Fletcher, K. A. and Pandey, S., Surfactant aggregation within room-temperature ionic liquid 1-ethyl-3-methylimidazolium bis(trifluoromethylsulfonyl)imide, *Langmuir* **20**, 33–36 (2004).
13. Evans, D. F. and Wennerstrom, H., *The Colloidal Domain: Where Physics, Chemistry, Biology and Technology Meet*, VCH, New York, 1994.
14. Gordon, C. M., Holbrey, J. D., Kennedy, A. R., and Seddon, K. R., Ionic liquid crystals: hexafluorophosphate salts, *J. Mater. Chem.* **8**, 2627–2636 (1998).
15. Bonhôte, P., Dias, A.-P., Armand, M., Papageorgiou, N., Kalyanasundaram, K., and Grätzel, M., Hydrophobic, highly conductive ambient-temperature molten salts, *Inorg. Chem.* **35**, 1168–1178 (1996).
16. Lakowicz, J., *Principles of Fluorescence Spectroscopy*, 2nd ed., Kluwer, New York, 1999.
17. Moroi, Y., *Micelles: Theoretical and Applied Aspects*, Kluwer, New York, 1992.

18. Vanyúr, R., Biczók, L., and Miskolczy, Z., Micelle formation of 1-alkyl-3-methylimidazolium bromide ionic liquids in aqueous solution, *Colloids Surfaces A: Physicochem. Eng. Aspects* **299**, 256–261 (2007).
19. Anacker, E. W. and Ghose, H. M., Counterions and micelle size. II. Light scattering by solutions of cetylpyridinium salts, *J. Am. Chem. Soc.* **90**, 3161–3166 (1968).
20. Zana, R., *Surfactant Solutions: New Methods of Investigation*, Marcel Dekker, New York, 1986.
21. (a) Pádua, A. A. H. and Canongia Lopes, J. N., Nanostructural organization in ionic liquids, *J. Phys. Chem. B* **110**, 3330–3335 (2006). (b) Canongia Lopes, J. N., Costa Gomes, M. F., and Pádua, A. A. H., Nonpolar, polar, and associating solutes in ionic liquids, *J. Phys. Chem. B* **110**, 16816–16818 (2006).

4 Mass Spectrometric Studies on Ionic Liquid Aggregates

R. LOPES DA SILVA

Departamento de Química, Universidade de Aveiro, Aveiro, Portugal

I. M. MARRUCHO and J. A. P. COUTINHO

CICECO, Departamento de Química, Universidade de Aveiro, Aveiro, Portugal

ANA M. FERNANDES

QOPNA, Departamento de Química, Universidade de Aveiro, Aveiro, Portugal

Abstract

The association of ionic liquids to form aggregates was studied in detail by electrospray ionisation mass spectrometry (ESI-MS) and tandem mass spectrometry (ESI-MS-MS). The ESI-MS results show the formation of aggregates $[C_{n+1}(BF_4)_n]^+$, where C = $[C_{4-10}mim]^+$, $[C_2OHmim]^+$, and $[C_4dmim]^+$ with $n = 1$ up to 13 for all the ionic liquids investigated. The differences observed between the numbers of neutral molecules that will form the most stable aggregate ion were correlated with structural differences in the cations. The ESI-MS-MS experiments performed for each mass–selected aggregate at different collision energies (breakdown graphs) showed that their dissociation, in the gas phase, may not necessarily occur always by sequential loss of $n(CBF_4)$ units, but also by intact $(CBF_4)_n$ neutrals. The relative order of hydrogen bond strengths between cation and anion, in each ion pair of ionic liquids studied, was established using Cooks's kinetic method. The results indicate that the order of hydrogen bond strength to $[BF_4]^-$ is $[C_4dmim]^+ < [C_{10}mim]^+ < [C_8mim]^+ < [C_4mim]^+ < [C_2OHmim]^+$, and to $[PF_6]^-$ the order is $[C_4dmim]^+ < [C_8mim]^+ < [C_6mim]^+ < [C_4mim]^+$.

Molten Salts and Ionic Liquids: Never the Twain? Edited by Marcelle Gaune-Escard and Kenneth R. Seddon
Copyright © 2010 John Wiley & Sons, Inc.

4.1 INTRODUCTION

Ionic liquids have been the subject of an increasing number of publications due to their unique physicochemical properties: negligible vapour pressure, relatively low viscosity, high thermal and chemical stability, and distinct solubility in both polar and nonpolar solvents inter alia [1]. The wide range of applications of ionic liquids, as well as the possibility of their use as "green" alternatives to volatile organic solvents, justifies the interest of researchers on these compounds [2]. The knowledge of their structural organization in the liquid phase is of fundamental importance in the understanding of their properties [3]. On the basis of spectroscopic analysis [4–8] and ab initio calculations [9, 10], particularly on imidazolium-based ionic liquids, they have been described as polymeric supramolecules formed through hydrogen bonds of the imidazolium cation with the anion [11].

Electrospray ionisation mass spectrometry (ESI-MS) is a major technique for the analysis of positive and negative ions present in solution. A high voltage is applied to a metallic capillary through which the solution is flowing. If the counterelectrode is large and planar, the strength of the electric field at the capillary tip is typically about 10^6–10^7 V m^{-1}. In the positive ion mode, anions migrate in the direction of the capillary, whereas cations migrate in the direction of the counterelectrode. At sufficiently high electric field strengths, a dynamic cone of liquid (referred to as a "Taylor cone") will form at the tip of the capillary. When the surface tension of the liquid does not compensate for the high charge density, a stream of droplets will emerge from the tip of the jet. As the solvent evaporates from the droplets with the assistance of warm nitrogen gas, the size of the droplets decreases until the electrostatic repulsion between the ions becomes equal to the surface tension and droplet fission occurs. Successive fissions will ultimately lead to the formation of gas phase ions that can be analysed within the mass spectrometer. A unique feature of

TABLE 4.1 Ionic Liquids Used in the Study

Ionic Liquid	R_1	R_2	A
[C$_4$mim][BF$_4$]	C$_4$H$_9$	H	BF$_4$
[C$_4$mim][PF$_6$]	C$_4$H$_9$	H	PF$_6$
[C$_6$mim][PF$_6$]	C$_6$H$_{13}$	H	PF$_6$
[C$_8$mim][BF$_4$]	C$_8$H$_{17}$	H	BF$_4$
[C$_8$mim][PF$_6$]	C$_8$H$_{17}$	H	PF$_6$
[C$_{10}$mim][BF$_4$]	C$_{10}$H$_{21}$	H	BF$_4$
[C$_4$dmim][BF$_4$]	C$_4$H$_9$	CH$_3$	BF$_4$
[C$_2$OHmim][BF$_4$]	C$_2$H$_4$OH	H	BF$_4$

this technique is the direct transfer to the gas phase of loosely bonded clusters of ions present in solution or formed in the droplets during the electrospray process [12, 13].

The aim of the present work is to use electrospray ionisation mass spectrometry and tandem mass spectrometry to investigate the influence of structural features of the cations on the formation of imidazolium-based ionic liquid aggregates, their gas-phase dissociation, and the relative order of hydrogen-bond strengths between cation and anion in each ion pair of ionic liquids studied (Table 4.1).

4.2 EXPERIMENTAL

The ionic liquids shown in Table 4.1 were used as ethanenitrile solutions (1.5×10^{-4} M). Electrospray ionisation mass (ESI-MS) and tandem mass spectra (ESI-MS-MS) were acquired with a Micromass Q-ToF 2 operating in the positive ion mode. Source and desolvation temperatures were 80 °C and 100 °C, respectively. The capillary voltage was 2600 V and cone voltage was 25 V. ESI-MS-MS spectra were acquired by selecting the precursor ion with the quadrupole, performing collisions with argon at energies of 2–30 eV in the hexapole, followed by mass analysis of product ions by the ToF analyser. N_2 was used as the nebulisation gas. The ionic liquid solutions were introduced at a 10 μL·min^{-1} flow. The breakdown graphs were obtained by acquiring the ESI-MS-MS spectra of each ion investigated at increasing collision energies and plotting the relative abundance of precursor and fragment ions as a function of collision energy. The relative order of hydrogen-bond strength between cation and anion in each ion pair studied was obtained by acquiring the ESI-MS-MS spectra, at 10 eV collision energy, of the cluster ions $[C_1 \cdots A \cdots C_2]^+$, and measuring the relative abundances of the two fragment ions observed.

4.3 RESULTS AND DISCUSSION

4.3.1 ESI Mass Spectra

The electrospray ionisation mass spectrum in the positive ion mode of the ionic liquids [C_{4-10}mim][BF_4], [C_4dmim][BF_4], and [C_2OHmim]BF_4] showed peaks corresponding to the formation of aggregates $[C_{n+1}(BF_4)_n]^+$ with $n = 1$ up to 13, in agreement with previously published results [11, 14]. The formation of an aggregate whose intensity does not follow the normal decreasing pattern has been observed for all the ionic liquids investigated. The spectrum of [C_{10}mim][BF_4], shown in Figure 4.1, exemplifies this behaviour with the out-of-range abundance of ion $[(C_{10}mim)_5(BF_4)_4]^+$ (m/z 1464), which corresponds to $n = 4$ in the general formula $[C_{n+1}(BF_4)_n]^+$. An increased gas phase stability of this particular ion as compared with its neighbours explains its greater relative abundance in the mass spectrum. Whereas the size of the N-alkyl side chain does not seem to have any effect on the composition of the most stable aggregate, which was $[(C_5A_4)]^+$ when $C = [C_{4-10}mim]^+$, for the ionic liquids with a methyl group at position C_2 of the imidazolium ring, or with a functionalised side chain with OH, the value of n decreased to three.

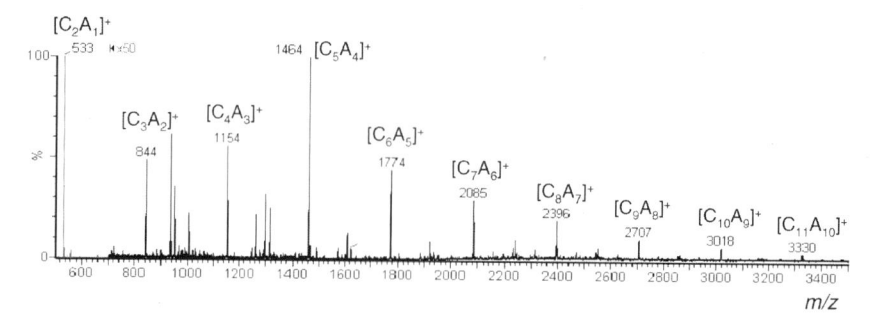

FIGURE 4.1 Electrospray ionisation mass spectrum of [C$_{10}$mim][BF$_4$].

4.3.2 ESI-MS-MS Spectra and Breakdown Graphs

The ESI-MS-MS spectra of all mass-selected aggregates [C$_{n+1}$(BF$_4$)$_n$]$^+$ activated by collision with argon show peaks at mass intervals corresponding to a neutral "molecule" of ionic liquid. This observation can be interpreted as a sequential loss, by the precursor ion, of n(CBF$_4$) units, or as loss of individual (CBF$_4$)$_n$ units. In order to obtain a better understanding of the fragmentation dynamics of ionic liquid aggregates, the dependence of the ESI-MS-MS spectra on collision energy was investigated. Treatment of data by plotting the relative abundances of precursor and fragment ions as a function of collision energy (eV) affords the breakdown graphs [15] exemplified in Figures 4.2–4.4 for [(C$_4$mim)$_4$(BF$_4$)$_3$]$^+$, [(C$_4$mim)$_5$(BF$_4$)$_4$]$^+$, and [(C$_4$mim)$_6$(BF$_4$)$_5$]$^+$, respectively.

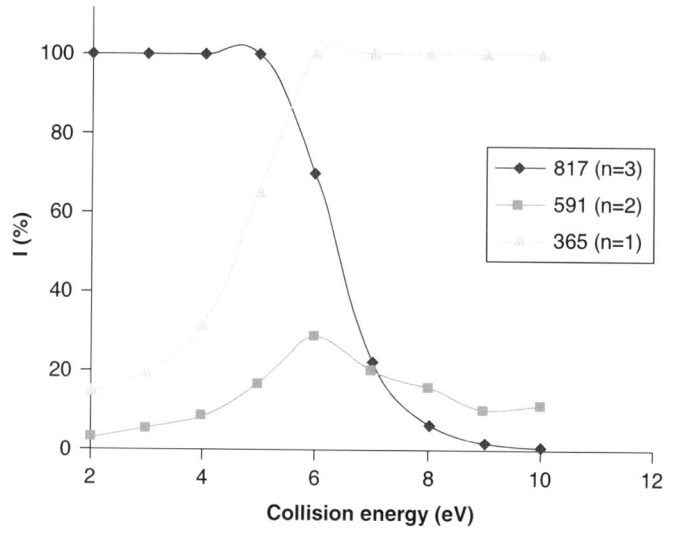

FIGURE 4.2 Breakdown graphs for [(C$_4$mim)$_4$(BF$_4$)$_3$]$^+$.

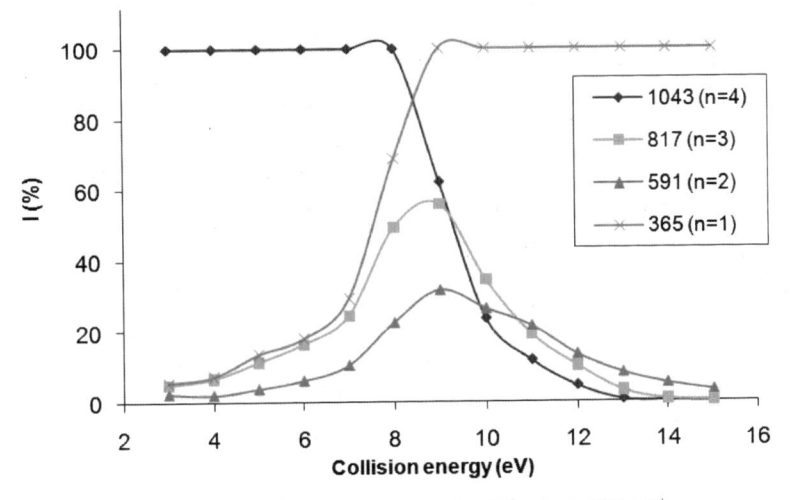

FIGURE 4.3 Breakdown graphs for $[(C_4mim)_5(BF_4)_4]^+$.

The breakdown pattern of ion $[(C_4mim)_4(BF_4)_3]^+$ (m/z 817; Fig. 4.2) shows a decay of its relative abundance and, at relatively low collision energies, coformation of ions m/z 591 and m/z 365, by loss of one or two neutral molecules, respectively, indicating that a direct loss of $([C_4mim][BF_4])_2$ from the precursor ion can occur. As the collision energy increases, the relative abundances of m/z 817 and 591 decrease, and ions m/z 365 become more abundant.

Figure 4.3 provides further evidence for the loss of neutral $([C_4mim][BF_4])_n$ with $n = 1-3$ from $[(C_4mim)_5(BF_4)_4]^+$ (m/z 1043). Since the maximum abundance for ions m/z 817 and m/z 591 occurs at roughly the same collision energy, it can be

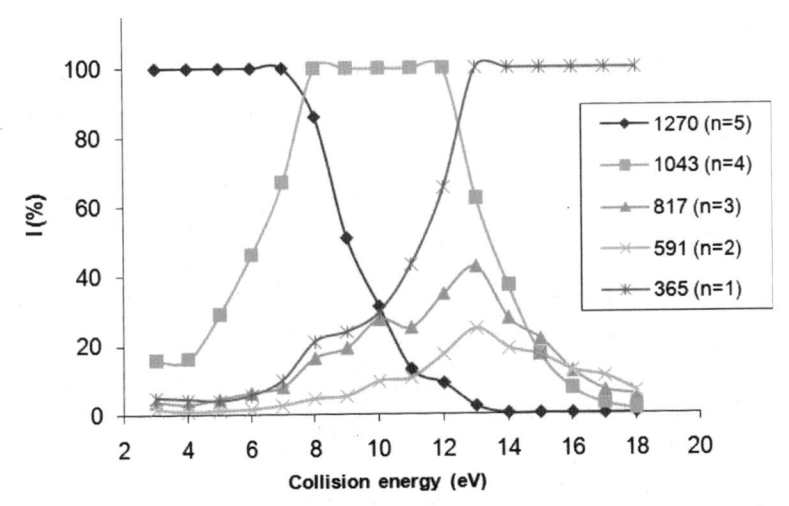

FIGURE 4.4 Breakdown graphs for $[(C_4mim)_6(BF_4)_5]^+$.

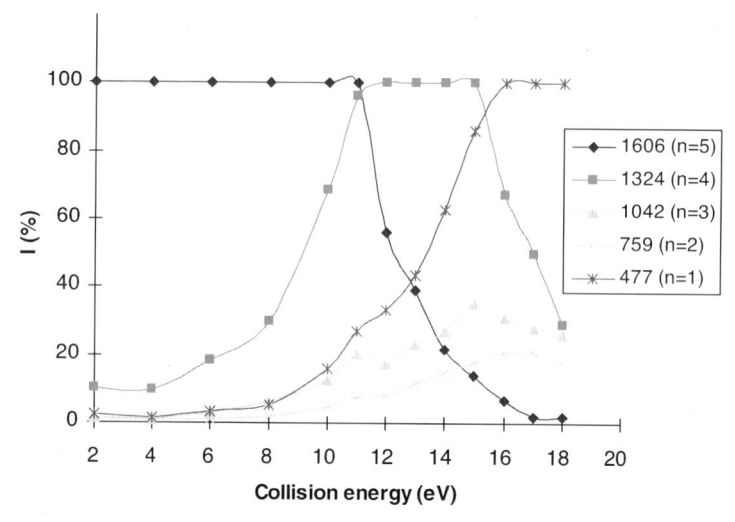

FIGURE 4.5 Breakdown graphs for $[(C_8mim)_6(BF_4)_5]^+$.

assumed that both will decompose, at higher collision energies, to $[(C_4mim)_2(BF_4)]^+$ (m/z 365) by loss of two or one neutral molecules, respectively, in agreement with the observations of Figure 4.2.

Interpretation of Figure 4.4 indicates a relative abundant formation of ion $[(C_4mim)_5(BF_4)_4]^+$ (m/z 1043), together with initial coformation of the low abundant ions m/z 817, 591, and 365 consistent with losses of $([C_4mim][BF_4])_n$ neutrals with n = 1 to 4. The predominance of ion m/z 1043 with 100% of relative abundance between 8 and 12 eV collision energy is noteworthy and consistent with the ESI

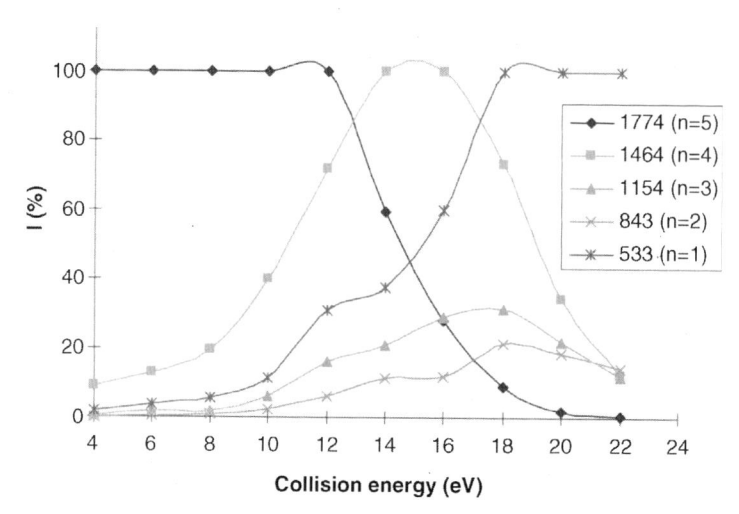

FIGURE 4.6 Breakdown graphs for $[(C_{10}mim)_6(BF_4)_5]^+$.

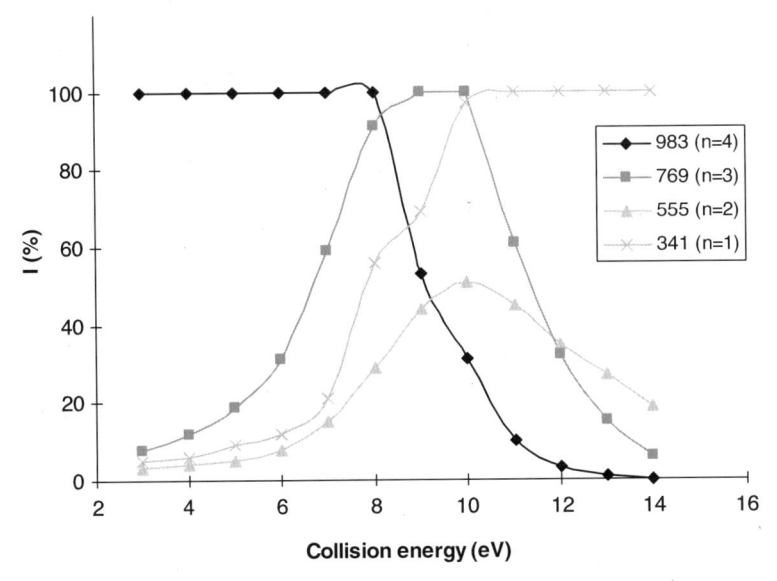

FIGURE 4.7 Breakdown graphs for $[(C_2OHmim)_5(BF_4)_4]^+$.

spectrum of [C$_4$mim][BF$_4$], where that ion was identified as the most stable aggregate. At higher collision energies, the decay and formation of the ions shown in Figure 4.4 are almost a replica of those shown in Figure 4.3.

Since the most favoured assemblies of the ionic liquids [C$_8$mim][BF$_4$] and [C$_{10}$mim][BF$_4$] also correspond to $n=4$ in the general formula $[C_{n+1}A_n]^+$, the breakdown graphs of ions $[(C_8mim)_6(BF_4)_5]^+$ and $[(C_{10}mim)_6(BF_4)_5]^+$ for these two ionic liquids are shown on Figures 4.5 and 4.6 for comparison with Figure 4.4. The pattern of decay and formation of the ions for these three ionic liquids are very similar: coformation of low abundant ions corresponding to losses of ([C$_4$mim][BF$_4$])$_n$ neutrals with $n=2$–4, and a predominant and more stable ion corresponding to $n=4$.

For the two ionic liquids where $n=3$, the comparison that has to be done is between the breakdown graphs of ions $[(C_2OHmim)_5(BF_4)_4]^+$ and $[(C_4dmim)_5(BF_4)_4]^+$ (Figs. 4.7 and 4.8) and those of Figures 4.4–4.6. The conclusions drawn are identical to the ones mentioned in the previous paragraph, particularly in what concerns the behaviour of the ions corresponding to the most stable aggregates for the five ionic liquids studied.

The interpretation of the above breakdown graphs represents a strong indication that the collision-induced dissociation of ionic liquid aggregates can proceed via loss of ([C$_4$mim][BF$_4$])$_n$ neutral molecules, where n can vary from 1 to 4, and not as a sequential decay via n([C$_4$mim][BF$_4$]) units.

4.3.3 Relative Hydrogen Bond Strengths by Cooks's Kinetic Method

The kinetic method of making thermochemical determinations proposed by Cooks [16–18] is based on the relative rates of competitive dissociations of a cluster

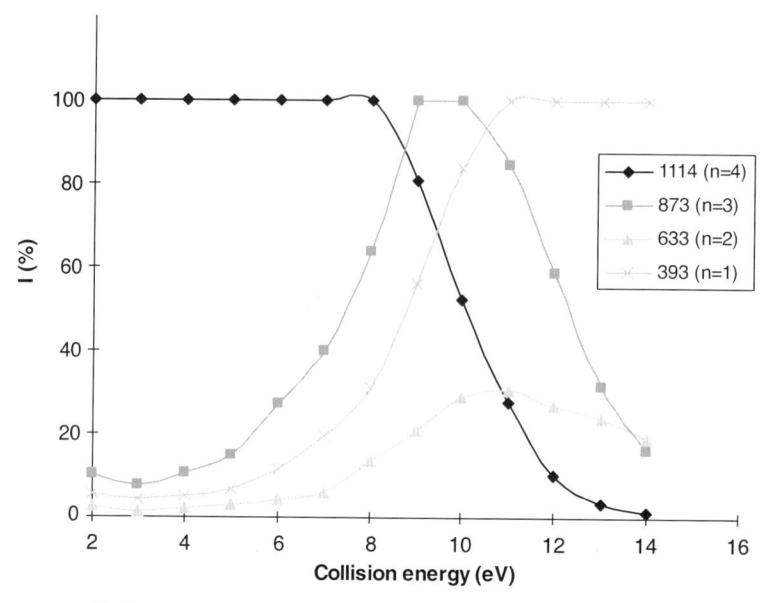

FIGURE 4.8 Breakdown graphs for $[(C_4dmim)_5(BF_4)_4]^+$.

ion comprised of the compound of interest and a reference compound. For example, the cluster ion $[B_1 \cdots H^+ \cdots B_2]$ dissociates as shown in Eq. (4.1):

$$B_1 \cdots H^+ \cdots B_2 \underset{k_2}{\overset{k_1}{\rightleftarrows}} \begin{array}{l} B_1H^+ + B_2 \\ B_2H^+ + B_1 \end{array} \quad (4.1)$$

where k_1 and k_2 are the rate constants for the competitive dissociations of the cluster ion to yield $[B_1H]^+$ and $[B_2H]^+$, respectively. The major assumptions of the method are the following:

- Negligible differences in the entropy requirements for the competitive channels
- Negligible reverse activation energies
- Absence of isomeric forms of the activated cluster ion

When these conditions are satisfied, the rates of the dissociations are controlled by the relative activation energies of each reaction channel, the difference in which is equivalent to the differences in proton affinities, $\Delta(\Delta H)$, of the two bases B_1 and B_2. Experimentally, the method can be performed using any tandem mass spectrometer by mass selection of the cluster ion $[B_1 \cdots H^+ \cdots B_2]$ and analysis of its fragmentation products. If no secondary reactions occur, the relative abundances of the fragment ions $[B_1H]^+$ and $[B_2H]^+$ will be determined by the rate constants k_1 and

k_2, respectively. Hence the ratio of the fragment ion abundances will be related to $\Delta(\Delta H)$:

$$\ln\frac{k_1}{k_2} = \ln\left[\frac{B_1 H^+}{B_2 H^+}\right] \approx \frac{\Delta(\Delta H)}{RT_{\text{eff}}} \qquad (4.2)$$

where T_{eff} is the effective temperature of the activated dimer. A plot of $\ln(k_1/k_2)$ versus $\Delta(\Delta H)$ for a set of reference compounds of known proton affinity will produce a straight line of slope $1/RT_{\text{eff}}$. From such a plot, the proton affinity of an unknown compound can be obtained by dissociation of the cluster ion formed by the unknown and a reference compound. A qualitative application of the method will produce a relative order of proton affinities (or other thermochemical quantities) in the following way: first the cluster ion $[B_1 \cdots H^+ \cdots B_2]$ produced in the ionisation source of the mass spectrometer will be selected in the first analyser, dissociated in a collision cell, and the fragmentation products analysed in the second mass analyser. If, for example, the proton affinity of B_1 is greater than the proton affinity of B_2, the loss of the neutral molecule B_2 will be favoured, and the relative abundance of ion $[B_1 H]^+$ will be greater than that of ion $[B_2 H]^+$. The analysis of several combinations of B_1 and B_2 will establish a relative order of proton affinities for the bases under study.

In the present study, the kinetic method was used to establish a relative order of hydrogen-bond strengths in the cation–anion ion pairs of imidazolium-based ionic liquids. Tetrafluoroborate, $[BF_4]^-$, and hexafluorophosphate, $[PF_6]^-$, were used as anions, $[A]^-$; and $[C_4\text{mim}]^+$, $[C_6\text{mim}]^+$, $[C_8\text{mim}]^+$, $[C_{10}\text{mim}]^+$, $[C_4\text{dmim}]^+$, and $[C_2\text{OHmim}]^+$ were used as cations, $[C]^+$. Equimolar mixtures of two ionic liquids in ethanenitrile were introduced into the electrospray ion source, the clusters $[C_1 \cdots A \cdots C_2]^+$ were selected with the quadrupole and dissociated by 10 eV collisions with argon, and the product ions were analysed with the ToF. Tables 4.2 and 4.3 present the results obtained for the binary mixtures of ionic liquids used to form the clusters. The ESI-MS-MS spectrum of the cluster $[C_4\text{mim} \cdots BF_4 \cdots C_8\text{mim}]^+$ (m/z 479; Fig. 4.9) shows that a stronger hydrogen bond of the ion pair $[C_4\text{mim}]^+ \cdots [BF_4]^-$ will favour the loss of the neutral molecule $[C_4\text{mim}][BF_4]$ from the cluster, which results in a greater abundance of ion $[C_8\text{mim}]^+$ (m/z 195) as compared with $[C_4\text{mim}]^+$ (m/z 139).

The analysis of the data in Tables 4.2 and 4.3 assumes, as explained earlier, that a stronger hydrogen bond of the ion pair will generate a less abundant cation in the ESI-MS-MS spectrum of the cluster. Thus the order of hydrogen-bond strength to $[BF_4]^-$ is

$$[C_4\text{dmim}]^+ < [C_{10}\text{mim}]^+ < [C_8\text{mim}]^+ < [C_4\text{mim}]^+ < [C_2\text{OHmim}]^+$$

and to $[PF_6]^-$ is

$$[C_4\text{dmim}]^+ < [C_8\text{mim}]^+ < [C_6\text{mim}]^+ < [C_4\text{mim}]^+$$

TABLE 4.2 Relative Abundances[a] of the Imidazolium Cations in the ESI-MS-MS Spectra of the Clusters $[C_1 \cdots BF_4 \cdots C_2]^+$

Cation	$[C_4mim]^+$	$[C_8mim]^+$
Relative abundance / %	31.3	100
Cation	$[C_4mim]^+$	$[C_{10}mim]^+$
Relative abundance / %	23.9	100
Cation	$[C_4mim]^+$	$[C_4dmim]^+$
Relative abundance / %	7.1	100
Cation	$[C_4mim]^+$	$[C_2OHmim]^+$
Relative abundance / %	64.7	14.9
Cation	$[C_8mim]^+$	$[C_{10}mim]^+$
Relative abundance / %	82.1	100
Cation	$[C_8mim]^+$	$[C_4dmim]^+$
Relative abundance / %	14.7	100
Cation	$[C_8mim]^+$	$[C_2OHmim]^+$
Relative abundance / %	100	4.4
Cation	$[C_{10}mim]^+$	$[C_4dmim]^+$
Relative abundance / %	17.4	100
Cation	$[C_{10}mim]^+$	$[C_2OHmim]^+$
Relative abundance / %	100	3.4
Cation	$[C_4dmim]^+$	$[C_2OHmim]^+$
Relative abundance / %	100	0.6

[a] Percentage % of the base peak.

These trends can be explained in terms of chain length increase on the N^1-alkyl group of the imidazolium, presence of a methyl group at C^2, and an OH-functionalised alkyl chain. A recent publication concerning the optimised structures of cation–anion ion pairs of 1,3-dialkylimidazolium-based ionic liquids at the B3LYP/6-31 + G level of

TABLE 4.3 Relative Abundances[a] of the Imidazolium Cations in the ESI-MS-MS Spectra of the Clusters $[C_1 \cdots PF_6 \cdots C_2]^+$

Cation	$[C_4mim]^+$	$[C_6mim]^+$
Relative abundance / %	44.2	100
Cation	$[C_4mim]^+$	$[C_8mim]^+$
Relative abundance / %	25.4	100
Cation	$[C_4mim]^+$	$[C_4dmim]^+$
Relative abundance / %	6.8	100
Cation	$[C_6mim]^+$	$[C_8mim]^+$
Relative abundance / %	47.3	64.4
Cation	$[C_6mim]^+$	$[C_4dmim]^+$
Relative abundance / %	16.5	100
Cation	$[C_8mim]^+$	$[C_4dmim]^+$
Relative abundance / %	22.1	100

[a] Percentage (%) of the base peak.

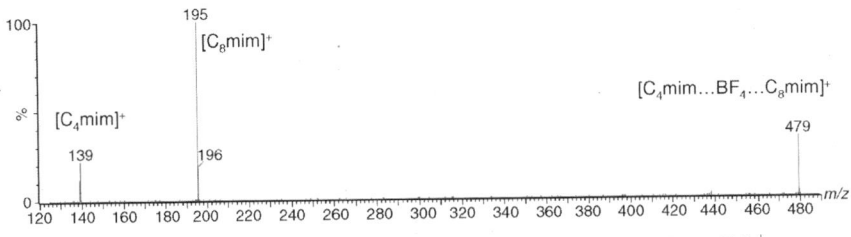

FIGURE 4.9 ESI-MS-MS spectrum of the cluster $[C_1 \cdots BF_4 \cdots C_2]^+$.

DFT theory [19] concluded that there were two stable hydrogen bonds, between the C^2–H of the imidazolium ring and one fluorine atom of the anion, and between a C–H of the side chain and another fluorine atom of the anion. Calculations of bond distances, bond angles, and interaction energies for the ion pairs [C$_{1-4}$mim][BF$_4$] and [C$_{1-4}$mim][PF$_6$] showed that with the increase in the size of the N-alkyl side chains, the distances of both H...F interactions increase, and the interaction energies between the cation and anion decrease. These conclusions are in complete agreement with the experimental results for the relative order of hydrogen-bond strength of cations to [BF$_4$]$^-$ (or [PF$_6$]$^-$) presented here. In [C$_4$dmim][BF$_4$] or [C$_4$dmim][PF$_6$], the C^2 position is blocked by a methyl group that not only decreases the hydrogen-bond donor ability [19, 20] of the cation but may represent a steric factor to the approach of either anion to the cation. Hence this gives the weakest hydrogen bond for these ion pairs. The strongest is, as expected, in [C$_2$OHmim][BF$_4$], since a shorter chain length with a terminal OH group will favour the interaction between the cation and the two fluorine atoms.

4.4 CONCLUSION

Aggregates of imidazolium-based ionic liquids of the general formula $[C_{n+1}(BF_4)_n]^+$, where C = [C$_{4-10}$mim]$^+$, [C$_2$OHmim]$^+$, and [C$_4$dmim]$^+$ with $n = 1$ up to 13, were identified by electrospray ionisation mass spectrometry, which also shows the formation of assemblies with extra stability for $n = 4$ or 3 depending on the cation. The dependence of [C$_{4-6}$(BF$_4$)$_{3-5}$]$^+$ aggregates gas phase dissociation with collision energy gives a strong indication that they decompose via losses of intact (CBF$_4$)$_n$ neutrals where $n = 1-4$. The relative hydrogen-bond strengths between the imidazolium cation and the anion, postulated for the formation of polymeric structures in the liquid state, was established using Cooks's kinetic method. For the six imidazolium cations studied, the relative order of hydrogen bond strengths to [BF$_4$]$^-$ is [C$_4$dmim]$^+$ < [C$_{10}$mim]$^+$ < [C$_8$mim]$^+$ < [C$_4$mim]$^+$ < [C$_2$OHmim]$^+$ and to [PF$_6$]$^-$ it is [C$_4$dmim]$^+$ < [C$_8$mim]$^+$ < [C$_6$mim]$^+$ < [C$_4$mim]$^+$. The chain length increase on the N^1-alkyl group of the imidazolium, the presence of a methyl group at C^2, and an OH-functionalised alkyl chain explain these trends.

REFERENCES

1. Chiappe, C. and Pieraccini, D., Ionic liquids: solvent properties and organic reactivity, *J. Phys. Org. Chem.* **18**, 275–297 (2005).
2. Anderson, J. L., Armstrong, D. W., and Wei, G., Ionic liquids in analytical chemistry, *Anal. Chem.* **78**, 2892–2902 (2006).
3. Hardacre, C., McMath, S. E. J., Nieuwenhuyzen, M., Bowron, D. T., and Soper, A. K., Liquid structure of 1,3-dimetyhylimidazolium salts, *J. Phys. Condens. Matter* **15**, S159–S166 (2003).
4. Lee, K. M., Chang, H., Jiang, J., Lu, L. Hsiao, C., Lee, Y., Lin, S. H., and Lin, I. J. B., Probing C–H··· X hydrogen bonds in amide-functionalized imidazolium salts under high pressure, *J. Chem. Phys.* **120**, 8645–8650 (2004).
5. Dupont, J., Suarez, P. A. Z., De Sousa, R. F., Burrow, R. A., and Kintzinger, J., C–H-p interactions in 1-*n*-butyl-3-methylimidazolium tetraphenylborate molten salt: solid and solution structures, *Chem. Eur. J.* **6**, 2377–2381 (2000).
6. Hardacre, C., Holbrey, J. D., McMath, S. E. J., Bowron, D. T., and Soper, A. K., Structure of molten 1,3-dimethylimidazolium chloride using neutron diffraction, *J. Chem. Phys.* **118**, 273–278 (2003).
7. Consorti, C. S., Suarez, P. A. Z., De Souza, R. F., Burrow, R. A., Farrar, D. H., Lough, A. J, Loh, W., Silva, L. H. M., and Dupont, J., Identification of 1,3-dialkylimidazolium salt supramolecular aggregates in solution, *J. Phys. Chem. B.* **109**, 4341–4349 (2005).
8. Dorbritz, S., Ruth, W., and Krafl, U., Investigation on aggregate formation of ionic liquids, *Adv. Synth. Catal.* **347**, 1273–1279 (2005).
9. Hunt, P. A., Kirchner, B., and Welton, T., Characterising the electronic structure of ionic liquids: an examination of the 1-butyl-3-methylimidazolium chloride ion pair, *Chem Eur. J.* **12**, 6762–6775 (2006).
10. Del Pópolo, M. G., Lynden-Bell, R. M., and Kohanoff, J., Ab initio molecular dynamics simulation of a room temperature ionic liquid, *J. Phys. Chem.* **109**, 5895–5902 (2005).
11. Abdul-Sada, A. K., Elaiwi, A. E., Greenway, A. M., and Seddon, K. R., Evidence for the clustering of substituted imidazolium salts via hydrogen bonding under the conditions of fast atom bombardment mass spectrometry, *Eur. Mass Spectrom.* **3**, 245–247 (1997); Dupont, J., On the solid, liquid and solution structural organization of imidazolium ionic liquids, *J. Braz. Chem. Soc.* **15**, 341–350 (2004).
12. Cole, R. R. B. (Ed.), *Electrospray Ionization Mass Spectrometry: Fundamentals, Instrumentation and Applications*, John Wiley & Sons, Hoboken, NJ, 1997.
13. Meng, C. K. and Fenn, J. B., Formation of charged clusters during electrospray ionization of organic solute species, *Org. Mass Spectrom.* **26**, 542–549 (1991).
14. Gozzo, F. C., Santos, L. S., Augusti, R., Consorti, C. S, Dupont, J., and Eberlin, M. N., Gaseous supramolecules of imidazolium based ionic liquids: "magic" numbers and intrinsic strengths of hydrogen bonds, *Chem. Eur. J.* **10**, 6187–6193 (2004).
15. Herzschuh, R. and Drewello, T., The fragmentation dynamics of small $Cs(CsI)_n{}^+$ cluster ions under low-energy multiple collision conditions, *Int. J. Mass Spectrom.* **233**, 355–359 (2004).
16. McLuckey, S. A., Cameron, D., and Cooks, R. G., Proton affinities from dissociations of proton-bound dimmers, *J. Am. Chem. Soc.* **103**, 1313–1317 (1981).

17. Cooks, R. G. and Wong, P. S. H., Kinetic method of making thermochemical determinations: advances and applications, *Acc. Chem. Res.* **31**, 379–386 (1998).
18. Cooks, R. G., Koskinen, J. T., and Thomas, P. D., The kinetic method of making thermochemical determinations, *J. Mass Spectrom.* **34**, 85–92 (1990).
19. Dong, K., Zhang, S., Wang, D., and Yao, X., Hydrogen bonds in imidazolium ionic liquids, *J. Phys. Chem. A* **110**, 9775–9782 (2006).
20. Crowhurst, L., Mawdsley, P. R., Perez-Arlandis, J. M., Salter, P. A., and Welton, T., Solvent–solute interactions in ionic liquids, *Phys. Chem. Chem. Phys.* **5**, 2790–2794 (2003).

5 Study of Sm–Al Alloy Formation in the Molten LiCl–KCl Eutectic

G. DE CÓRDOBA and C. CARAVACA

CIEMAT, Nuclear Fission Division/Radioactive Waste Unit, Madrid, Spain

Abstract

This chapter describes the preliminary results obtained on the electrochemical formation of Sm–Al alloys in the molten LiCl–KCl–SmCl$_3$ system. Cyclic voltammetry and open-circuit chronopotentiometry techniques have been applied by using solid aluminium electrodes to identify the Sm–Al intermetallic compounds. When inert tungsten is used as the working electrode, the cyclic voltammogram shows a single wave corresponding to the soluble Sm^{3-}/Sm^{2+} system at a potential of −0.99 V versus AgCl/Ag at 723 K. However, the reduction of Sm^{2+} to metallic samarium is not observed since its reduction potential is very negative and close to that of the solvent Li(I)/Li(0) (>−2.7 V vs. AgCl/Ag). When an aluminium substrate is used, a well-defined signal is found at a potential of -1.90 V versus AgCl/Ag at 723 K placed between the Sm^{3+}/Sm^{2+} system and the cathodic solvent limit. This signal is associated with the formation of an Sm–Al alloy. The shift of the deposition potential toward more positive values than that of pure metallic samarium is explained in terms of a lowering of the activity of samarium in the aluminium phase. The open-circuit chronopotentiometry technique has also been applied, which allows identifying more clearly the number of intermetallic compounds formed and the potentials at which these compounds are transformed one into another. Electrolysis, both potentiostatic and galvanostatic, has been conducted to characterise the Sm–Al compounds. Deposits obtained have been analysed by XRD and SEM-EDX techniques. Thermodynamic properties such as the activity of samarium in the aluminium phase and the relative Gibbs free energy, enthalpy, and entropy of formation of the Sm–Al intermetallics have been estimated from the chronopotentiometry measurements.

Molten Salts and Ionic Liquids: Never the Twain? Edited by Marcelle Gaune-Escard and Kenneth R. Seddon
Copyright © 2010 John Wiley & Sons, Inc.

5.1 INTRODUCTION

The operation of nuclear fission reactors gives rise to long-lived radionuclides, which may be considered a potential risk borne by future generations that must be minimised. The partitioning and transmutation (P&T) strategy is one of the alternatives that are currently under investigation for the management of highly radioactive nuclear waste. The aim of P&T is to transform the long-lived radionuclides, formed during nuclear fuel irradiation, into shorter or even stable ones, by means of fission or neutron capture reactions. This would lead to the reduction of the long-term radiological impact that involves its final disposal [1].

To achieve this goal, it is necessary to separate the radionuclides to be transmuted, mainly plutonium and some minor actinides (Np, Am, and Cm), from the fission products (FPs) contained in the irradiated nuclear fuel. Nowadays, different separation methods are being explored, both hydrometallurgical and pyrometallurgical ones. Pyrometallurgical processes in molten salt media present some characteristics that convert them in particularly attractive ways in the P&T context. The most significant ones are (1) the radiation stability of the salt and metallic phases used, which allows the handling of fuels with shorter cooling times, and (2) their inherent proliferation resistance [2]. Hence these processes would be appropriate to treat advanced nuclear fuels proposed for transmutation, which will have considerable plutonium and minor actinide content, and will reach very high burn-ups.

The main step of a pyrometallurgical reprocess is the grouped separation of actinides from the rest of the radionuclides contained in the irradiated fuel, mainly the lanthanide elements. The objective is to reach high actinide recovery yields (>99%) and, at the same time, to minimise the lanthanide content in the final product, mainly because of their neutronic poison effect. The chemical similarity between these groups of elements and the high mass content of lanthanides in the irradiated fuel make their separation difficult. Electrorefining and electrolysis in a molten metal chloride, and salt/metal reductive extraction in a molten fluoride media, are the selected techniques to achieve this goal.

The design of a pyrochemical reprocessing scheme requires a good knowledge of the basic thermodynamic and kinetic properties of the elements for separation, mainly actinides and lanthanides, in the different salt and metallic media proposed. Thermodynamic calculations as well as experimental studies performed have shown that, for the time being, aluminium is one of the most promising metallic solvents to support the selective recovery of actinides [3, 4].

The present work is part of an investigation that is being carried out at CIEMAT concerning the behaviour of samarium in the molten LiCl–KCl eutectic. Several studies have been conducted on the electrochemistry of this element in molten chlorides by using different types of solid and liquid metallic substrates, with [5–8] and without [9–13] alloy formation. However, no information about the behaviour of samarium on solid aluminium cathodes has been found in the literature.

The purpose of this investigation is to study Sm–Al alloy formation by using solid reactive aluminium electrodes in molten LiCl–KCl. Basic thermodynamic properties of the intermetallic compounds formed have been estimated. The electrochemical techniques used as well as the preliminary results obtained are described in the following sections.

5.2 EXPERIMENTAL PROCEDURE

Handling of chemicals was carried out inside a glove box under an inert argon atmosphere (Air Liquide, $H_2O < 0.5$ ppm, $O_2 < 0.1$ ppm). The experiments were performed in a quartz cell under argon gas flow, which was additionally purified by passing it through both a moisture and an oxygen trap (Agilent, S.A.). The thermocouple and electrodes were positioned in the cell, which also supported a glassy carbon crucible (Sofacel, S.A.) containing the electrolyte.

The electrolyte used was the eutectic LiCl–KCl (59 : 41 mol %) (Aldrich, 99%; Merck, 99.999%). Preparation of the molten bath was performed by heating up the mixture at 473 K under vacuum (2×10^{-3} atm) for approximately 20 h, by fusing, and afterwards by raising it to atmospheric pressure using inert argon. The bath was purified by $HCl_{(g)}$ treatment (Air Liquide, $H_2O < 10$ ppm, $Cl_2 + Br_2 < 20$ ppm) for 45–60 min, followed by argon sparging in order to remove the dissolved hydrogen chloride.

Solutions of samarium(III) chloride were prepared by direct addition of anhydrous $SmCl_3$ (Aldrich, 99.999%) without further treatment.

Cyclic voltammetry and chronopotentiometry measurements were conducted by an Autolab PGSTAT30 (Eco-Chimie) potentiostat-galvanostat controlled with Autolab GPES software v.4.9.

The cathodes used for the electrochemical study consisted of 1 mm diameter tungsten (Aldrich, 99.9 + %) and aluminium (Aldrich, 99.999%) metallic wires. The lower end of the electrodes was polished thoroughly by using sandpaper and then cleaned with ethanol using ultrasound. The surface area was determined after each experiment by measuring the immersion depth of the electrode in the melt. As counterelectrode, either a 1 mm tungsten wire or a 3 mm glassy carbon rod (Sofacel, S.A.) was used. The reference electrode was made of a 1 mm silver wire (Aldrich, 99.9%) dipped into a closed-end Pyrex tube containing a solution of AgCl (Merck, >99.4%) in LiCl–KCl (0.75 mol kg^{-1}). For the electrolysis, 1 mm thick aluminium plates (Aldrich, 99.999%) were used as cathodes and a 3 mm graphite rod (Sofacel, S.A.) was used as the anode.

Since samarium alloys are easily oxidised [5], deposits obtained by electrolysis were not washed with water but with ethylene glycol (Aldrich, 99 + %) by ultrasound to remove residual salt and probable metallic lithium.

Surface analysis of the deposits was performed by X-ray diffraction (XRD) (Philips, X'PERT MPD) to identify the intermetallic Sm–Al compounds formed. SEM (scanning electron microscopy)-EDX (energy-dispersive X-ray) (HITACHI, S-2500) techniques were also used to analyse the surface morphology and to measure the concentration profiles of samarium and aluminium. Cross-section of the samples allowed determination of the thickness of the deposits by measuring it with an optical microscope (Leika, DM 4000 M).

The concentration of samarium in solution was determined by taking samples from the melt (about 200 mg), which were dissolved and diluted in 1 M aqueous nitric acid and then analysed by ICP-MS (Perkin Elmer Elan 6000).

5.3 RESULTS AND DISCUSSION

According to the Sm–Al phase diagram, which is shown in Figure 5.1 [14], samarium can form five solid intermetallic compounds with the more noble metal aluminium ($AlSm_2$, $AlSm$, Al_2Sm, Al_3Sm, and $Al_{11}Sm_3$). At the present working temperature range (723–823 K), $Al_{11}Sm_3$, the most aluminium rich compound, is not present. Cyclic voltammetry (CV) and open-circuit chronopotentiometry (OCP) techniques have been applied to identify the Sm–Al intermetallics in the molten LiCl–KCl eutectic.

5.3.1 Cyclic Voltammetry Study

Figure 5.2 shows the cyclic voltammograms obtained at inert tungsten and reactive solid aluminium electrodes in a LiCl–KCl–$SmCl_3$ solution (0.053 mol·kg^{-1}). The electroactivity domain of pure LiCl–KCl on aluminium is also plotted, which is significantly smaller than that on inert tungsten. This is due to the formation of Al–Li alloys [15] that shifts the cathodic limit of the solvent toward more positive values

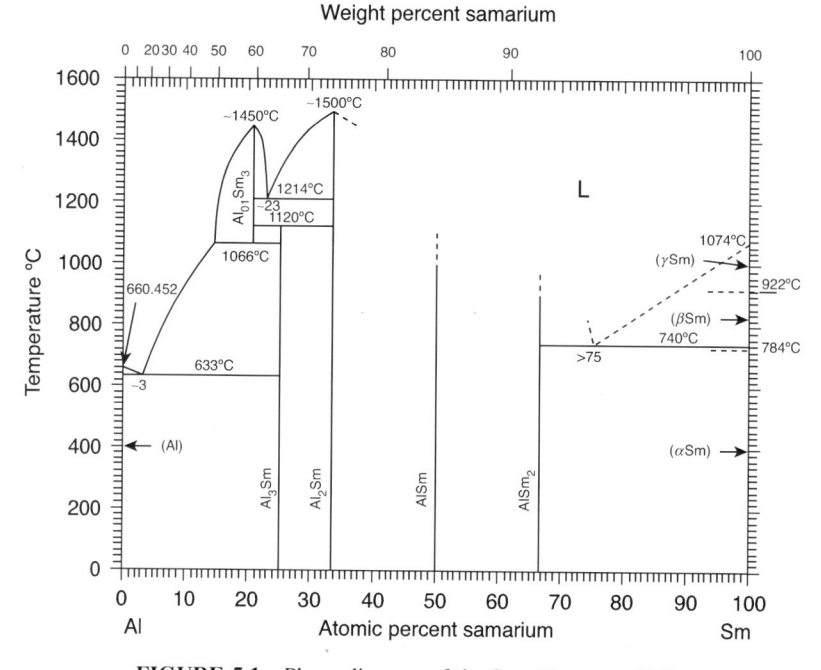

FIGURE 5.1 Phase diagram of the Sm–Al system [14].

[Figure 5.2: Cyclic voltammograms with axes i/A cm⁻² vs $E_{AgCl/Ag}$/V, showing curves labeled A, A', B, B', Li, Li-Al, Al → Al³⁺, for LiCl-KCl-SmCl₃ at W, LiCl-KCl at Al, and LiCl-KCl-SmCl₃ at Al.]

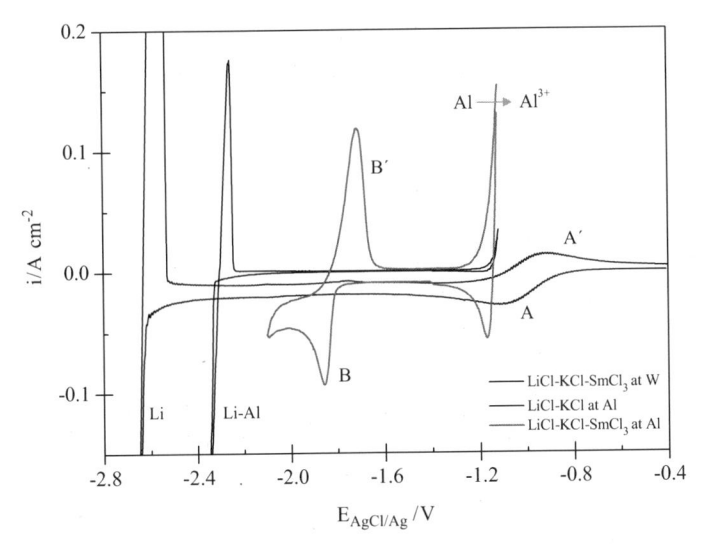

FIGURE 5.2 Cyclic voltammograms obtained in LiCl–KCl–SmCl₃ (0.053 mol·kg⁻¹) at tungsten and aluminium electrodes. $T = 723$ K, $v = 0.1$ V·s⁻¹.

(−2.35 V vs. AgCl/Ag at 723 K), and to the oxidation of aluminium (−1.15 V vs. AgCl/Ag at 723 K), which constitutes the anodic limit of the electrochemical window.

When inert tungsten is used (see Fig. 5.2), a cathodic wave A is observed at a potential of ∼ −0.99 V versus AgCl/Ag and its corresponding anodic wave A' (E ∼ −0.85 V vs. AgCl/Ag). This signal, with a shape characteristic of a soluble–soluble system, has previously been identified as the Sm(III)/Sm(II) redox couple [13]. However, the Sm(II)/Sm(0) redox system cannot be studied because Sm(II) ions are reduced to pure metallic samarium at a potential more negative than that of the solvent reduction, Li(I)/Li(0). Therefore, at a tungsten substrate, the reduction of Sm(III) ions takes place in the following two steps:

$$Sm^{3+} + 1e^- \leftrightarrow Sm^{2+} \tag{5.1}$$

$$Sm^{2+} + 2e^- \leftrightarrow Sm^0 \tag{5.2}$$

On a solid aluminium electrode, a well-defined cathodic peak B is observed at a potential of −1.86 V versus AgCl/Ag (see Fig. 5.2). Since this potential is more positive than the potential of pure samarium deposition, this peak is considered to be associated with the formation of an Sm–Al alloy. When the sweep direction is reversed, an anodic peak B' is observed that corresponds to the samarium dissolution from the alloy. The shift of the samarium reduction potential toward more positive values is explained in terms of a reduction of the activity of metallic samarium in the aluminium phase. However, the number of cathodic–anodic signals observed is less than the number of alloy phases shown on the Sm–Al phase diagram (see Fig. 5.1). This suggests that certain alloys form rapidly while the formation of other phases is much slower [16].

Taking into account that the Al(III)/Al(0) reduction potential is more negative than the Sm(III)/Sm(II) one in LiCl–KCl [17], the reduction of Sm(III) ions by metallic aluminium to Sm(II) at the electrode surface is expected to occur. Moreover, from published thermochemical data for pure compounds [18], it has been found that this reaction is thermodynamically favourable ($\Delta G_r^0 = -106.69$ kJ · mol^{-1} at 723 K). Therefore, when solid aluminium is used as cathode, the expected reduction mechanism consists of the chemical reduction of Sm(III) to Sm(II) by aluminium, followed by the electrochemical reduction of Sm(II) to metallic samarium, and then the Sm–Al alloy formation, as follows:

$$3Sm^{3+} + Al + 3e^- \leftrightarrow 3Sm^{2+} + Al^{3+} \tag{5.3}$$

$$yAl + xSm^{2+} + 2e^- \leftrightarrow Sm_xAl_y \tag{5.4}$$

5.3.2 Open-Circuit Chronopotentiometry Study

The open-circuit chronopotentiometry (OCP) technique has been used to identify the formation of Sm–Al alloys. It consists of applying short cathodic polarisations (between 60 and 360 s) at potentials more positive than the pure samarium deposition. Then the open-circuit potential of the aluminium electrode is measured as a function of time. In order to confirm the reproducibility of the experiments, the same measurement was repeated several times for each temperature tested (723–823 K). The uncertainty is given by the standard deviation for each measurement.

When metallic samarium is deposited on the aluminium electrode, samarium and aluminium interdiffuse to form different Sm$_x$Al$_y$ phases. The potential gradually shifts to more positive values until the rest potential of the aluminium substrate is reached. During this process, the E–t curve shows successive plateaux, each of them being typical of the equilibrium in the solid state between two intermetallic compounds [19].

Figure 5.3 shows the OCP curve obtained after applying a potential of -2.35 V versus AgCl/Ag for 30 s at an aluminium cathode in LiCl–KCl–SmCl$_3$ (0.05 mol·kg^{-1}) at 773 K. Three potential plateaux are observed at 1 (-2.3 V), 2 (-1.95 V), and 3 (-1.75 V), potentials being versus the AgCl/Ag reference system.

We have compared this curve with the one obtained in pure LiCl–KCl (broken line, Fig. 5.3), where a very stable plateau at a potential close to -2.3 V is observed. This plateau is thought to correspond to the formation of Al–Li alloys, as it was also mentioned in the cyclic voltammetry study (vide supra). Plateaus numbers 2 (-1.95 V) and 3 (-1.75 V) are associated with Sm–Al alloy formation.

Since the plateau at -1.95 V versus AgCl/Ag was not clearly observed, probably due to the fast transformation of one phase into another, this equilibrium is not going to be considered for the determination of the Sm–Al compounds' thermodynamic properties.

Before determining the thermodynamic properties it is necessary to identify the Sm–Al intermetallics formed. This is possible from the characterisation of the deposits obtained by electrolysis, as described in the next section.

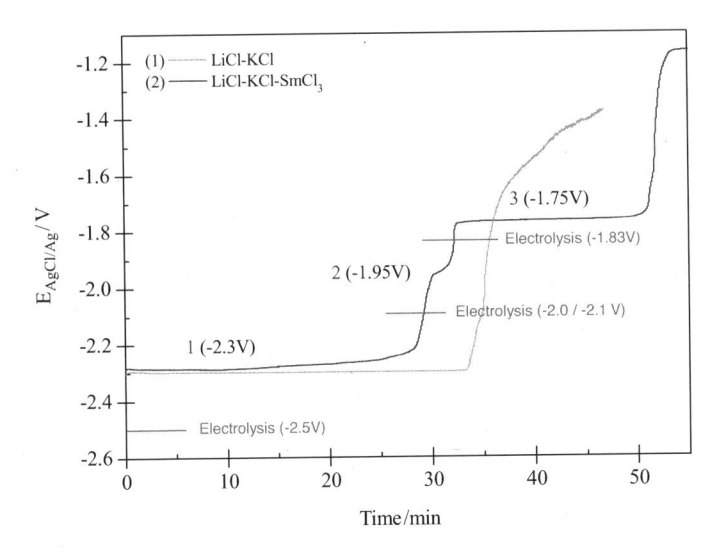

FIGURE 5.3 Open-circuit potential curve for an aluminium electrode after applying (1) −2.4 V for 60 s in pure LiCl–KCl at 723 K; and (2) −2.35 V for 30 s in the LiCl–KCl–SmCl$_3$ system (0.05 mol· kg^{-1}) at 773 K.

5.3.3 Electrolysis: Characterisation of Sm–Al Alloys

Potentiostatic electrolyses have been performed to identify the Sm–Al intermetallic compounds. The working temperature to perform electrolysis was 723 K. Deposits obtained were analysed by XRD and SEM-EDX techniques. The potentials applied were selected by considering those between the plateaux in the open-circuit chronopotentiograms (see Fig. 5.3). These potentials were (1) −1.83 V and −2.1 V versus AgCl/Ag, values more positive than pure samarium and lithium deposition potentials; and (2) −2.5 V versus AgCl/Ag, by assuming the metal lithium codeposition. After the electrolysis, the samples were washed with ethylene glycol and stored inside a desiccator until their analysis.

The X-ray diffraction analysis of the samples obtained after applying a potential of −1.83 V versus AgCl/Ag showed the existence of the SmAl$_4$ phase, as the main component, and metallic aluminium. This intermetallic compound is not present in the Sm–Al phase diagram but is in the XRD index cards (JCPDS 00-019-0047. ICDD 2004). It is also known from the XRD results that this phase has been formed at high temperature, and it has a lower symmetry. According to this, peaks found in the diffractogram may correspond to another composition phase but this is not registered in the ICDD 2004 database. Further investigation concerning this phase formation process is needed.

Figure 5.4 shows the surface morphology of the deposit obtained after applying −2.1 V at an aluminium plate for 285 min. The composition profile of the deposit obtained from the EDX analysis shows the presence of samarium and aluminium. The XRD diffractogram that is also plotted in Figure 5.4 confirms the Sm–Al alloy formation. The observed peaks were identified as SmAl$_3$ and metallic aluminium; the

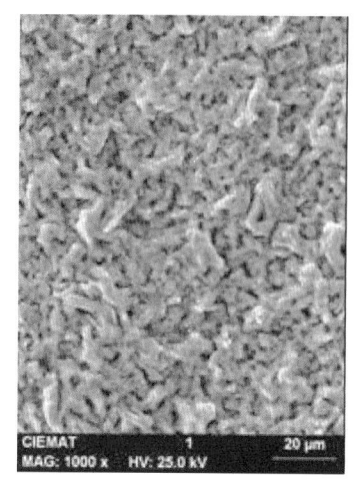

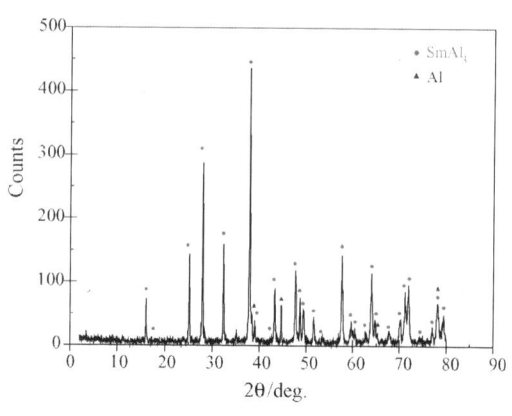

FIGURE 5.4 SEM image and X-ray diffractogram of an aluminium plate covered by an SmAl$_3$ deposit obtained by potentiostatic electrolysis (E_{appl}: -2.1 V vs. AgCl/Ag, time: 285 min, $Q = 190$ C) in LiCl–KCl–SmCl$_3$ at 723 K.

SmAl$_3$ phase is the main component. The cross section of the sample allowed measuring the thickness of the deposit with an optical microscope, being equal to ~10 μm. In some cases, XRD analysis of the deposits showed peaks corresponding to SmAl$_3$, as the main component, but the SmAl$_4$ phase was also present.

According to the studies of Iida et al. [5] to form Sm–Ni alloys in LiCl–KCl, it seems that codeposition of lithium considerably improves the SmNi$_2$ formation rate. Based on this, the potential of -2.5 V versus AgCl/Ag (see Fig. 5.3) was selected to carry out the electrolysis. Only one sample obtained by applying this potential has been analysed by XRD; hence results presented here are just preliminary. The diffractogram shows the presence of the Al–Li phase, the main component, and the SmAl$_2$ phase. Deposition of lithium at the aluminium surface would hinder the interdiffusion of aluminium and samarium to form alloys, which may explain the small SmAl$_2$ deposit and the absence of other Sm–Al phases. This was not the case for the Iida et al. [5] study, where no Li–Ni alloys exist. Therefore plateau number 1 at -2.3 V versus AgCl/Ag (see Fig. 5.3) would correspond to an equilibrium in which Al–Li and SmAl$_2$ are involved. The mechanism of this equilibrium has not clearly been identified. It should also be indicated that all samples obtained at -2.5 V versus AgCl/Ag presented cracks all over the aluminium plate. This could be due to some kind of interaction between the deposited metallic lithium and aluminium.

Galvanostatic electrolyses ($I = cte$) have been conducted to compare them with those performed at constant potential. Samples were prepared by applying constant cathodic currents (~-10 mA) on aluminium plates at 723 K. XRD analyses of the deposits showed the formation of the SmAl$_4$ intermetallic phase. The cathode potential evolution was followed during the electrolysis, which was almost constant and equal to ~-1.85 V versus AgCl/Ag. This value is similar to that applied under potentiostatic conditions at which SmAl$_4$ was also formed. Under these conditions,

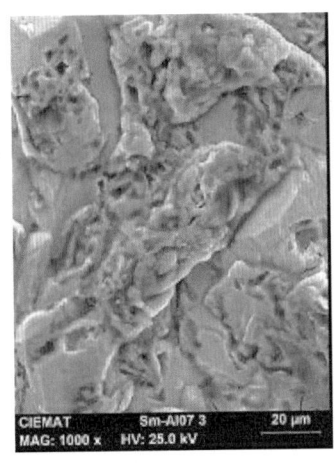

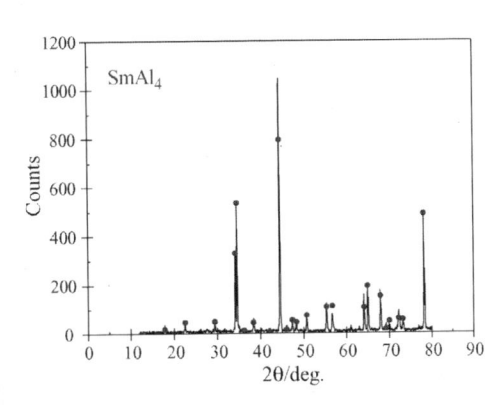

FIGURE 5.5 SEM image and XRD spectrum of an aluminium plate covered by a SmAl$_4$ deposit obtained under galvanostatic conditions (-10 mA) for 2 h. $T = 723$ K, $Q = 70$ C.

thicker deposits than those formed by potentiostatic electrolysis were obtained in less time.

Figure 5.5 shows the SEM image and the XRD spectrum of a 60 μm thick deposit obtained under galvanostatic conditions (-10 mA). The EDX composition profile of the sample shows the presence of samarium and aluminium in the deposit.

Once the different Sm$_x$Al$_y$ phases have been identified, it is necessary to associate them to the two-phase equilibriums found in the open-circuit chronopotentiograms (Fig. 5.3). According to the results obtained, the following equilibria between solid Sm–Al phases are proposed:

Plateau X (< -2.0 V vs. AgCl/Ag) $2\text{SmAl}_3 + \text{Sm}^{2+} + 2e^- \leftrightarrow 3\text{SmAl}_2$

Plateau 2 (-1.95 V vs. AgCl/Ag) $3\text{SmAl}_4 + \text{Sm}^{2+} + 2e^- \leftrightarrow 4\text{SmAl}_3$

Plateau 3 (-1.75 V vs. AgCl/Ag) $4\text{Al} + \text{Sm}^{2+} + 2e^- \leftrightarrow \text{SmAl}_4$

As shown above, the presence of an additional plateau (X) at a potential <-2.0 V versus AgCl/Ag is proposed. The fact that this two-phase equilibrium plateau is not observed in the OCP curve (Fig. 5.3) could be due to a very fast transformation of one phase into another.

For the determination of the thermodynamic properties we have only considered the most stable plateau, which seems to correspond to the equilibrium between SmAl$_4$ and aluminium, according to the XRD results.

5.3.4 Thermodynamic Property Determination

When the potentials measured versus AgCl/Ag reference electrodes in the chronopotentiograms are converted into values versus the Sm(II)/Sm(0) system, they correspond to electromotive forces (emfs) for the two-phase coexisting state. Then, it is possible to calculate the thermodynamic properties of the Sm–Al intermetallic compounds such as the activity of samarium in the aluminium phase and the relative molar Gibbs free energy of formation.

Under our experimental conditions, the Sm(II)/Sm(0) reduction potential cannot be measured; therefore it was estimated by using data from the literature. Lebedev et al. [20] calculated the apparent standard potential of the Sm(II)/Sm(0) system in LiCl–KCl by measuring the emf of the galvanic cell: Sm-Zn$_{liq}$/KCl–LiCl–SmCl$_2$// LiCl–KCl–MCl$_3$ (M = La or Ce)/M–Zn$_{liq}$ in the temperature range 973–1273 K. The following expression for the apparent standard potential dependence with temperature has been found:

$$E'^{\circ}_{Sm(II)/Sm}(\text{V vs. Cl}_2/\text{Cl}^-) = -4.282 + 9.066 \times 10^{-4}\, T(K) \quad (5.5)$$

The equilibrium potential for the reduction of Sm(II) ions into pure metallic samarium can be estimated as

$$E_{eq} = E'^{\circ}_{Sm(II)/Sm} + \frac{2.3RT}{2F}\log X_{Sm(II)} \quad (5.6)$$

where $X_{Sm(II)}$ is the concentration of Sm(II) ions in molar fraction, assuming that $X_{Sm(II)} = X_{Sm(III)}$ at the electrode surface, R is the ideal gas constant (8.314 J·mol^{-1}·K^{-1}), F is the Faraday constant (96500 C·mol^{-1}), and E'° is the apparent standard potential of the Sm(II)/Sm(0) redox system from Lebedev. Calculated equilibrium potentials versus a Cl$_2$/Cl$^-$ reference electrode have been converted into potentials versus AgCl/Ag (see Table 5.1) [21].

From the estimated emf corresponding to a chemical composition of the alloy formed at the surface of the electrode, it is possible to calculate the activity of samarium in the aluminium phase by the following expression:

$$emf = -\frac{RT}{2F}\ln a_{Sm} \quad (5.7)$$

The activity of samarium in the aluminium phase at different temperatures is quoted in Table 5.1. These values, around 10^{-14} to 10^{-12}, give an idea of the low solubility of samarium in aluminium.

TABLE 5.1 Thermodynamic Properties of the SmAl$_4$ Compound in the SmAl$_4$ + Al Two-Phase Coexisting State at Different Temperatures

T (K)	E (V) versus Sm(II)/Sm(0)	ΔG/kJ mol^{-1}	ΔG/ kJ mol^{-1} atm^{-1}	a_{Sm}
723	0.976 ± 0.005	−188.34 ± 0.97	−37.67 ± 0.19	(2.47 ± 0.85) × 10^{-14}
748	0.970 ± 0.003	−187.25 ± 0.58	−37.45 ± 0.12	(8.38 ± 0.91) × 10^{-14}
773	0.964 ± 0.006	−186.10 ± 1.16	−37.22 ± 0.23	(2.66 ± 0.84) × 10^{-13}
798	0.959 ± 0.002	−184.99 ± 0.39	−37.00 ± 0.08	(7.78 ± 0.94) × 10^{-13}
823	0.950 ± 0.004	−183.30 ± 0.77	−36.66 ± 0.15	(2.32 ± 0.89) × 10^{-12}

ΔG(kJ·mol^{-1}) = −224.18(±2.56) − 0.0494(±0.0023)(T/(K))

Moreover, the relative partial molar Gibbs free energy of formation for the SmAl$_4$ intermetallic compound can be calculated from the emf values as follows:

$$\Delta G_\text{f}(\text{SmAl}_4) = -2 \cdot F \cdot emf \tag{5.8}$$

The temperature dependence of the Gibbs free energy of formation allows calculation of the relative partial molar entropy and enthalpy of formation of the SmAl$_4$ compound. By applying the Gibbs–Helmholtz equation, a linear relationship was obtained. Values determined are included in Table 5.1.

5.4 CONCLUSION

Sm–Al alloy formation has been investigated in the molten LiCl–KCl eutectic by using solid reactive aluminium electrodes at several temperatures (723–823 K). Cyclic voltammetry and open-circuit chronopotentiometry measurements have shown that samarium alloys with solid aluminium. Potentiostatic and galvanostatic electrolyses have been performed to identify the Sm–Al intermetallics formed. Preliminary results indicate the presence of three phases: SmAl$_2$, SmAl$_3$, and SmAl$_4$.

The intermetallic SmAl$_4$ phase that is not present in the Sm–Al phase diagram is formed at applied potentials close to pure aluminium. Further investigation concerning the formation of this phase is needed. The Gibbs energy of formation of the SmAl$_4$ intermetallic compound in the two-phase equilibrium state SmAl$_4$ + Al, the most stable, has been estimated at different temperatures. In addition, the activity of samarium in the aluminium phase has been determined.

ACKNOWLEDGEMENTS

This work is part of the activities performed at CIEMAT within the EUROPART project (FI6W-508854) of the European Commission. The authors thank ENRESA for financial support within the agreement CIEMAT-ENRESA (No. 7800087, Annex X) on separation of long-lived radionuclides. We also thank V. Smolenski for technical discussions, and L. Gutiérrez, A. Del Río, A. M. Lancha, Y. Álamo, M. J. Tomás, M. Rosado, J. Quiñones, J. M. Cobo, and J. Alcaide for technical assistance.

REFERENCES

1. OECD/NEA, *Accelerator Driven Systems (ADS) and Fast Reactors (FR) in Advanced Fuel Cycles. A Comparative Study*, 2002, pp. 1–349.
2. OECD/NEA, *Pyrochemical Separations in Nuclear Applications. A Status Report*, 2004.
3. Conocar, O., Douyere, N., and Lacquement, J., Extraction behavior of actinides and lanthanides in a molten fluoride/liquid aluminum system, *J. Nucl. Mater.* **344**, 136–141 (2005).

4. Serp, J., Allibert, M., LeTerrier, A., Malmbeck, R., Ougier, M., Rebizant, J., and Glatz, J., Electroseparation of actinides from lanthanides on solid aluminum electrode in LiCl–KCl eutectic melts, *J. Electrochem. Soc.* **152**(3), C167–C172 (2005).
5. Iida, T., Nohira, T., and Ito, Y., Electrochemical formation of Sm–Ni alloy films in a molten LiCl–KCl–SmCl$_3$ system, *Electrochim. Acta* **46**, 2537–2544 (2001).
6. Iida, T., Nohira, T., and Ito, Y., Electrochemical formation of Sm–Co alloy films by Li codeposition method in a molten LiCl–KCl–SmCl$_3$ system, *Electrochim. Acta* **48**, 901–906 (2003).
7. Iida, T., Nohira, T., and Ito, M., Electrochemical formation of Sm–Co alloys by codeposition of Sm and Co in a molten LiCl–KCl–SmCl$_3$–CoCl$_2$ system, *Electrochim. Acta* **48**, 2517–2521 (2003).
8. Lebedev, V. A., *Selectivity of Liquid Metal Electrodes in Molten Halides* (in Russian), Chelyabinsk, Metallurgiya, 1993.
9. Johnson, K. E. and Mackenzie, J. R., Samarium, europium and ytterbium electrode potentials in LiCl–KCl eutectic melt, *J. Electrochem. Soc.* **116**(12), 1697–1703 (1969).
10. Novosielova, A., Shishkin, V., and Khokhlov, V., Redox potentials of samarium and europium in molten lithium chloride, *Z. Naturforsch.* **56(A)**, 754–756 (2001).
11. Nikolaeva, E. V., Chemezov, O. V., and Chkochklov, V. A., Study of the process of electroreduction of Sm^{3+} in the NaCl–KCl melt, in *XIII Russian Conference on Physical Chemistry and Electrochemistry of Molten Salts and Solid Electrolytes*, Ekaterinburg, Russia, 2004, Vol. 1.
12. Kushkov, H. B. and Vindizeva, M. K., The investigation of electroreduction of samarium ions in halide ions, in *XIII Russian Conference on Physical Chemistry and Electrochemistry of Molten Salts and Solid Electrolytes*, Ekaterinburg, Russia, 2004, Vol. 1.
13. Cordoba, G. and Caravaca, C., An electrochemical study of samarium ions in the molten eutectic LiCl + KCl, *J. Electroanal. Chem.* **572**, 145–151 (2004).
14. Gschneidner, K. A. J. and Calderwood, F. W., Al–Sm phase diagram, *Bull. Alloy Phase Diagrams* **10**(1), 37–39 (1989).
15. McAlister, A. J., Al–Li phase diagram, *Bull. Alloy Phase Diagrams* **3**(2), 172–177 (1982).
16. Konishi, H., Nohira, T., and Ito, M., Formation and phase control of Dy alloy films by electrochemical implantation and displantation, *J. Electrochem. Soc.* **148**(7), C506–C511 (2001).
17. Plambeck, J. A., Encyclopedia of Electrochemistry of the Elements, Volume X, *Fused Salt Systems*, Marcel Dekker, New York, 1976.
18. HSCChemistry, Outokumpu Research Oy, Version 4.1.
19. Massot, L., Chamelot, P., and Taxil, P., Cathodic behaviour of samarium(III) in LiF–CaF$_2$ media on molybdenum and nickel electrodes, *Electrochim. Acta* **50**(28), 5510–5517 (2005).
20. Lebedev, V. A., Kovalevsky, A. B., Nichkov, I. F., and Raspopin, S. P., Electrochemical behaviour of Sm in fused LiCl-KCl eutectic, *Russian J. Electrochem.* **10**(9), 1342–1344 (1974).
21. Mottot, Y. E., Propiétés chimiques et électrochimiques des chlorures de lanthanides cériques en milieux chlorures fondus. Stabilité thermodynamique des alliages La-Ni et Nd-Fe à haute température, Université Pierre et Marie Curie (Paris VI), Paris, 1986.

6 Alumina Solubility and Electrical Conductivity in Potassium Cryolites with Low Cryolite Ratio

ALEXANDER DEDYUKHIN, ALEXEI APISAROV, OLGA TKATCHEVA, YURII ZAIKOV, and ALEXANDER REDKIN

Institute of High Temperature Electrochemistry, Yekaterinburg, Russia

Abstract

Electrical conductivity and alumina solubility in molten systems KF–AlF_3, KF–AlF_3–Al_2O_3, KF–AlF_3–LiF, and KF–AlF_3–LiF–Al_2O_3 with cryolite ratio (CR) 1.3 in the temperature range 700–800 °C have been investigated. Cells of two types—with a boron nitride capillary and with two parallel molybdenum electrodes—were used for conductivity measurements. Electrical conductivity of the KF–AlF_3 system was found to be $1.21\ \Omega^{-1}\cdot cm^{-1}$ at 795 °C. An electrical conductivity decrease in the presence of alumina can be compensated by lithium fluoride additions. Alumina solubility was determined by a potentiometric titration method in the temperature range 700–760 °C. Alumina solubility in potassium cryolite (CR = 1.3) with LiF additives up to 5 mass % was in the range 3.9–4.7 mass % at 700 °C. LiF additives over 5 mass % essentially decrease Al_2O_3 solubility. The values of potassium and aluminium solubility in melts under investigation were determined.

6.1 INTRODUCTION

The modern aluminium industry needs new technologies in order to reduce energy consumption and to meet ecological requirements. Use of low melting cryolites with a low cryolite ratio (CR; the ratio of the number of moles of MF to the number of moles of AlF_3) is one of the possible ways to solve this problem. Electrical conductivity and alumina solubility are the crucial properties of any electrolyte for

---

Molten Salts and Ionic Liquids: Never the Twain? Edited by Marcelle Gaune-Escard and Kenneth R. Seddon
Copyright © 2010 John Wiley & Sons, Inc.

aluminium electrolysis. These properties are strongly temperature dependent, so while the operating temperature of the process is reduced by 200–300 °C, both electrical conductivity and alumina solubility also drop. It is known from literature data [1] that alumina solubility in MF–AlF$_3$ melts is strongly affected by the nature of the cation and increases in the sequence Li < Na < K. Thus the best alumina solvents are electrolytes based on potassium cryolite, but Al$_2$O$_3$ solubility in acid electrolytes (CR = 1.3–1.5) at temperatures below 800 °C has not been investigated yet. In contrast, electrical conductivity in MF–AlF$_3$ rises in reverse order (Li > Na > K). Nevertheless, electrical conductivity decrease can be compensated with lithium fluoride additives [2]. The investigation of electrical conductivity and alumina solubility in the molten mixtures KF–AlF$_3$, KF–AlF$_3$–Al$_2$O$_3$, KF–AlF$_3$–LiF, and KF–AlF$_3$–LiF–Al$_2$O$_3$ with a cryolite ratio of 1.3 in the temperature range 700–800 °C was the focus of this present work. Since these melts could be used as electrolytes for aluminium electrowinning, the solubility of aluminium and potassium in KF–AlF$_3$–Al$_2$O$_3$ and KF–AlF$_3$–LiF–Al$_2$O$_3$ systems was also estimated.

6.2 EXPERIMENTAL

6.2.1 Electrolyte Preparation

Electrolytes investigated were prepared from AlF$_3$, LiF, and KF (added as K[HF$_2$]). The mixture was heated in a glassy carbon container by raising the temperature up to 700 °C for 3 hours, and then keeping at this temperature for 4 hours, while HF was removed from the melt due to thermal decomposition ($T = 238.7$ °C). The starting materials certainly contained oxygen in different forms, at concentrations up to several tenths of a percent. During the electrolyte preparation, the fluorination of oxides occurred due to their interaction with HF, and the oxygen concentration in the electrolyte prepared did not exceed 0.1 mass %. Compositions of electrolytes used in this research are listed in Table 6.1. The cryolite ratio of all melts was kept at 1.3.

6.2.2 Electrical Conductivity

Oxide-fluoride melts are complicated systems for investigation, due to their high chemical aggression. Therefore the construction and material selection for an electrical conductivity investigation are very important. The experimental cell designs could be divided into two types. The first one is a completely metallic cell: as usual,

TABLE 6.1 Electrolyte Composition in Mass %

Number	System	KF	AlF$_3$	LiF	Al$_2$O$_3$
1	KF–AlF$_3$	47.35	52.65	—	—
2	KF–AlF$_3$–LiF	47.35	52.65	0–10	—
3	KF–AlF$_3$–Al$_2$O$_3$	47.35	52.65	—	0–4.8
4	KF–AlF$_3$–LiF–Al$_2$O$_3$	39.09	55.91	5	0–4.4

crucible and electrodes are made of platinum [3, 4]. Such cells have a relatively simple design, but the low cell constant is to their considerable disadvantage and greatly affects measurement precision. The capillary type cells avoid these shortcomings [5]. Pyrolytic boron nitride (PBN) is now widely used as a capillary material [6]. PBN is characterised by a high corrosion resistance in cryolite melts and a small thermal expansion coefficient, and it is an electrical insulator even at 1000 °C. However, our preliminary tests showed that this material is not universal either, since it is exposed to destruction in lithium fluoride containing melts.

To determine the electrical conductivity of the electrolytes under study, cells of two types were used: a capillary cell and a cell with two parallel electrodes. A schematic of the capillary type cell is shown in Figure 6.1. A glassy carbon crucible filled with a weighed quantity of the salt mixture was put at the bottom of the quartz test tube, which was tightly closed by a vacuum rubber plug with holes for the electrodes, thermocouple, and the inert gas inlet and outlet. Small tubes for the gas inlet and outlet and the cases for the thermocouple and electrodes were prepared from alumina. The platinum–platinum-rhodium thermocouple without any casing was immersed directly into the molten salt. The difference between the cell with the

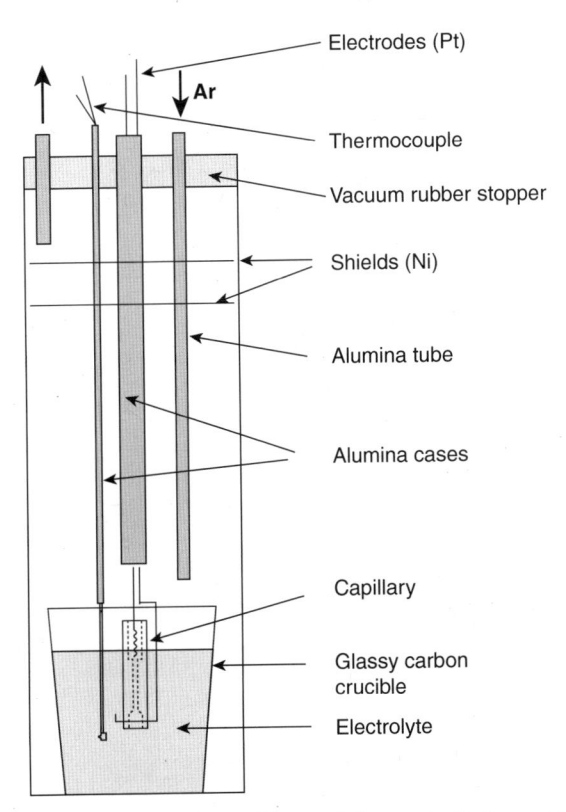

FIGURE 6.1 Scheme of the capillary cell for conductivity measurements.

parallel electrodes and the capillary type cell is the fact that two rigidly fixed parallel molybdenum electrodes were immersed into the melt. A unit for the load of additives in the flow of inert gas was attached to the alumina gas-feeder tube. The test tube was put into the furnace, evacuated, filled with inert gas, and heated to a temperature of 795 °C. The impedance measurements were performed by using a Zahner electric IM6-E in the AC frequency range between 1 Hz and 1 MHz, with an AC amplitude of 5 mV. The electrolyte resistance was determined from the impedance diagrams. The capillary type cells were used for measuring the electrical conductivity of the molten potassium cryolite $KF–AlF_3$ (CR = 1.3). The electrical conductivity of the $KF–AlF_3–Al_2O_3$, $KF–AlF_3–LiF$, and $KF–AlF_3–LiF–Al_2O_3$ melts was measured in the cells with the parallel electrodes. In the experiments with the $KF–AlF_3–LiF$ system, the lithium fluoride was added to the mixture with the potassium cryolite in the quantity providing an unchanged cryolite ratio. After each addition, the submersion depth of the electrodes was corrected with allowance for the changed electrolyte volume, and the change in the electrolyte resistance was recorded.

The capillary unit was calibrated against molten potassium chloride [7] in a temperature range of 790–860 °C. In a temperature range of 690–795 °C, the cell constant was $12.36 \pm 0.03\,\text{cm}^{-1}$. The constant of the cell with two parallel molybdenum electrodes was determined from the electrical conductivity value obtained for potassium cryolite with CR = 1.3 in the capillary type cell. The electrical conductivity measured for the electrolytes under study in the cells with the parallel electrodes was calculated with allowance for the temperature dependence of the constant.

6.2.3 Alumina Solubility

To determine alumina solubility, a potentiometric titration method was used [8]. The potentiometric titration technique is easy enough in technical performance and allows carrying out visual observation of the alumina behaviour. Figure 6.2 represents the

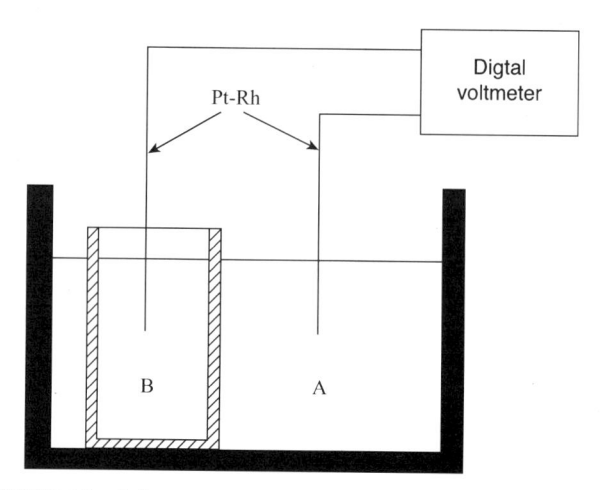

FIGURE 6.2 Scheme of the cell for determining alumina solubility.

laboratory apparatus scheme. Initially, alumina saturated electrolyte was loaded into the alundum container (A). A weighed amount of electrolyte under investigation was put in the carbon crucible (B) used as a diaphragm (thickness 1.5 mm). Pt-Rh electrodes were immersed into compartments A and B after electrolyte melting. The electromotive force (EMF) of the concentration element, which occurs as a result of the oxygen ion concentration difference in the compartments, was registered with a digital voltmeter. Alumina addition was continued until the A–B section EMF difference became zero, which meant that the alumina concentration was equal. Based on known masses of melt and additives, the solubility was calculated.

6.2.4 Metal Solubility

Metal solubility determination is a difficult technical task. There are some known techniques for this investigation—thus it is easy to practically realise the mass loss method, but it neglects the metal–moisture and metal–crucible interaction [6]. The frozen samples analysis method is more difficult to realise, but it has less experimental error. The tin extraction method was described by Wang et al. [10]. A modified technique of this method was used for the metal solubility determination in this potassium cryolite work. The experimental unit is depicted in Figure 6.3.

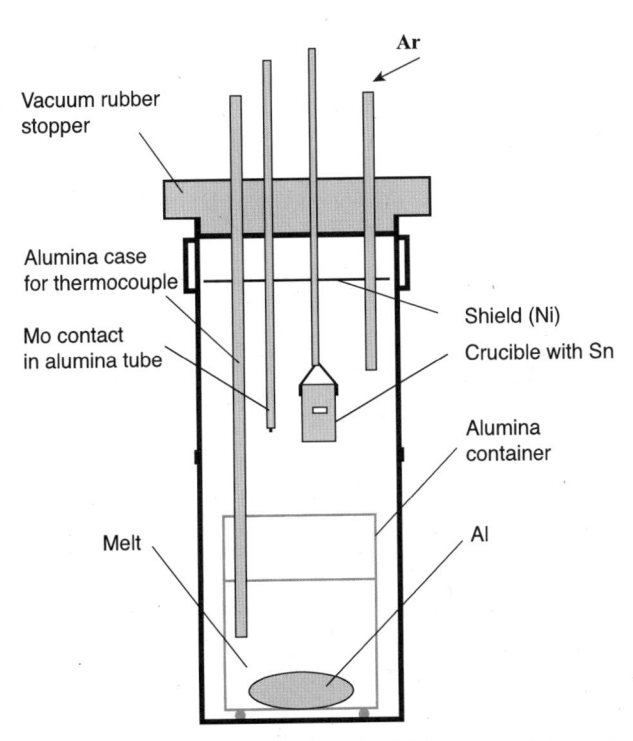

FIGURE 6.3 Scheme of the metal solubility determining cell.

The measured amount of pure aluminium ($m = 40$ g) and the electrolyte ($m = 160$ g) was loaded into the alumina container. A small alumina crucible suspended on a nichrome wire contained pure tin ($m = 8$ g). The small crucible has a slot for electrolyte sampling and could be shifted vertically. During the experiment, the cell was evacuated, filled with pure argon, and heated up to the test temperature (700 or 750 °C). After 4 hours at this temperature, the small crucible was immersed into the melt for electrolyte sampling. The submersion depth was controlled by the electrical contact between the nichrome wire and the molybdenum one. As a result, the ladling crucible contained approximately 8 g of electrolyte. After 3 hours alloying of dissolved metals and tin, the small crucible was lifted up to the cool zone of the cell. The samples obtained were analysed for the presence of individual metals by spectral emission analysis.

6.3 RESULTS AND DISCUSSION

6.3.1 Electrical Conductivity

The electrical conductivity data for the molten KF–AlF$_3$ (CR = 1.3) system measured in capillary type cells at a temperature of 690–795 °C are given in Table 6.2. The electrical conductivity is substantially lower than that of conventional electrolyte based on sodium cryolite [6]. The comparison of electrical conductivity for potassium cryolite (CR=1.22) obtained by Hives and Thonstad [9] and the present work is shown in Figure 6.4. All results are in good agreement within the experimental error.

The measurement of the KF–AlF$_3$ electrolyte (CR = 1.3) electrical conductivity, depending on the alumina content, was performed in the cells with the parallel molybdenum electrodes in a temperature range of 680–770 °C. The polythermal curves are given in Figure 6.5. As one would expect, the electrical conductivity decreases with decreasing temperature and with increasing alumina concentration. The curve of the temperature dependence of the electrical conductivity in the KF–AlF$_3$ melt with 4.8 mass % alumina exhibits a small bend in a temperature range of 700–720 °C. This means that the saturation of melt by alumina occurs at temperatures below 700 °C. The solubility of Al$_2$O$_3$ at 700 °C, determined from the conductivity data, is about 4.6 mass %.

The electrical conductivity measurement of the KF–AlF$_3$ electrolyte with LiF additives in a temperature range of 700–770 °C was performed in the cells with the parallel molybdenum electrodes. It is evident from Figure 6.6 that the electrical conductivity decreases with decreasing temperature, and substantially increases with

TABLE 6.2 Electrical Conductivity of the KF–AlF$_3$ (CR = 1.3) Melt

Temperature/°C	692	717	737	754	774	795
Conductivity/$\Omega^{-1} \cdot$cm^{-1}	0.94	1.01	1.06	1.11	1.16	1.21

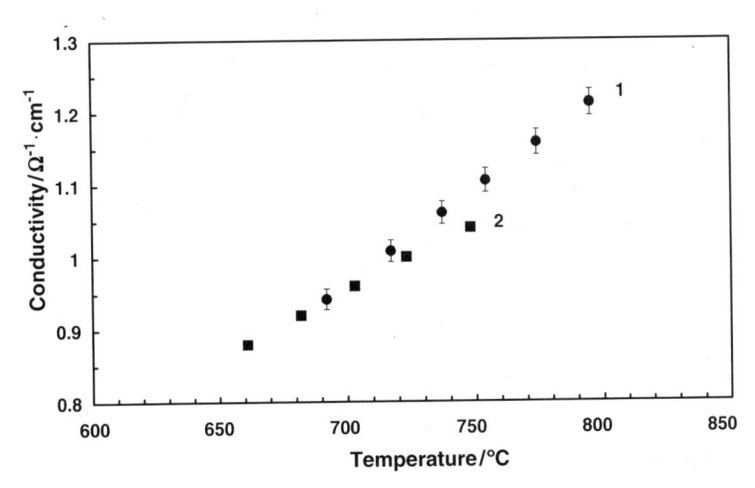

FIGURE 6.4 Electrical conductivity of the KF–AlF$_3$ system versus temperature. Data set (1) ●, CR = 1.3 [present work]; data set (2) ■, CR = 1.22 [9].

increasing content of added LiF. The addition of 5.7 mass % LiF led to an increase in electrical conductivity of approximately 15%.

Figure 6.7 shows the electrical conductivity dependence of the KF–AlF$_3$–LiF system containing 5 mass % LiF on the Al$_2$O$_3$ concentration at 700 and 750 °C. The solubility of alumina in these melts is somewhat lower than its solubility in the electrolytes free from lithium fluoride. It is about 3.8 mass % at 700 °C.

6.3.2 Alumina Solubility

Alumina solubility experimental data obtained in KF–AlF$_3$ and KF–AlF$_3$–LiF melts in the temperature range 700–760 °C are listed in Table 6.3. Alumina solubility in

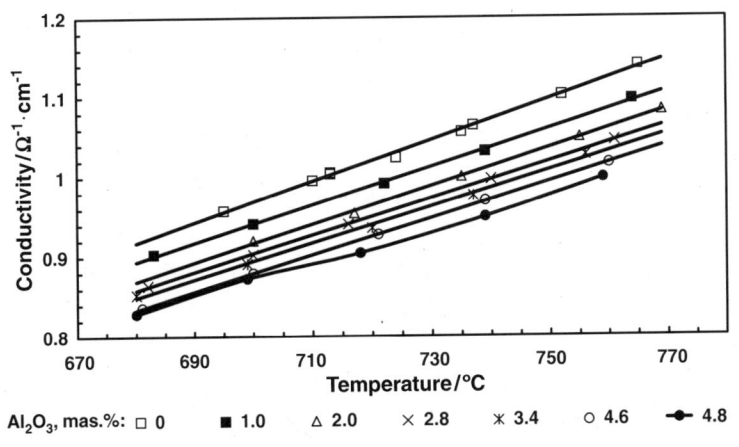

FIGURE 6.5 Electrical conductivity of the KF–AlF$_3$ + Al$_2$O$_3$ system versus temperature.

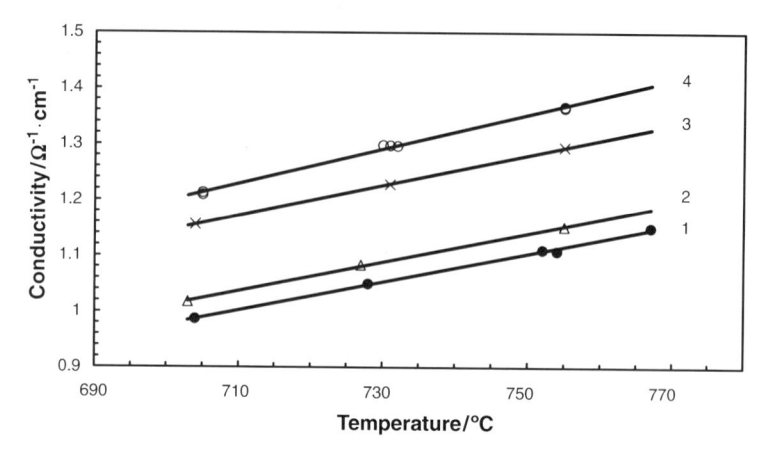

FIGURE 6.6 Electrical conductivity of the KF–AlF$_3$–LiF system versus temperature. LiF content, in mass %: (1) 0; (2) 1.0; (3) 5.7; (4) 10.0.

potassium cryolite (CR = 1.3) with LiF additives up to 5 mass % was in the range 3.9–4.7 mass % at 700 °C, which is suitable for commercial electrolysis. LiF additives over 5 mass % essentially decrease Al$_2$O$_3$ solubility.

Alumina solubility values obtained in the molten systems under investigation by the potentiometric technique, and found from conductivity data, are in good agreement.

6.3.3 Metal Solubility

Metal solubility data, calculated based on the condition of the equilibrium in the tin–electrolyte system, are presented in Table 6.4, where C_{Al} and C_K are the concentrations of aluminium and potassium, respectively.

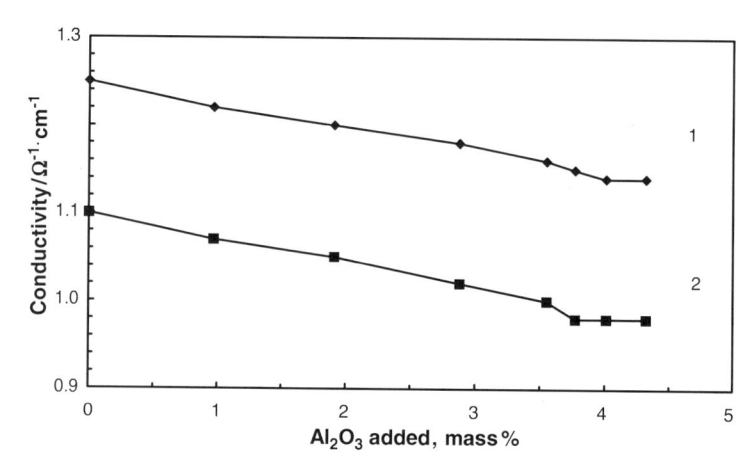

FIGURE 6.7 Conductivity of KF–AlF$_3$–LiF–Al$_2$O$_3$ versus alumina added at (1) 750 °C and (2) 700 °C.

TABLE 6.3 Alumina Solubility in KF–AlF$_3$–LiF Melts

Number	LiF/mass %	[Al$_2$O$_3$]$_{satur}$/mass %		
		700 °C	730 °C	760 °C
1	0	4.7	5.2	5.7
2	1.5	4.4	4.7	5.3
3	5	3.9	4.1	4.5
4	10	2.0	—	—

Based on these data, the metal solubility increases considerably with rising temperature. LiF additives (up to 5 mass %) do not significantly influence solubility at 700 °C and lithium in the melt was not detected, within the limits of the chemical analysis accuracy. LiF additives reduce potassium and aluminium solubility at 750 °C. It is necessary to note that the measured data might be undervalued due to the open surface of the melt during the experiment. Unfortunately, there are no available data for metal solubility in potassium cryolite to make a comparison.

6.4 CONCLUSION

The investigation of the properties of the low-melting KF–AlF$_3$ (CR = 1.3) electrolyte with alumina additions in a temperature range of 680–770 °C showed that its electrical conductivity is substantially lower than that of a conventional electrolyte based on sodium cryolite; however, the solubility of alumina at these temperatures is sufficiently high. A decrease in the electrical conductivity in the presence of alumina can be compensated for by lithium fluoride additions, which significantly increase it. Metal solubility in the KF–AlF$_3$ (CR = 1.3) system at 700–750 °C are essentially lower than in conventional sodium cryolite.

TABLE 6.4 Metal Solubility in KF–AlF$_3$–LiF Melts

T/°C	Electrolyte Composition/mass %			
	AlF$_3$–52.6; KF–47.4; + Al$_2$O$_3$ (4 mass %)		AlF$_3$–55.9; KF–39.1; LiF–5; + Al$_2$O$_3$ (4 mass %)	
	C_{Al}/ mass % × 10^3	C_K/ mass % × 10^3	C_{Al}/ mass % × 10^3	C_K/ mass % × 10^3
700	0.44	1.27	0.40	1.49
750	4.16	5.20	2.74	2.74

REFERENCES

1. Robert, E., Olsen, J. E., Danek, V., Tixhon, E., Ostvold, T., and Gilbert, B., Structure and thermodynamics of alkali fluoride–aluminium fluoride–alumina melts. Vapour pressure, solubility, and raman spectroscopic studies, *J. Phys. Chem. B* **101**, 9447–9457 (1997).
2. Matiasovsky, K., Danek, V., and Malinovsky, M., Effect of LiF and Li_3AlF_6 on electrical conductivity of cryolite–alumina melts, *J. Electrochem. Soc.* **116**, 1381–1383 (1969).
3. Edwards, J. D., Taylor, C. S., Russell, A. S., and Maranville, L. F., Electrical conductivity of molten cryolite and potassium, sodium, and lithium chlorides, *J. Electrochem. Soc.* **99**, 527–535 (1952).
4. Matiasovsky, K., Malinovsky, M., and Danek, V., Specific electrical conductivity of molten fluorides, *Electrochim. Acta* **15**, 25–32 (1970).
5. Smirnov, M. V., Shumov, Y. A., and Khohlov, V. A., Electrical conductivity of molten alkali metal fluorides, Electrochemistry of Molten and Solid Electrolytes, in *Proceedings of the Electrochemistry Institute*, Sverdlovsk, No. 18, 3–9 (1972) (in Russian).
6. Grjotheim, K., Krohn, C., Malinovsky, M., Matiasovsky, K., and Thonstad, J., *Aluminium Electrolysis. Fundamentals of the Hall–Heroult Process*, 2nd ed., Verlag, Dusseldorf, 1982.
7. Janz, J., Thermodynamic and transport properties for molten salts: correlation equation for critically evaluated density, surface tension, electrical conductance, and viscosity data, *J. Phys. Chem. Ref. Data* **17** (Suppl. 2), (1988).
8. Apisarov, A. P., Shurov, N. I., Zaikov, Yu. P., and Khramov, A. P., Investigation of alumina solubility and solubility rate in low-melting fluoride electrolytes by potentiometric titration method, *Vestnik UGTU-UPI*, **14** (44), 13–16 (2004) (in Russian).
9. Hives, J. and Thonstad, J., Electrical conductivity of low-melting electrolytes for aluminium smelting, *Light Metals* **49**, 5111 (2004); Danelik, V. and Hives, J., Low-melting electrolyte for aluminium smelting, *J. Chem. Eng. Data* **49**, 1414–1417 (2004).
10. Wang, X., Peterson, R. D., and Richards, N. E., Dissolved metals in cryolitic melts, *Light Metals*, 323–330 (1991).

7 Ionic Liquids as Solvents for the Variable Temperature Electrodeposition of Metals and Semiconductors: A Short Introduction

S. ZEIN EL ABEDIN

Electrochemistry and Corrosion Laboratory, National Research Centre, Dokki, Cairo, Egypt

F. ENDRES

Institute of Particle Technology, Clausthal University of Technology, Clausthal-Zellerfeld, Germany

Abstract

Ionic liquids have attracted considerable attention since they have extraordinary physical properties. In the field of electrochemistry, there is an increasing interest in ionic liquids due to their extremely large electrochemical windows, often more than 5 V. Therefore many metals and semiconductors can be electrodeposited using ionic liquids, for example, elements that are not accessible from aqueous solutions, such as tantalum, aluminium, and the rare earth elements. Moreover, the high thermal stability of ionic liquids allows the direct electrodeposition of crystalline metals and semiconductors at elevated temperatures, and in this context ionic liquids can be regarded as the link to molten salts. In this chapter, we present a short review on the electrodeposition of tantalum, aluminium, silicon, and selenium in water - and air-stable ionic liquids. Our results show that uniform, adherent layers of tantalum can be obtained in the ionic liquid 1-butyl-1-methylpyrrolidinium bis(trifluoromethylsulfonyl)amide [C$_4$mpyrr][NTf$_2$] containing TaF$_5$ at 200 °C, under the right conditions. Nano- and microcrystalline aluminium can be made in the ionic liquids [C$_4$mpyrr][NTf$_2$] and [C$_2$mim][NTf$_2$], respectively. We also present some results on the electrodeposition of silicon and hexagonal grey selenium in the ionic liquid [C$_4$mpyrr][NTf$_2$].

Molten Salts and Ionic Liquids: Never the Twain? Edited by Marcelle Gaune-Escard and Kenneth R. Seddon
Copyright © 2010 John Wiley & Sons, Inc.

7.1 INTRODUCTION

In recent years, ionic liquids have attracted considerable attention since they have extraordinary physical properties superior to those of water or organic solvents. As ionic liquids are basically constituted of ions, they behave very differently than conventional molecular liquids when they are used as solvents. In contrast to conventional molecular solvents, ionic liquids are usually nonvolatile, nonflammable, and less toxic, are good solvents for both organics and inorganics, and can be used over a wide temperature range. Moreover, ionic liquids can have pretty large electrochemical windows, more than 5 V, and hence they give access to elements that cannot be electrodeposited from aqueous or organic solutions. Another advantage of ionic liquids is that problems associated with hydrogen ions in conventional protic solvents can be eliminated in ionic liquids because most of the commercially available ionic liquids are aprotic. Therefore ionic liquids, especially air- and water-stable ones, are considered promising solvents for a wide variety of applications including electrodeposition, batteries, catalysis, separations, and organic synthesis [1, 2]. Moreover, the development of ionic liquids has opened the door for the electrodeposition of reactive elements, which cannot be obtained from aqueous solutions (viz., aluminium, magnesium, titanium, and tantalum) at moderate temperatures, although magnesium and titanium have not yet been electrodeposited as elements in the bulk phase. To get more information on the electrodeposition of metals and semiconductors in ionic liquids, we refer the reader to recently published review articles [3, 4].

In this chapter, we present a short review on the electrodeposition of some metals, such as tantalum and aluminium, and semiconductors, such as silicon and selenium, in air- and water-stable ionic liquids. We show that ionic liquids can be regarded as possible alternatives to high temperature molten salts. Currently, ionic liquids cannot replace molten salts in the electrowinning of reactive and refractory metals. Nevertheless, they can be used efficiently for the coating of other metals with thin layers of tantalum or aluminium. From a practical point of view, molten salts are hardly suited for the coating of sensitive materials like the NiTi shape memory alloy with other metals, since the electrolysis process is performed at elevated temperatures between 500 and more than 1000 °C, under aggressive chemical conditions. With ionic liquids, a technical electroplating process might be performed at moderate, or even room, temperature.

7.2 EXPERIMENTAL

The ionic liquids 1-butyl-1-methyl-pyrrolidinium bis(trifluoromethylsulfonyl)amide ([C$_4$mpyrr][NTf$_2$]) and 1-ethyl-3-methylimidazolium bis(trifluoromethylsulfonyl) amide [C$_2$mim][NTf$_2$] were purchased in ultrapure quality. The liquids were dried under vacuum for 12 h at a temperature of 100 °C to yield water contents below 3 ppm (by Karl Fischer titration), and stored in an argon-filled glove box with water and dioxygen below 1 ppm (OMNI-LAB from Vacuum-Atmospheres). Anhydrous aluminium(III) chloride grains (Fluka, 99%), tantalum(V) fluoride (Alfa, 99.99%), selenium(IV) chloride (Alfa, 99.5%), and silicon(IV) chloride (Alfa, 99.999%) were used without further purification.

All electrochemical measurements were performed in the glove box using a Parstat 2263 Potentiostat/Galvanostat (Princeton Applied Research) controlled by PowerCV and PowerStep software. Au(111) on mica (Molecular Imaging), platinum sheets (Alfa, 99.99%), and highly oriented pyrolytic graphite (HOPG) were used as working electrodes. In the case of aluminium electrodeposition, aluminium wires (Alfa, 99.999%) were used as reference and counterelectrodes. However, in tantalum, selenium, and silicon electrodeposition, platinum wires (Alfa, 99.99%) were used as reference and counterelectrodes, respectively. For measurements at room temperature, the electrochemical cell was made of polytetrafluoroethylene and clamped over a Teflon-covered Viton Oring onto the substrate, thus yielding a geometric surface area of 0.3 cm^2. For variable temperature measurements, a quartz round-bottomed flask was used as an electrochemical cell. Prior to use, all parts in contact with the solution were thoroughly cleaned in a mixture of 50 : 50 vol% H_2SO_4 : H_2O_2, followed by heating under reflux in doubly distilled water. A high resolution field emission scanning electron microscope (Carl Zeiss DSM 982 Gemini) was utilised to investigate the surface morphology of the deposited films and energy dispersive X-ray analysis (EDX) was used to determine the film composition. The X-ray diffractograms of the deposited samples were recorded using a Siemens D-500 diffractometer with CoK_α radiation.

The STM experiments were performed with in-house built STM heads and scanners under inert gas conditions (H_2O and O_2 < 1 ppm) with a Molecular Imaging Pico Scan 2500 STM controller in feedback mode. STM tips were prepared by electrochemical etching of tungsten wires (0.25 mm diameter) and electrophoretically coated with an electropaint (BASF ZQ 84-3225 0201). During the STM experiments the electrode potential was controlled by the PicoStat from Molecular Imaging. For the current/voltage tunnelling spectroscopy, the tip was positioned on the site of interest and the tip voltage was scanned between an upper and a lower limit. During this procedure, the feedback is switched off.

7.3 RESULTS AND DISCUSSION

7.3.1 Electrodeposition of Tantalum

The unique properties of tantalum, such as high melting point, ductility, toughness, and excellent corrosion resistance, make it an attractive coating material for components exposed to high temperature, wear, and severe chemical environments. Because of its good thermal stability at elevated temperatures, thin layers of tantalum are applied as a diffusion barrier on silicon in ultralarge-scale integrated (ULSI) circuits to prevent copper atoms from diffusing into dielectrics or a silicon substrate [5].

Many efforts have been made to develop an electroplating process for the electrodeposition of tantalum. High temperature molten salts were found to be efficient baths for the electrodeposition of tantalum [6–11]. Senderoff and Mellors have reported the first results on the electrodeposition of tantalum using the ternary eutectic mixture LiF–NaF–KF as a solvent, and K_2TaF_7 as a source of tantalum at temperatures between 650 and 850 °C [6, 7]. However, these baths have many technical and economic problems, such as the loss in the current efficiency of the electrolysis process due to the dissolution of metal after its deposition [12], and the expected corrosion problems at high temperatures. Recently, we have reported for the first time that tantalum can be electrodeposited as thin layers in the water- and air-stable ionic liquid 1-butyl-1-methylpyrrolidinium bis(trifluoromethylsulfonyl) amide at 200 °C using TaF_5 as a source of tantalum [13]. The quality of the deposit was found to be improved on addition of LiF to the deposition bath.

Figure 7.1a shows the cyclic voltammogram of [C_4mpyrr][NTf_2] containing 0.5 M TaF_5 on Au(111) at room temperature. As shown, two reduction processes are recorded in the forward scan. The first one starts at a potential of -0.5 V with a peak at -0.75 V; it might be correlated to the electrolytic reduction of Ta(V) to Ta(III). The second process starts at a potential of -1.5 V and is accompanied by the formation of a black deposit on the electrode surface. This can be attributed to the reduction of Ta(III) to tantalum metal simultaneously with the formation of insoluble tantalum compounds on the electrode surface. The anodic peak recorded on the backward scan is due to the dissolution of the electrodeposit, which, however, does not seem to be complete. At $E > 1.5$ V, the anodic current increases as a result of gold dissolution. The deposit obtained is less adherent and can easily be removed by washing with propanone.

The electrodeposition of tantalum was also investigated at different temperatures up to 200 °C. It was found that the mechanical quality and the adherence of the electrodeposits improve at 200 °C. Moreover, the quality and the adherence of the electrodeposit were found to be improved upon addition of LiF to the electrolyte. The SEM micrograph of the tantalum electrodeposit, Figure 7.1b, made potentiostatically at -1.8 V in [C_4mpyrr][NTf_2] containing 0.25 M TaF_5 and 0.25 M LiF on a platinum electrode at 200 °C for 1 h, shows a smooth, coherent, and dense layer. XRD patterns of the electrodeposit clearly show the characteristic patterns of crystalline tantalum (Fig. 7.1c).

Clearly, the electrodeposition of tantalum is not a straightforward process. According to our experience, there is a limiting current density. Below, tantalum is obtained; above, mainly nonstoichiometric tantalum subhalides are formed. We suggest as one explanation the following mechanism: when TaF_5 or $[TaF_7]^{2-}$ are reduced to tantalum, five to seven fluoride ions are liberated during electrodeposition. At low current densities, fluoride diffuses to the bulk of the liquid. If the current density is too high, the liquid is saturated at the electrode surface with fluoride ions. In this case, the deposition proceeds, but rather polynuclear tantalum complexes are formed due to the presence of the fluoride. The role of LiF is not yet clear in this system. On one hand, it might reduce local viscosity; on the other hand, it might alter the (unknown) structure of the double layer in this system, thus facilitating tantalum deposition. In any case there is a dependence on the tantalum source. If $TaCl_5$ is employed under similar conditions, thick deposits can be made, too, but they contain practically no crystalline tantalum. Apart from the stability of the subvalent tantalum

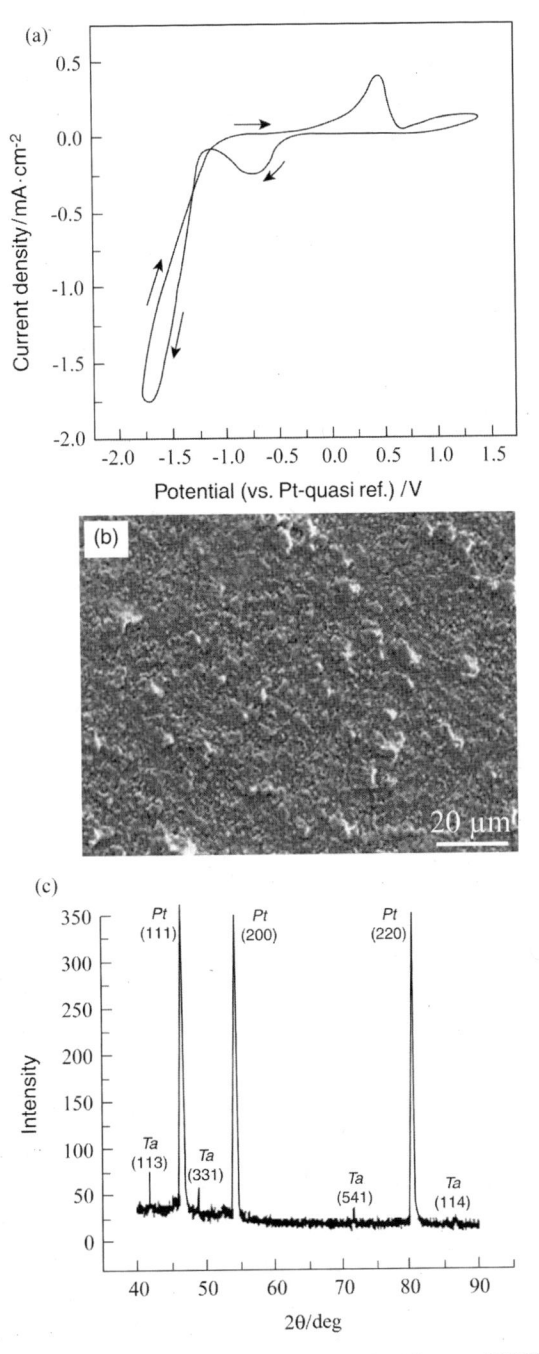

FIGURE 7.1 (a) Cyclic voltammogram of 0.5 M TaF$_5$ in [C$_4$mpyrr][NTf$_2$] on Au(111) at room temperature: scan rate, 10 mV·s^{-1}. (b) SEM micrograph of the electrodeposit formed potentiostatically on Pt in [C$_4$mpyrr][NTf$_2$] containing 0.25 M TaF$_5$ and 0.25 LiF at a potential of -1.8 V for 1 h at 200 °C. (c) XRD patterns of the deposited layer obtained potentiostatically on Pt in [C$_4$mpyrr][NTf$_2$] containing 0.25 M TaF$_5$ and 0.25 LiF at a potential of -1.8 V for 1 h (see Reference [13]).

chlorides, the diffusion of chloride ions to the bulk of the solution might be slower than that of fluoride due to its larger radius. We have started to investigate the tantalum deposition with in situ scanning tunnelling microscopy, with the aim to shed more light on the difficult process of tantalum deposition.

7.3.2 Electrodeposition of Aluminium

Aluminium and aluminium alloys have a remarkable economic importance and, nowadays, the main end-use market for aluminium is the automobile industry. The light weight of aluminium and its alloys makes them attractive materials to use in car manufacturing, leading to reduction in the car weight, which, in turn, lessens fuel consumption. The Audi A8 and the Jaguar XJ chassis are completely made of an aluminium alloy. It is well known that the production of aluminium is carried out at elevated temperatures (ca. 1000 °C) by electrolysis of molten cryolite (Na_3AlF_6) in which aluminium oxide is dissolved [14]. This has been the main industrial method for primary aluminium production for the past century. However, it is hardly suited to the coating of other metals with a layer of aluminium, since the electrolysis is performed under aggressive chemical conditions, and aluminium is obtained in the liquid phase in the Hall–Heroult process. Nowadays, there are a wide variety of methods for aluminium coatings, such as hot dipping, thermal spraying, sputter deposition, vapour deposition, and electroplating. The electroplating process is commercially attractive since it can be performed at low temperatures, as the experimental setup is less demanding. Because of its high reactivity ($E° = -1.7$ V vs. NHE), the electrodeposition of aluminium in aqueous solutions is impossible. Only hydrogen would be formed at the cathode. Therefore the electrolytes must be aprotic, such as organic solvents or ionic liquids. Currently, the commercial processes for electroplating of aluminium are based on organic solvents, such as REAL (room-temperature electroplated aluminium) [15, 16] and SIGAL (Siemens galvano-aluminium) processes [17, 18]. A main disadvantage of both processes is the flammability of the aluminium-containing species and/or of the organic solvents. Therefore ionic liquids are regarded as suitable electrolytes for aluminium electrodeposition. Many papers have been published on the electrodeposition of aluminium from chloroaluminate (III) ionic liquids, which are considered the first generation of ionic liquids [19–30]. Although high quality aluminium deposits can be obtained using chloroaluminate ionic liquids, a main disadvantage of them is that they are extremely hygroscopic and they must be strictly handled under inert gas conditions. Therefore the electrodeposition of aluminium in air- and water-stable ionic liquids is of great interest and stimulates further routes for aluminium electrodeposition.

Quite recently, we reported for the first time that nano- and microcrystalline aluminium can be electrodeposited in three different air- and water-stable ionic liquids, namely, 1-butyl-1-methylpyrrolidinium bis(trifluoromethylsulfonyl)amide [C_4mpyrr][NTf_2], 1-ethyl-3-methylimidazolium bis(trifluoromethylsulfonyl)amide [C_2mim][NTf_2], and trihexyltetradecylphosphonium bis(trifluoromethylsulfonyl) amide ([P_{66614}][NTf_2]) [31, 32]. It was found that the ionic liquids [C_4mpyrr][NTf_2]

and [C$_2$mim][NTf$_2$] form biphasic mixtures in the concentration range of AlCl$_3$ from 1.6 to 2.5 mol·L^{-1} and 2.5 to 5 mol·L^{-1}, respectively [32]. Moreover, the electrodeposition of aluminium at room temperature occurs only from the upper phase at AlCl$_3$ concentrations $\geq$1.6 mol·L^{-1} and $\geq$5 mol·L^{-1} in the ionic liquids [C$_4$mpyrr][NTf$_2$] and [C$_2$mim][NTf$_2$], respectively. Brausch et al. [33] reported the first results on the biphasic behaviour of new highly acidic ionic liquids based on [cation][N(SO$_2$CF$_3$)$_2$]/AlCl$_3$. They assumed that the lower phase is formed by neutral, mixed chloro-[bis(trifluoromethylsulfonyl)amide]-aluminium species, whereas the upper phase contains the organic cation and a mixture of chloro-[bis(trifluoromethylsulfonyl)amide]-aluminate ions. This means that the reducible aluminium-containing species exists only in the upper phase of the biphasic mixtures, and hence the electrodeposition of aluminium occurs only from the upper phase.

Figure 7.2 shows high resolution SEM micrographs of about 10 μm thick layers of aluminium on gold substrates electrodeposited potentiostatically at 100 °C in the ionic liquids [C$_4$mpyrr][NTf$_2$] and [C$_2$mim][NTf$_2$], respectively. As seen in the SEM micrograph of Figure 7.2a, the deposit contains very fine crystallites with sizes in the nanometre regime. On the other hand, coarse cubic-shaped aluminium particles of sizes in the micrometer range were obtained in the ionic liquid [C$_2$mim][NTf$_2$], Figure 7.2b. Presumably, the [C$_4$mpyrr]$^+$ cation acts as a grain refiner that is adsorbed on the substrate and on the growing nuclei, thus hindering strongly the growth of crystallites. Recent STM studies have shown that aluminium shows an underpotential deposition (UPD) in [C$_2$mim][NTf$_2$] followed by three-dimensional growth, whereas in [C$_4$mpyrr][NTf$_2$] there are no hints of UPD.

7.3.3 Electrodeposition of Silicon

Silicon is one of the most important semiconductors, as it is the basis of any computer chip. There were several approaches in the past to electrodeposit silicon in organic solvents [34–36]. However, the authors report a disturbing effect by water that can hardly be avoided in organic solvents. Furthermore, there were studies on the electrodeposition of silicon in high temperature molten salts [37]. It was reported by Katayama et al. [38] that silicon can also be electrodeposited in a low temperature molten salt. In that study, the authors employed 1-ethyl-3-methylimidazolium hexafluorosilicate, and at 90 °C they could deposit a thin layer of silicon. A disadvantage of imidazolium cations is that silicon deposition occurs at the onset of irreversible cation reduction. However, this film reacted with water to form SiO$_2$, so that evidence whether the deposited silicon species was elemental or even semiconducting is missing. Recently, we have shown that silicon can be electrodeposited on the nanoscale from 1-butyl-1-methylpyrrolidinium bis(trifluoromethylsulfonyl)amide saturated with silicon(IV) chloride [39].

Figure 7.3a shows the cyclic voltammogram of the ionic liquid 1-butyl-1-methylpyrrolidinium bis(trifluoromethylsulfonyl)amide saturated with silicon(IV) chloride on highly oriented pyrolytic graphite (HOPG). As seen, a strong reduction current is recorded in the forward scan. After having passed the lower switching potential, the anodic scan crosses the cathodic one at −2000 mV versus Fc/[Fc]$^+$, which is typical

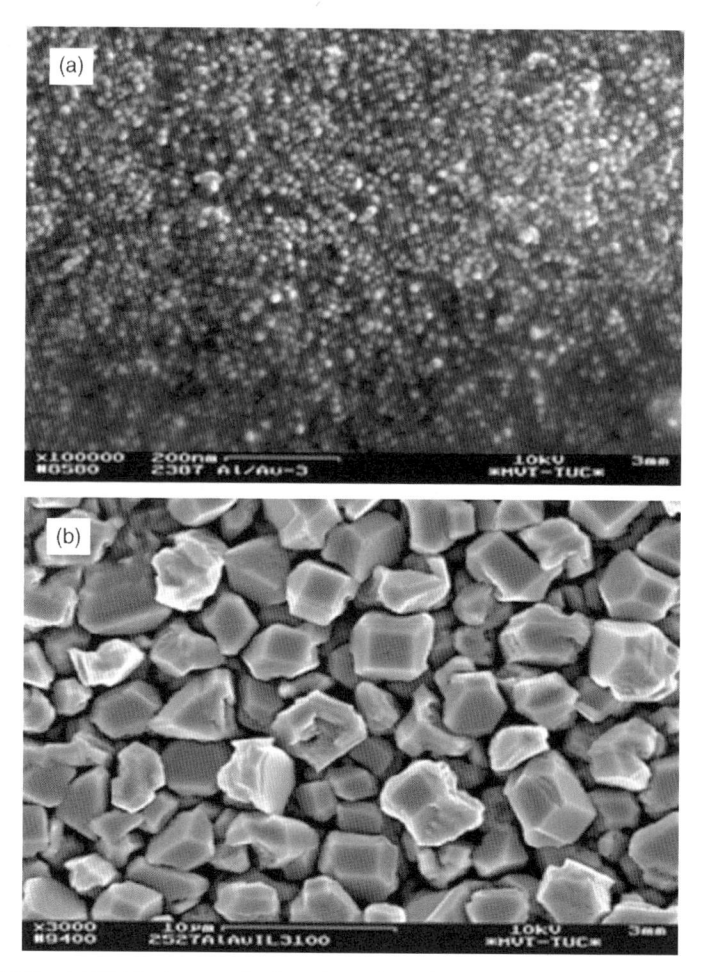

FIGURE 7.2 (a) SEM micrograph of an electrodeposited aluminium film on gold formed in the upper phase of the mixture AlCl$_3$/[C$_4$mpyrr][NTf$_2$] after potentiostatic polarisation at −0.45 V (vs. Al) for 2 h at 100 °C. (b) SEM micrograph of an electrodeposited aluminium film on gold made in the upper phase of the AlCl$_3$/[C$_2$mim][NTf$_2$] mixture after potentiostatic polarisation at −0.05 V (vs. Al) for 2 h at 100 °C.

for nucleation. Approaching an electrode potential of +400 mV versus Fc/[Fc]$^+$, a strong oxidation current starts, which is in part correlated to the silicon(IV) chloride reduction process beginning at −1600 mV versus Fc/[Fc]$^+$ and in part correlated to HOPG oxidation, as with SiCl$_4$ in the liquid a similar oxidation behaviour is observed if the scan is started from the open-circuit potential toward positive potentials.

Figure 7.3b shows a high-resolution SEM picture of an electrodeposited silicon layer on a gold substrate. As seen, the deposit contains small crystallites with sizes of about 50 nm and smaller. The deposit can keep its dark appearance even under air,

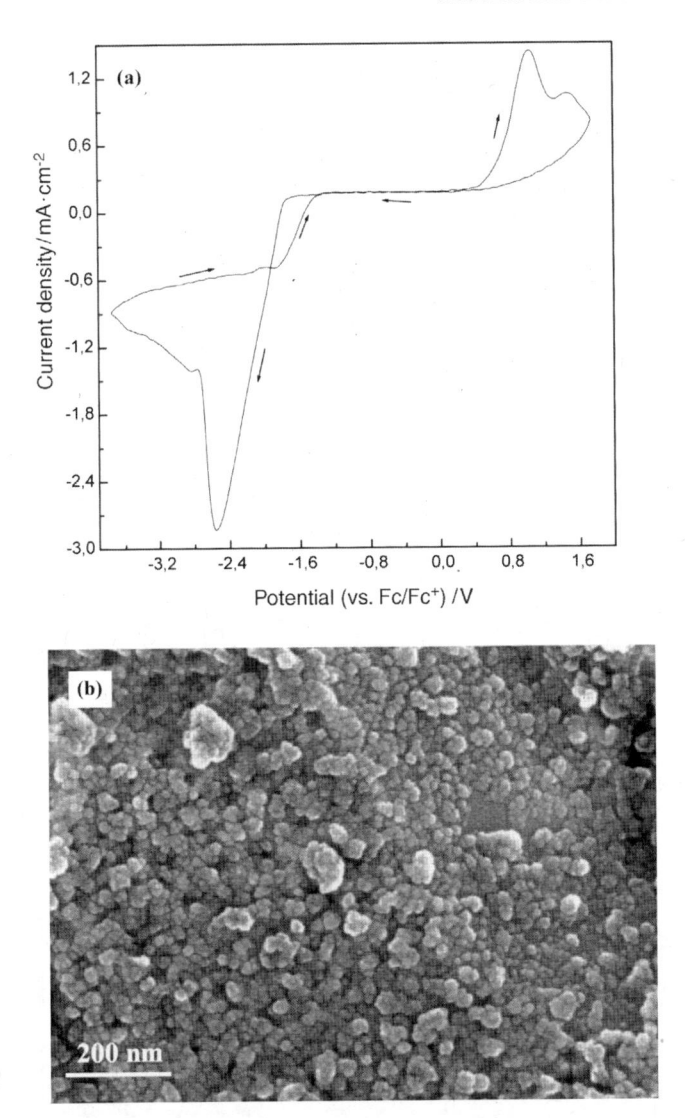

FIGURE 7.3 (a) Cyclic voltammogram of SiCl$_4$ saturated in the ionic liquid [C$_4$mpyrr][NTf$_2$]: scan rate, 10 mV·s^{-1} (see Ref. 39). (b) SEM micrograph of electrodeposited silicon, made potentiostatically at −2.7 V versus Fc/[Fc]$^+$ (Fc = ferrocene) (see Reference [1]).

but sometimes it is rapidly transformed into white powder (i.e., SiO$_2$). The EDX analysis showed only gold from the substrate and silicon, but no detectable chlorine. This proves that elemental silicon was electrodeposited, which can be subject to some oxidation under environmental conditions.

Figure 7.4a shows the STM picture of a ca. 100 nm thick silicon layer that was electrodeposited at −1600 mV versus Fc/[Fc]$^+$ on HOPG, probed under potential control with the in situ STM. Its surface is smooth on the nanometre

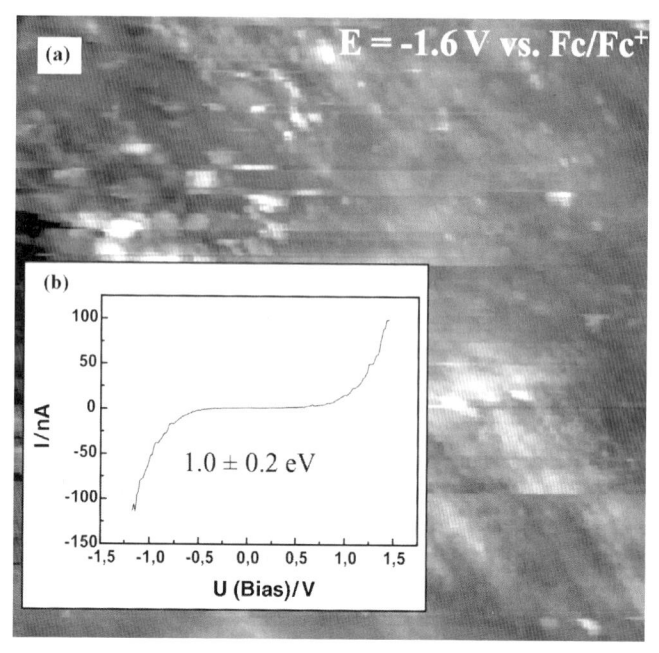

FIGURE 7.4 (a) In situ STM picture of a ca. 100 nm thick film (600 nm × 600 nm). (b) In situ current/voltage tunnelling spectrum of the silicon electrodeposit on HOPG.

scale. Furthermore, it is clear from the STM image that the tip tends to push away the deposits, which is typical for HOPG as a substrate. In order to show the semiconducting behaviour of the deposited film, current/voltage tunnelling spectroscopy was performed. It has already been shown by us that the I-U tunnelling spectroscopy is a valuable technique for in situ characterisation of electrodeposited semiconductors [40, 41] and metals. We could show with the in situ I-U tunnelling spectroscopy that germanium with layer thicknesses of 20 nm and more is semiconducting with a symmetric bandgap of 0.7 ± 0.1 eV. On the other hand, we have found that very thin layers of germanium with thicknesses of several monolayers exhibit clearly metallic behaviour [40, 41]. Figure 7.4(b) shows an in situ current/voltage tunnelling spectrum of the 100 nm thick silicon layer. The spectra are of the same quality from any point on the surface. As seen in the spectrum, a typical bandgap of 1.0 ± 0.2 eV is observed. This value is quite similar to the value that we observed for hydrogen terminated n-doped Si(111) in an ionic liquid [42]. The value of microcrystalline silicon in the bulk phase at room temperature is 1.1 eV. In light of these results, it can be concluded that elemental, intrinsically semiconducting silicon was electrodeposited from the employed ionic liquid as a thin layer. An XPS study (where nanoparticles were made electrochemically in the glove box, and rinsed and transferred under an inert gas to the UHV chamber of our XPS device) showed that the particles consist of an elemental silicon core and a silicon(IV) oxide shell [43].

7.3.4 Electrodeposition of Selenium

Selenium exhibits both photovoltaic action, where light is converted directly into electricity, and photoconductive properties, where the conductivity increases with increased illumination. These properties make selenium useful in the production of photocells and solar cells. Moreover, compound semiconductors containing selenium, such as InSe, CdSe, or CuInSe$_2$ (CIS), have many optoelectronic applications, including advanced solar cells, IR detectors, and solid-state lasers [44].

The electrodeposition of selenium in aqueous solutions was studied extensively; see, for example, References [45–48]. However, the exclusive electrodeposition of grey selenium has not yet been successful in aqueous solutions. The electrodeposition of selenium in aqueous solutions is usually complicated by the formation of amorphous red and black selenium, which exhibit bad electronic conductivity. Grey selenium can only be deposited quantitatively at elevated temperatures of more than 100 °C, which is practically impossible to achieve in aqueous galvanic baths. It is known that, thermodynamically, a phase transition from amorphous red to crystalline grey selenium occurs at a temperature of about 80 °C. Therefore the use of ionic liquids, especially the air- and water-stable ones, in the electrodeposition of selenium is of great benefit owing to their high thermal stability and their low vapour pressures. Consequently, the direct electrodeposition of grey selenium and its compounds is possible in ionic liquids since the deposition process can be performed efficiently at elevated temperatures. We have recently shown that semiconducting grey selenium can be exclusively deposited in the ionic liquid [C$_4$mpyrr][NTf$_2$] at temperatures of more than 100 °C [49].

The cyclic voltammograms of [C$_4$mpyrr][NTf$_2$], containing 0.1 mol·L^{-1} SeCl$_4$ on a Pt substrate at 25 °C and at 150 °C, are displayed in Figure 7.5. The potential started at the open-circuit potential and was initially swept in the negative direction at a scan rate of 10 mV·s^{-1}. In these experiments, we had no other choice but to employ a platinum quasi-reference. The peaks for Fc/[Fc]$^+$ were difficult to identify, and due to the complexity of the voltammograms, it was decided to avoid any interference with ferrocene. Fortunately, the platinum quasi-reference electrode was fairly stable in the course of the experiment. At 25 °C, the cyclic voltammogram is characterised by a main cathodic peak at −0.75 V (vs. Pt), c$_3$, and two shoulders, c$_2$ and c$_1$, preceding c$_3$. As a dark red deposit is obtained at −0.75 V after deposition for 2 h, the main reduction peak, c$_3$, might be attributed to the reduction of Se(IV) to elemental selenium, presumably the red phase, and the two shoulders c$_2$ and c$_1$ might be associated with two different underpotential deposition processes.

The anodic peak a$_3$ recorded in the reverse scan is ascribed to partial dissolution of the deposited selenium. The anodic counterparts to the cathodic peaks c$_1$ and c$_2$ (a$_1$ and a$_2$) are recorded at potentials higher than 0.0 V. From the experiments at 25 °C, it can be concluded that a thin film of X-ray amorphous red selenium is obtained. This result is not at all surprising, as thermodynamically a phase transition from amorphous red selenium to crystalline grey selenium occurs only at temperatures ≥80 °C. Thus the exclusive deposition of grey selenium should occur at temperatures well above 80 °C. Hence we have performed the cyclic voltammetry

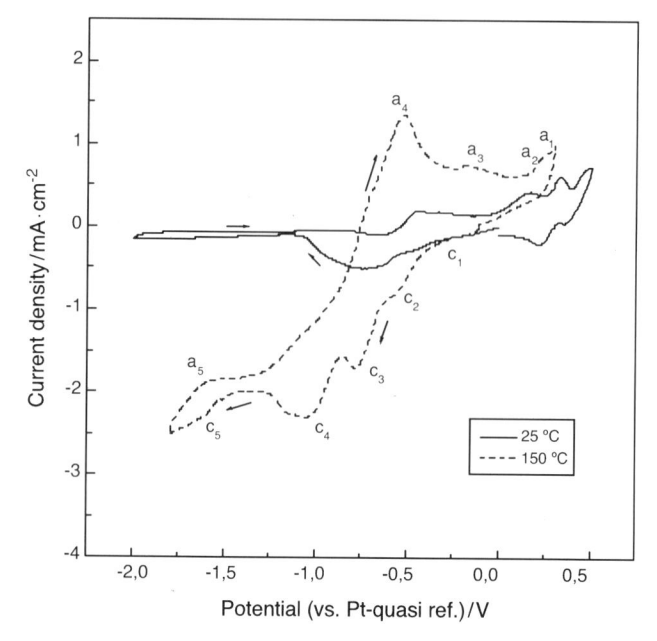

FIGURE 7.5 Cyclic voltammograms of the ionic liquid [C$_4$mpyrr][NTf$_2$] containing 0.1 M SeCl$_4$ on platinum at 25 °C and 150 °C: scan rate, 10 mV·s^{-1} (see Reference [39]).

measurements at 150 °C. As seen in Figure 7.5, two new cathodic processes are observed in the cathodic branch, c$_4$ and c$_5$, with corresponding anodic counterparts, a$_4$ and a$_5$, respectively. At c$_3$, a red deposit was evident on the electrode surface with the naked eye, which turned grey at c$_4$. Thus the peaks c$_3$ and c$_4$ are likely related to the deposition of red and grey selenium, respectively. When the electrode potential is scanned further in the negative direction, after formation of the grey selenium deposit, a further cathodic peak c$_5$ is observed, with corresponding anodic peak a$_5$. This pair of peaks might be associated with further reduction of the deposited selenium to Se^{2-}, as the selenium film can disappear completely at this electrode potential. The reduction of selenium to Se^{2-} has been reported in aqueous solutions [45–53]. Our results show that selenium electrochemistry in [C$_4$mpyrr][NTf$_2$] is fairly difficult and requires a lot of fundamental studies.

In order to perform a material analysis of selenium, the deposits were made at −1.1 V (vs. Pt) for 2 h at 150 °C on platinum substrates, and subsequently characterised by means of SEM and XRD to explore morphology and composition. After the deposition experiments, the deposits were rinsed by isopropanol and then cleaned with propanone in an ultrasonic bath for 10 min. Visually, a dark grey, well adherent deposit was obtained after potentiostatic deposition. As seen in the SEM micrograph shown in Figure 7.6a, the deposit made at 150 °C appears to be dense with coarse crystallites. The XRD patterns of the electrodeposit show the characteristic peaks of the crystalline grey selenium, Figure 7.6b. Therefore it can be concluded that grey selenium can be deposited exclusively in the employed ionic liquid at 150 °C.

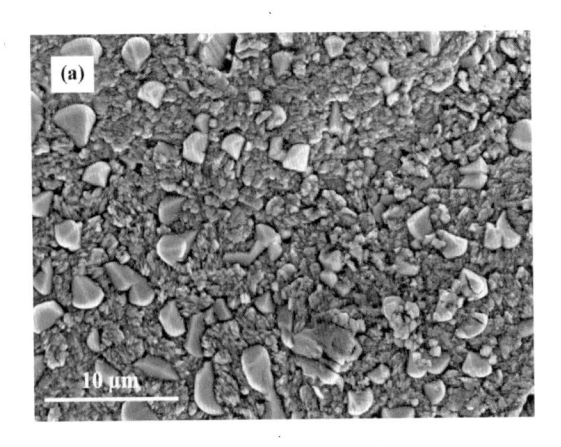

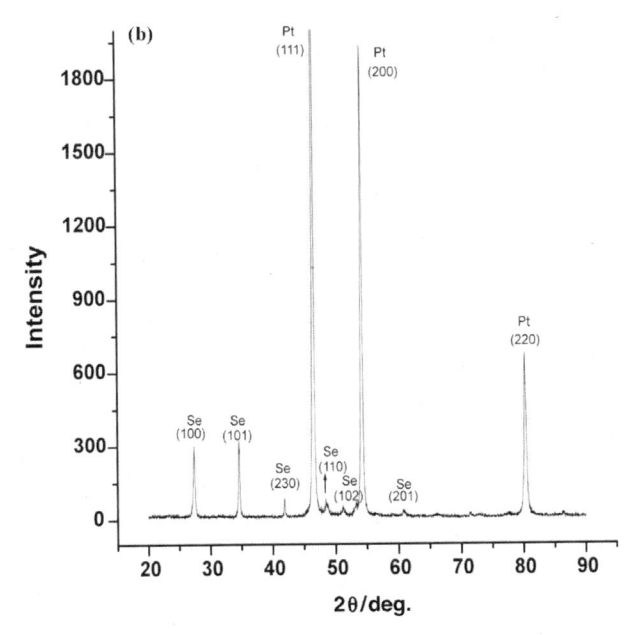

FIGURE 7.6 (a) An SEM micrograph of an electrodeposited selenium layer obtained potentiostatically on Pt in [C$_4$mpyrr][NTf$_2$] containing 0.1 M SeCl$_4$ at a potential of −1.1 V (vs. Pt) for 2 h at 150 °C. (b) XRD patterns of the electrodeposited selenium layer.

7.4 CONCLUSION

We have shown that air- and water-stable ionic liquids allow the variable temperature electrodeposition of reactive elements such as tantalum, aluminium, silicon, and selenium, as some examples. Due to their extraordinary physical properties, such as wide electrochemical windows, high thermal stability, and negligible vapour pressure, ionic liquids give a novel approach for the electrodeposition of many metals and semiconductors at elevated temperatures. In this context, we have summarised some of our key results on the electrodeposition of tantalum, aluminium, silicon, and selenium in air- and water-stable ionic liquids. Our results show that uniform, adherent layers of tantalum can be obtained in the ionic liquid [C$_4$mpyrr][NTf$_2$] containing TaF$_5$ and LiF at 200 °C. Nano- and microcrystalline aluminium can be made from the ionic liquids [C$_4$mpyrr][NTf$_2$] and [C$_2$mim][NTf$_2$], respectively. It was also shown that a silicon layer with thickness of about 100 nm, electrodeposited from the ionic liquid [C$_4$mpyrr][NTf$_2$], exhibits a bandgap of 1.0 ± 0.2 eV, indicating that semiconducting silicon was electrodeposited. Furthermore, it was found that adherent, dense layers of semiconducting grey selenium can be made exclusively in the ionic liquid [C$_4$mpyrr][NTf$_2$] at 150 °C.

In our opinion, the variable temperature chemistry/electrochemistry in ionic liquids should be investigated in more detail. The viscosity of ionic liquids is of no concern at elevated temperature and, furthermore, kinetic limitations are much easier to overcome. In this context ionic liquids might be regarded as the missing link to high temperature molten salts.

REFERENCES

1. Endres, F. and Zein El Abedin, S., *Phys. Chem. Chem. Phys.* **8**, 2101 (2006).
2. Buzzeo, M. C., Evans, R. G., and Compton, R. G., *ChemPhysChem.* **5**, 1106 (2004).
3. Abbott, A. P. and McKenzie, K. J., *Phys. Chem. Chem. Phys.* **8**, 4265 (2006).
4. Zein El Abedin, S., and Endres, F., *ChemPhysChem.* **7**, 58 (2006).
5. Chen, X., Peterson, G. G., Goldberg, C., Nuesca, G., Frisch, H. L., Arkles, B., and Sullivan, J., *J. Mater. Res.* **14**, 2043 (1999).
6. Mellors, G. W., and Senderoff, S., *J. Electrochem. Soc.* **112**, 840 (1965).
7. Senderoff, S. and Mellors, G. W., *Science* **153**, 1475 (1966).
8. Dutra, A. J. B., Vazquez, J. C., and Espinola, A., *Minerals Eng.* **6**, 663 (1993).
9. Chamelot, P., Taxil, P., and Lafage, B., *Electrochim. Acta* **39**, 2571 (1994).
10. Chamelot, P., Palau, P., Massot, L., Savall, A., and Taxil, P., *Electrochim. Acta* **47**, 3423 (2002).
11. Lantelme, F., Barhoun, A., Li, G., and Besse, J. P., *J. Electrochem. Soc.* **139**, 1255 (1992).
12. Haarberg, G. M. and Thonstad, J., *J. Appl. Electrochem.* **119**, 789 (1989).
13. Zein El Abedin, S., Farag, H. K., Moustafa, E. M., Welz-Biermann, U., and Endres, F., *Phys. Chem. Chem. Phys.* **7**, 2333 (2005).
14. Brukin, A. R., *Production of Aluminum and Alumina, Critical Reports in Applied Chemistry*, Vol. **20**, John Wiley & Sons, Chichester, UK, 1987.
15. Berg, J. V. D., Daenen, T., Krijl, G., and Leest, R. V. D., *Metalloberfläche* **35**, 218 (1985).
16. Altgeld, W., *Metalloberfläche* **40**, 253 (1986).
17. Ziegler, K. and Lehmkuhl, H., *Chemie* **283**, 414 (1956).
18. Kautek, W. and Birkle, S., *Electrochim. Acta* **34**, 1213 (1989).
19. Zhao, Y. and VanderNoot, T. J., *Electrochim. Acta* **42**, 3 (1997).
20. Lipsztajn, M. and Osteryoung, R. A., *J. Electrochem. Soc.* **130**, 1968 (1983).

REFERENCES

21. Melton, T. J., Joyce, J., Maloy, J. T., Boon, J. A., and Wilkes, J. S., *J. Electrochem. Soc.* **137**, 3865 (1990).
22. Welch, B. J. and Osteryoung, R. A., *J. Electrochem. Soc.* **118**, 455 (1981).
23. Carlin, R. T. and Osteryoung, R. A., *J. Electrochem. Soc.* **136**, 1409 (1989).
24. Moffat, T. P., *J. Electrochem. Soc.* **141**, L115 (1994).
25. Ali, M. R., Nishikata, A., and Tsuru, T., *Electrochim. Acta* **42**, 2347 (1997).
26. Stafford, G. R., *J. Electrochem. Soc.* **141**, 945 (1994).
27. Tsuda, T., Hussey, C. L., and Stafford, G. R., *J. Electrochem. Soc.* **151**, C379 (2004).
28. Stafford, G. R., *J. Electrochem. Soc.* **136**, 635 (1989).
29. Endres, F., Bukowski, M., Hempelmann, R., and Natter, H., *Angew. Chem.* **115**, 3550 (2003); *Angew. Chem. Int. Ed.* **42**, 3428 (2003).
30. Liu, Q. X., Zein El Abedin, S., and Endres, F., *Surf. Coat. Tech.* **201**, 1352 (2006).
31. Zein El Abedin, S., Moustafa, E., Hempelmann, R., Natter, H., and Endres, F., *Electrochem. Commun.* **7**, 1111 (2005).
32. Zein El Abedin, S., Moustafa, E., Hempelmann, R., Natter H., and Endres, F., *ChemPhysChem.* **7**, 1535 (2006).
33. Brausch, N., Metlen, A., and Wasserscheid, P., *Chem. Commun.* **13**, 1552 (2004).
34. Agrawal, A. K. and Austin, A. E., *J. Electrochem. Soc.* **128**, 2292 (1981).
35. Gobet, J. and Tannenberger, H., *J. Electrochem. Soc.* **133**, C322 (1986).
36. Gobet, J. and Tannenberger, H., *J. Electrochem. Soc.* **135**, 109 (1988).
37. Matsuda, T., Nakamura, S., Ide, K., Nyudo, K., Yae, S. J., and Nakato, Y., *Chem. Lett.* **7**, 569 (1996).
38. Katayama, Y., Yokomizo, M., Miura, T., and Kishi, T., *Electrochemistry* **69**, 834 (2001).
39. Zein El Abedin, S., Boressinko, N., and Endres, F., *Electrochem. Commun.* **6**, 510 (2004).
40. Endres, F. and Zein El Abedin, S., *Phys. Chem. Chem. Phys.* **4**, 1640 (2002).
41. Endres, F. and Zein El Abedin, S., *Phys. Chem. Chem. Phys.* **4**, 1649 (2002).
42. Freyland, W., Zell, C. A., Zein El Abedin, S., and Endres, F., *Electrochim. Acta* **48**, 3053 (2003).
43. Bebensee, F., Borisenko, N., Frerichs, M., Höftt, A., Maus-Friedrichs, W., Zein El Abedin, S., and Endres, F., *Z. Phys. Chem.*, **222**, 671 (2008).
44. Schock, H. W., *Appl. Surf. Sci.* **92**, 606 (1996).
45. Huang, B. M., Lister, T. E., and Stickney, J. L., *Surf. Sci.* **392**, 27 (1997).
46. Sorenson, T. A., Lister, T. E., Huang, B. M., and Stickney, J. L., *J. Electrochem. Soc.* **146**, 1019 (1999).
47. Alanyalioglu, M., Demir, U., and Shannon, C., *J. Electroanal. Chem.* **561**, 21 (2004).
48. Zhang, X. Y., Cai, Y., Miao, J. Y., Ng, K. Y., Chan, Y. F., Zhang, X. X., and Wang, N., *J. Cryst. Growth.* **276**, 674 (2005).
49. Zein El Abedin, S., Saad, A. Y., Farag, H. K., Borisenko, N., Liu, Q. X., and Endres, F., *Electrochim. Acta*, **52**, 2746 (2007).
50. Santos, M. C. and Machado, S. A. S., *J. Electroanal. Chem.* **567**, 203 (2004).
51. Furuya, N. and Motoo, S., *J. Electroanal. Chem.* **98**, 189 (1979).
52. Modolo, R., Traore, M., and Vittori, O., *Electrochim. Acta* **31**, 859 (1986).
53. Lister, T. E. and Stickney, J. L., *J. Phys. Chem.* **100**, 19568 (1996).

8 Predicting the Thermodynamic Behaviour of Water + Ionic Liquids Systems Using COSMO-RS

M. G. FREIRE

CICECO, Departamento de Química, Universidade de Aveiro, Aveiro, Portugal

L. M. N. B. F. SANTOS

CIQ, Departamento de Química, Faculdade de Ciências da Universidade do Porto, Porto, Portugal

I. M. MARRUCHO and J. A. P. COUTINHO

CICECO, Departamento de Química, Universidade de Aveiro, Aveiro, Portugal

Abstract

Ionic liquids are a novel class of chemical compounds with interesting properties that are driving much research in several fields. The complete understanding of the phase behaviour of ionic liquids with water is an important issue, yet there is little data on their phase equilibria. In this work, the predictive capability of COSMO-RS, a predictive model based on unimolecular quantum chemistry calculations, was evaluated on the description of the liquid–liquid equilibria and the vapour–liquid equilibria of water and several imidazolium-based ionic liquid binary mixtures. The performance of the different calculation procedures, the basis set parameterisations, and the effect of the ion conformers on the quality of the predictions were evaluated. COSMO-RS calculations were performed at the following levels: BP/TZVP, BP/SVP/AM1, and B88-VWN/DNP. It was found that the calculation procedure based on the quantum chemical COSMO calculation at the BP/TZVP level derived from the lower energy ion conformations provides the best prediction capacity of the model. Good agreement between the model predictions and experimental VLE and LLE data was obtained. The COSMO-RS proved to be very helpful for scanning the growing set of already known ionic liquids in order to find suitable candidates for a certain task or to design new ionic liquids for specific applications.

Molten Salts and Ionic Liquids: Never the Twain? Edited by Marcelle Gaune-Escard and Kenneth R. Seddon
Copyright © 2010 John Wiley & Sons, Inc.

8.1 INTRODUCTION

Room-temperature ionic liquids are salts commonly composed of relatively large organic cations and inorganic or organic anions that cannot form an ordered crystal and thus remain liquid at or near room temperature. Unlike molecular liquids, the ionic nature of these liquids results in a unique combination of intrinsic physical properties such as high thermal stability, large liquidus range, high ionic conductivity, negligible vapour pressures, nonflammability, and a highly solvating capacity for both polar and nonpolar compounds. Among the several applications foreseeable for ionic liquids in the chemical industry, such as solvents in organic synthesis, as homogeneous and biphasic transfer catalysts, and in electrochemistry, there has been considerable interest in the potential of ionic liquids for separation processes as extraction media where, among others, ionic liquids have shown promise in the liquid–liquid extraction of organics from water [1–3]. Nevertheless, for the extraction of organic products from chemical reactions that proceed in aqueous media and for liquid–liquid extractions from aqueous phases, ionic liquids with lower solubility in water are preferred, and while they cannot contribute to air pollution due to their negligible vapour pressure, they do have in fact a significant solubility in water and, as a result, this is the most likely medium through which ionic liquids will enter the environment. Moreover, the loss of ionic liquids into the aqueous phase may be an important factor in estimating the cost of the ionic liquid used and the cost of water treatments. Furthermore, it was already shown that the presence of water in the ionic liquid phase can dramatically affect their physical properties [4, 5].

Another intrinsic attribute of ionic liquids is the potential of tuning their physical and chemical properties by varying different features of the ionic liquid, including the alkyl chain length and number of alkyl groups of the cation and the anion identity. At present, measurements on the solubility and phase equilibria of ionic liquids and water are limited. To our knowledge, only a few studies on these kinds of solubilities have been reported [6–12]. Thus, in our group, a systematic study of the liquid–liquid equilibria was conducted by variation of the cation chain length, in the presence of additional alkyl substitution on the cation, as well as in the anion identity. The main goal is to determine the impact of the different ionic liquids characteristics and the possibility to design a solvent for a specific application, in this case to fine-tune a particular ionic liquid with known mutual solubilities with water.

As it is infeasible to experimentally measure all possible combinations of anions and cations in ionic liquids vapour–liquid equilibria (VLE) and liquid–liquid equilibria (LLE) systems, it is essential to make measurements on selected systems to provide results that can be used to develop correlations and to test predictive methods. Several models have been used for correlating experimental data of phase equilibria with ionic liquids systems. Based on excess free Gibbs energy models, Wilson, UNIFAC, and UNIQUAC equations have been applied to correlate solid–liquid equilibria and VLE of ionic liquids [11, 13–16]. Another local composition model that proved capable of correlating data for ionic liquids systems was the nonrandom two-liquid (NRTL) model that was applied to VLE and LLE systems [16–19]. A different approach was proposed by Rebelo [20] that used a "polymer-like" G^E model to correlate the LLE of ionic liquids solutions, because of the similarity between the LLE phase diagrams of polymer solutions and those of ionic liquid solutions. Nevertheless, correlations and group contribution methods are not a good alternative, due to the lack of experimental data at present. On the other hand, the use of equations of state (EoS) requires critical parameters of the ionic liquid, which are not directly measurable and have been obtained indirectly [21, 22]. However, on the basis of unimolecular quantum calculations of the individual molecules, the Conductor-like Screening Model for Real Solvents, COSMO-RS, appears to be a novel method for the prediction of thermophysical properties of fluids and can be considered as an alternative to the structure-interpolating group-contribution methods (GCMs) [23, 24]. The COSMO-RS is also based on a physically founded model but, unlike GCMs, uses only atom-specific parameters. This method is therefore, at least qualitatively, able to describe structural variations correctly. To our knowledge, two applications of COSMO-RS for ionic liquids LLE systems were found in the literature, concerning ionic liquids and alcohols, ethers or ketone systems [25, 26]. Conversely, no application of COSMO-RS to ionic liquid/water systems was previously attempted.

Thus the goal of the present study is to evaluate the COSMO-RS potential for prediction of the thermodynamic behaviour for systems of ionic liquids and water for imidazolium-based ionic liquids, since no data for other family-based ionic liquids are currently available.

8.2 PHASE EQUILIBRIA PREDICTION OF SYSTEMS INVOLVING IONIC LIQUIDS AND WATER

Traditional approaches for correlating or predicting the properties of fluid mixtures such as EoS methods and schemes based primarily on dividing the molecules into various groups (GCMs) require a large bank of experimental data prior to their application. At present, the major requirement is a predictive method that could screen the huge number of possible combinations of ionic liquids and their mixtures, in this study with water, prior to making extensive experimental measurements.

The COSMO-RS is a unique method for predicting the thermodynamic properties of mixtures on the basis of unimolecular quantum chemical calculations for the individual molecules [23–27]. COSMO-RS combines the electrostatic advantages and the computational efficiency of the quantum chemical dielectric continuum solvation model COSMO with a statistical thermodynamics approach, based on the results of the quantum chemical calculations [28, 29]. The standard procedure of COSMO-RS calculations consists of two steps: quantum chemical COSMO calculations for all the molecular species involved, and COSMO-RS calculations [28, 29].

The quantum chemical model COSMO is an efficient variant of dielectric continuum solvation methods. In these calculations, the solute molecules are assumed to be in a virtual conductor environment, where the solute molecule induces a polarisation charge density σ on the interface between the molecule and the conductor, that is, on the molecular surface. These charges act back on the solute and generate a more polarised electron density than in a vacuum. During the quantum chemical self-consistency cycle, the solute molecule is thus converged to its energetically optimal state in a conductor with respect to electron density, and the molecular geometry can be optimised using conventional methods for calculations in a vacuum. Although time consuming, one advantage of this procedure is that the quantum chemical calculations have to be performed just once for each molecule of interest.

The COSMO-RS calculation procedure, performed using the COSMOtherm program [28, 29], describes all the interactions between molecules as contact interactions of the molecular surfaces, and these interactions can be related to the screening charge densities σ and σ' of the interacting surface pieces. For the statistical mechanical calculation, the molecular surface is split into small effective areas, a_{eff}, and contact with each of these areas is considered to be independent. The difference between the two screening charge densities of a particular pair causes a specific interaction energy that, in combination with the energy contributions of hydrogen bonds and the van der Waals interactions, allows the interactions of the components of a particular mixture to be determined. The application of statistical thermodynamics gives the chemical potential of each component, and from these LLE, VLE, and other properties of any mixture are derived. It should be noted that these calculations do not depend on any properties of the ionic liquids mixtures. Furthermore, the mixture properties are calculated from an entropy term based on the surface area and volume of the molecules, and an enthalpy term that is calculated from the energies of interaction between the surfaces of the molecules.

Within the COSMOtherm program, a *pseudo-binary* approach was used to calculate the LLE and VLE of a mixture composed of an ionic liquid and water, with the cation and anion of the ionic liquid input as separate compounds with the same mole fraction. The chemical potentials are calculated for the ternary system (anion + cation + water) with the chemical potential of the ionic liquid as the sum of the chemical potentials of both the cation and anion. For the case of LLE, a numerical approach is used to find the two compositions having equal chemical potentials of the three components (in the pseudo-binary system) in the two phases at a particular temperature.

In order to evaluate the performance of the different calculation procedures and basis set parameterisations on the quality of the predictions, COSMO-RS calculations were done at the following levels:

- BP/TZVP (Turbomole [30], DFT/COSMO calculation with the BP functional and TZVP [31] basis set using the optimised geometries at the same level of theory)—parameter file: BP_TZVP_C21_0105.
- BP/SVP/AM1 (Turbomole [30], DFT/COSMO single-point calculation with the BP functional and SVT [31] basis set upon geometries optimised at semiempirical MOPAC-AM1/COSMO level)—parameter file: BP_SVP_AM1_C21_0105.
- B88-VWN/DNP (DMOL3 [32], DFT/COSMO calculation with the B88-VWN functional and numerical DNP [32] basis using the optimised geometries at the same level of theory)—parameter file: DMOL3_C21_0105.

8.3 LLE AND VLE EXPERIMENTAL DATABASE

Liquid–liquid equilibria experimental measurements between water and imidazolium-based ionic liquids were studied for the following ionic liquids: 1-butyl-3-methylimidazolium hexafluorophosphate ([C$_4$mim][PF$_6$]), 1-butyl-3-hexylimidazolium hexafluorophosphate ([C$_6$mim][PF$_6$]), 1-octyl-3-methylimidazolium hexafluorophosphate ([C$_8$mim][PF$_6$]), 1-butyl-2,3-dimethylimidazolium hexafluorophosphate ([C$_4$C$_1$mim][PF$_6$]), 1-ethyl-3-methylimidazolium bis(trifluoromethylsulfonyl)amide ([C$_2$mim][NTf$_2$]), 1-butyl-3-methylimidazolium bis(trifluoromethylsulfonyl)amide ([C$_4$mim][NTf$_2$]), 1-hexyl-3-methylimidazolium bis(trifluoromethylsulfonyl)amide ([C$_6$mim][NTf$_2$]), 1-heptyl-3-methylimidazolium bis(trifluoromethylsulfonyl)amide ([C$_7$mim][NTf$_2$]), 1-octyl-3-methylimidazolium bis(trifluoromethylsulfonyl)amide ([C$_8$mim][NTf$_2$]), and 1-octyl-3-methylimidazolium tetrafluoroborate ([C$_8$mim][BF$_4$]). The ionic liquid content in the water-rich phase was analysed using UV-Vis spectroscopy, and the water content in the ionic liquid-rich phase was analysed by Karl Fischer titration. The temperature range of the experimental analysis was between 288 and 318 K and at atmospheric pressure. Further details about the experimental procedure and respective results can be found elsewhere [33, 34].

Vapour–liquid equilibria experimental isothermal measurements between water and imidazolium-based ionic liquids were taken from Anthony et al. [6] for [C$_4$mim][PF$_6$], [C$_8$mim][PF$_6$], and [C$_4$mim][BF$_4$] and from Kato and Gmehling [11] for [C$_2$mim][NTf$_2$], [C$_4$mim][NTf$_2$], and 1-methyl-3-methylimidazolium dimethylphosphate ([C$_1$mim][(CH$_3$)$_2$PO$_4$]).

8.4 RESULTS AND DISCUSSION

Prior to extensive comparisons between COSMO-RS predictions and the experimental data available, some studies concerning the COSMO-RS calculations were carried out. Different calculation procedures, basis set parameterisations, and the use of different energy conformers were evaluated to assess the performance of the predictive results. After defining the best conditions for the COSMO-RS procedure, comparisons between modelling results and experimental data were conducted. The effect of the alkyl chain length, methyl inclusion, anion identity, and temperature dependence in both LLE and VLE systems are presented and discussed.

8.5 PARAMETERISATION INFLUENCE ON THE PREDICTIONS

The quality, accuracy, and systematic errors of the electrostatic energies resulting from the underlying quantum chemical calculations depend on the quantum chemical method as well as on the basis set/parameterisation combinations. For that reason, three parameterisations coupled with the COSMOtherm program were tested, both on the LLE and VLE predictions using different calculation procedures and basis set

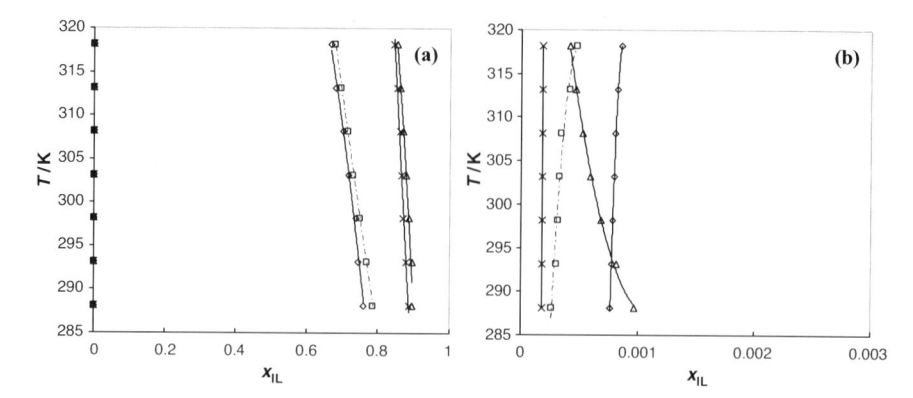

FIGURE 8.1 Liquid–liquid phase diagram for water and [C$_4$mim][NTf$_2$]: □, experimental; ◊, BP/TZVP; Δ, BP/SVP/AM1; ×, B88-VWN/DNP. The dashed and the solid lines represent, respectively, the experimental data and the prediction by the COSMO-RS calculation.

parameterisations: namely; BP/TZVP, BP/SVP/AM1, and B88-VWN/DNP, as described previously. The calculations for the three parameterisations tested were performed with the minimum energy conformers for one LLE example and one VLE example, and the results are presented in Figures 8.1 and 8.2.

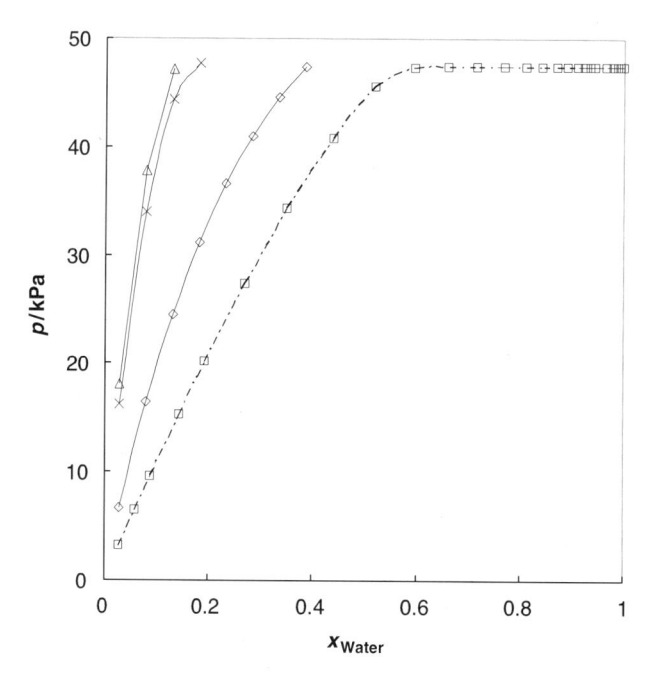

FIGURE 8.2 Vapour–liquid phase diagram for water and [C$_2$mim][NTf$_2$]: □, experimental; ◊, BP/TZVP; Δ, BP/SVP/AM1; ×, B88-VWN/DNP. The dashed and the solid lines represent, respectively, the experimental data and the prediction by the COSMO-RS calculation.

The question of which quantum chemistry method and basis set and also which parameterisation of COSMOtherm to use for an application depends on the required quality and the later usage of the predictions. The results obtained here show, in accordance with the suggestions of Eckert [29], that the BP/TZVP procedure is the best choice because it is the result of a high-quality quantum chemistry method in combination with a large basis set and therefore is able to capture the strong polarity of ionic species. In fact, for the LLE and VLE prediction of the binary mixtures, in general, the best results in description of the experimental data were achieved with the BP/TZVP procedure.

8.6 CONFORMERS INFLUENCE ON THE PREDICTIONS

A molecule prefers to occupy the levels of the minimum potential energy and arranges its atoms accordingly. By rotation around single bonds, molecules with the same molecular formula can form geometrical isomers by arranging their atoms in different, nonequivalent positions to each other, the so-called minimum energy conformations or stable conformations. There are different energy states for the various conformers in the alkyl chains of $[C_2mim]^+$ to $[C_8mim]^+$ cations and in the $[NTf_2]^-$ and $[(CH_3)_2PO_4]^-$ anions studied. Thus it is important from a theoretical point of view to evaluate the effect of the various conformers on the predicted LLE and VLE systems. To study the influence of the ionic liquids conformations on the COSMO-RS predictions, the stable conformations with the minimum and the maximum COSMO energy have been tested. The minimum energy conformations of the ionic liquids consist of the geometrically optimised minimum energy structure of the cation and the anion and the maximum energy conformations consist of the opposite analogy; besides the $[NTf_2]^-$ and $[(CH_3)_2PO_4]^-$ anions that were found to have two and three stable conformers each, in some cases just the effect of the cation is studied, because only one structure of the anions $[PF_6]^-$ and $[BF_4]^-$ exists. Some examples of the conformer influence results are depicted in Figures 8.3–8.5, both in the LLE and VLE phase diagrams.

The results presented in Figure 8.3 show that the different cation energy conformations have a small effect on the predicted mutual solubilities, and similar behaviours in the LLE study for other alkyl chain length examples and also in combination with the $[BF_4]^-$ anion were obtained. Higher deviations were found in the $[C_4mim][NTf_2]$ ionic liquid phase diagram presented in Figure 8.4, where besides the presence of the cation conformers, the anion also presents two different minimum energy geometrical isomers. There is a combination of minimal energy conformations of both cation and anion, showing that the anion plays an important role in the interactions with water. The $[NTf_2]^-$ anion has four atoms of oxygen and the interaction between these and hydrogen from water will be stronger and hydrogen bonded, compared to the $[BF_4]^-$ and $[PF_6]^-$ anions.

One example of the diverse energy conformations influence on the VLE behaviour is presented in Figure 8.5. In both cases of the $[C_2mim][NTf_2]$ and $[C_4mim][NTf_2]$

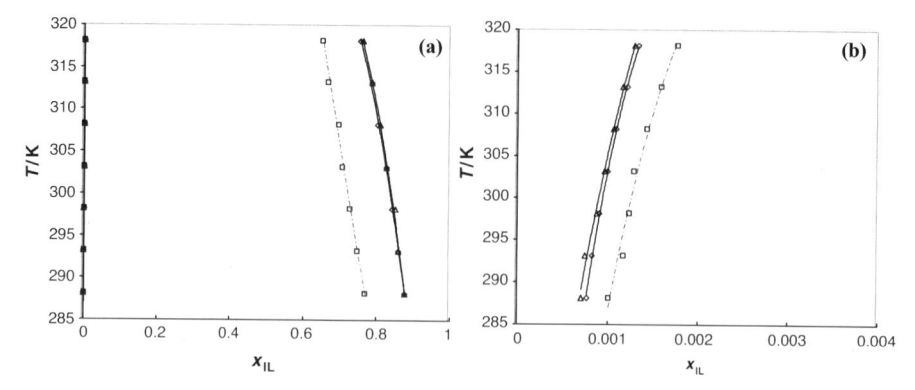

FIGURE 8.3 Complete liquid–liquid phase diagrams for (a) water and [C$_4$mim][PF$_6$] and (b) in the water-rich phase: □, experimental; ◊, lower energy conformation; Δ, higher energy conformation. The dashed and the solid lines represent, respectively, the experimental data and the prediction by the COSMO-RS calculation using the BP/TZVP procedure.

with water vapour–liquid phase diagrams, the positive deviation from Raoult's law is predicted, and the best results with respect to experimental data are that with ions of lower energy conformation.

In general, the LLE of systems with imidazolium-based ionic liquids and water seem to be more affected by the presence of the different conformers of the anion than the VLE phase diagram, where the higher deviations were found with the cation multiple conformations.

In both the LLE and VLE studies, the best predictions were obtained with the lower energy conformations of both cation and/or anion.

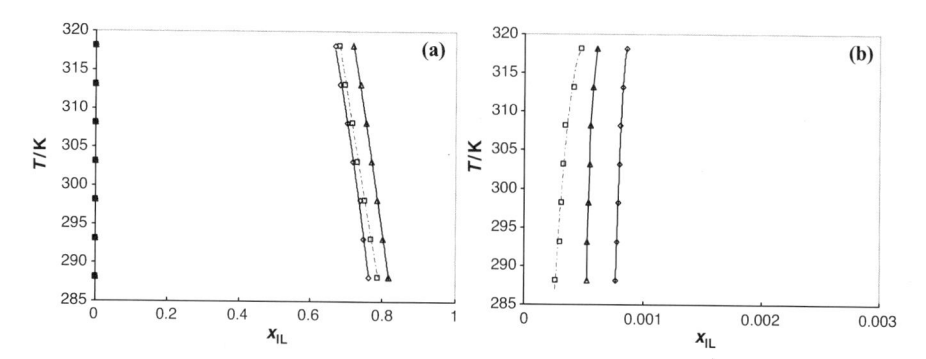

FIGURE 8.4 Complete liquid–liquid phase diagrams for (a) water and [C$_4$mim][NTf$_2$] and (b) in the water-rich phase: □, experimental; ◊, lower energy conformation; Δ, higher energy conformation. The dashed and the solid lines represent, respectively, the experimental data and the prediction by the COSMO-RS calculation using the BP/TZVP procedure.

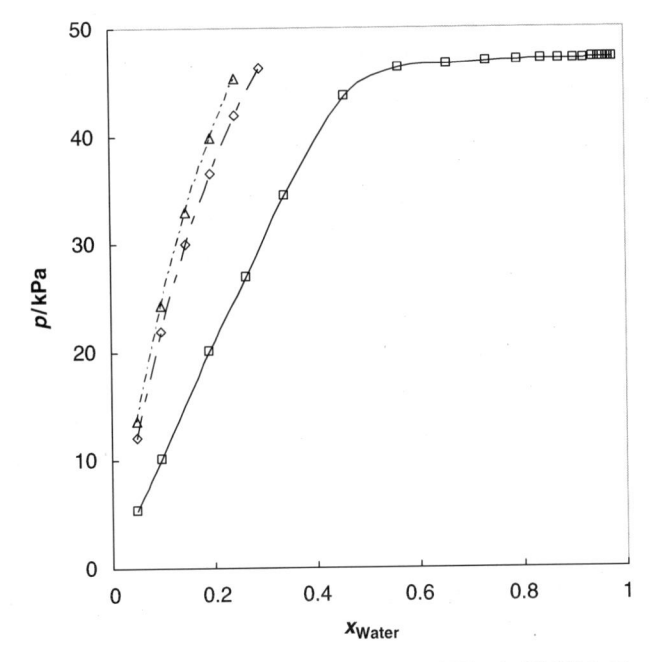

FIGURE 8.5 Vapour–liquid phase diagram for water and [C₄mim][NTf₂]: □, experimental; ◊, lower energy conformation; Δ, higher energy conformation. The dashed and the solid lines represent, respectively, the experimental data and the prediction by the COSMO-RS calculation using the BP/TZVP procedure.

8.7 LIQUID–LIQUID EQUILIBRIA MODELLING

Having established the best parameterisation and the optimal conformers, the phase diagram for a number of binary ionic liquid–water systems was predicted. In this part of the work, the quantum chemical COSMO calculations for the ionic liquids under study were performed with the Turbomole program package [30] using the BP density functional theory and the triple-ζ valence polarised large basis set (TZVP) [31]. These calculations were made for a true three-component mixture, where the cation and anion of equal concentrations are treated as separate species, and were performed with the optimised minimum energy conformations of both the cation and/or the anion. Experimental data in the form of T–x data for each binary mixture investigated and the results obtained with the COSMO-RS calculations are compared in Figures 8.6–8.10.

The ionic liquids–water systems studied presented an asymmetric LLE behaviour due to the high solubility of water in ionic liquids while their solubility in water is very limited, ranging from 10^{-5} to 10^{-2} mol·L^{-1}. Imidazolium-based ionic liquids can act as both hydrogen bond acceptors (anion) and donors (cation) and would be expected to interact with solvents with both accepting and donating sites, such as water. Water is

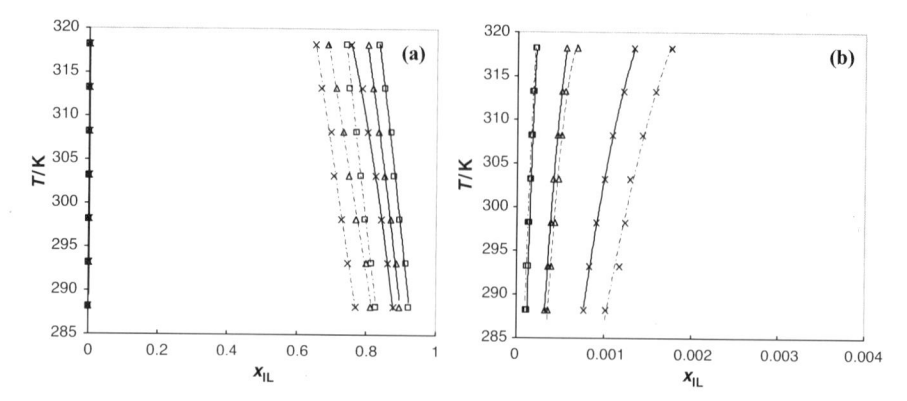

FIGURE 8.6 Comparison of the liquid–liquid phase diagrams for (a) water and ionic liquids and (b) the water-rich phase: ×, [C$_4$mim][PF$_6$]; △, [C$_6$mim][PF$_6$]; □, [C$_8$mim][PF$_6$]. The dashed and the solid lines represent, respectively, the experimental data and the prediction by the COSMO-RS calculation using the BP/TZVP procedure.

well known to form hydrogen-bonded networks with both high enthalpies and constants of association, and it is expected to stabilise with hydrogen-bond donor sites. However, the existence of the liquid–liquid phase equilibria in these mixtures is the evidence that the interaction between the ionic liquid and water is not very significant, and that the ion–ion and water–water interactions were stronger than water–ion interactions in the observed mixtures.

The results obtained are discussed below from different perspectives to evaluate the impact of the ionic liquids structural variations and the COSMO-RS predictive capability for mutual solubilities.

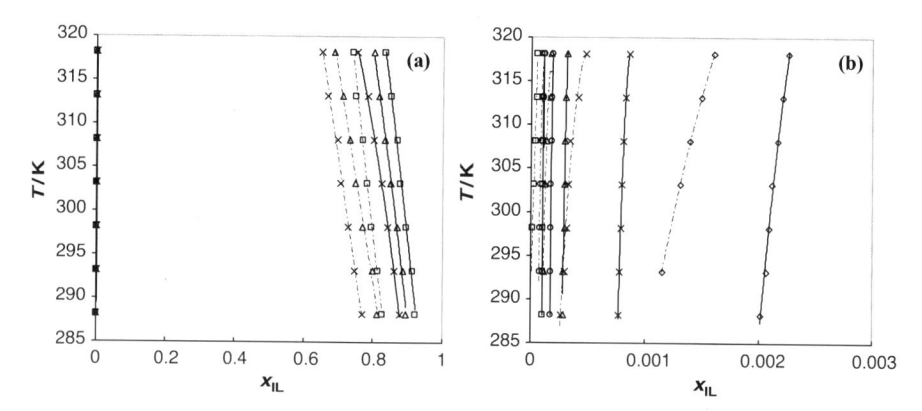

FIGURE 8.7 Comparison of the liquid–liquid phase diagrams for (a) water and ionic liquids and (b) the water-rich phase: ◊, [C$_2$mim][NTf$_2$]; ×, [C$_4$mim][NTf$_2$]; △, [C$_6$mim][NTf$_2$]; ○, [C$_7$mim][NTf$_2$]; □, [C$_8$mim][NTf$_2$]. The dashed and the solid lines represent, respectively, the experimental data and the prediction by the COSMO-RS calculation using the BP/TZVP procedure.

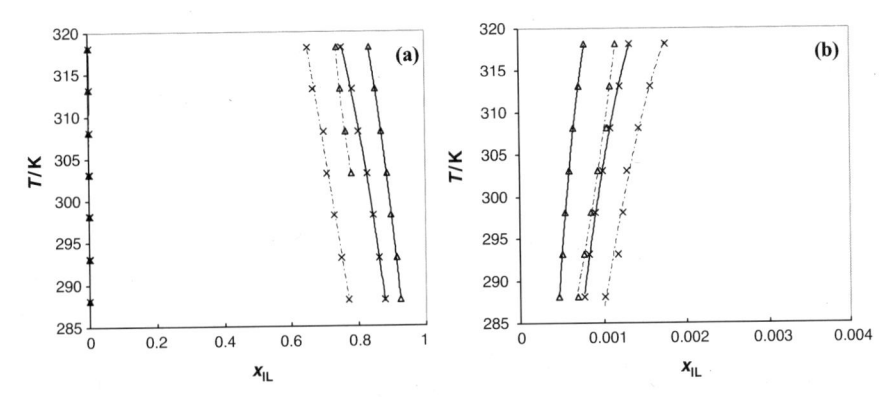

FIGURE 8.8 Comparison of the liquid–liquid phase diagrams for (a) water and ionic liquids and (b) the water-rich phase: ×, [C$_4$mim][PF$_6$]; Δ, [C$_4$C$_1$mim][PF$_6$]. The dashed and the solid lines represent, respectively, the experimental data and the prediction by the COSMO-RS calculation using the BP/TZVP procedure.

8.7.1 Cation Alkyl Chain Length

Figures 8.6 and 8.7 show the liquid–liquid phase behaviour for two anions, [PF$_6$]$^-$ and [NTf$_2$]$^-$, in combination with different alkyl chain length imidazolium-based ionic liquids. Due to the asymmetrical character of the LLE behaviour, the figures present both the general LLE diagrams and the water-rich side of the equilibrium.

The influence of the cation alkyl chain length in the mutual solubilities was verified to follow the same trend with both the anions, where there is a hydrophobicity increase concomitant with the cation alkyl chain length increase. That hydrophobic tendency occurs at both sides of the equilibrium, but plays a major role on the water-rich side,

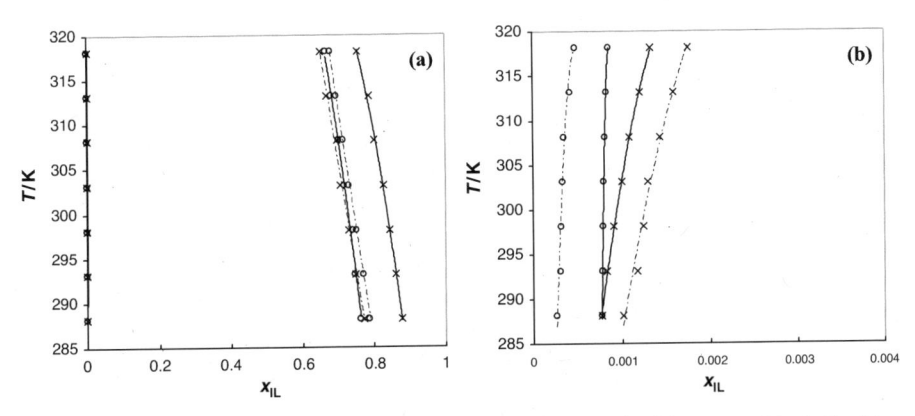

FIGURE 8.9 Comparison of the liquid–liquid phase diagrams for (a) water and ionic liquids and (b) the water-rich phase: ○, [C$_4$mim][NTf$_2$]; ×, [C$_4$mim][PF$_6$]. The dashed and the solid lines represent, respectively, the experimental data and the prediction by the COSMO-RS calculation using the BP/TZVP procedure.

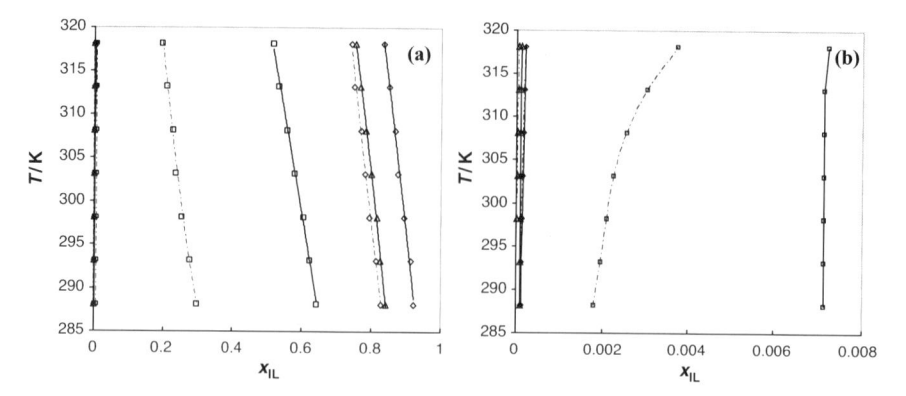

FIGURE 8.10 Comparison of the liquid–liquid phase diagrams for (a) water and ionic liquids and (b) the water-rich phase: △, [C$_8$mim][NTf$_2$]; ◊, [C$_8$mim][PF$_6$]; □ [C$_8$mim][BF$_4$]. The dashed and the solid lines represent, respectively, the experimental data and the prediction by the COSMO-RS calculation using the BP/TZVP procedure.

where differences of one order of magnitude appear when comparing the solubility of [C$_4$mim][PF$_6$] with [C$_8$mim][PF$_6$], and even of two orders of magnitude when comparing [C$_2$mim][NTf$_2$] with [C$_8$mim][NTf$_2$] in water.

The results obtained from COSMO-RS calculations show an acceptable agreement with the experimental data available, and follow the same hydrophobicity tendency to increase with the cation alkyl chain length increase, depicting the qualitatively good prediction capacity of this model. Higher relative deviations were found in the water-rich phase, but it should be mentioned that there is a relatively low solubility of ionic liquids in water, and the predictions were always of the same order of magnitude as the experimental data; that is, if the solubilities were reported as log(x), as is usual in this field, the difference between the experimental data and the predictions is well below one log unit.

8.7.2 Cation Methyl Inclusion

Figure 8.8 presents the comparison between water and [C$_4$mim][PF$_6$] or [C$_4$C$_1$mim][PF$_6$] mutual solubilities.

By replacing the hydrogen of the [C$_4$mim]$^+$ cation at the C2 position with a methyl group (forming [C$_4$C$_1$mim]$^+$), the ability of the cation to hydrogen bond with water is greatly diminished, resulting in a decrease in the mutual solubilities between the previous imidazolium-based ionic liquid and water. Clearly, hydrogen bonding of water with the acidic hydrogen of the imidazolium cation has some influence in controlling liquid–liquid phase behaviour between imidazolium-based ionic liquids and water. COSMO-RS calculations agree well with experimental results for both sides of the equilibrium, predicting correctly the increased hydrophobicity and the variation in mutual solubilities due to a methyl inclusion in the imidazolium cation.

8.7.3 Anion Identity

Figures 8.9 and 8.10 show the comparison between the experimental data and COSMO-RS predictions using the BP/TZVP procedure for the liquid–liquid phase behaviour for two independent cations, $[C_4mim]^+$ and $[C_8mim]^+$, in combination with different anions, $[NTf_2]^-$, $[PF_6]^-$, and $[BF_4]^-$.

An excellent predictive capacity was found for the case of the $[C_4mim][NTf_2]$ entire liquid–liquid phase diagram. Also, in the water-rich phase, COSMO-RS proved to predict the hydrophobic tendency increase due to the anion identity from $[BF_4]^- < [PF_6]^- < [NTf_2]^-$, following the experimental mutual solubilities decrease with water. However, from Figure 8.9, in the ionic liquid-rich phase, the hydrophobic tendency between $[PF_6]^-$ and $[NTf_2]^-$ is not well described when compared to the experimental data. COSMO-RS predicts a higher solubility of water in $[C_4mim][NTf_2]$ than in $[C_4mim][PF_6]$, which may be due to the fact that $[NTf_2]^-$ is a stronger Lewis base than $[PF_6]^-$. Conversely, $[PF_6]^-$ has a greater charge density than $[NTf_2]^-$ because it is smaller, so it can have stronger Coulombic interactions, and clearly the phase behaviour is the result of several competing interactions in the solution.

Moreover, it was demonstrated that both the cation and anion affect the mutual solubilities, but from Figures 8.9 and 8.10, it is the anion that plays the major role on the phase behaviour of imidazolium-based ionic liquids with water. Thus changing the anion is the easiest way to adjust the liquid–liquid equilibrium with water.

8.8 VAPOUR–LIQUID EQUILIBRIA MODELLING

The vapour-phase behaviour for six ionic liquid–water systems was available from the literature [6, 11] and the comparison of the COSMO-RS predictions, using the BP/TZVP procedure and the lower energy ion conformations, with the experimental data was performed. The results are presented in Figures 8.11–8.14 in the form of p–x data for each binary mixture investigated. Again, the COSMO-RS calculations were made for a true three-component mixture where the cation and anion of equal concentrations are treated as separate species. The results obtained for the isotherm p–x phase diagrams are discussed below from different views to evaluate the influence of the ionic liquids structural variations and their dependence on temperature and the COSMO-RS predictive capability.

8.8.1 Cation Alkyl Chain Length

Figures 8.11 and 8.12 show the vapour–liquid phase behaviour for two anions, $[PF_6]^-$ and $[NTf_2]^-$, in combination with different alkyl chain length imidazolium-based ionic liquids.

From Figures 8.11 and 8.12, COSMO-RS provided a good description of the p–x phase diagrams with respect to the cation alkyl chain length variation with both the anions, $[PF_6]^-$ and $[NTf_2]^-$, when compared to the experimental data [6, 11]. There is an increase of the positive deviation from Raoult's law and an increase of the ionic

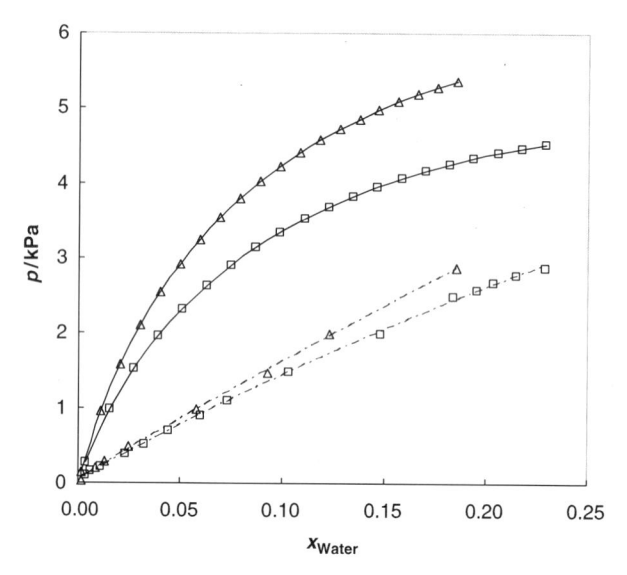

114 PREDICTING THE THERMODYNAMIC BEHAVIOUR OF WATER

FIGURE 8.11 Comparison of the vapour–liquid phase diagrams for water and ionic liquids at 298.15 K: □, [C$_4$mim][PF$_6$]; Δ, [C$_8$mim][PF$_6$]. The dashed and the solid lines represent, respectively, the experimental data [6] and the prediction by the COSMO-RS calculation using the BP/TZVP procedure.

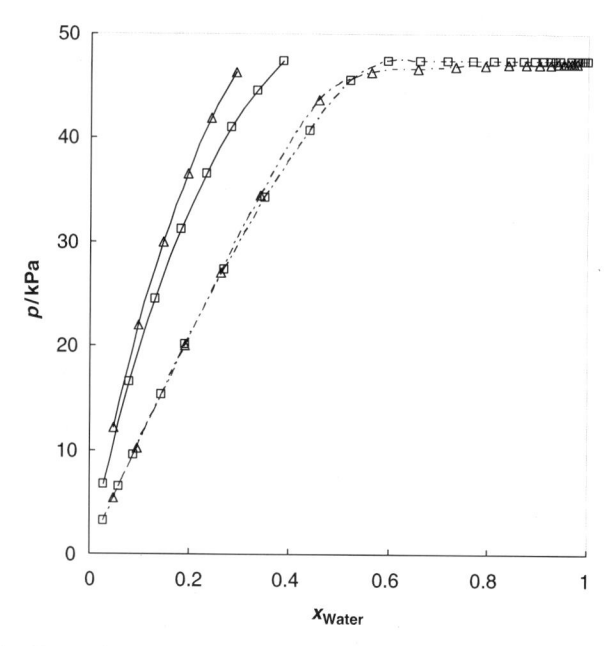

FIGURE 8.12 Comparison of the vapour–liquid phase diagrams for water and ionic liquids at 353.15 K: □, [C$_2$mim][NTf$_2$]; Δ, [C$_4$mim][NTf$_2$]. The dashed and the solid lines represent, respectively, the experimental data [11] and the prediction by the COSMO-RS calculation using the BP/TZVP procedure.

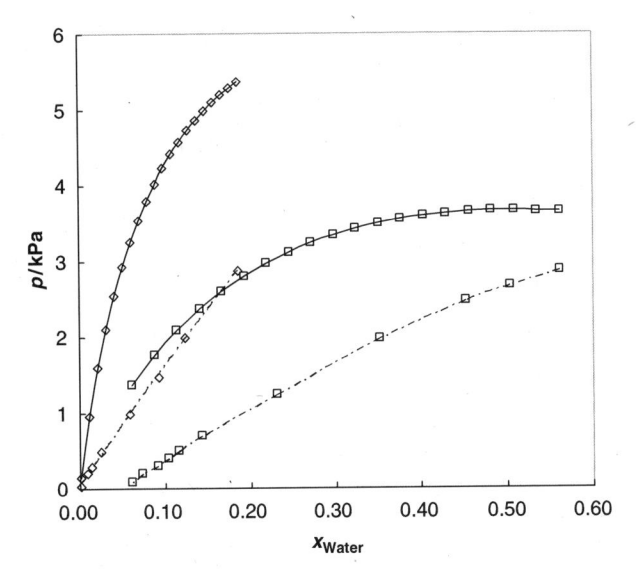

FIGURE 8.13 Comparison of the vapour–liquid phase diagrams for water and ionic liquids at 298.15 K: ◊, [C$_8$mim][PF$_6$]; □, [C$_8$mim][BF$_4$]. The dashed and the solid lines represent, respectively, the experimental data [6] and the prediction by the COSMO-RS calculation using the BP/TZVP procedure.

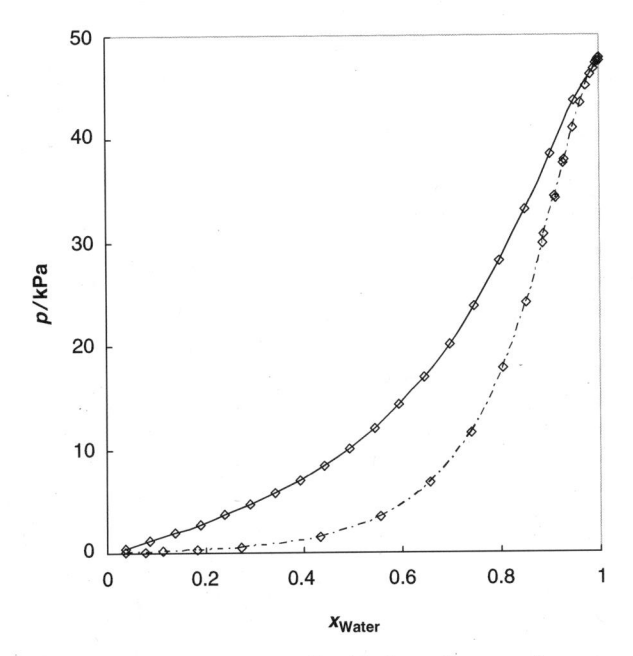

FIGURE 8.14 Comparison of the vapour–liquid phase diagrams for water and [C$_1$mim][(CH$_3$)$_2$PO$_4$] at 353.15 K. The dashed and the solid lines represent, respectively, the experimental data [6] and the prediction by the COSMO-RS calculation using the BP/TZVP procedure.

liquids' hydrophobicity with the alkyl chain length increase. In Figure 8.12 the occurrence of a miscibility gap with the $[NTf_2]^-$ based ionic liquids is shown; this same behaviour is expected with the $[PF_6]^-$ based ionic liquids, although the mole fraction range presented is not enough to prove this fact experimentally. Besides the positive deviation from Raoult's law that is predicted for the ionic liquids based on these two anions, COSMO-RS was able to give at least a priori good qualitative predictions.

8.8.2 Anion Identity

Figure 8.13 shows the comparison between experimental data [6] and COSMO-RS predictions using the BP/TZVP procedure for the vapour–liquid phase behaviour for the $[C_8mim]^+$ cation, in combination with different anions $[NTf_2]^-$ and $[PF_6]^-$. Figure 8.14 presents the results obtained for water with the $[C_1mim][(CH_3)_2PO_4]$ ionic liquid vapour–liquid phase equilibria [11].

COSMO-RS proved to predict the hydrophobic tendency increase in the vapour–liquid phase equilibria due to the anion identity from $[BF_4]^- < [PF_6]^-$, following the same trend of the liquid–liquid equilibria.

Moreover, it was demonstrated that both the cation and anion affect the vapour–liquid phase equilibria but, comparing Figures 8.11–8.13, it is the anion that plays the major role in the vapour-phase behaviour of imidazolium-based ionic liquids with water, as in the LLE systems.

From Figure 8.14, COSMO-RS demonstrated that it is also able to describe phase diagrams with strong negative deviations from Raoult's law, as experimentally evidenced.

8.8.3 Temperature Dependence

Figures 8.15–8.17 show the comparison between the experimental data [6] and COSMO-RS predictions using the BP/TZVP procedure for the vapour–liquid phase behaviour at several isotherms for the following ionic liquids: $[C_4mim][PF_6]$, $[C_8mim][PF_6]$, and $[C_8mim][BF_4]$.

From Figures 8.15–8.17, COSMO-RS showed the ability to describe the vapour–liquid phase diagram behaviour as a function of temperature, increasing pressure with the increase of temperature, as verified experimentally. Although the quantitative predictions are not so excellent, COSMO-RS was able to describe the qualitative variations due to temperature in all the ionic liquids analysed which were composed of different cations and/or anions.

8.9 CONCLUSION

Ionic liquids have been suggested as potential "green" solvents to replace volatile organic solvents in reaction and separation processes due to their negligible vapour pressure. To develop ionic liquids for these applications, it is important to gain a

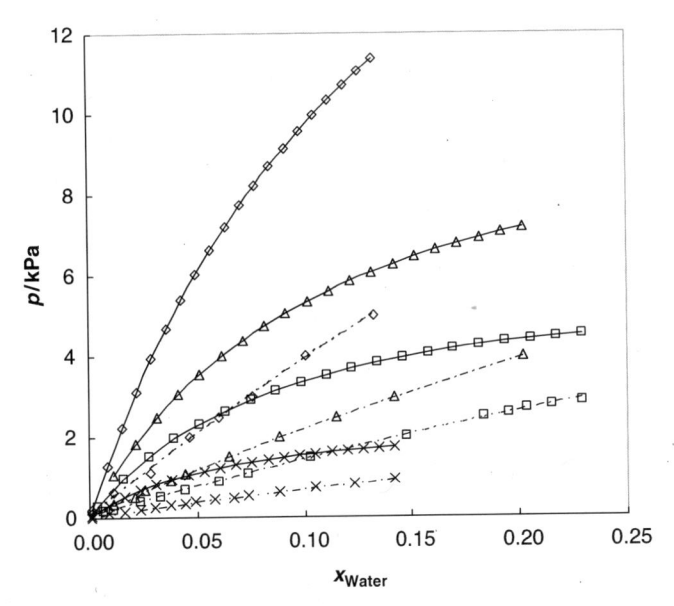

FIGURE 8.15 Comparison of the vapour-liquid phase diagrams for water and [C$_4$mim][PF$_6$] at isotherms: ×, 283.15 K; □, 298.15 K; Δ, 308.15 K; ◊, 323.15 K. The dashed and the solid lines represent, respectively, the experimental data and the prediction by the COSMO-RS calculation.

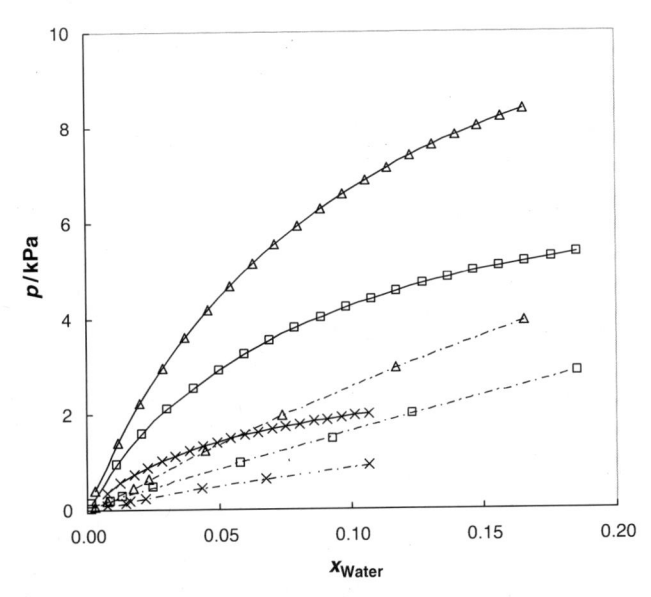

FIGURE 8.16 Comparison of the vapour–liquid phase diagrams for water and [C$_8$mim][PF$_6$] at isotherms: ×, 283.15 K; □, 298.15 K; Δ, 308.15 K. The dashed and the solid lines represent, respectively, the experimental data and the prediction by the COSMO-RS calculation.

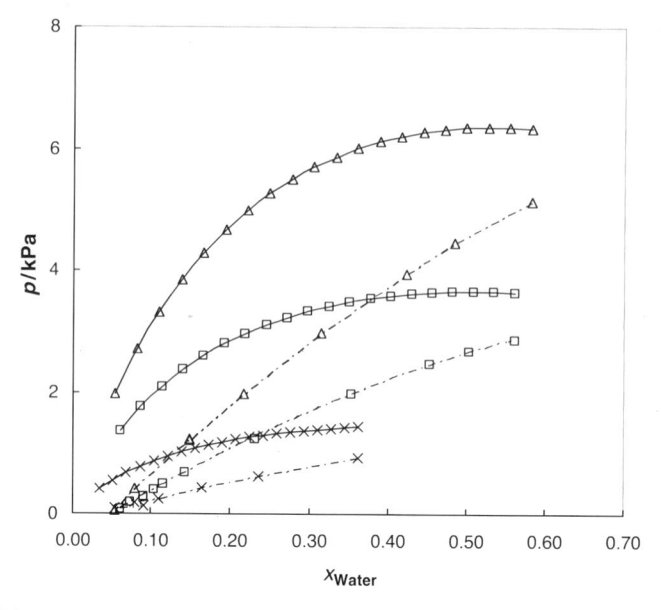

FIGURE 8.17 Comparison of the vapour–liquid phase diagrams for water and [C$_8$mim][BF$_4$] at isotherms: ×, 283.15 K; □, 298.15 K; Δ, 308.15 K. The dashed and the solid lines represent, respectively, the experimental data and the prediction by the COSMO-RS calculation.

fundamental understanding of the factors that control the phase behaviour of ionic liquids with other liquids, including polar solvents such as water. Since it is not feasible to experimentally determine all possible combinations with ionic liquids, a predictive method capable of describing the phase behaviour of such systems is extremely important. Quantum chemical calculations based on the σ profiles of the cation, the anion, and water were used for the prediction of LLE and VLE systems incorporating ionic liquids and water. COSMO-RS and its implementation in the program COSMOtherm was shown to be capable of giving satisfactory a priori predictions of the thermodynamics of systems involving ionic liquids, which may be of considerable value for the exploration of suitable ionic liquids for practical and specific applications.

ACKNOWLEDGEMENTS

The authors acknowledge financial support from Fundação para a Ciência e a Tecnologia (Project POCI/EQU/58152/2004) and a Ph.D. grant (SFRH/BD/14134/2003) to Mara G. Freire. They also acknowledge F. Eckert and A. Klamt, COSMOtherm, Version C2.1, Release 01.05, COSMOlogic GmbH & Co. KG, Leverkusen, Germany, 2005, and M. Diedenhofen of COSMOlogic for advice and assistance in the use of COSMOtherm.

REFERENCES

1. Huddleston, J. G., Willauer, H. D., Swatloski, R. P., Visser, A. E., and Rogers, R. D., Room temperature ionic liquids as novel media for clean liquid-liquid extraction, *Chem. Commun.* **44**, 1765–1766 (1998).
2. Fadeev, A. G. and Meagher, M. M., Opportunities for ionic liquids in recovery of biofuels, *Chem. Commun.* **44**, 295–296 (2001).
3. McFarlane, J., Ridenour, W. B., Luo, H., Hunt, R. D., DePaoli, D. W., and Ren, R. X., Room temperature ionic liquids for separating organics from produced water, *Sep. Sci. Technol.* **40**, 1245–1265 (2005).
4. Seddon, K. R., Stark, A., and Torres, M.-J., Influence of chloride, water, and organic solvents on the physical properties of ionic liquids, *Pure Appl. Chem.* **72**, 2275–2287 (2000).
5. Huddleston, J. G., Visser, A. E., Reichert, W. M., Willauer H. D., Broker, G. A., and Rogers, R. D., Characterisation and comparison of hydrophilic and hydrophobic room temperature ionic liquids incorporating the imidazolium cation, *Green Chem.* **3**, 156–164 (2001).
6. Anthony, J. L., Maggin, E. J., and Brennecke J. F., Solution thermodynamics of imidazolium-based ionic liquids and water, *J. Phys. Chem. B* **105**, 10942–10949 (2001).
7. Wong, D. S. H., Chen, J. P., Chang, J. M., and Chou, C. H., Phase equilibria of water and ionic liquids [emim][PF$_6$] and [bmim]PF$_6$], *Fluid Phase Equilib.* **194–197**, 1089–1095 (2002).
8. Crosthwaite, J. M., Aki, S. N. V. K., Maggin, E. J., and Brennecke, J. F., Liquid phase behavior of imidazolium-based ionic liquids with alcohols, *J. Phys. Chem. B* **108**, 5113–5119 (2004).
9. Rebelo, L. P. N., Najdanovic-Visak, V., Visak, Z. P., Nunes da Ponte, M., Szydlowski, J., Cerdeiriña, C. A., Troncoso, J., Romaní, L., Esperança, J. M. S. S., Guedes, H. J. R., and de Sousa, H. C., A detailed thermodynamic analysis of [C$_4$mim][BF$_4$] + water as a case study to model ionic liquid aqueous solutions, *Green Chem.* **6**, 369–381 (2004).
10. Najdanovic-Visak, V., Esperança, J. M. S. S., Rebelo, L. P. N., Nunes da Ponte, M., Guedes, H. J. R., Seddon, K. R., and Szydlowskiy, J., Phase behaviour of room temperature ionic liquid solutions: an unusually large co-solvent effect in (water + ethanol), *Phys. Chem. Chem. Phys.* **4**, 1701–1703 (2002).
11. Kato, R. and Gmehling, J., Measurement and correlation of vapour–liquid equilibria of binary systems containing the ionic liquids [EMIM][(CF$_3$SO$_2$)$_2$N], [MMIM][(CF$_3$SO$_2$)$_2$N], [MMIM][(CH$_3$)$_2$PO$_4$] and oxygenated organic compounds respectively water, *Fluid Phase Equilib.* **231**, 38–43 (2005).
12. Calvar, N., González, B., Gómez, E., and Domínguez, A., Vapor–liquid equilibria for the ternary system ethanol + water + 1-butyl-3-methylimidazolium chloride and the corresponding binary systems at 101.3 kPa, *J. Chem. Eng. Data* **51**, 2178–2181 (2006).
13. Domańska, U., Bogel-Łukasik, E., and Bogel-Łukasik, R., 1-Octanol/water coefficients of 1-alkyl-3-methyllimidazolium chloride, *Chem. Eur. J.* **9**, 3033–3041 (2003).
14. Domańska, U., Thermophysical properties and thermodynamic phase behavior of ionic liquids, *Thermochim. Acta* **448**, 19–30 (2006).
15. Domańska, U. and Bogel-Łukasik, E., Measurements and correlation of the (solid + liquid) equilibria of [1-decyl-3-methylimidazolium chloride + alcohols (C2–C12), *Ind. Eng. Chem. Res.* **42**, 6986–6992 (2003).

16. Doker, M. and Gmehling, J., Measurement and prediction of vapor–liquid equilibria of ternary systems containing ionic liquids, *Fluid Phase Equilib.* **227**, 255–266, (2005).
17. Hu, X., Yu, J., and Liu, H., Liquid–liquid equilibria of the system 1-(2-hydroxyethyl)-3-methylimidazolium tetrafluoroborate or 1-(2-hydroxyethyl)-2,3-dimethylimidazolium tetrafluoroborate + water + 1-butanol at 293.15 K, *J. Chem. Eng. Data* **51**, 691–695 (2006).
18. Letcher, T. M., Deenadayalu, N., Soko, B., Ramjugernath, D., and Naicker, P. K., Ternary liquid–liquid equilibria for mixtures of 1-methyl-3-octylimidazolium chloride + an alkanol + an alkane at 298.2 K and 1 bar, *J. Chem. Eng. Data* **48**, 904–907 (2003).
19. Letcher, T. M. and Reddy, P., Ternary liquid–liquid equilibria for mixtures of 1-hexyl-3-methylimidazolium (tetrafluoroborate or hexafluorophosphate) + ethanol + an alkene at T = 298.2 K, *Fluid Phase Equilib.* **219**, 107–112 (2004).
20. Rebelo, L. P. N., Simple g^E-model for generating all basic types of binary liquid–liquid equilibria and their pressure dependence. Thermodynamic constraints at critical loci, *Phys. Chem. Chem. Phys.* **1**, 4277–4286 (1999).
21. Shariati, A. and Peters, C. J., High-pressure phase behavior of systems with ionic liquids: measurements and modeling of the binary system fluoroform + 1-ethyl-3-methylimidazolium hexafluorophosphate, *J. Supercrit. Fluids* **25**, 109–111 (2003).
22. Rebelo, L. P. N., Lopes, J. N. C., Esperança, J. M. S. S., and Filipe, E., On the critical temperature, normal boiling point, and vapor pressure of ionic liquids, *J. Phys. Chem. B* **109**, 6040–6043 (2005).
23. Klamt, A. and Eckert, F., COSMO-RS: a novel and efficient method for the a priori prediction of thermophysical data of fluids, *Fluid Phase Equilib.* **172**, 43–72 (2000).
24. Wu, C.-T., Marsh, K. N., Deev, A. V., and Boxal, J. A., Liquid–liquid equilibria of room-temperature ionic liquids and butan-1-ol, *J. Chem. Eng. Data* **48**, 486–491 (2003).
25. Domańska, U., Pobudkowska, A., and Eckert, F., (Liquid + liquid) phase equilibria of 1-alkyl-3-methylimidazolium methylsulfate with alcohols, or ethers, or ketones, *J. Chem. Thermodyn.* **38**, 685–695 (2006).
26. Klamt, A., Jonas, V., Bürger, T., and Lohrenz, J. C. W., Refinement and parametrisation of COSMO-RS, *J. Phys. Chem. A* **102**, 5074–5085 (1998).
27. Klamt, A. and Schüürmann, G., COSMO—a new approach to dielectric screening in solvents with explicit expressions for the screening energy and its gradient, *J. Chem. Soc. Perkins Trans.* **2**, 799–805 (1993).
28. Eckert, F. and Klamt, A., *COSMOtherm*, Version C2.1, Release 01.05; COSMOlogic GmbH & Co. KG, Leverkusen, Germany, 2005.
29. Eckert, F., *COSMOtherm User's Manual*, Version C2.1, Release 01.05, COSMOlogic GmbH & Co. KG, Leverkusen, Germany, 2005.
30. Schäfer, A., Klamt, A., Sattel, D., Lohrenz, J. C. W., and Eckert, F., COSMO implementation in TURBOMOLE: extension of an efficient quantum chemical code towards liquid systems, *Phys. Chem. Chem. Phys.* **2**, 2187–2193 (2000).
31. This density functional method and basis set combination is equivalent to the Turbomole method. Thus the COSMOtherm parameter set optimised for the Turbomole DFT method can be used with COSMO files produced by this quantum chemical program package.

32. Andzelm, J., Kölmel, C., and Klamt, A., Incorporation of solvent effects into density functional calculations of molecular energies and geometries, *J. Chem. Phys.* **103**, 9312–9320 (1995).
33. Freire, M. G., Carvalho, P. J., Gardas, R. L., Santos, L. M. N. B. F., Marrucho, I. M., and Coutinho, J. A. P., Mutual solubilities of imidazolium-based ionic liquids and water and the Hofmeister series, oral communication presented at EUCHEM Conference on Molten Salts and Ionic Liquids, 2006.
34. Freire, M. G., Neves, C., Carvalho, P. J., Gardas, R. L., Fernandes, A. M., Marrucho, I. M., Santos, L., and Coutinho, J. A. P., Mutual solubilities of water and hydrophobic ionic liquids, *J. Phys. Chem. B* **111** (45), 13082–13089 (2007).

9 Metallic Inert Anodes for Aluminium Electrolysis

I. GALASIU and R. GALASIU
Institute of Physical Chemistry "I.G. Murgulescu" of The Romanian Academy, Bucharest, Romania

C. NICOLESCU
Faculty of Industrial Chemistry, University Polytechnia of Bucharest, Bucharest, Romania

Abstract

Metallic inert anodes based on iron alloys were tested during electrolysis in cryolite–alumina melts. The corrosion resistance as a function of alloy composition was measured during electrolysis. For some inert anodes, the content of impurities in the cathodic aluminium was lower than that in the aluminium obtained by the conventional method. SEM analyses showed modifications during electrolysis into the layer of oxides on the anode surfaces.

9.1 INTRODUCTION

As is well known, industrial production of aluminium is based on the electrolysis of cryolite–alumina melts, at temperatures of 950–970 °C using carbon anodes. During the electrolysis, oxygen is formed at the carbon anodes and reacts with the carbon from the anodic material forming CO_2 gas, which is evolved. At low alumina content in the electrolyte, during the so-called anodic effects, CF_4 is formed at the anode.

Moreover, CO_2 and CF_4 are gases that generate greenhouse effects that negatively impact the environment and humanity (e.g., droughts, storms, floods, ice melting, increase of sea level, major modifications of the Earth's climate). If the CO_2 remains in the atmosphere for "only" 50–200 years, CF_4 as an inert gas is accumulated in the atmosphere for 10^4–10^6 years. In addition, the greenhouse effects of CF_4 are 6500 times greater than those of CO_2.

Molten Salts and Ionic Liquids: Never the Twain? Edited by Marcelle Gaune-Escard and Kenneth R. Seddon
Copyright © 2010 John Wiley & Sons, Inc.

Because of the major disadvantages of the actual technology for industrial aluminium production, it is imperative to replace it with another technology having a less negative impact on the environment and, when possible, a lower electrical consumption.

This technology should use inert anodes that resist the corrosive actions of both the melted cryolite and metallic aluminium dissolved in the melt. At the same time, the anodes have to be resistant to the action of oxygen at 1000 °C.

Even from this brief introduction, the difficulties in realising this kind of technology are obvious. However, a solution has to be found in the shortest possible time. In the last 30–40 years, researchers in the field of aluminium electrolysis have made significant efforts to obtain this essential new technology.

Laboratory and prototype experiments have been performed, and both Alcoa (in the United States) and Moltech (in Switzerland) have announced that they will apply at the industrial level the results they obtained in their own researches. Also, other big companies producing aluminium at the industrial level are working in secret to obtain inert anodes for aluminium electrolysis.

The most important issue to be followed and checked with these new anodes is their corrosion during electrolysis and, consequently, the purity of the obtained aluminium. Up to now, three important categories of inert anodes have been used:

1. *Ceramic inert anodes*, especially based on SnO_2 doped with CuO and Sb_2O_3, were tested at the laboratory scale in the 1980s, all over the world. They have the advantage that, being prepared from oxides in their highest oxidation state, they do not oxidise with the formed oxygen during electrolysis.

2. *Cermet anodes* consist of two phases: one ceramic phase made of nickel ferrites, and one metallic phase, made of copper, nickel, iron, or their alloys. A nickel ferrite has been chosen as the ceramic material. Unfortunately, the nickel ferrite has a high electrical resistivity, powders of metal or alloys being added to increase the electrical conductivity. The mixture obtained is sintered in an inert atmosphere at 1300 °C. This type of anodic material was proposed by Alcoa in the early 1980s, and research studies with it continue to the present. This material has a major disadvantage: during electrolysis the metallic part oxidises.

3. *Metallic inert anodes* started to be tested after the year 2000, mainly in laboratory scale experiments. Although the majority of the metals and metal alloys dissolve during the anodic process of electrolysis, in the cryolite–alumina melts, some alloys undergo a surface oxidation and a lower dissolution in the melt. If the resulting oxides have a low solubility in the electrolyte, the metals enter a passivity process, similar to the air passivity of metals/metallic alloys. With the anodes used in aluminium electrolysis, the surface layer has to be electrically conductive and impermeable to oxygen (which forms in the electrolysis process). By oxygen diffusion, the undesirable "metallic core" oxidation can take place. Also, the metallic ions should be prevented from migrating into the melted electrolyte, where they can oxidise.

9.2 EXPERIMENTAL

Experiments consisted of testing steel materials having different compositions (low and high alloyed) as anodes in the electrolysis process of cryolite–alumina melts. The experimental arrangement is shown in Figure 9.1.

A vertical muffle furnace with Kanthal resistance was used. The graphite crucible containing the electrolyte mixture was placed in the furnace to melt the electrolyte. On the bottom of the crucible, 100 g of melted aluminium was deposited and functioned as the cathode. The anode was a metallic bar with a diameter of 15 mm. The immersion depth of the anode in the molten electrolyte was kept at about 25 mm with a vertical-positioning mechanical system.

The electrolysis conditions were as follows:

- Temperature: 980 °C
- Current density: 0.8 A·cm^{-2}
- Electrolyte composition: 81% Na_3AlF_6 + 4% CaF_2 + 7% AlF_3 + 8% Al_2O_3
- Electrolysis time: 4 h

Alumina content was kept close to the saturated value, taking into consideration that the closer the alumina content in the electrolyte is to this value, the lower is the solubility of other metallic oxides, so corrosion is minimal.

After the electrolysis experiment, the anode was mechanically cleaned and the diameter was measured in order to evaluate the corrosion that took place during the process. The cathodic aluminium composition was analysed in order to determine the level of impurities after electrolysis under the experimental conditions. Iron

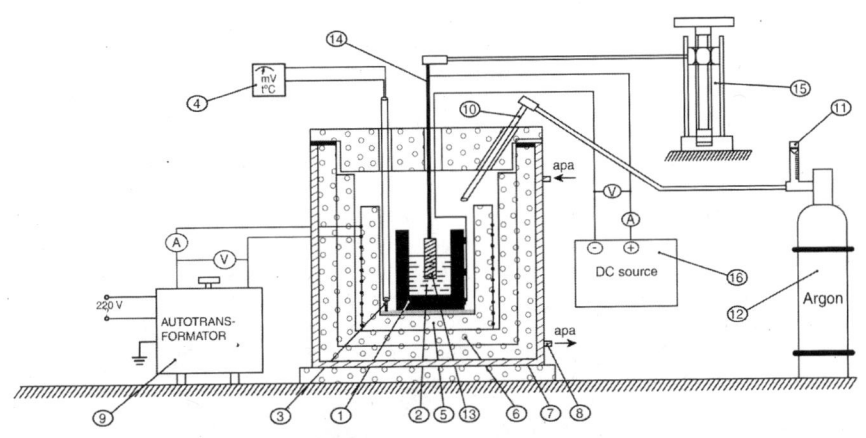

FIGURE 9.1 Experimental arrangement: ① graphite crucible (counterelectrode), ② melt, ③ thermocouple Pt-PtRh, ④ mV-thermometer, ⑤ muffle with Kanthal resistance, ⑥ muffle protecting vessel, ⑦ metallic case, ⑧ water cooling circuit, ⑨ alternating current transformer, ⑩ tube for passing argon gas, ⑪ flow-metre, ⑫ argon bomb, ⑬ metallic anode, ⑭ metallic rod, ⑮ mechanic positioning system, ⑯ direct current source.

TABLE 9.1 Composition of the Tested Metallic Inert Anodes, the Balance Being Iron

Anode Number	Composition/wt %					
	Mn	Si	Cr	Ni	Cu	Mo
1	0.75	0.5	15.01	7.05	1.40	0.70
2	1.20	0.57	18.75	9.33	0.20	0.25
3	1.07	1.99	25.50	20.55	0.15	0.22
4	0.33	0.55	1.25	3.33	0.35	0.17
5	0.30	0.25	2.20	0.3	0.25	0.1

content was determined in the aluminium, as iron is the main component of the alloys that were tested as anodic material. The initial content of iron in aluminium was 0.14%. The results in Table 9.1 showed an increase in iron content in the aluminium during the electrolysis using the described inert anodes.

9.3 RESULTS AND DISCUSSION

The chemical compositions of the tested anodes made of metallic alloys are presented in Table 9.1.

The first three samples are made of highly alloyed steels and the last two are made of low alloyed steels. Figures 9.2, 9.4, 9.6 and 9.8 show the photos of the first four tested anodes, and Figures 9.3, 9.5, 9.7 and 9.9 show the metallographic microscopy photos of the same samples. The diameter variations after electrolysis are presented in Table 9.2.

The iron content in the cathodic aluminium after 4 h of electrolysis is presented in Table 9.3, in comparison with the initial iron content in aluminium used in the experiments. The concentration of iron in cathodic aluminium (Table 9.3) represents the concentration of iron in aluminium initially used as the cathode, plus iron coming

FIGURE 9.2 Macroscopic aspect of sample 1 after 4 h of electrolysis.

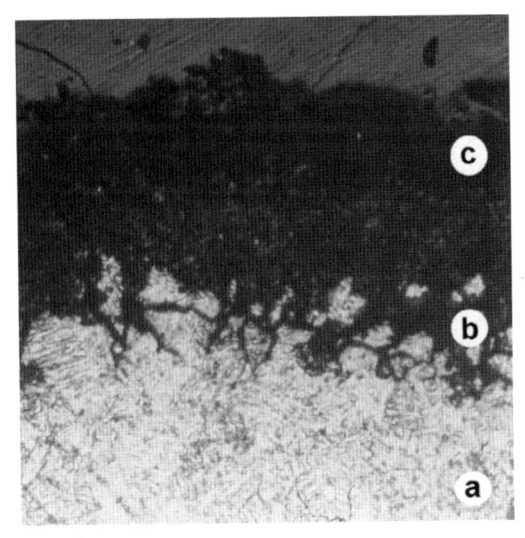

FIGURE 9.3 Micrograph (×100) on a zone from the transverse section of sample 1 after electrolysis; electrochemical attack with 10% chromium(VI) oxide: (a) metal, (b) zone of intergrain corrosion, and (c) layer of oxides.

from the anode corrosion. The last column in the table represents the accumulation of iron in the aluminium during electrolysis with inert anodes.

For the highly alloyed steels, the samples with a larger content of nickel had better resistance to corrosion. For samples 1–3, one can see that while the nickel content is increasing, the rate of corrosion is decreasing (Table 9.2) as well as the content of iron in the aluminium (Table 9.3).

We note a surprising behaviour of the low alloyed steels. To date, everybody believed that only noble metals could resist electrolysis in the cryolite–alumina melts. At the start of our work, the most corrosion resistant alloys were investigated, which are obviously expensive. Recent literature reports [1–6], as well as our

FIGURE 9.4 Macroscopic aspect of sample 2 after 4 h of electrolysis.

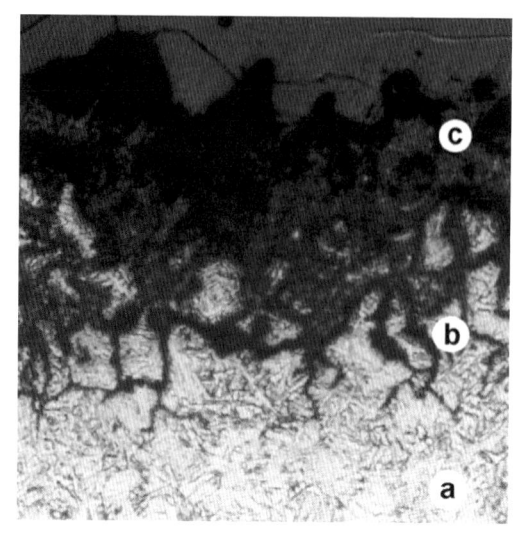

FIGURE 9.5 Micrograph (×100) on a zone from the transverse section of sample 2 after electrolysis; electrochemical attack with 10% chromium(VI) oxide: (a) metal, (b) zone of intergrain corrosion, and (c) layer of oxides.

experiments [7, 8], have shown that low alloyed steels (cheap materials) exist which can be used with success to produce inert anodes. Surprisingly, some of these steels seem to be even better than some types of stainless steel. The secret is that during electrolysis, those steels are covered by layers of oxides, which coat the metal and protect it from further oxidation.

Metallographic optical microscopy studies made on those samples have shown that, on the low alloyed steel, a layer of intergrain corrosion is formed on the surface, having a fine structure and a depth of 40–60 μm, and on the external zone, the one in direct contact with the electrolyte, a layer of oxides exists having a thickness of 120–170 μm. In the case of high alloyed steels, a higher content of nickel determines

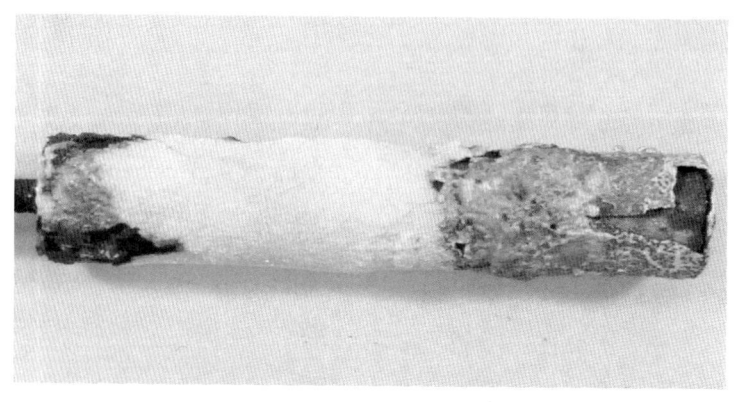

FIGURE 9.6 Macroscopic aspect of sample 3 after 4 h of electrolysis.

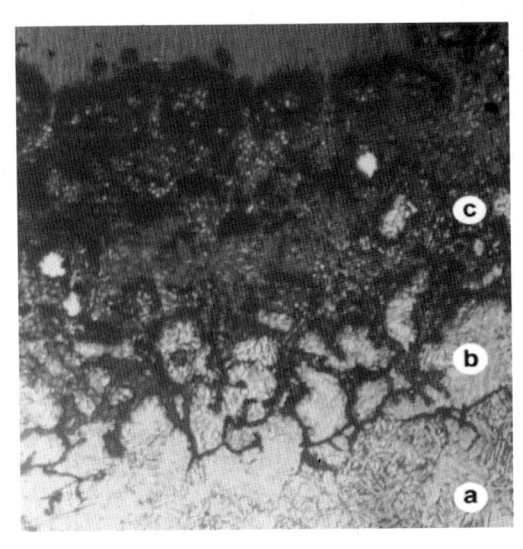

FIGURE 9.7 Micrograph (×100) on a zone from the transverse section of sample 3 after electrolysis; electrochemical attack with 10% chromium(VI) oxide: (a) metal, (b) zone of intergrain corrosion, and (c) layer of oxides.

an increase of the layer's thickness, both in the intergrain corrosion zone and in the oxide zone. A clear difference regarding the behaviour of low alloyed steels compared with high alloyed steels is given by the structure of the intergrain corrosion layer. A finer structure of this layer stops the removal of the oxide layer from the anode surface. For the high alloyed steel, because of the structure of the intergrain corrosion layer, parts of the external layer are removed from the anode surface and are passed into the electrolyte.

The maximum content of iron in commercial aluminium is stated at the moment to be 0.35% [9]. Aluminium obtained by us in the case of anodes 2–5 has an iron content below this limit. The electrolysis time in our laboratory experiments was 4 h, a very

FIGURE 9.8 Macroscopic aspect of sample 4 after 4 h of electrolysis.

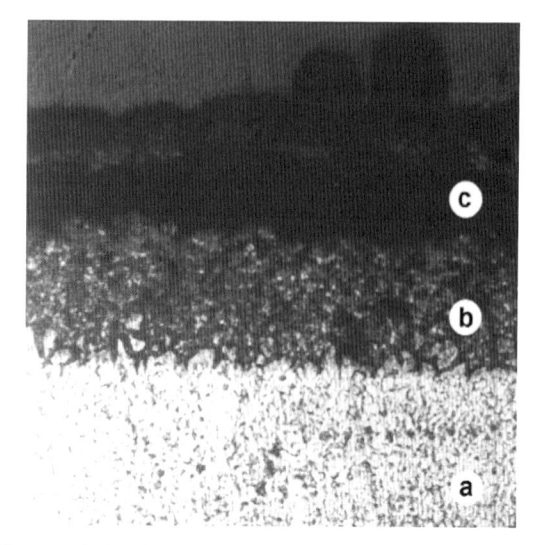

FIGURE 9.9 Micrograph (×100) on a zone from the transverse section of sample 2 after electrolysis; electrochemical attack with 10% chromium(VI) oxide: (a) metal, (b) zone of intergrain corrosion, and (c) layer of oxides.

short time. But, in industrial cells during electrolysis, the quantity of aluminium deposited on the cathode increases and thus the probable content of impurities will not increase as much. Moreover, the current efficiency for aluminium deposition in a laboratory scale installation, such as ours, hardly ever goes over 50%, while in industrial cells the cathodic current efficiency is over 90%.

TABLE 9.2 Diameter Variation of the Anode Bars During Electrolysis

Anode Number	Diameter Variation/mm
1	1.1
2	0.6
3	0.0
4	0.0
5	0.3

TABLE 9.3 Iron Content (wt %) in Cathodic Aluminium After 4 h of Electrolysis[a]

Al Sample	% Fe	%Fe$_{total}$ − %Fe$_{initial}$
1	0.65	0.51
2	0.35	0.21
3	0.15	0.01
4	0.15	0.01
5	0.20	0.05

[a] The initial content in the added aluminium is 0.14 wt %.

In conclusion, our results regarding the use of inert steel anodes for aluminium electrolysis are promising, showing that problems could be solved in the near future.

REFERENCES

1. Sekhar, J. A., Liu, J., Deng, H., Duruz, J. J., and de Nora, V., in *Light Metals*, B. J. Welch (Ed.), TMS, Warrendale, PA, 1998, p. 597.
2. Duruz, J. J. and de Nora, V., MOLTECH, International Patent, WO 00/6800, February 2000.
3. Duruz, J. J., de Nora, V., and Crotaz, O., MOLTECH, International Patent, WO OO/6802, February 2000.
4. Duruz, J. J., de Nora, V., and Crotaz, O., MOLTECH, International Patent, WO 00/6803, February 2000.
5. Crotaz, O. and Duruz, J. J., MOLTECH, International Patent, WO 00/6804, February 2000.
6. de Nora, V. and Duruz, J. J., MOLTECH, International Patent, WO 00/6805, February 2000.
7. Galasiu, I., Galasiu, R., Nicolescu, C., and Istodor, V., in *Proceedings of the International Symposium on Ionic Liquids*, Carry le Rouet, France, June 2003, p. 179.
8. Galasiu, I., Galasiu, R., Nicolescu, C., Vlaicu, G., and Istodor, V., in *Proceedings of the 12th International Symposium on Aluminum Electrowinning*, Bratislava, Slovakia, 2003, p. 233.
9. Galasiu, I. and Galasiu, R., *Aluminum Electrolysis*, Technical Publishing House, Bucharest, 2000.

10 The Behaviour of Phosphorus and Sulfur in Cryolite-Alumina Melts: Thermodynamic Considerations

I. GALASIU and R. GALASIU
Institute of Physical Chemistry, "I. G. Murgulescu" of The Romanian Academy, Bucharest, Romania

C. NICOLESCU
Faculty of Industrial Chemistry, University Polytechnia of Bucharest, Bucharest, Romania

J. THONSTAD and G. M. HAARBERG
Department of Material Sciences Engineering, Norwegian University of Science and Technology, Trondheim, Norway

Abstract

The behaviour of impurities containing phosphorus and sulfur in cryolite–alumina melts was studied. Mixtures containing $Na_3AlF_6 + AlF_3 + CaF_2 + Al_2O_3$ in different proportions were prepared. The mixtures were melted in a graphite crucible together with carbon dust, metallic aluminium, and the impurity (phosphorus was added as Na_3PO_4 and sulfur as Na_2SO_4). The phosphorus and the sulfur contents were determined in fractions separated from each sample after solidification. The quantity of adsorbed impurities depends on the electrolyte composition, and the presence/absence of metallic aluminium in the electrolyte. The HSC Chemistry for Windows—Outokumpu program was used to make thermodynamic calculations regarding the existence of phosphorus or sulfur compounds and their possible chemical reactions. Phase stability diagrams have been drawn for the conditions of the experiments, and Gibbs energy calculations have been made. The presence of metallic aluminium in the studied mixtures was analysed from this thermodynamic perspective.

Molten Salts and Ionic Liquids: Never the Twain? Edited by Marcelle Gaune-Escard and Kenneth R. Seddon
Copyright © 2010 John Wiley & Sons, Inc.

10.1 INTRODUCTION

The production of aluminium with low contents of impurities is a continuous challenge. In addition to the metal quality perspective, the accumulation of certain impurities (such as phosphorus) in the electrolyte is known to have a significant deleterious effect on current efficiency. Also, the environmental perspective should be considered—impurities like sulfur representing a serious pollution hazard in aluminium electrolysis [1]. Sulfur has the same deleterious influence on the current efficiency when inert anodes are used for the aluminium electrolysis [2].

In recent years, research workers in the primary aluminium field have shown constant interest in the behaviour of phosphorus [3–5] and sulfur [5, 6] in the cryolite–alumina melts, and also during electrolysis in the Hall–Héroult process.

10.2 EXPERIMENTAL

Mixtures containing Na_3AlF_6–AlF_3–CaF_2–Al_2O_3 in different proportions were mixed mechanically at room temperature, together with carbon powder from an aluminium factory and impurity (phosphorus added as Na_3PO_4 and sulfur added as Na_2SO_4). The samples, mixed with the respective impurity, were introduced in a graphite crucible that was placed in a furnace heated to 1000 °C. A vertical muffle furnace was used and argon was passed during the experiments. After melting, the samples were held at a constant temperature of 970 °C for 1 h. For some samples, metallic aluminium was added to the molten electrolyte; the holding time in these cases was also 1 h.

Several compositions were studied, with phosphorus as impurity, added as Na_3PO_4. Different concentrations of Al_2O_3 and also different molar ratios $NaF:AlF_3$ were used. CaF_2 was added in order to reduce the melting temperature of the mixtures. Samples with and without metallic aluminium addition were studied, to see the influence of metallic aluminium on the adsorption of impurities on the surface of the carbon powder. Some results have been published earlier [5].

After the holding time, the crucibles were removed from the furnace, covered, and allowed to cool down to room temperature. The solidified samples were examined, and a separation into several fractions was noticed:

- Fraction F1, "electrolyte," having a low content of carbon
- Fraction F2, "carbon dust" (industrial quality)
- Fraction F3, metallic aluminium

10.3 RESULTS AND DISCUSSION

By the use of the program HSC Chemistry for Windows—Outokumpu, the thermodynamic possibility of the presence of phosphorus compounds in the melts was

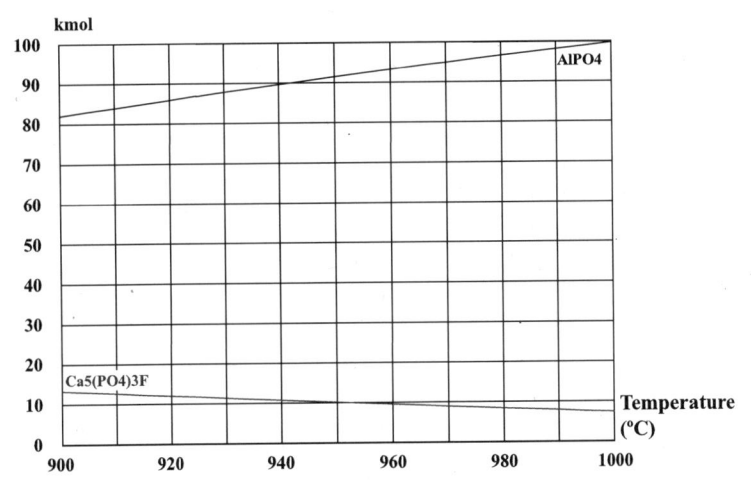

FIGURE 10.1 A phase stability diagram for a sample without addition of metallic aluminium, in the 0–100 kmol range.

examined. For the calculations, the experimental conditions were set as initial data in the program, and the activity coefficients were supposed to be constant and equal to unity. The obtained phase stability diagrams were interpreted and conclusions were drawn. In the diagrams the various compounds are supposed to be in the pure state, liquid or solid.

In Figures 10.1–10.4, we present some phase stability diagrams obtained for samples with the following compositions (wt %): 81.1% Na_3AlF_6, 6.6% AlF_3, 4.7%

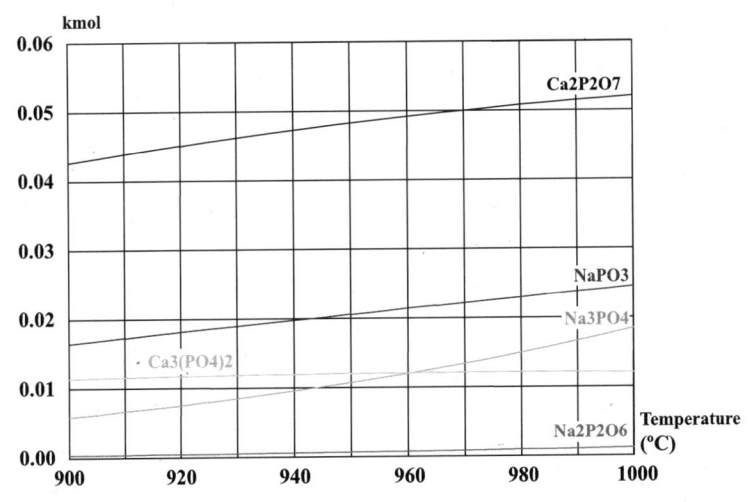

FIGURE 10.2 Phase stability diagram for a sample without addition of metallic aluminium, in the 0–0.06 kmol range.

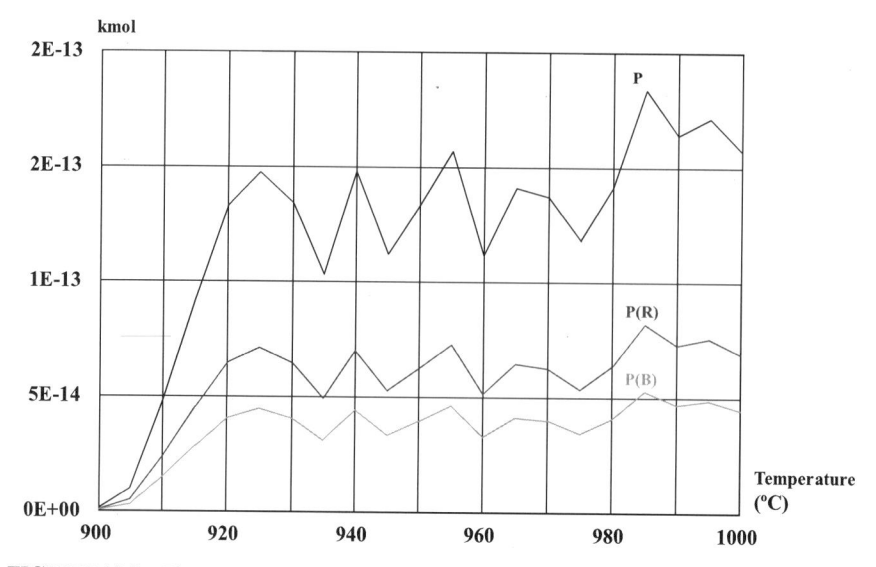

FIGURE 10.3 Phase stability diagram for a sample without addition of metallic aluminium, in the $0–2 \times 10^{-13}$ kmol range. P(R)—red phosphorus; P(B)—black phosphorus.

CaF_2, 1.9% Al_2O_3, 4.7% C + 1% Na_3PO_4, without and with metallic aluminium addition, respectively. The calculations of thermodynamic equilibria were made for the 900–1000 °C temperature range. The quantities were multiplied by 10^7 (proportions remained constant) to obtain better clarity for the diagrams.

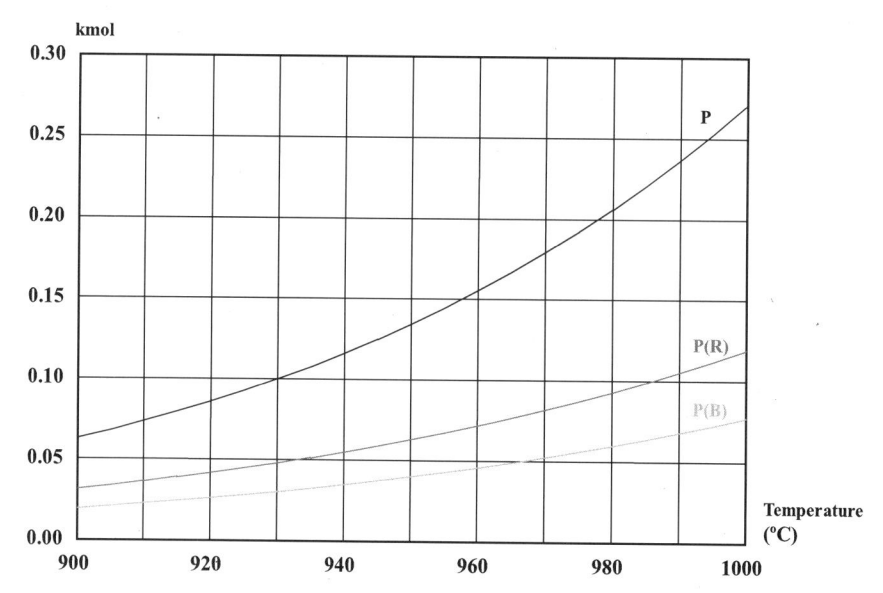

FIGURE 10.4 Phase stability diagram for a sample with addition of metallic aluminium, in the 0–0.3 kmol range. P(R)—red phosphorus; P(B)—black phosphorus.

RESULTS AND DISCUSSION

The thermodynamic calculations performed showed a rather different behaviour for samples with and without metallic aluminium addition, mainly regarding the nature and quantity of phosphorus compounds that may form.

The following observations were considered for samples having the compositions as mentioned previously, without metallic aluminium addition:

1. There is a thermodynamic possibility for the formation and existence in the melt of $Ca_5(PO_4)_3F$ and $AlPO_4$ in quantities of the same order.
2. The existence in the melt of dissolved compounds of phosphorus with calcium and sodium, such as $Ca_3(PO_4)_2$, $Ca(PO_3)_2$, $Ca_2P_2O_7$, $NaPO_3$, Na_3PO_4, and $Na_2P_2O_6$, is thermodynamically possible, but in quantities 10^4 lower than the ones from No. 1.
3. The formation at equilibrium, and in very low concentrations, of the compounds P_2O_3, P_2O_5, P_4O_6, and P_4O_{10} is also possible.
4. The thermodynamic possibility for the formation of a phosphorus compound with carbon was not detected.

The following reactions involving phosphorus compounds were considered:

$$5\,CaF_2 + 3\,Na_3PO_4 = Ca_5(PO_4)_3F + 9\,NaF \tag{10.1}$$

$$\Delta G(1000\,°C) = -19.269\,\text{kcal}$$

$$5\,CaF_2 + 2\,Na_3PO_4 = Ca_3(PO_4)_2 + 6\,NaF \tag{10.2}$$

$$\Delta G(900\,°C) = -0.942\,\text{kcal}$$

For samples with aluminium addition (15.8 g Al to 212 g sample) and a wt% composition similar to the one discussed earlier (81.1% Na_3AlF_6, 6.6% AlF_3, 4.7% CaF_2, 1.9% Al_2O_3, 4.7% C + 1% Na_3PO_4, the following conclusions were reached (the calculations were made under the same conditions):

1. The following phosphorus compounds may be present in the melt: elemental phosphorus (red, black) and small quantities of $AlPO_4$, Ca_3P_2, and P_2O_3.
2. The thermodynamic possibility of a phosphorus compound with carbon was not detected in these systems.

This thermodynamic study showed that, for samples with the given composition and temperature range, several types of phosphorus compounds may form in the melts without metallic aluminium addition. These substances remain dissolved in the electrolyte, and after solidification (by cooling down to room temperature), phosphorus is concentrated in fraction F1 ("electrolyte," having a low content of carbon). In samples with metallic aluminium addition, gaseous phosphorus forms preferentially. The gases, with their natural tendency to escape from the melt, may adsorb on carbon particles. Thus it is supposed that higher quantities of phosphorus will be found in fraction F2 ("carbon dust"). Results presented in a previous paper [5] confirm this assumption.

When comparing Figures 10.3 and 10.4, it may be noted that in samples with metallic aluminium addition, gaseous elemental phosphorus forms in far higher quantities than in samples with the same composition, but without aluminium addition. In the latter samples, phosphorus compounds remain preferentially in the melt. This observation is based on the distribution coefficients for carbon dust/electrolyte calculated from experimental data and presented in a previous paper (higher distribution coefficients in samples with metallic aluminium addition) [5].

In the following, a thermodynamic study is presented for samples with composition (wt %) close to the electrolyte in industrial aluminium electrolysis and containing sulfur as an impurity, added as Na_2SO_4. By the use of the HSC Chemistry for Windows program, the calculations of thermodynamic equilibria were made for the 900–1000 °C temperature range. The quantities were multiplied by 10^7 (proportions remained constant) to obtain better clarity of the diagrams. The purpose of this calculation was to find sulfur compounds that may exist at equilibrium, in the given temperature range, and their possible chemical reactions (showing a negative Gibbs energy).

Samples with the following proportions (wt %) were studied: 72.3% Na_3AlF_6, 10.7% AlF_3, 4.5% CaF_2, 1.8% Al_2O_3, 4.5% C, and 6.2% Na_2SO_4, obtained by the experimental procedure described earlier. A comparison was made between samples with this composition (Fig. 10.5) and samples having the same proportions (wt %), but with metallic aluminium addition (16 g Al to 224 g sample; Fig. 10.6).

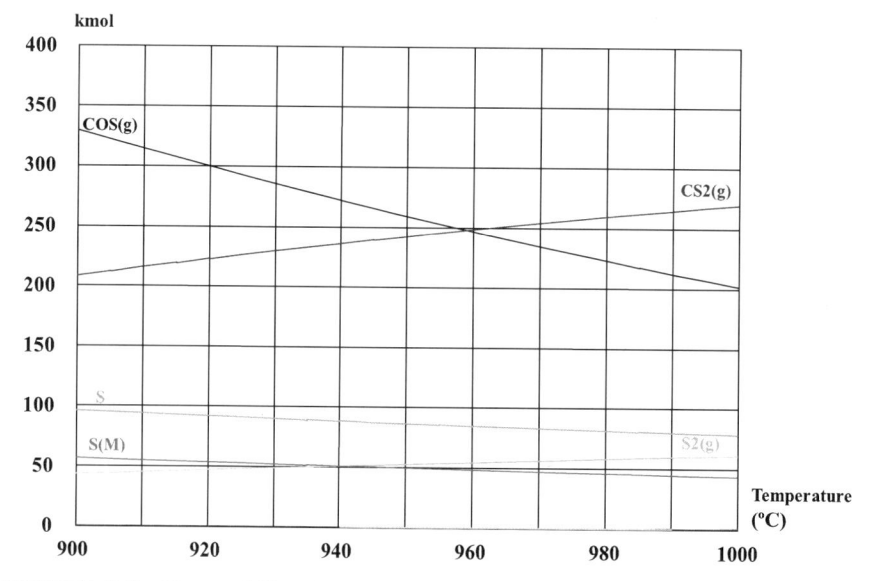

FIGURE 10.5 Phase stability diagram for a sample without addition of metallic aluminium, in the 0–400 kmol range. S(M)—monoclinic sulfur.

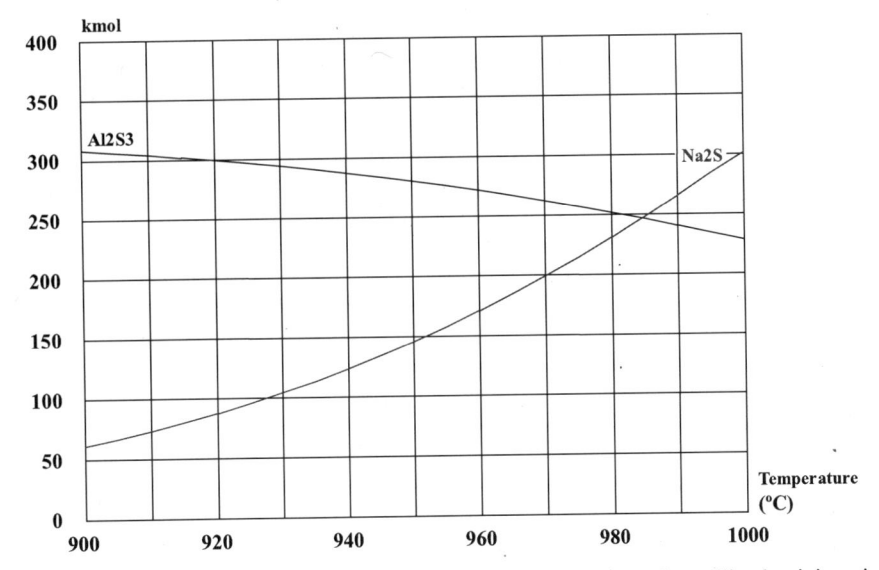

[Figure 10.6: Phase stability diagram showing Al2S3 and Na2S curves vs Temperature (°C) from 900 to 1000, with kmol on y-axis from 0 to 400.]

FIGURE 10.6 Phase stability diagram for a sample with addition of metallic aluminium, in the 0–400 kmol range.

It was concluded that the following sulfur compounds may exist at equilibrium, in samples without metallic aluminium addition, in the 900–1000 °C temperature range:

1. Sulfur compounds with carbon, oxygen, and fluorine: CS_2, COS, CS, S_2O, SO, SO_2, SF, S_2F_2, SF_2, SO_3, SOF, SOF_2, SO_2F_2, SF_3, SF_4, and SF_6.
2. Elemental sulfur as S, S_2, S_3, S_4, S_5, S_6, S_7, and S_8.
3. Sulfur compounds with carbon, sodium, calcium, and aluminium: Na_2SO_3, $Na_2S_2O_3$, AlS, AlS_2, $CaSO_3$, Na_2SO_4, $CaSO_4$, CaS, $(AlS)_2$, Al_2S, and Al_2S_3.

The following reactions that are thermodynamically possible have been considered:

$$Na_2SO_4 + C = Na_2SO_3 + CO(g) \tag{10.3}$$
$$\Delta G(1000\,°C) = -6.571\,\text{kcal}$$

$$2\,Na_2SO_4 + 5\,C = Na_2S_2O_3 + 5\,CO(g) + 2\,Na(g) \tag{10.4}$$
$$\Delta G(1000\,°C) = -0.942\,\text{kcal}$$

$$2\,Na_2SO_4 + 8\,C = S_2(g) + 5\,CO(g) + 4\,Na(g) \tag{10.5}$$
$$\Delta G(1000\,°C) = -13.473\,\text{kcal}$$

$$3\,Na_2SO_4 + 12\,C = S_3(g) + 12\,CO(g) + 6\,Na(g) \qquad (10.6)$$
$$\Delta G(1000\,°C) = -10.381\,\text{kcal}$$

$$4\,Na_2SO_4 + 16\,C = S_4(g) + 16\,CO(g) + 8\,Na(g) \qquad (10.7)$$
$$\Delta G(1000\,°C) = -10.413\,\text{kcal}$$

$$5\,Na_2SO_4 + 20\,C = S_5(g) + 20\,CO(g) + 10\,Na(g) \qquad (10.8)$$
$$\Delta G(1000\,°C) = -7.114\,\text{kcal}$$

$$6\,Na_2SO_4 + 24\,C = S_6(g) + 24\,CO(g) + 12\,Na(g) \qquad (10.9)$$
$$\Delta G(1000\,°C) = -11.642\,\text{kcal}$$

$$7\,Na_2SO_4 + 28\,C = S_7(g) + 28\,CO(g) + 14\,Na(g) \qquad (10.10)$$
$$\Delta G(1000\,°C) = -13.699\,\text{kcal}$$

$$8\,Na_2SO_4 + 32\,C = S_8(g) + 32\,CO(g) + 16\,Na(g) \qquad (10.11)$$
$$\Delta G(1000\,°C) = -12.434\,\text{kcal}$$

The phase stability diagrams (Figs. 10.5 and 10.6) showed important differences between samples with and without addition of metallic aluminium. In the presence of metallic aluminium, sodium, aluminium sulfides (Na_2S, NaS, Na_2S_2, NaS_2, Na_2S_3, Al_2S_3), and elemental sulfur (S) may form; compounds of sulfur with carbon, oxygen, and fluorine were not found.

10.4 CONCLUSION

Several phase stability diagrams for the 900–1000 °C temperature range have been presented (other diagrams were also used) and Gibbs energy (ΔG) calculations for possible chemical reactions have been performed. We concluded that some phosphorus compounds and sulfur compounds may exist at equilibrium in the studied systems. Also, several chemical reactions gave a negative Gibbs energy, thus showing that they are thermodynamically possible. All calculations were performed with the use of the HSC Chemistry for Windows— Outokumpu program. The limitations of the program calculations should be considered for a further study.

REFERENCES

1. Thonstad, J., Fellner, P., Haarberg, G. M., Hives, J., Kvande, H., and Sterten, A., *Aluminum Electrolysis. Fundamentals of the Hall–Héroult Process*, 3rd ed., Aluminum-Verlag, Dusseldorf, 2001.
2. La Camera et al., U.S. Patent, 2004/002786 A1.

3. Daněk, V., Chrenková, M., Silný, A., Haarberg, G. M., and Stas, M., *Can. Met. Q.* **38**, 149 (1999).
4. Haarberg, G. M., Thisted, E., Thonstad, J., and Haugland, E., in *Proceedings of 11th International Aluminum Symposium*, Trondheim, September 2001, p. 243.
5. Galasiu, I., Galasiu, R., Nicolescu, C., Thonstad, J., and Haarberg, G. M., in *CD Proceedings of the 12th International Aluminum Symposium*, Slovakia, September 2003.
6. Ambrova, M., Fellner, P., Gabcova, J., and Thonstad, J., in *CD Proceedings of the 12th International Aluminum Symposium*, Slovakia, September 2003.

11 Ionic Liquid–Ionic Liquid Biphasic Systems

DIRK GERHARD, FRIEDRICH FICK, and PETER WASSERSCHEID

Lehrstuhl für Chemische Reaktionstechnik, Universität Erlangen–Nürnberg, Erlangen, Germany

Abstract

New examples of ionic liquid–ionic liquid biphasic systems are presented. For selected combinations, the ion distribution in the two liquid phases has been determined, and first hints for the use of these systems in separation applications have been obtained.

11.1 INTRODUCTION

Ionic liquids are low melting (<100 °C) salts that represent a new class of non-molecular, ionic solvents [1–3]. Nowadays, ionic liquids are widely applied as alternatives to classical organic solvents in chemical transformations and separation technologies. In past years, the range of known and available ionic liquids has been expanded, so that many different candidates are accessible and even commercially available today.

One of the key aspects that has attracted scientists to the application of ionic liquids concerns their highly tunable, and often unusual, solubility properties [4]. Ionic liquid-organic biphasic systems have proved to be highly versatile, especially in catalysis [5–7] and in extraction technologies [8–10]. In catalysis, the liquid–liquid biphasic operation using ionic liquids as the catalyst immobilisation phase allows not only the recycling of the precious metal complex dissolved but also the recycling of the ionic liquid itself. Moreover, the ionic liquid–organic miscibility gap provides a certain ionic environment for the catalyst, even in those cases where only a small

Molten Salts and Ionic Liquids: Never the Twain? Edited by Marcelle Gaune-Escard and Kenneth R. Seddon
Copyright © 2010 John Wiley & Sons, Inc.

TABLE 11.1 Examples of Ionic Liquid Combinations (IL1 and IL2) Showing a Miscibility Gap at 20 °C

Number	IL1	IL2
1	[C$_2$mim][EtOSO$_3$]	[N$_{1888}$][NTf$_2$]
2	[C$_1$mim][MeOSO$_3$]	[N$_{1888}$][NTf$_2$]
3	[C$_2$mim][SCN]	[N$_{1888}$][NTf$_2$]
4	[C$_2$mim][N(CN)$_2$]	[N$_{1888}$][NTf$_2$]
5	[SMe$_3$][N(CN)$_2$]	[N$_{1888}$][NTf$_2$]

amount of ionic liquid is used in the presence of a large amount of organic reaction mixture.

First examples of mutually immiscible ionic liquids were recently published [11] while the systems reported here were still under detailed investigation in our laboratories. Arce et al. [11] described the phase behaviour of mixtures containing 1-alkyl-3-methylimidazolium and trihexyltetradecylphosphonium based ionic liquids. They determined temperature–composition diagrams as well as ion distributions for those systems. In this chapter, we report additional examples of mixtures of two ionic liquids that show pronounced miscibility gaps (and thus forming ionic liquid–ionic liquid biphasic systems) without any organic solvent being present. The systems presented here were discovered in our laboratories in context with systematic studies aimed at the modification of physicochemical properties by mixing two or more different ionic liquids. Table 11.1 displays some systems that show a miscibility gap at 20 °C. The upper phase is always methyl-trioctylammonium bis(trifluoromethylsulfonyl)amide ([N$_{1888}$][NTf$_2$]), an ionic liquid that has been reported to be hydrophobic and of relatively low polarity in nature.

In order to further elucidate the chemical nature of the two phases, ^{1}H NMR studies were carried out in which the amounts of the specific ions in the two individual phases were quantitatively determined (see experimental section). The amount and molar ratio of the [NTf$_2$]$^-$ ion (which does not contain protons) were calculated based on the necessary charge compensation in each of the two liquid phases. The results for the system [N$_{1888}$][NTf$_2$]/[C$_2$mim][EtOSO$_3$] are given in Table 11.2.

The data indicate that for this specific case the phase discrimination is more clearly developed on the cation side compared to the anion side, with the lower phase hardly containing any [N$_{1888}$]$^+$ ions. On the anion side, the denser phase shows surprisingly an excess of ethylsulfate, whereas in the phase of lower density, the [NTf$_2$]$^-$ is the dominating ion. This result indicates that the specific interaction between the [N$_{1888}$]$^+$ ion and the [NTf$_2$]$^-$ ion overcompensated the density driven tendency of the [NTf$_2$]$^-$ ion to form a part of denser phase.

To illustrate this important aspect further, the densities of all pure components that can form from the ions [N$_{1888}$]$^+$, [C$_2$mim]$^+$, [NTf$_2$]$^-$, and [EtOSO$_3$]$^-$ and the densities found for the two immiscible liquid phases formed at 20 °C are compared in Table 11.3.

INTRODUCTION

TABLE 11.2 Composition[a] of the Two Liquid Phases Formed in the System [N$_{1888}$][NTf$_2$]/[C$_2$mim][EtOSO$_3$] at 20 °C

	X_{cation}	X_{anion}	X_{total}
Upper phase			
[C$_2$mim]$^+$	0.24		0.12
[N$_{1888}$]$^+$	0.76		0.38
[EtOSO$_3$]$^-$		0.34	0.17
[NTf$_2$]$^-$		0.66	0.33
Lower phase			
[C$_2$mim]$^+$	0.98		0.49
[N$_{1888}$]$^+$	0.02		0.01
[EtOSO$_3$]$^-$		0.66	0.33
[NTf$_2$]$^-$		0.34	0.17

[a] X_{ion} = molar ratio of the specific ion with respect to all cations/anions in the specific phase; X_{total} = molar ratio of the specific ion with respect to all ions present in the specific phase.

Using the determined densities of the two immiscible phases and the volumes of both phases, it was also possible to close the mass balance for the two ionic liquids applied in the experiment, proving the reliability of the data obtained from the NMR measurements in combination with the calculated [NTf$_2$]$^-$ ion content (see above).

In another set of experiments, a small amount of tetrahydrofuran (10 mol % with respect to the overall molar amount of ions) was added to the mixture, to see how the phase behaviour of the two ionic liquids would change. In order to determine the partition coefficient of the tetrahydrofuran between the two phases, a toluene insert was used as internal standard in the respective ^{1}H NMR measurements. The partition coefficient was measured to be K (tetrahydrofuran in upper phase/tetrahydrofuran in lower phase) = 1.3, indicating that the major part of the added tetrahydrofuran was found in the upper phase. The compositions of the different phases with added tetrahydrofuran were determined applying again ^{1}H NMR spectroscopy

TABLE 11.3 Densities of the Pure Ionic Liquids Formed from the Ions [N$_{1888}$]$^+$, [C$_2$mim]$^+$, [NTf$_2$]$^-$, and [EtOSO$_3$]$^-$ and of the Separated Phases of the System [N$_{1888}$][NTf$_2$]/[C$_2$mim][EtOSO$_3$] at 20 °C

System	ρ/g cm^{-3}
Upper phase	1.090
Lower phase	1.298
[C$_2$mim][EtOSO$_3$]	1.230[a] [12]
[N$_{1888}$][NTf$_2$]	1.106 [13]
[C$_2$mim][NTf$_2$]	1.559 [14]
[N$_{1888}$][EtOSO$_3$]	Solid

[a] Measured at 25 °C.

TABLE 11.4 Composition[a] of the Two Immiscible Phases Formed by the System [N$_{1888}$][NTf$_2$]/[C$_2$mim][EtOSO$_3$] with 10 mol % THF added at 20 °C

	X_{cation}	X_{anion}	X_{total}
Upper phase			
[C$_2$mim]$^+$	0.30		0.13
[N$_{1888}$]$^+$	0.70		0.31
[EtOSO$_3$]$^-$		0.40	0.18
[NTf$_2$]$^-$		0.60	0.26
THF			0.12
Lower phase			
[C$_2$mim]$^+$	0.90		0.48
[N$_{1888}$]$^+$	0.01		0.01
[EtOSO$_3$]$^-$		0.76	0.37
[NTf$_2$]$^-$		0.24	0.11
THF			0.03

[a]X_{ion} = molar ratio of the specific ion with respect to all cations/anions in the specific phase; X_{total} = molar ratio of the specific ion with respect to all ions present in the specific phase.

(see Table 11.4). Obviously, the general trend is not strongly influenced by the addition of tetrahydrofuran. The upper phase is still [N$_{1888}$]$^+$ and [NTf$_2$]$^-$ rich, while the denser phase consists almost entirely of the [C$_2$mim]$^+$ ion with an excess of the [EtOSO$_3$]$^-$ anion.

In an attempt to expand the system to molecules of interest (e.g., catalytic applications), we also studied the phase behaviour of 1-hexene and 1-octene in the system [C$_2$mim][EtOSO$_3$]/[N$_{1888}$][NTf$_2$] (10 mol % to the overall molar amount of ions). Again, toluene was used as an internal standard for the corresponding ^{1}H NMR experiments. The partition coefficients were measured to be K (upper phase/lower phase) = 13.8 for oct-1-ene and K (upper phase/lower phase) = 6.8 for hex-1-ene, respectively, indicating that almost the whole added olefin is found in the upper, less polar phase. Remarkably, a relatively big difference in partition coefficient was found for these two olefins of relatively similar polarity (composition of phases: see Tables 11.5 and 11.6).

A question of high relevance for both fundamental and practical aspects is the upper solution temperature for the systems under investigation. To determine this, all five systems displayed in Table 11.1 were heated to 120 °C and left there for 24 h to fully equilibrate. Only two mixtures, [C$_1$mim][MeOSO$_3$]/[N$_{1888}$][NTf$_2$] and [SMe$_3$][N(CN)$_2$]/[N$_{1888}$][NTf$_2$], still showed two immiscible phases under these conditions. The upper temperatures of complete miscibility of the other mixtures were determined by slowly heating the systems from 60 to 120 °C, and the temperature range at which the formation of one clear phase was observed was determined (Table 11.7).

It should also be mentioned that addition of very nonpolar organic solvents leads to the formation of unusual triphasic systems. For example, the addition of cyclohexane to the mixture of [N$_{1888}$][NTf$_2$]/[C$_2$mim][EtOSO$_3$] results in such an ionic liquid–ionic liquid–organic system.

TABLE 11.5 Phase Composition of [C₂mim][EtOSO₃]/[N₁₈₈₈][NTf₂]/Hex-1-ene

	X_{cation}	X_{anion}	X_{total}
Upper phase			
[C₂mim]⁺	0.19		0.091
[N₁₈₈₈]⁺	0.81		0.400
[EtOSO₃]⁻		0.29	0.144
[NTf₂]⁻		0.71	0.347
Hex-1-ene			0.018
Lower phase			
[C₂mim]⁺	0.98		0.489
[N₁₈₈₈]⁺	0.02		0.010
[EtOSO₃]⁻		0.64	0.340
[NTf₂]⁻		0.36	0.159
Hex-1-ene			0.002

In conclusion, we have demonstrated that selected ionic liquids show pronounced miscibility gaps, even at elevated temperatures when mixed with each other. This, once more, demonstrates the wide scope of the ionic liquid concept. Indeed, the concept includes ion combinations that are very different in their physicochemical properties—even to an extent that they do not completely mix. The composition of the different phases, determined by proton NMR spectroscopy, revealed interesting effects about the formation of preferred ion–ion clustering. The hydrophobic anion [NTf₂]⁻ is preferably found in the upper phase with the [N₁₈₈₈]⁺ ion, despite the fact that its highly fluorous nature implies a relatively high density of this ion. We anticipate that multiphasic ionic liquid–ionic liquid systems will find applications in both the areas of multiphase catalysis and separation technologies.

TABLE 11.6 Phase Composition of [C₂mim][EtOSO₃]/[N₁₈₈₈][NTf₂]/Oct-1-ene

	X_{cation}	X_{anion}	X_{total}
Upper phase			
[C₂mim]⁺	0.21		0.103
[N₁₈₈₈]⁺	0.79		0.389
[EtOSO₃]⁻		0.29	0.143
[NTf₂]⁻		0.71	0.349
Oct-1-ene			0.016
Lower phase			
[C₂mim]⁺	0.98		0.489
[N₁₈₈₈]⁺	0.02		0.010
[EtOSO₃]⁻		0.64	0.318
[NTf₂]⁻		0.36	0.181
Oct-1-ene			0.002

TABLE 11.7 Upper Solution Temperature for Some Ionic Liquid–Ionic Liquid Biphasic Mixtures

Number	IL1	IL2	Upper Solution Temperature/°C
1	[C$_2$mim][EtOSO$_3$]	[N$_{1888}$][NTf$_2$]	95–100
2	[C$_1$mim][MeOSO$_3$]	[N$_{1888}$][NTf$_2$]	>120
3	[C$_2$mim][SCN]	[N$_{1888}$][NTf$_2$]	105–110
4	[C$_2$mim][N(CN)$_2$]	[N$_{1888}$][NTf$_2$]	115–120
5	[SMe$_3$][N(CN)$_2$]	[N$_{1888}$][NTf$_2$]	>120

11.2 EXPERIMENTAL

11.2.1 Chemicals

The ionic liquids used in this research—[C$_2$mim][EtOSO$_3$], [C$_2$mim][N(CN)$_2$], [N$_{1888}$][NTf$_2$] (Solvent Innovation), tetrahydrofuran (Acros Organics), hex-1-ene (Aldrich), oct-1-ene (Aldrich), and toluene (Aldrich)—were purchased from the companies given in parentheses. The ionic liquid [SMe$_3$][N(CN)$_2$] was synthesised following the synthetic approaches reported earlier [15]. [C$_1$mim][MeOSO$_3$] was obtained by reaction of 1-methylimidazole with equimolar amounts of dimethyl sulfate under neat conditions. [C$_2$mim][SCN] was synthesised by anion exchange starting from [C$_2$mim]Cl and Ag[SCN]. All materials were characterised by ^{1}H and ^{13}C NMR spectroscopy and were used without further purification.

11.2.2 Measurements

^{1}H and ^{13}C NMR spectra were recorded on a JEOL ECX 400 MHz spectrometer, in d_6-DMSO, or without solvent (see below). Density was measured at room temperature in a BLAUBRAND pyknometer with a defined volume of 5.330 cm^3 according to DIN ISO 3507.

11.2.3 Determination of Phase Compositions by NMR

Equimolar amounts of the two ionic liquids were mixed for 3 h in a Schlenk tube using a magnetic stirring bar. Afterwards, the mixture was kept at ambient conditions for 7 days to reach equilibrium. Samples of each phase were taken with a syringe, followed by addition of dry d_6-DMSO, and an ^{1}H NMR spectrum was recorded.

The molar ratios of [C$_2$mim]$^+$, [N$_{1888}$]$^+$, and [EtOSO$_3$]$^-$ were determined by using the H_3C-N signal of [C$_2$mim]$^+$ as an internal reference. The amount of the [NTf$_2$]$^-$ anion was calculated based on the charge compensation in each phase.

11.2.4 Determination of Partition Coefficients by NMR Spectroscopy

Equimolar amounts of the two ionic liquids and the third component under investigation (tetrahydrofuran, 1-octene, or 1-hexene) were mixed for 3 h in a Schlenk tube, using a magnetic stirring bar. Afterwards, the mixture was kept under ambient conditions for 7 days to reach equilibrium. Samples of each phase were taken with a syringe, followed by an ^{1}H NMR experiment with a coaxial insert containing toluene.

The partition coefficient of the third component was determined by using the H_3C signal of toluene as an internal standard. The phase compositions were measured using the described method.

11.2.5 Molar Balance of the Biphasic System [C$_2$mim][EtOSO$_3$]/[N$_{1888}$][NTf$_2$]

Equimolar amounts (0.04 mol) of the two ionic liquids were mixed in a plain cylinder. Afterwards, the mixture was kept at ambient conditions for 7 days to reach equilibrium. The volume of each phase was determined. Samples of both phases were taken with a syringe, followed by ^{1}H NMR experiments (using d_6-DMSO as solvent) and density measurements using a BLAUBRAND pyknometer. Using the NMR experiments, an average molar mass was obtained. Using the known densities, volumes, and average molar masses of each phase, the molar amount of each species in both phases could be calculated.

11.2.6 Determination of Upper Solution Temperature

All mixtures were heated to 60 °C and kept there for 1 h using a stirred heating bath. The temperature was then increased in 5 °C steps, keeping every target temperature stable for 30 min. The visible formation of one clear phase was set as the upper solution temperature.

REFERENCES

1. Welton, T., Room-temperature ionic liquids. Solvents for synthesis and catalysis, *Chem. Rev.* **99**, 2071–2084 (1999).
2. Wasserscheid, P. and Welton, T. (Eds.), *Ionic Liquids in Synthesis*, 2nd ed. Wiley-VCH, Weinheim, 2007.
3. Dupont, J. and Spencer, J., On the noninnocent nature of 1,3-dialkylimidazolium ionic liquids, *Angew. Chem. Int. ed.* **43**, 5296–5297 (2004).
4. Holbrey, J. D., Visser, A. E., and Rogers, R. D., in *Ionic Liquids in Synthesis*, P. Wasserscheid and T. Welton (Eds.), Wiley-VCH, Weinheim, 2003, pp. 68–81.
5. Wasserscheid, P. and Keim, W., Ionic liquids–new "solutions" for transition metal catalysis, *Angew. Chem. Int. Ed. Engl.* **39**, 3773–3789 (2000).
6. Solinas, M., Pfaltz, A., Cozzi, P. G., and Leitner, W., Enantioselective hydrogenation of imines in ionic liquid/carbon dioxide media, *J. Am. Chem. Soc.* **126**, 16142–16147 (2004).
7. Dyson, P. J., Biphasic chemistry utilising ionic liquids, *Chimia* **59**, 66–71 (2005).
8. Eßer, J., Wasserscheid, P., and Jess, A., Deep desulfurisation of oil refinery streams by extraction with ionic liquids, *Green Chem.* **6**, 316–322 (2004).

9. Pereiro, A. B., Tojo, E., Rodriguez, A., Canosa, J., and Tojo, J., HMImPF$_6$ ionic liquid that separates the azeotropic mixture ethanol plus heptane, *Green Chem.* **8**, 307–310 (2006).
10. Vallee, C., Biard, A., and Olivier-Bourbigou, H., *Fr. Demande* **23** (2006).
11. Arce, A., Earle, M. J., Katdare, S. P., Rodriguez, H., and Seddon, K. R., Mutually immiscible ionic liquids, *Chem. Commun.*, 2548–2550 (2006).
12. Yang, J.-Z., Lu, X.-M., Gui, J.-S., and Xu, W.-G., A new theory for ionic liquids—the Interstice Model Part 1. The density and surface tension of ionic liquid EMISE, *Green Chem.* **6**, 541–543 (2004).
13. Solvent-Innovation GmbH, Cologne, Germany (www.solvent-innovation.de).
14. Bônhote, P., Dias, A. P., Papageorgiou, N., Kalyanasundaram, K., and Grätzel, M., Hydrophobic, highly conductive ambient-temperature molten salts, *Inorg. Chem.* **35** (5), 1168–1178 (1996).
15. Gerhard, D., Alpaslan, S. C., Gores, H. J., Uerdingen, M., and Wasserscheid, P., Trialkylsulfonium dicyanamides—a new family of ionic liquids with very low viscosities, *Chem. Commun.*, 5080–5082 (2005).

12 Recent Developments in the Reprocessing of Spent Fuel by Catalyst Enhanced Molten Salt Oxidation (CEMSO)

TREVOR R. GRIFFITHS

Redston Trevor Consulting Limited, Leeds, United Kingdom

VLADIMIR A. VOLKOVICH

Department of Rare Metals, Ural State Technical University–UPI, Ekaterinburg, Russia

W. ROBERT CARPER

Department of Chemistry, Wichita State University, Wichita, Kansas

Abstract

A novel process, catalyst enhanced molten salt oxidation (CEMSO), for high temperature pyrochemical reprocessing of spent ceramic nuclear fuels is described. It employs alkali carbonate melts containing small amounts of alkali nitrate as catalyst and air as an oxidant. Uranium(IV) oxide is converted into alkali metal uranates(VI) (essentially insoluble in the melt) and the dissolution behaviour of the fission products depends on their nature and chemical state in the spent fuel. The mechanism of the process, involving the catalytic action of nitrate ions, is discussed and the structure of the prospective fuel cycle is proposed. Preliminary results of mathematical modelling of the activity of superoxide ions support the oxidation explanation given here. The CEMSO process can also be used for the destruction of various wastes including CFCs.

12.1 INTRODUCTION

At the EUCHEM Molten Salts meeting in Poland in 2004, we introduced a potential alternative technique for reprocessing spent nuclear fuel using molten salt oxidation

---

Molten Salts and Ionic Liquids: Never the Twain? Edited by Marcelle Gaune-Escard and Kenneth R. Seddon
Copyright © 2010 John Wiley & Sons, Inc.

(MSO). We now report an updated and more detailed critical analysis. First, the process is outlined, and then recent developments are described in more detail. As a consequence of our improved understanding of the process, it is now termed catalyst enhanced molten salt oxidation (CEMSO). It had been shown previously [1] that the reason why the early, original and successful oxidation experiments using molten carbonate and air sparging had been discontinued was because they were somewhat slow and incomplete when attempting to oxidise paper, plastic, and cotton. This was due to an incorrect understanding of molten salt oxidation.

Molten salt oxidation is commonly described as a thermal, nonflame process that oxidises and destroys completely many organic compounds and simultaneously retains nonoxidisable and nonvolatile inorganic species in the molten carbonate-based salt: hydrocarbons are converted to carbon dioxide and steam. Applications of MSO included the complete destruction of hazardous waste, propellants, and explosives and metal recovery [2, 3]. The early literature describes the various successes but does not attempt to explain the mechanism, while referring to the catalytic role of carbonate. This is because, essentially, the process of combustion was achieved at a lower temperature in molten carbonate, a natural conclusion. To speed up the process, catalysts typical of the oil refining industry were added to the carbonate melt, such as transition metal oxides. These had a minor effect on oxidation rates but the amounts required, around 40%, caused problems due to increased viscosity of the system. It was also thought that additives intended to increase the solubility of oxygen into the melt would be beneficial, but no description of their mechanism was offered.

In 1998, after some $40,000,000 had been expended on molten salt oxidation research, an enquiry essentially concluded that further funding was unlikely to yield further improvements in oxidation rates or efficiency. The attempt at describing the process mentioned peroxide and superoxide as potential participants, but went no further. The only explanation we can find in the literature and various reports was written in 1993 by Stelman and Gay [4] and was based on studies on the reaction between graphite and molten sodium carbonate [5–7]. They describe MSO as a system that consists of waste particles and air bubbles suspended in molten sodium carbonate, and that both the oxidiser and the waste react separately with the molten salt to form intermediates, which in turn react to complete the process. They stated that there is a bidirectional mass transfer of gaseous species between the bulk gas in the bubble in the melt and the gas–liquid interface at the surface of the bubble. At this interface, oxygen reacts with the salt to form the peroxide (Na_2O_2), superoxide (NaO_2), peroxycarbonate (Na_2CO_4), and peroxydicarbonate ($Na_2C_2O_6$), and the authors noted that all these are strong oxidising agents. They then proposed a bidirectional mass transfer between the surface of the liquid (presumably the liquid in contact with the bubble surface) and the bulk of the liquid, followed by turbulent mixing that disperses the oxidising species from the bubble side and the reducing species from the waste side throughout the bulk liquid, where they annihilate each other. They thus also have bidirectional mass transfer between the bulk liquid and the interface between the liquid salt and the waste. The final stage in their explanation was that chemical reactions occur on the surface of the waste. They stated that these may include the direct reaction of the "oxy-salts" and the waste, as well as reduction of sodium carbonate by the waste to form sodium metal and CO. They concluded that the elementary reactions involve a large network of intermediate salt compounds, and the overall reaction is a catalytic cycle consuming no sodium carbonate (hence, presumably, the concept that sodium carbonate is the catalyst in this process).

They also stated that sodium sulfate is another catalyst and that, when added to sodium carbonate melt at up to 10%, a new catalytic reaction path is involved, and cited three, now expired, patents [8–10]. They described the catalytic effect as due to a $[SO_4]^{2-}/S^{2-}$ cycle and stated that the mechanism involves reduction of sulfate in several steps to sulfide by reaction of the salt with the waste (citing two of Stelman's papers in support [11, 12]). They then claimed that the sulfide is reoxidised back to sulfate by reaction of the salt with air. They stated that this is what takes place with carbon, and that analogous reactions occur for the hydrogen content of the waste, with steam as the product. No supporting thermodynamic data were given.

Previously, we have shown that, while carbonate participated in the oxidation process, it did not act as a catalyst. The key feature is that when oxygen dissolves in molten carbonate it is *not* largely present as molecular oxygen, but is present as a mixture of peroxide and superoxide ions. Frangini and Scaccia [13] have maintained that the oxygen content in carbonate melts consists of molecular oxygen, peroxide, superoxide, and percarbonate. They used a standard Cr(III) technique to determine the oxygen solubility [14, 15], but unless performed on the molten carbonate and not the quenched melt, results are not reliable. In addition, they used the simple Henry's law and did not take into account the partial pressure of CO_2 and the presence of peroxides and superoxides. We have developed a precise technique that achieves this [16]. Furthermore, the existence of percarbonate is, in our view, dubious. As far as we have been able to ascertain, it has only been proposed as a means of explaining electrochemical results, and no identification by an independent technique, such as spectroscopy, has been advanced. We have commenced theoretical calculations concerning the reactivity of peroxides and superoxides in carbonate melts.

12.2 EXPERIMENTAL

All the experimental details employed are given in our published papers referenced herein.

12.3 ENHANCED MOLTEN SALT OXIDATION

Oxygen reacts with molten carbonate, dissolving chemically and not physically, an uncommon but essential property. There are three possible reactions:

Peroxide formation: $\quad O_2 + 2\,[CO_3]^{2-} \rightarrow 2\,[O_2]^{2-} + 2\,CO_2 \quad$ (12.1)

Superoxide formation: $\quad 3\,O_2 + 2\,[CO_3]^{2-} \rightarrow 4\,[O_2]^{-} + 2\,CO_2 \quad$ (12.2)

Percarbonate formation: $\quad O_2 + 2\,[CO_3]^{2-} \rightarrow 2\,[CO_4]^{2-} \quad$ (12.3)

but we neglect reaction (12.3) for the previously given reasons.

To improve the oxidative capability of carbonate melts, the addition of sodium peroxide and potassium superoxide would obviously but momentarily help, but they quickly decomposed when added at around 500 °C and thus this is not a long-term solution. Oxygen is slightly soluble in molten sulfates, and more so in molten chlorides, doubtless as molecular oxygen in these melts. Thus their addition to carbonate melts did increase slightly, but not significantly, the oxidative capability, the molecular oxygen entering the solution now reacting with carbonate, as in Eqs. (12.1) and (12.2). We measured and defined the oxidative capacity in terms of the percentage of insoluble uranium(IV) oxide that could be oxidised to insoluble uranate(VI) under the experimental conditions [17]: the mechanism does not involve reduction of sulfate, as previously suggested [4].

Addition of nitrate or nitrite was more effective [18]. Figure 12.1 summarises the reactions involved. We evaluated, for example, a mixture of peroxide and nitrate for the optimum ratio of peroxide-to-nitrate in carbonate melts [19] for the oxidation of UO_2. This ratio, under the conditions employed, was 3.0, and no oxygen was

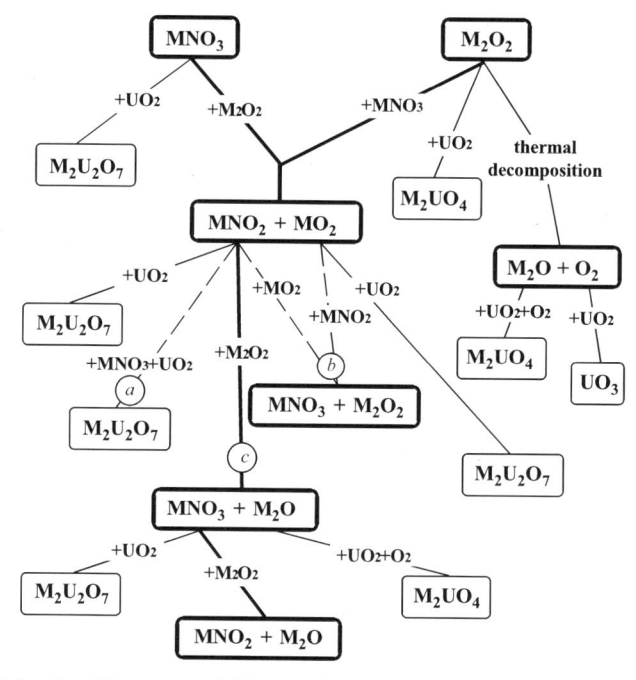

FIGURE 12.1 Possible routes of UO_2 oxidation in melts containing added alkali metal peroxide (M_2O_2) and nitrate (MNO_3). Thick solid lines show formation of oxidising agents, thin lines show formation of alkali metal uranates(VI), and dashed lines indicate reactions not favoured. Reaction ⓐ requires a three times excess of MNO_2 to MNO_3 by $3MNO_2 + MNO_3 + 3UO_2 \rightarrow M_2UO_4 + M_2U_2O_7 + 4NO$ and is therefore unlikely to occur in the system. In reaction ⓑ, since KO_2 reacts rapidly with UO_2, in the presence of excess UO_2 the reaction of MO_2 with uranium(IV) oxide is preferred. Reaction ⓒ is only possible in the presence of an excess of peroxide over nitrate [19].

additionally bubbled through the melt. The proportion of UO_2 oxidised increased with temperature in the range 450–750 °C, but the reaction always ceased before the oxidising species reached the centre of the powder particles, and usually after 2–7 h.

The carbonate melts used were the eutectics $(Li-Na-K)_2[CO_3]$, $(Li-Na)_2[CO_3]$, and $(Li-K)_2[CO_3]$ and the low melting mixture $(Na-K)_2[CO_3]$, which have the corresponding melting points, 393, 500, 500, and 710 °C.

A speculative experiment was undertaken with potassium chlorate(VII) [20], which normally starts to decompose above 400 °C. The minimum usable temperature of our carbonate mixtures was that of the ternary carbonate eutectic, at 450 °C, yet we found that >90% of the added chlorate(VII) was still present in the melt after 30 minutes: however, its effect was gone after 2–3 h. This melt was the most efficient at oxidising UO_2 powder, and oxidising completely ceramic UO_2, but its oxidation rate did not increase at 500 °C, due to increased thermal decomposition of the oxidising species: the mechanism of its decomposition was discussed [20]. Obviously, using added chlorate(VII) is not suitable or advised for industrial use.

12.4 CATALYST ENHANCED MOLTEN SALT OXIDATION

We developed catalyst enhanced molten salt oxidation (CEMSO) after finding that both nitrite and nitrate ions were stable in carbonate melts and no oxides of nitrogen were observed or detected in the off-gases, particularly since these acid gases would react in an alkaline melt to form nitrite and nitrate. Furthermore, the addition of peroxide to a nitrate melt generates superoxide [20]:

$$2[NO_3]^- + [O_2]^{2-} \rightarrow 2[NO_2]^- + 2[O_2]^- \tag{12.4}$$

Thus continuously bubbling air into a carbonate melt containing nitrate and nitrite ions would maintain the peroxide ions formed (Eq. (12.1)) and hence increase the concentration of superoxide ions. The peroxide initially formed in an oxygen sparged carbonate + nitrate melt will be converted into the more active superoxide in the presence of nitrate, and the nitrite thereby formed will be oxidised back to nitrate by fresh peroxide now continuously being formed, thereby increasing the oxidising power of the melt and enabling the cycle to be repeated.

The rate and extent of oxidation in these melts is thus increased, and the system becomes viable for reprocessing spent nuclear fuel. The catalytic cycle of CEMSO thus can be represented as follows:

Oxygen to peroxide: $\quad O_2 + 2[CO_3]^{2-} \rightarrow 2[O_2]^{2-} + 2CO_2$

Peroxide to superoxide and nitrate to nitrite: $\quad 2[NO_3]^- + [O_2]^{2-} \rightarrow 2[NO_2]^- + 2[O_2]^-$

Nitrite back to nitrate $\quad [NO_2]^- + [O_2]^{2-} \rightarrow [NO_3]^- + O^{2-}$

Nitrite back to nitrate : $\quad 2[NO_2]^- + O_2 \rightarrow 2[NO_3]^-$

12.4.1 Electron Transfer

The four above reactions can only be overall summaries: intermediates may be involved. It is not usual for two same-charged ions to collide and react. In the above catalytic cycle, this would appear to happen, with electron transfer resulting. Collisions between same-charged ions can arise, but whether at temperatures above 500 °C in an ionic melt they will do so with sufficient energy such that they separate after electron transfer probably requires confirmation with computer calculations, which are in progress. The role of cations present in the melt as potential participants in intermediate species would similarly have to be tested.

12.4.2 Conditions for Complete Oxidation of Ceramic UO_2

Further testing showed that CEMSO could achieve complete oxidation of ceramic UO_2 in 2 h at 600 °C in an air-bubbled ternary carbonate melt to which potassium nitrate had been added, and that potassium diuranate(VI) was the main product [21]. The rate and extent of reaction of ceramic UO_2 were not altered, within experimental error, when air was replaced by oxygen, an economic advantage.

When the reaction is complete and the air-bubbling stopped, the diuranate(VI) settles to the bottom of the melt and thus the melt (or the precipitate of diuranates(VI)) can readily be removed and the fission products therein treated. Knowledge of the behaviour, and particularly the solubility, of uranates(VI) in molten carbonates is important for the effective separation of uranium and plutonium from molten carbonate.

12.4.3 Solubility of Uranates(VI) in Molten Carbonates

Extensive studies were undertaken of the chemistry of all possible uranates(VI) of the alkali metals [19, 22–26] in molten carbonates [23, 24, 26], including solubilisation enthalpies and solubility in the various carbonate melts. These were studied as a function of temperature, and the effect of changing the basicity of the carbonate melts by altering the partial pressure of carbon dioxide above the melt was also evaluated [24, 26]. The general result was to show a V-shaped dependence of solubility with increase in temperature and CO_2 partial pressure; Figure 12.2 is an example of the extensive data produced [27].

The solubilities were particularly small, but our most important datum was that the solubility of $K_2U_2O_7$ in the ternary eutectic at 600 °C in air was only 70 wt ppm (29 mol ppm); see Table 12.1. The maximum solubility at 800 °C for any mono- or diuranate(VI) of lithium, sodium, or potassium in the four carbonates used rarely exceeded 200 wt ppm. Table 12.1 also shows that the 12 repeated low values, mostly below 50 wt ppm., were all within 7 units. Thus, significantly, the amount of uranium remaining in the carbonate melt is reliably known and exceedingly small. We note that the ANL electrochemical separation process generally leaves around 10% in the melt.

CATALYST ENHANCED MOLTEN SALT OXIDATION   157

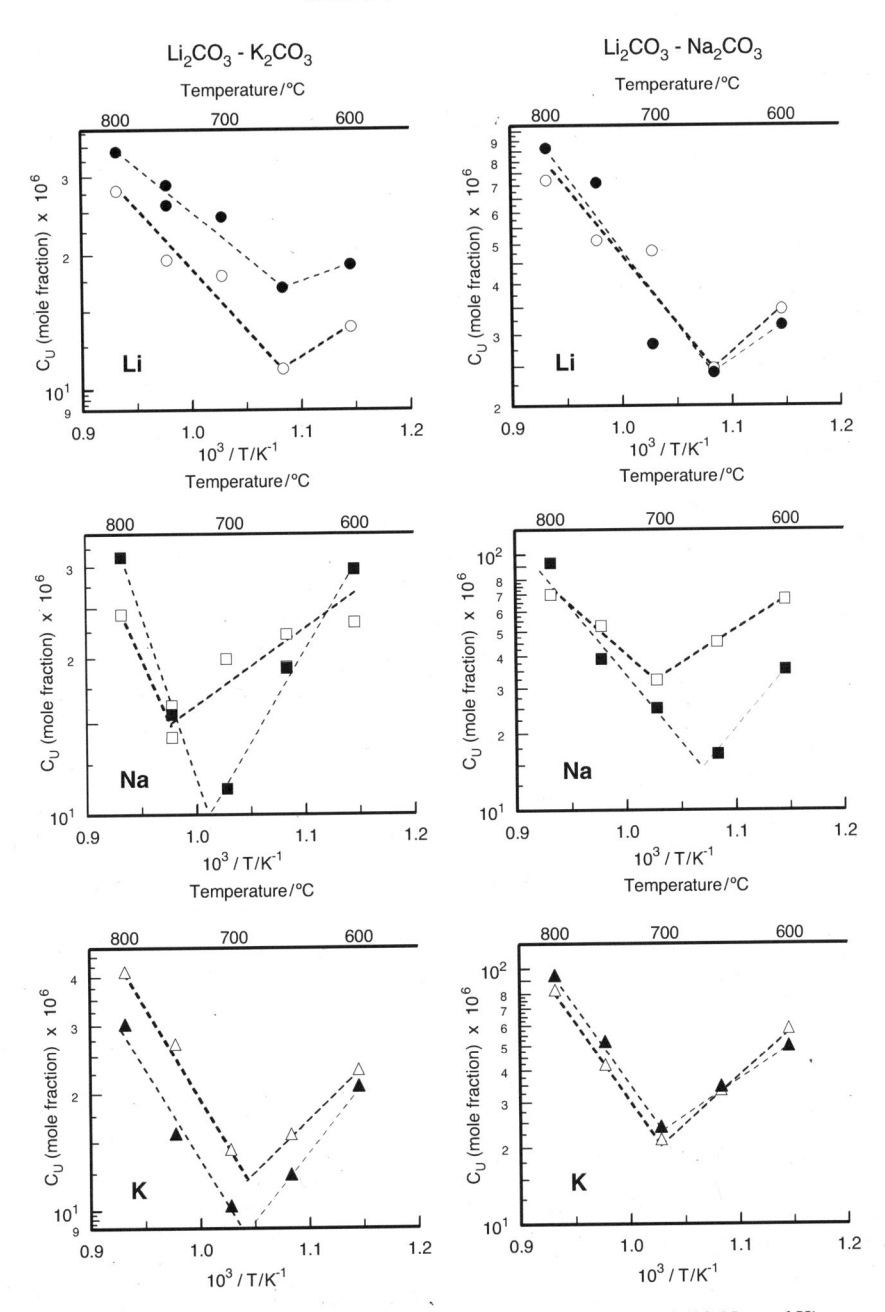

FIGURE 12.2 Effect of temperature on the solubilities of alkali metal (Li, Na, and K) mono- and diuranates(VI) in (Li–Na)$_2$CO$_3$ and (Li–K)$_2$CO$_3$ melts. Melt compositions are shown above the plots; alkali metal symbols indicate the cations of the uranates(VI) added. Results for monouranates(VI) are shown by open symbols and dashed lines, and for diuranates(VI) by solid symbols and dashed lines [27].

TABLE 12.1 Solubilities of Alkali Metal Mono- and Diuranates(VI) in Molten Carbonate Mixtures, Atmosphere Air[a]

Temperature/°C	Li₂UO₄ wt ppm	Li₂UO₄ mol ppm	Li₂U₂O₇ wt ppm	Li₂U₂O₇ mol ppm	Na₂UO₄ wt ppm	Na₂UO₄ mol ppm	Na₂U₂O₇ wt ppm	Na₂U₂O₇ mol ppm	K₂UO₄ wt ppm	K₂UO₄ mol ppm	K₂U₂O₇ wt ppm	K₂U₂O₇ mol ppm
\multicolumn{13}{c}{(Li–K)₂CO₃ Eutectic, 44.5 mol% Li₂CO₃–55.5 mol % K₂CO₃}
600	30	14	41	19	51	23	64	30	50	23	45	21
650	24	11	37	17	48	22	42	19	34	16	27	12
650 (rpt)					42	19						
700	39	18	53	24	43	20	24	11	31	14	22	10
750	42	20	56	26	31	14	34	16	59	27	34	16
750 (rpt)			62	29	36	16						
800	61	28	74	34	53	24	68	31	90	41	66	30
\multicolumn{13}{c}{(Li–Na–K)₂CO₃ Eutectic, 44.1 mol % Li₂CO₃–29.8 mol % Na₂CO₃–26.1 mol % K₂CO₃}
450	202	85	146	62	103	44	149	63	171	72	171	72
500	244	103	65	28	106	44	78	33	173	73	123	52
550	156	66	72	30	81	34	70	29	158	66	71	30
550 (rpt)			65	27								
600	109	46	39	16	41	18	33	14	81	34	70	29
650	109	46	31	13	47	20	30	13	74	31	41	17
700	70	30	51	22	48	20	41	17	29	12	24	10
700 (rpt)	73	31	54	23								

750	89	38	77	32	55	23	55	23	54	23	57	24
800	152	64	144	61	132	56	118	50	127	53	155	65

(Li–Na)₂CO₃ Eutectic, 53.3 mol % Li₂CO₃–46.7 mol % Na₂CO₃

600	93	35	84	32	179	67	95	35	156	58	132	50
650	66	25	65	24	122	45	45	17	90	33	93	35
700	129	48	76	28	86	32	67	25	57	22	64	24
750	137	51	190	71	141	53	104	39	112	42	139	52
800	193	72	231	86	185	70	248	93	219	83	253	94

(Na–K)₂CO₃ melt, 56 mol % Na₂CO₃–44 mol % K₂CO₃

750	181	91	193	97	175	88	201	102	142	72	190	96
800	264	133	266	136	312	158	276	141	316	160	275	139

[a]See Reference [8].

12.4.4 Limited Plutonium Data

We have not, for statutory reasons, been permitted to conduct similar tests with PuO_2. However, it is understood from the information available that PuO_2 will also yield plutonates under the above conditions, but these may be slightly more soluble in molten carbonates, but have yet to be measured or published. Solubilities near 10%, even for plutonates, are not anticipated.

We note that, in the 1990s, there were reports of plans and experiments for the application of MSO to plutonium and other radioactive materials [28–33], but unfortunately they all used sodium carbonate at around 900 °C and misrepresented the mechanism by which MSO technology operates. A test using solid combustible waste containing plutonium at TRU levels examined the off-gas at five points sequentially downstream, at the off-gas line, water trap, condensate, prefilter, and HEPA filter. Measurable amounts of plutonium were found only in the prefilter and 99.9% of the plutonium remained in the sodium carbonate [33].

We can conclude that plutonium will remain in our lower temperature CEMOS system: unfortunately, none of the publications stated if the plutonium in the melt was in the form of plutonates. We fully anticipate that other transuranics in spent fuel will remain in the melt, and any leaving in the off-gas will be trapped downstream.

12.4.5 Reprocessing Cycle

It is now possible to propose an efficient reprocessing cycle. First, the fuel rods would be left in cooling ponds for around 100 days so that the short-lived, intensely radioactive species lose their activity, and the generation of heat has subsided. Second, the spent fuel would be treated by CEMOS, and insoluble UO_2 (and PuO_2) would be completely oxidised to insoluble alkali uranates(VI), and separated off and more rods treated (the low solubility of uranates(VI) means that they do not accumulate in carbonate melts). But, since the fission products dissolve in the melt and accumulate, they will be removed by phosphate precipitation, the carbonate melt being recycled after appropriate clean-up.

12.4.6 Mathematical Modelling

As an initial report, we describe the mathematical modelling by one of the authors (WRC) of a severe test of the oxidising ability of superoxide on the strongest covalent bond—the carbon fluorine bond. We have investigated the potential reactions between superoxide ($[O_2]^-$) and selected chlorofluorocarbons including CCl_2F_2, CCl_3F, $CClF_2H$, CCl_2FCF_2Cl, and $CClF_2CClF_2$. Semiempirical and ab initio methods were used to determine stable (gas phase) superoxide–chlorofluorocarbons that might represent intermediate complexes in degradation reactions. This was accomplished by interacting the various chlorofluorocarbons with superoxide at potential points of attack, that is, the bonds between the central carbon atoms and the attached halogens. It was established that the C–Cl bond would accept the insertion of superoxide to form a stable superoxide–chlorofluorocarbon adduct for all five chlorofluorocarbons. This was confirmed by using the semiempirical methods, AM1(1), PM3(2), and PM5,

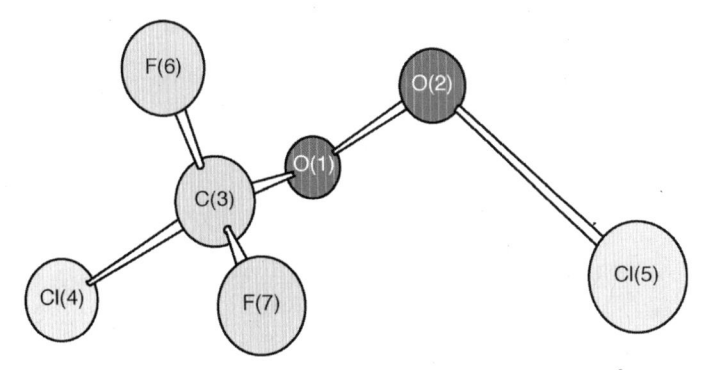

FIGURE 12.3 Calculated structure of $[C(O_2Cl)ClF_2]^-$. Bond distances (Å) are predicted as: Cl4–C3 = 1.844; F6–C3 = 1.348; F7–C3 = 1.331; C3–O1 = 1.341; O1–O2 = 1.427; O2–Cl5 = 2.482. Bond angles (°) are: O2–O1–C3 = 107.45; O1–O2–Cl5 = 112.99; O1–C3–Cl4 = 105.83; O1–C3–F6 = 113.79; O1–C3–F7 = 115.15; Cl4–C3–F6 = 106.94; Cl4–C3–F7 = 107.75; F6–C3–F7 = 106.94.

contained in the CACHE Program (Version 5.04) to generate enthalpies of formation and reaction, all of which are considerably exothermic. Unfortunately, the insertion of superoxide between carbon and fluorine to form a stable species was not successful in the majority of the semiempirical calculations (regardless of method). The AM1 structures of the five superoxide adducts were obtained, and that of $[C(O_2Cl)ClF_2]^-$ is shown in Figure 12.3. These results suggest that attack by superoxide in molten carbonate on chlorofluorocarbons is possible and, since these calculations relate to room temperature, attack on Freons having no chlorine bonds would seem possible; further calculations are in progress.

These modelling results support our contention that CEMOS is a powerful oxidising technique for converting uranium and plutonium oxides into insoluble uranates(VI) and plutonates, while bringing fission products into solution.

12.4.7 Separation of Fission Products

In many respects, the task of separating the fission products remaining in the carbonate melt is similar to that of identifying the components of a mixture of unknowns that used to be part of an undergraduate's inorganic practical chemistry course. If the mixture was not soluble in dilute acids, the next step was to boil it with a strong solution of sodium carbonate. The cations would be precipitated as insoluble carbonates and the anions would remain in solution with the excess carbonate ions. In the present situation, the cations remain in solution because their concentration is low and water is absent.

The only possible insoluble compounds that could be formed are the oxides, but the concentration of oxide ions in the carbonate melt will be low and remain low—the higher the partial pressure of carbon dioxide, the lower the concentration of oxide ions. The temperatures involved are also well below the decomposition temperature of alkali metal carbonates.

Some metals, particularly noble metals, may not be oxidised and thus not go into solution. Spent fuel contains minor amounts of the noble metals ruthenium, rhodium, and palladium, in the form of alloy inclusions. When these noble elements are formed by high temperature decomposition on matrices in low concentration (for use as catalysts), they migrate together to form microcrystals. Such pellets in ceramic fuel will be released into the carbonate melt, if on or near the fuel rod surface, and the remainder will remain occluded within the uranate(VI) formed. The uranate(VI) separated from the molten carbonate dissolves readily in dilute acid and the alloy pellets will be left behind. The conditions of the CEMSO process are not, in our view, sufficient to attack these pellets.

12.4.8 Relative Abundance of Components

The most abundant fission product elements, per kilogram of uranium, are caesium at around 3000 mg·kg^{-1}, then barium and strontium at around 2000 and 1000 mg·kg^{-1}, respectively, and then the early lanthanides, lanthanum, cerium, and praseodymium at around 1500, 2800, and 1500 mg·kg^{-1}, respectively. The number of kilograms of spent fuel that need to be treated by CEMSO before it is necessary to remove fission products will depend on many factors: however, it has been demonstrated that over 97% of fission products can be precipitated out of molten chloride solutions [34].

Many compounds of barium and strontium are insoluble, but the solubility of caesium, strontium, and barium oxides in molten carbonate can be expected to be high, and we have determined [35] that the concentrations of the oxides of the lanthanides lanthanum, praseodymium, and neodymium in the ternary carbonate at 600 °C are 0.41, 0.29, and 0.71 wt %, respectively. These, too, will not be coprecipitated with the uranates(VI) formed, since the latter do not enter solution but are formed by the inward movement of the oxidising species, which releases the soluble species.

In practice, separation can be achieved by batch or continuous methods. The latter is feasible since alkali metal carbonates are essentially impervious to radiation, and the former will depend on an agreed maximum concentration accumulated of the fission products or the radiation levels reached.

12.5 PRECIPITATION OF FISSION PRODUCTS

There are now two routes for precipitating the fission products from the molten carbonate remaining after the uranates(VI) and plutonates have been removed. One involves quenching the melt and treating it with water, and the other uses the addition of precipitants, largely phosphates.

12.5.1 Water Treatment

Addition of water will enable the insoluble carbonates to precipitate out, and these can be removed by filtration, dissolved in dilute acid, and subjected to conventional aqueous separation techniques. The fission product remaining in aqueous solution will be largely the highly radioactive ^{137}Cs. It can be removed by various methods. If it is used industrially, it can be recovered by ion exchange. Alternatively, being particularly soluble, the water content can be reduced to deposit the other alkali metals and the caesium will be concentrated in the filtrate. It can then be precipitated quantitatively as caesium tetraphenylborate, $Cs[B(C_6H_5)_4]$, a standard procedure, but since potassium ions may also be present in significant concentration, the precipitate may also contain some of these cations.

Upon complete removal of the remaining water, the carbonate mixture can be recycled, after any adjustment of the cation ratios and satisfactory residual radiation levels.

12.5.2 Phosphate Precipitation

Most phosphates are insoluble in water, and there is evidence that this also applies in molten salts. To our knowledge, there are no common molten salts, or mixtures or eutectics thereof, in which phosphates are even slightly soluble, with the exception of alkali metal phosphates. Our proposal is therefore that alkali metal phosphates be added to molten carbonate and all the fission products filtered or separated off and sent for disposal. Since many phosphates form glasses, vitrification may be a particularly suitable treatment. The remaining liquid now consists of carbonate with an excess of phosphate, since an excess is required to ensure essentially complete precipitation. Before it can be recycled, this excess phosphate must be removed. A procedure for doing this has not been developed yet, but we suggest that if it can be established that a particular phosphate will be precipitated essentially completely when stoichiometric amounts are present, then this would be the simplest method. Related preliminary experiments are described next. Alternatively, molten salt ion exchange could be employed. If these high temperature methods are insufficient, then aqueous methods would be needed.

12.5.3 Phosphate Precipitation from Molten Chlorides

Molten carbonates can safely be contained in suitable ceramic ware. They are not very amenable toward the transparent containers preferred in the laboratory due to their aggressive nature and attack on optical silica. We therefore have made detailed studies of phosphate precipitation from molten chlorides of the majority of the fission product elements formed in spent nuclear fuel, namely, caesium, magnesium, strontium, barium, lanthanides (lanthanum to dysprosium), zirconium, chromium, molybdenum, manganese, rhenium (to simulate technetium), iron, ruthenium, nickel, cadmium, bismuth, and tellurium [34]. This choice of a change to working with molten chlorides instead of carbonates was also because Gen IV molten salt reactors would probably employ molten chlorides, and we were experienced with molten chloride systems. We expect almost identical results when comparable experiments are performed in molten carbonates.

To date, we have examined these fission product elements individually plus some mixtures. We have not investigated multication mixtures of all the major fission product elements, but we have established that a fivefold excess of phosphate is normally sufficient to precipitate >99% of the majority of these elements, and conditions for >97% for almost all the rest have been identified. The cations present in the chloride melt are often critical to the nature or extent of precipitation. The following are some of the major features found:

- Complete reaction normally required a mole ratio of phosphate-to-lanthanide of 5.
- In LiCl–KCl-based melts, lanthanide phosphate was generally formed in around 100% yield [36].
- In NaCl–KCl-based melts, a mixture of $Na_3Ln(PO_4)_2$ and lanthanide phosphate was precipitated.
- For caesium, using Li_3PO_4 as precipitant with LiCl–KCl–CsCl melt ($[PO_4]^{3-}$: Cs^+ mole ratio 3.55), an insoluble white precipitate was formed and X-ray powder diffraction showed only four phases, namely, LiCl, KCl, Li_3PO_4, and Cs_3PO_4, with no remaining CsCl, and thus complete removal of caesium.
- If Na_3PO_4 is added to NaCl–KCl–CsCl melt ($[PO_4]^{3-}$: Cs^+ mole ratio 3.07), again there is a mixture of products, but after 3 h the reaction had not proceeded to completion as some CsCl remained. To remove all the caesium from this melt, an additional treatment is required, such as ion exchange.
- Strontium and barium can be precipitated together. Using NaCl–KCl-based melts and Na_3PO_4 at 750 °C, the crystalline double phosphate, $NaMPO_4$ (M = Sr, Ba), separated out, with 99% and 93% efficiency, respectively. When lithium replaced sodium, at 550 °C, only Li_3PO_4 was precipitated: ion exchange might be required if LiCl–KCl-based melts are used.
- With bismuth, cadmium, chromium, iron, manganese, molybdenum, nickel, rhenium, ruthenium, and zirconium [34], studies in the lithium-containing melts showed that not all these elements formed individual well-defined phosphate phases, but with zirconium 100% was precipitated. Generally, only Li_3PO_4 was found by X-ray powder diffraction of the final precipitate.
- With the sodium-containing melts, rather different results were obtained. With zirconium only 83% was precipitated. Some melts, chromium(III), iron(II), molybdenum(III), rhenium(V), and ruthenium(III), formed oxide phases, identified by X-ray diffraction powder analysis: the source of the oxygen is presumed to be from the air, as chloride melts can dissolve limited amounts of oxygen [18].
- Cadmium and bismuth are 100% separated in the sodium-containing and lithium-containing melts, respectively, and do not precipitate at all in the other melts.
- Ruthenium is fully precipitated in the sodium-containing melt (not tested in the lithium-containing melt).
- With a mixture of strontium, barium, cerium, and zirconium, and using Li_3PO_4 as precipitant, all the cerium and zirconium were deposited as $CePO_4$ and $(Li/K)_2Zr(PO_4)_2$, and over 90% of the alkaline earths remained in the melt.

12.6 CONCLUSION

Catalyst enhanced molten salt oxidation, (CEMSO) is a highly effective oxidising system for converting the uranium (and plutonium) in spent fuel rods into insoluble uranates(VI) (and plutonates). These can be completely removed, leaving just the fission products in the carbonate melt, which can be recycled until fission product levels have reached concentrations requiring their removal (which can be achieved by conventional chemistry, if the quenched melt is dissolved in water or by addition of phosphate to the carbonate melt). This latter has not been researched, but studies in chloride melts have shown features to expect, including the role of cations on the identity and extent of the phosphates precipitated. Some radioactive isotopes can be removed selectively and made available for industrial or medical use. Since the fission products are largely separated as phosphates, they are particularly suitable for disposal by vitrification.

Industrial large-scale application of CEMSO will not require much research, since the original research on MSO successfully employed pilot plant and larger apparatus, and thus the details are in the literature.

The rate at which CEMSO reacts is also much faster than existing techniques. Ceramic UO_2 was oxidised to uranate(VI) in 3 h and complete precipitation of fission product phosphates was also achieved in 3 h.

CEMSO is also an effective technique for oxidising completely dangerous waste and reducing dramatically the volume of low and intermediate level radioactive waste [1].

ACKNOWLEDGEMENTS

We thank British Nuclear Fuels plc for providing the UO_2 and V.A.V. additionally thanks them for a post doctoral research fellowship at Leeds University, UK, and a research fellowship at Manchester University, UK.

REFERENCES

1. Griffiths, T. R., Volkovich, V. A., and Anghel, E. M., Molten salt oxidation: a reassessment of its supposed catalytic mechanism and hence its development for the disposal of waste automotive tires, in *Molten Salts XIII, Proceedings of the Thirteenth International Symposium on Molten Salts,* Philadelphia, PA, USA, May 12–17, 2002, H. C. Delong, R. W. Bradshaw, M. Matsunaga, G. R. Stafford, and P. C. Trulove (Eds.), The Electrochemical Society Proceedings Series, Vol. PV 2002-19, The Electrochemical Society, Pennington, NJ, 2002, pp. 306–317.
2. Navratil, J. D., and Stewart, A. E., *Nukleonika* **41**, 57 (1996).
3. Cooley, C. R., EM-50, Memorandum, Department of Energy, U.S. Government, December 17, 1998.
4. Stelman, D. and Gay, R. L., Fundamental chemical and process differences between molten salt oxidation and incineration, Report for Rockwell International, California, USA, August 1993.

5. Dunks, G. B., Stelman, D., and Yosim, S. J., *Carbon* **18**, 363 (1980).
6. Dunks, G. B. and Stelman, D., *Inorg. Chem.* **22**, 2168 (1983).
7. Dunks, G. B., *Inorg. Chem.* **23**, 828 (1984).
8. Lefrancoiss, P. A. and Barclay, K. M., U.S. Patent No. 3,567,412 (1971).
9. Birk, J. R. and Huberr, D. A., U.S. Patent No. 3,708,270 (1973).
10. Birk, J. R., U.S. Patent No. 3,710,737 (1973).
11. Stelman, D., Darnell, A. J., Christie, J. R., and Yosim, S. J., in *Molten Salts*, The Electrochemical Society, Princeton, N J, 1976, p. 299.
12. Dunks, G. B. and Stelman, D., *Inorg. Chem.* **21**, 108 (1982).
13. Frangini, S. and Scaccia, S., *J. Electrochem. Soc.* **152**, A2155 (2005).
14. Frangini, S. and Scaccia, S., *J. Electrochem. Soc.* **152**, A1251 (2004).
15. Scaccia, S. and Frangini, S., *Talanta* **64**, 791 (2004).
16. Volkovich, V. A., Griffiths, T. R., Fray, D. J., and Thied, R. C., *J. Nucl. Mater.* **282** (2–3) 152 (2000).
17. Volkovich, V. A., Griffiths, T. R., Fray, D. J., Fields, M., and Wilson, P. D., *J. Chem. Soc. Faraday Trans.* **92**, 5059 (1996).
18. Volkovich, V. A., Griffiths, T. R., Fray, D. J., and Fields, M., *J. Chem. Soc. Faraday Trans.* **93**, 3819 (1997).
19. Volkovich, V. A., Griffiths, T. R., Fray, D. J., and Fields, M., *J. Chem. Soc. Faraday Trans.* **93**, 3819 (1997).
20. Volkovich, V. A., Griffiths, T. R., Fray, D. J., Fields, M., and Thied, R. C., *J. Chem. Soc. Faraday Trans.* **94**, 2623 (1998).
21. Volkovich, V. A., Griffiths, T. R., Fray, D. J., and Fields, M., *J. Nucl. Mater.* **256**, 131 (1998).
22. Griffiths, T. R., and Volkovich, V. A., *J. Nucl. Mater.* **274**, 229 (1999).
23. Volkovich, V. A., Griffiths, T. R., Fray, D. J., and Fields, M., *Vib. Spectrosc.* **17**, 83 (1998).
24. Volkovich, V. A., Griffiths, T. R., Fray, D. J., and Fields, M., *Dyes Pigments* **39**, 139 (1998).
25. Volkovich, V. A., Griffiths, T. R., and Thied, R. C., *Vib. Spectrosc.* **25** (2), 223 (2001).
26. Volkovich, V. A., Griffiths, T. R., Fray, D. J., and Thied, R. C., *Phys. Chem. Chem. Phys.* **3** (23), 5182 (2001).
27. Volkovich, V. A., Griffiths, T. R., Fray, D. J., and Thied, R. C., *Phys. Chem. Chem. Phys.* **1**, 3297 (1999).
28. Volkovich, V. A., May, I., Griffiths, T. R., Charnock, J. M., and Lewin, R. G., *J. Nucl. Mater.* **344**, 73 (2005).
29. Bell, J. T., Hass, P. A., and Rudolph, J. C., *Sep. Sci. Technol.* **30**, 1755 (1995).
30. Ismagilov, Z. R., Kerzhentsev, M. A., and Adamson, M., in *Scientific Advances in Alternative Demilitarisation Technologies*, F.W. Holm (Ed.), Kluwer Academic Publishers, Holland, 1996, p. 29.
31. Wishau, R., Montoya, A., and Ramsey, K. B., in 32nd Annual Midyear Meeting, *Creation and Future Legacy of Stockpile Stewardship Isotope Production, Applications, and Consumption*, Sealed Sources, Recycling, and Transmutation Session, January 24–27, 1999.
32. Ismagilov, Z. R., Kerzhentsev, M. A., Shkrabina, R. A., Tsikoza, L. T., Lunyushkin, B. I., Ostrovski, Yu. V., Kostin, A. L., Abroskin, I. Ye., Malya, Ye. N., Matyukha, V. A.,

Adamson, M. G., Heywood, A. C., Zundelevich, Yu., Ismagilov, I. Z., Babko-Malyi, S., and Battleson, D. M. *Catalysis Today* **55**, 23 (2000).

33. Grantham, L. F., McKenzie, D. E., Oldenkamp, R. D., and Richards, W. L., Disposal of transuranic solid waste using Atomics International molten salt combustion process, Rockwell International Report Al ERDA-13169 prepared for ERDA, March 15, 1976.
34. Volkovich, V. A., Griffiths, T. R., and Thied, R. C., *J. Nucl. Mater.* **323**, 49 (2003).
35. Volkovich, V. A., Yakimov, S. M., Vasin, B. D., Polovov, I. B., Medvedev, E. O., Griffiths, T. R., and Rebrin, O. I., in *Proceedings of the 7th International Symposium on Molten Salts Chemistry and Technology,* August 29 to September 2, 2005, Toulouse, France, 2005, Vol. 2, p. 663.
36. Griffiths, T. R., Volkovich, V. A., and Thied, R. C., in *Proceedings of the International George Papatheodorou Symposium*, S. Boghosian, V. Dracopoulos, C. G. Kontoyannis, and G. A. Voyiatzis (Eds.), Institute of Chemical Engineering and High Temperature Chemical Processes, Patras, Greece, September 17–18, 1999, p. 78.

13 Plasma-Induced Molten Salt Electrolysis to Form Functional Fine Particles

YASUHIKO ITO

Department of Environmental Systems Science, Faculty of Science and Engineering, Doshisha University, Kyotanabe, Kyoto, Japan

TOKUJIRO NISHIKIORI

I'MSEP Co. Ltd., Shimogyo, Kyoto, Japan

TAKUYA GOTO

Department of Fundamental Energy Science, Graduate School of Energy Science, Kyoto University, Sakyo-ku, Kyoto, Japan

Abstract

Plasma-induced molten salt electrolysis is a nonconventional electrolysis that utilises a discharge generated between an electrode and a molten salt electrolyte. Even in a case where one electrode is positioned outside over the electrolyte surface, plasma-induced stationary discharge makes continuous electrolysis possible under some appropriate conditions, even at atmospheric pressure. Using a plasma-induced cathodic discharge electrolysis, for instance, various kinds of fine particles have been formed: Ag, Cu, Fe, Ni, Ti, Al, Nb, Ta, TiNi$_2$ alloy, and C. On the other hand, in the case of plasma-induced anodic discharge electrolysis, fine particles of TiO$_2$, WS$_2$, MoS$_2$, C, and C$_3$N$_4$ were obtained.

Bench-scale experiments have been conducted successfully to continuously obtain well-defined fine particles by this plasma-induced molten salt electrolysis.

13.1 INTRODUCTION

Recent technological developments have focused on high functionalisation of materials, achieving low environmental impact as well as low manufacturing cost.

Molten Salts and Ionic Liquids: Never the Twain? Edited by Marcelle Gaune-Escard and Kenneth R. Seddon
Copyright © 2010 John Wiley & Sons, Inc.

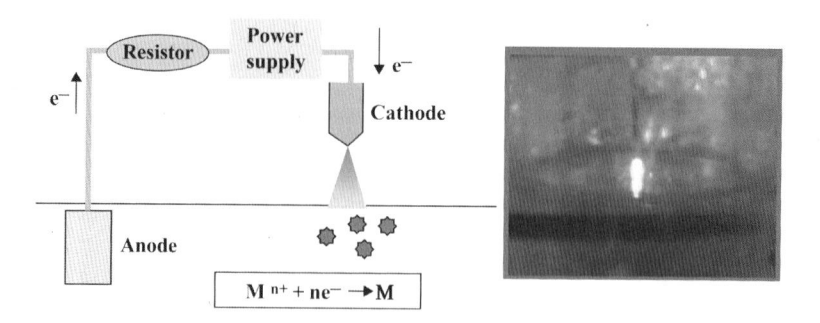

FIGURE 13.1 The principle and a photograph of plasma-induced molten salt electrolysis (cathodic discharge).

Among various types of functionalisation methods, molten salt electrochemical processes can play important roles due to their low environmental impact, low energy consumption, and good flexibility. Against this background, various novel molten salt electrochemical processes have been created and developed by the authors and their collaborators to obtain highly functionalised materials [1–4]. Here, the plasma-induced molten salt electrolysis method to produce functional particles is selected as an example and described in detail.

13.2 PRINCIPLE OF PLASMA-INDUCED MOLTEN SALT ELECTROLYSIS

Even when one electrode is positioned outside over an electrolyte surface, plasma-induced stationary discharge makes continuous electrolysis possible under some appropriate conditions, even at atmospheric pressure. The principle of this process is shown in Figure 13.1, selecting a case of cathodic discharge as an example.

In the case of cathodic discharge, it is generated between cathode and melt, which is maintained by electron emission from the cathode: argon gas at atmospheric pressure is partially dissociated to form a condensed plasma in the gas phase. Thus when a metal ion source is dissolved in the melt, fine metal particles are formed in the melt whose sizes are in the nano- or micrometre scale. On the other hand, in the case of anodic discharge, it can be maintained by the following two parallel and competitive charge transfer processes: cation emission from the anode and electron emission from the melt [5]. The principle of the anodic discharge is shown in Figure 13.2. Both cathodic and anodic discharge electrolyses have high potential as methods for producing fine functional particles, as will be demonstrated.

13.3 CATHODIC DISCHARGE

Concerning cathodic discharge electrolysis, the sizes of the obtained particles are strongly dependent on the electrolytic conditions, such as concentration of metal ion

CATHODIC DISCHARGE 171

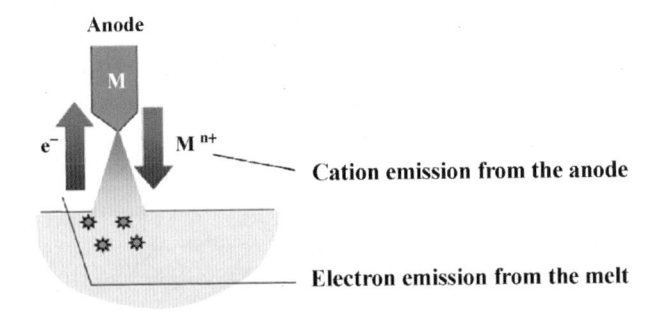

FIGURE 13.2 The principle of plasma-induced molten salt electrolysis (anodic discharge).

source, volumetric current density (current per electrolyte volume), quantity of electricity, and bath temperature. Some examples are shown in Figures 13.3 [6] and 13.4 [7].

Figure 13.5 shows an SEM image and XRD pattern for tantalum particles obtained from LiCl–KCl–K$_2$TaF$_7$ (0.1 mol %) at 723 K, where tantalum metal scrap was used as an anode. The formation of the tantalum particles was confirmed by EPMA. The XRD pattern of the particles shows that the lattice constant increased by about 3%.

Figure 13.6 shows titanium particles obtained by a cathodic discharge of LiCl–KCl–K$_2$TiF$_6$ (0.1 mol%) melt under an atmospheric pressure of argon at 723 K [8].

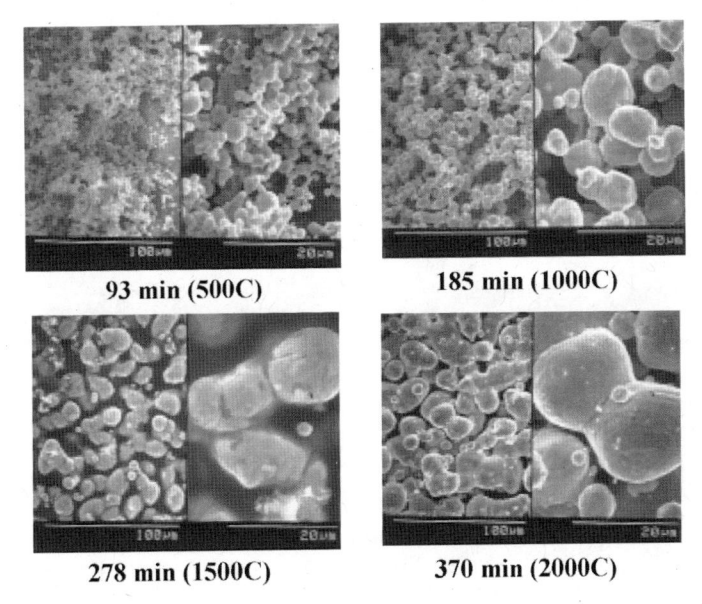

93 min (500C)   185 min (1000C)

278 min (1500C)   370 min (2000C)

FIGURE 13.3 The sizes of silver particles obtained by cathodic discharge electrolysis with various quantities of electricity: LiCl–KCl–AgCl (0.1 mol %), 90 mA, 723 K.

172 PLASMA-INDUCED MOLTEN SALT ELECTROLYSIS

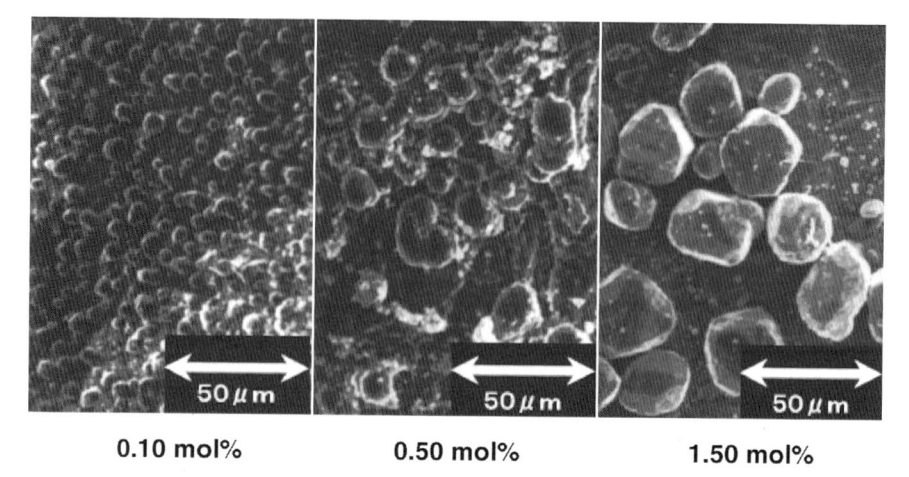

FIGURE 13.4 The sizes of nickel particles obtained by cathodic discharge electrolysis from an electrolyte containing various nickel ion concentrations: LiCl–KCl–NiCl$_2$, 723 K, 90 mA, 2000 C.

Results of EPMA, XRD, and XPS suggest that the particles consisted of metallic or partially oxidised titanium in the core and TiO$_2$ in the surface layer. Originally, the titanium particles are considered to have been formed in the melt by cathodic reduction of titanium ions, and the TiO$_2$ layer is considered to have been formed by exposure of the titanium particles to air or water after electrolysis.

The particle sizes were dependent on the electrolytic conditions, such as quantity of electricity, K$_2$TiF$_6$ concentration, bath temperature, and volumetric current density. As an example, the influence of volumetric current density on the particle size is shown in Figure 13.7.

By applying this cathodic discharge electrolytic process, various kinds of fine particles other than the above have been formed, for example, niobium, aluminium, iron, and carbon.

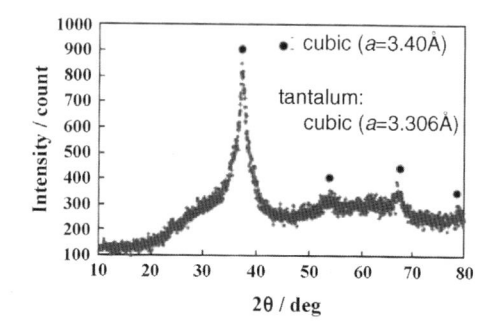

FIGURE 13.5 SEM image and XRD pattern of tantalum particles obtained by cathodic discharge: LiCl–KCl–K$_2$TaF$_7$ (0.1 mol %), 723 K, tantalum anode.

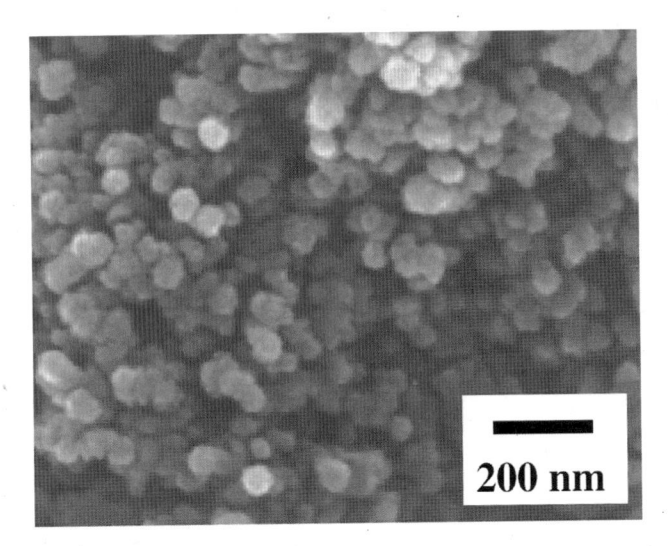

FIGURE 13.6 SEM image of well-defined titanium particles: LiCl–KCl–K$_2$TiF$_6$ (0.1 mol %), 723 K, 180 mA, 300 C, bath volume = 0.12 dm^3.

13.3.1 Bench Scale Experiments

As for the particle formation, there exist two possible reaction paths: direct cathodic reduction of the ions dissolved in the melt and indirect reduction in which cathodically formed alkali metal chemically reduces metallic ions dissolved in the melt to produce metal particles. As for the particle growth, on the other hand, there also exist two main possible mechanisms: one is the growth by direct cathodic reduction where suspended metal particles play roles as moving electrodes (for the electrodeposition on the particle surface), and the other is the growth by continuous collision and coalescence of nanoscale particles occurring in the melt.

In order to examine the possibility of the occurrence of these reaction and growth paths, the following experiments have been conducted. First, titanium particles were formed by cathodic discharge electrolysis of LiCl–KCl–K$_2$TiF$_6$. After a quantity

FIGURE 13.7 SEM images of the particles formed by cathodic discharge electrolysis of LiCl–KCl–K$_2$TiF$_6$ (0.1 mol %) melt at 723 K. Volumetric current density: (a) 1.5, (b) 3.0, and (c) 4.5 A · dm^{-3}, respectively.

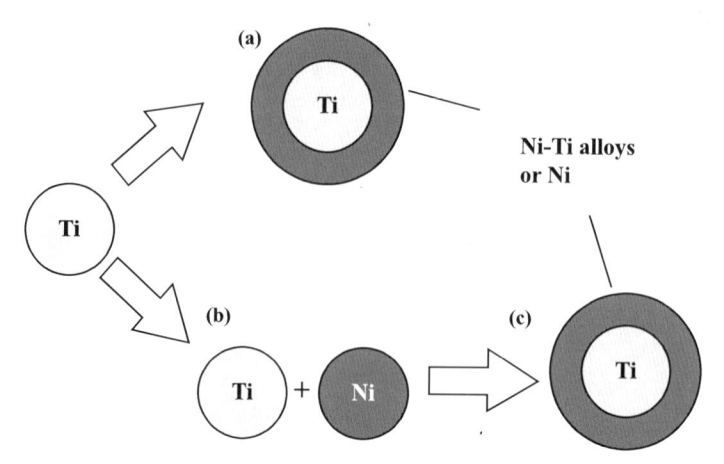

FIGURE 13.8 Conceptual drawing of the possible particle growth pathways: (a) direct Ni–Ti alloy formation, (b) independent formation of Ni and Ti, and (c) collision and coalescence between Ni and Ti particles.

of electricity sufficient for reducing all the titanium ions in the melt was passed, nickel(II) chloride was added to the melt and cathodic discharge electrolysis was again performed. The XRD result suggested the existence of $NiTi_2$, in addition to nickel and TiO_2. The TiO_2 is considered to be formed by exposure of the titanium particles to air or water after electrolysis as described previously. The formation of nickel particles independent of titanium particles suggests that cathodic reduction of nickel ions occurs also in a place other than the titanium particle surface. The formation of $NiTi_2$ together with Ni and Ti particles means that the $NiTi_2$ is formed by a cathodic reduction of nickel ions at the titanium particle surface (Fig. 13.8a), and/or by continuous collision and coalescence of nickel and titanium particles (Fig. 13.8b, c).

On the other hand, even when Ni and Ti particles coexisted in an LiCl–KCl melt for a long period, $NiTi_2$ and other types of Ni–Ti alloy were not obtained. This result suggests that $NiTi_2$ was formed only by direct cathodic reduction of nickel ions at the titanium particle surface. Consequently, it was confirmed that reduction of ions at the particle surface is involved in the particle growth mechanism during cathode discharge electrolysis.

Based on these experimental results, together with other experimental data and knowledge accumulated during those fundamental experiments, several different types of the bench scale electrolytic systems have been designed and constructed, where the mechanisms for the formation and growth of the particles were taken into account. Although it is rather difficult to intentionally choose one specified formation path and one specified growth mechanism at present, it is considered that the most effective way to obtain well-defined fine particles is, at least, to move the formed particles, as soon as possible, away from the place facing the cathode where current flow is most intensive and melt temperature is locally and extremely high, due to the continuously generated cathodic discharge. Based on the above

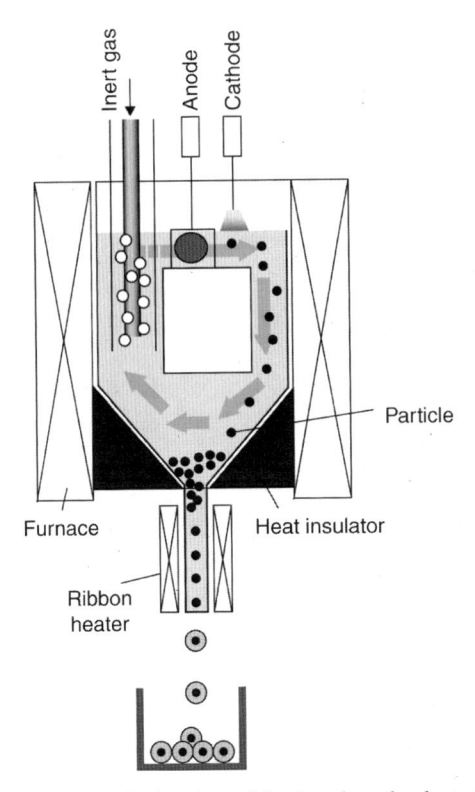

FIGURE 13.9 Schematic drawing of the bench scale electrolysis system.

considerations, melt circulation by means of the lifting power of argon gas bubbled into the melt was adopted. A scheme for the bench scale electrolysis system is shown in Figure 13.9.

The height of the total system is about 1 m. This system has been operated with success, and some results are briefly described in the following: $LiCl-KCl-NiCl_2$ (0.5 mol %) melt was used as an electrolyte at 723 K. A tungsten rod was used as a cathode and nickel plate was used as an anode, which anodically dissolved into the melt to continuously provide nickel(II) ions; the electrolytic current was 1 A. The particles transported to the bottom part of the system were collected through the channel located underneath, which is otherwise closed with a frozen valve. The yields of the collected nickel particles calculated on the basis of Faraday's law stayed around 95%. Details of the experiments will be disclosed in the near future.

13.4 ANODIC DISCHARGE

As described previously, in some cases of anodic discharge electrolysis, the discharge is predominantly maintained by cation emission from the anode. This cation emission

176 PLASMA-INDUCED MOLTEN SALT ELECTROLYSIS

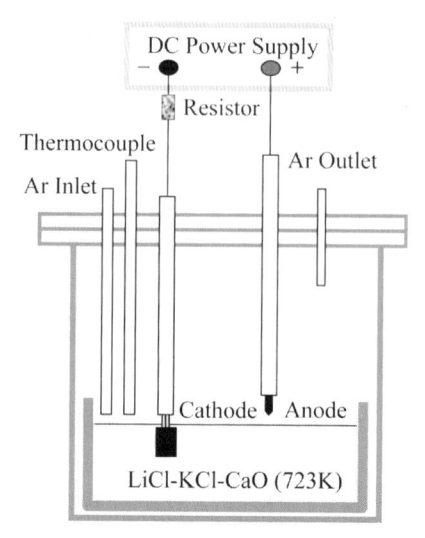

FIGURE 13.10 Scheme of an experimental cell for anodic discharge electrolysis.

process can be applied to metal oxide particle formation. That is, metal ions emitted from the anode can react with oxide ions dissolved in the melt to produce metal oxide particles.

13.4.1 Metal Oxide Particles

A titanium wire was used as the anode and anodic discharge electrolysis was conducted with the use of an LiCl–KCl melt containing CaO as an oxide ion source. Figure 13.10 shows a scheme of the experimental cell, and Figures 13.11 and 13.12

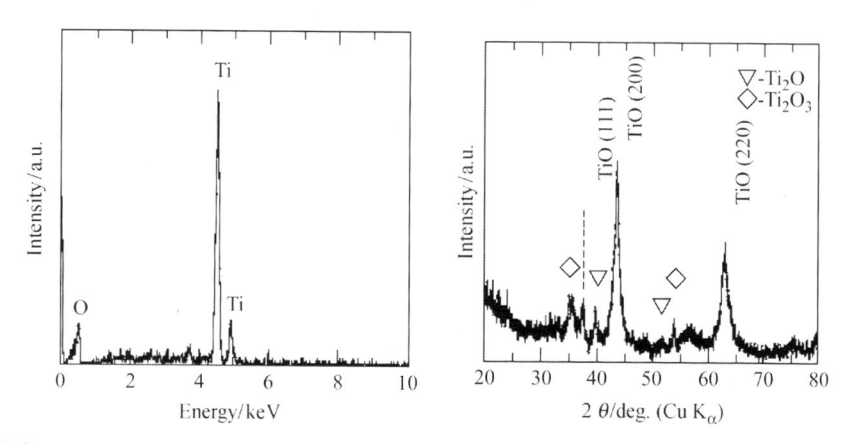

FIGURE 13.11 EDX spectrum and XRD pattern of the particle formed by anodic discharge electrolysis of LiCl–KCl–CaO (0.2 mol %) at 723 K.

ANODIC DISCHARGE 177

FIGURE 13.12 SEM images of the particles formed by anodic discharge electrolysis of LiCl–KCl–CaO (0.2 mol %) at 723 K. Quantity of electricity: (a) 100 C, (b) 175 C, and (c) 250 C, respectively.

show the results of product analysis and SEM images of the formed particles, respectively [9].

13.4.2 Metal Sulfide Particles

Anodic discharge electrolysis was conducted with the use of a molybdenum anode and LiCl–KCl melt containing KSCN as the sulfur source. Particles of MoS_2 have been obtained as shown in Figures 13.13 and 13.14 [10]. When a tungsten anode was used in place of the molybdenum anode, formation of WS_2 was achieved, as shown in Figure 13.15 [10].

13.4.3 Carbon Nitride Particles

As mentioned earlier, anodic discharge is maintained either by cation emission from the anode or electron emission from the melt. Accordingly, various kinds of particles can be formed by the electron emission reaction from the melt. If an electron emission reaction from the melt occurs predominantly, carbide ion dissolved in the melt should donate electrons to produce elemental carbon in the melt. This was

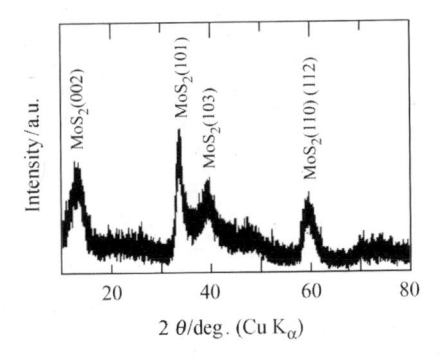

FIGURE 13.13 An XRD pattern and SEM image of the particles formed by anodic discharge electrolysis of LiCl–KCl–KSCN (0.1 mol %) at 723 K (Mo anode).

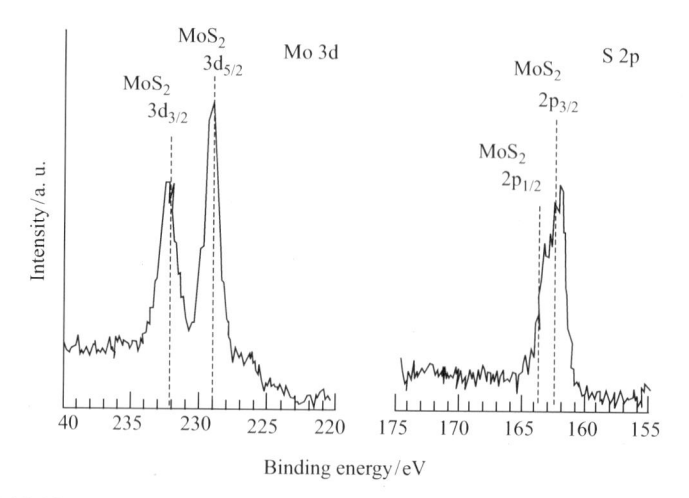

FIGURE 13.14 XPS spectra of the particles formed by anodic discharge electrolysis of LiCl–KCl–KSCN (0.1 mol %) at 723 K (Mo anode).

confirmed by the anodic discharge electrolysis of a molten LiCl–KCl–CaC$_2$ system, where elemental carbon particles were obtained after the electrolysis. Based on this result, the formation of carbon nitride compounds has been conducted by anodic discharge electrolysis of a LiCl–KCl–CaC$_2$–Li$_3$N system with the use of a glassy carbon anode. In this case, carbide ions and nitride ions are expected to donate electrons to be oxidised, and the formed carbon and nitrogen atoms would react to form carbon nitride according to the reaction

$$C_2^{2-} + 2xN^{3-} \rightarrow 2CN_x + (2+6x)e^-$$

After anodic discharge electrolysis, black particles were obtained from the melt. An XRD pattern of the obtained particles was similar to that of graphite [11]. An XPS spectrum for N(1s) indicated the existence of nitrogen atoms predominantly bonded

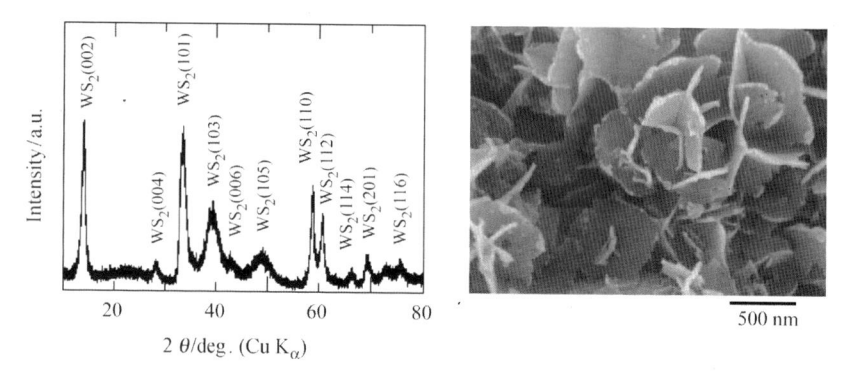

FIGURE 13.15 An XRD pattern and SEM image of the particles formed by anodic discharge electrolysis of LiCl–KCl–KSCN (0.1 mol %) at 723 K (W anode).

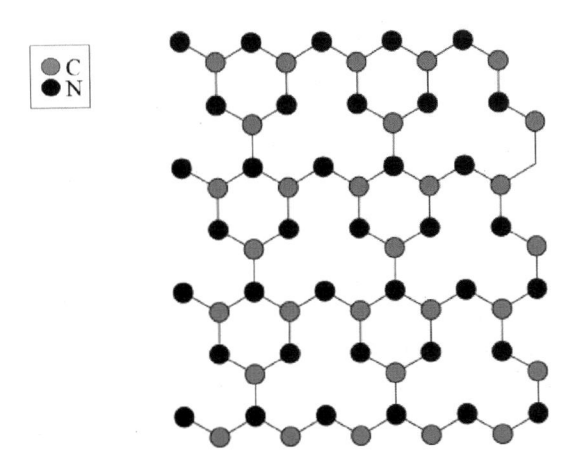

FIGURE 13.16 A conceptual drawing of the structure of particles formed by anodic discharge electrolysis of LiCl–KCl–CaC$_2$ (0.2 mol %)–Li$_3$N (0.5 mol %) at 723 K.

to sp^2-carbon. In addition, a part of the nitrogen seemed to be bonded to sp^3-carbon. Therefore almost all of nitrogen was supposed to be incorporated into the graphite ring, and a part of the nitrogen is considered to have formed other carbon nitride compounds including N–sp^{3}C bonds, such as α-C$_3$N$_4$ and β-C$_3$N$_4$. The N/C ratio was estimated to be 0.04 from the peaks of the XPS spectra. Even taking into consideration contaminated carbon, the N/C ratio is smaller than that of C$_3$N$_4$. From the results described above, the obtained particles are considered to be a carbon nitride compound embedded in a graphite matrix. A possible structure of the obtained particles is schematically shown in Figure 13.16.

13.5 CONCLUSION

Plasma-induced molten salt electrolysis to produce functional fine particles, which has been newly created and developed by the authors and their collaborators, is a very interesting and promising process from the viewpoint of practical applications. Further extensive R&D activities are currently underway at the newly established venture company I'MSEP Co. Ltd., a major target of which is the commercialisation of some of the mentioned highly functionalised materials, including various fine particles.

REFERENCES

1. Ito, Y., Some approaches to novel molten salt electrochemical processes, *Electrochemistry* **68**, 88–94 (2000).
2. Ito, Y., Advances in molten salt electrochemistry towards future energy systems, *Electrochemistry* **73**, 545–551 (2005).

3. Ito, Y., Novel molten salt electrochemical processes directed towards industrial applications, in *Proceedings of the 7th International Symposium on Molten Salts and Chemical Technology*, 2005, 33–39.
4. Ito, Y. and Nishikiori, T., Novel electrochemical reactions related to electrodeposition and electrochemical synthesis, *J. Min. Metallurgy* **39B**, 233–249 (2003).
5. Oishi, T., Goto, T., and Ito, Y., Anode discharge electrolysis of molten LiCl–KCl system, *J. Electrochem. Soc.* **150**, D13–D16 (2003).
6. Kawamura, H., Moritani, K., and Ito, Y., Discharge electrolysis in molten chloride: formation of fine silver particles, *Plasmas and Ions* **1**, 29–36 (1998).
7. Kawamura, H., Moritani, K., and Ito, Y., Discharge electrolysis in molten chloride: formation of fine nickel particles, *J. Jpn. Soc. Powder Metallurgy* **45**, 1142–1147 (1998).
8. Oishi, T., Kawamura, H., and Ito, Y., Formation and size control of titanium particles by cathode discharge electrolysis of molten chloride, *J. Appl. Electrochem.* **32**, 819–824 (2002).
9. Oishi, T., Goto, T., and Ito, Y., Formation of metal oxide particles by anode discharge electrolysis of molten LiCl–KCl–CaO system, *J. Electrochem. Soc.* **149**, D155–D159 (2002).
10. Oishi, T., Goto, T., and Ito, Y., Formation of transition metal sulfide particles by anode discharge electrolysis of molten LiCl–KCl–KSCN system, *Electrochemistry*, **70**, 697–700 (2002).
11. Oishi, T., Hattori, T., Goto, T., and Ito, Y., Formation of carbon nitride by anode discharge electrolysis of molten salt, *J. Electrochem. Soc.* **149**, D178–D181 (2002).

14 Liquid Electrolytes: Their Characterisation, Investigation, and Diverse Applications

KEITH E. JOHNSON

Department of Chemistry and Centre for Studies in Energy and Environment, University of Regina, Regina, Saskatchewan, Canada

Abstract

The origins of liquid electrolytes and their relationships to solids and vapours are discussed. The information available from diffraction, spectroscopic, density, and electrochemical measurements is outlined. It is shown that conductivity is the critical technique for establishing a spectrum of ionicity, ranging from liquid silicates and alkali halides through organic salts and aqueous solutions to the barely conducting ionising solvents. The dramatic effects of high pressure on conductivity are described. Electrochemical applications are distinguished from those for which ionicity is of secondary importance.

> Michael Faraday in the 1830s was the first to investigate systematically, the electrolysis of molten salts and used his results to assist in establishing the fundamental law of electrolysis which bears his name.
>
> It is now accepted that pure molten salts consist predominantly of ions. They differ, therefore, from all other classes of liquids in that they are the only group of pure liquids in which positively and negatively charged particles coexist and could therefore logically be called "liquid electrolytes" or "ionic liquids."
>
> —Harry Bloom, Liverpool 1961
> From the Eleventh Spiers Memorial Lecture in
> *The Structure and Properties of Ionic Melts:*
> *A General Discussion of the Faraday Society*

Molten Salts and Ionic Liquids: Never the Twain? Edited by Marcelle Gaune-Escard and Kenneth R. Seddon
Copyright © 2010 John Wiley & Sons, Inc.

182 LIQUID ELECTROLYTES

14.1 INTRODUCTION

The intent of this chapter is to compare and contrast the various liquid systems that show electrolytic conductance. In particular, we are interested in simple salts such as KCl, solutions of salts in nonconducting but ionising solvents, and more complex organic or semiorganic salts.

The early history of electricity and electrochemistry was discussed by MacInnes [1]. Table 14.1 includes a selection of pertinent work [1–7] prior to 1910. Wilkes [8] has discussed the evolution of the organic and semiorganic systems from ethylammonium nitrate to the popular 1,3-dialkylimidazolium salts. Reviews by Pagni [9] and Burwell [10] are also relevant.

The title of the EUCHEM 2006 Conference includes two subjects not one: Molten Salts and Ionic Liquids. "Molten Salt" implies a salt that is normally solid — it should mean solid in a standard state of 298 K and 1 bar. Most solid salts have ionic lattices and remain ionic upon melting, but, *under normal conditions*, there are exceptions: the molten mercury(II) halides [11] are not totally ionised, PCl_5 is ionic in the solid ($[PCl_4][PCl_6]$) but not the liquid [12] phase, and the molecular solid $HBF_4 \cdot H_2O$ becomes ionic ($[H_3O][BF_4]$) on melting [13]. "Ionic Liquid" implies a liquid composed of ions and thus a salt. It was suggested that this term be restricted to liquids with freezing points below 100 °C, and this notion has been accepted by many and has encouraged research and industrial interest regarding several organic and semiorganic salts that are "new materials" as opposed to cool relatives of KCl. What then should one call pyridinium chloride (mp 144 °C) or pyridinium ethanoate (mp < 25 °C)? Clearly, the specific temperature does not belong in the definition but rather, an ionic liquid is best described as one consisting of ions and ion pairs (or parent molecules); the dominant forces are ion–ion interactions, such that a relaxed ionic lattice with a *pseudo*-Madelung energy exists. The relationship to ionic solids and gaseous salt molecules is thus relevant.

TABLE 14.1 A Little History

Year	Investigator(s)	Activity	Reference
1600	Gilbert	Static electricity	1
1786	Galoani	Frog's legs	1
1795	Volta	Battery and emf series	1
1805	Grotthus	Ionic conduction mechanism	1
1807	Davy	K from liquid KOH	1
1835	Faraday	Law and various terms	1,2
1847	Hittorf	Liquid nitrate electrolysis	3
1886	Hall and Heroult	Al from Al_2O_3 in Na_3AlF_6	4
1887	Arrhenius	Modern theory of electrolytes	4
1890	Gardner	Carboxylic acid/base azeotropes	5
1900	Zawidzki	Acid/base vapour pressures	6
1909	Kaufler and Kunz	Halohydrogenates	7

14.2 MAKING LIQUID ELECTROLYTES

Conceptually, ionic systems arise (1) by electron transfer from a nonmetal to a metal (NaCl, CaO); (2) by proton transfer from acid to base [(C_2H_5)$_3$NH][HCOO]); (3) through self-ionisation of select polar molecules (ICl, H_3PO_4); (4) by dissolution of salts or ionophores in high dielectric constant solvents (aqueous CsCl, aqueous HCl); (5) by modifying basic anions through reaction with Lewis acids ([CO_3]$^{2-}$, [$AlCl_4$]$^-$ salts); and (6) through alkylation of certain protons of cations and anions ([C_4py]$^+$, [C_2mim]$^+$, [$EtSO_4$]$^-$).

It is noteworthy that the mixing of nonconducting molecular compounds can generate modest ionic conducting liquids (1-methylimidazole + HBr) [14]. Furthermore, the dissociative patterns of molecular/ionic liquids are quite varied (cf. H_3PO_4 [15], H_2SO_4 [15], CH_3COOH [16]) and their behaviour in aqueous solutions or mixtures may be quite different (e.g., [HF_2]$^-$ but not [HCl_2]$^-$ survives in H_2O).

14.3 RELATIONSHIP TO SOLIDS AND VAPOURS

Simple ionic solids such as NaCl are poor conductors of electricity. The only charge carriers are interstitial ions (Frenkel defects) and vacancies or holes (Schottky defects). More complex systems, such as mixed valency materials, have functional conductivities. Examples of these are the fluoride ion electrode (LaF_3 + EuF_2), yttria-stabilised zirconia, and $LiCoO_2$ in Li-ion batteries. The salts Ag_4MI_5 (M = K, Rb, Cs) are good Ag^+ conductors in both solid and liquid states [17].

The lattice or Madelung energy stabilises ionic solids and is responsible for the high melting points of the simple inorganic salts. This energy is less for the organic salts, where the charge density is dispersed so melting is easier. Concurrently, glass formation and supercooling of the latter, when liquid, are common.

In the gas or vapour phase, salts are molecular. Besides monomers, there is evidence for the formation of dimers, such as (NaCl)$_2$, (NiCl$_2$)$_2$ [18], and Al_2Cl_6, and even hexamers (the cluster Pd_6Cl_{12}) [19]. The vapours of ethanoic acid and hydrogen halides have been shown to include (CH_3COOH)$_2$ and the monocyclic (HF)$_6$, and possibly (HCl)$_6$ [20]. A salt such as pyridinium ethanoate dissociates into pyridine and ethanoic acid in the vapour [21] since the proton affinity (a gas phase property) of ethanoate exceeds that of pyridine [22]. Quaternary nitrogen and similar salts of select anions may survive in the vapour [23], particularly if their degree of ion pairing in the liquid is substantial. Purification by distillation of both of the latter salt types is feasible [24], although many of the acid/base systems form azeotropes of acid-rich composition [25], so these liquids will not necessarily distill in a straightforward manner [21].

14.4 THEORIES OF LIQUID ELECTROLYTES

There are a number of theories of ionic liquids including a thermodynamic one by Pitzer [26] and the hole theory of Fürth [27], well-documented in the book by Bockris and Reddy [28].

14.5 SPECTROSCOPY AND STRUCTURE

X-ray diffraction studies [29] of liquid alkali halides show two or three coordination spheres, rather broad radial distribution functions (RDFs), coordination numbers less than in the solids, and smaller distances of closest approach. Neutron diffraction of several 1,3-dialkylimidazolium salts shows charge ordering to two anion shells [30] and, in another study, a sharp drop in the Al–Cl–Al angle in [Al$_2$Cl$_7$]$^-$, when going from solid to liquid [30].

Numerous studies of the spectra of transition metal ions in liquid salts took place in the 1955–1975 period, especially in connection with corrosion possibilities with molten salt reactors [31]. In essence, the crystal field stabilisation energies (CFSEs) of the ions and the coordinate site availability must be reconciled. For cobalt(II), clear coordination numbers were observed—tetrahedral (T_d) for halides, dodecahedral (D_{2d}) for nitrates and sulfates—but the picture for nickel(II) was less defined, with *pseudo*-octahedral (O_h) and *pseudo*-tetrahedral (T_d) species together fitting some of the spectra of chloride mixtures [32]. The spectra of cobalt(II) [33] and nickel(II) [34] in Lewis basic organic chloroaluminates support the formation of tetrahedral complex ions, but in Lewis acidic systems octahedral or *pseudo*-octahedral coordination spheres arise, as confirmed by an EXAFS study [35]—coordination of the Lewis acidic transition metal ions by basic [AlCl$_4$]$^-$ rather than by acidic [Al$_2$Cl$_7$]$^-$ is preferred.

Raman spectroscopy has been used by several groups but, in particular, Gilbert has identified Na$_2$AlF$_5$ in cryolite melts [36] and several alkylhaloaluminates [37] in the systems developed by IFP (vide infra). Recently, Hamaguchi and Ozawa [38] recorded the Raman spectra of liquid and crystalline salts of the 1-butyl-3-methyl-1*H*-imidazolium cation, together with X-ray structures of the crystals; of particular interest is their finding of local structures in the liquids, with either *trans* or *gauche* butyl substituents.

Infrared absorption spectroscopy finds obvious applications in monitoring organic cations, hydride formation [39], functional groups of organic products, and so on, but IR emission spectroscopy [40] can also be applied at low as well as high temperatures. ^{1}H NMR spectra are used both to characterise solvent cations and to follow solute protons in combination with anions ([HCl$_2$]$^-$, [H$_2$Br$_3$]$^-$, [HAl$_2$Cl$_8$]$^-$, etc.) [41, 42] and in their reactions with strong bases ([pyH]$^+$, [Hmim]$^+$ [14], etc.). ^{1}H and ^{19}F spin-echo NMR spectra of salts with fluorine in only one of the two ions [43] have been employed to determine self-diffusion coefficients, which predict conductivities exceeding the measured values. The use of ^{13}C spectra to follow organic reactions is routine, except that the solvent shifts for ionic systems are notable [14]. It is also possible to analyse liquids and vapours at equilibrium in these systems by GC/MS [44].

14.6 ELECTROCHEMISTRY

It has become common to report an "electrochemical window" for a liquid electrolyte. This measures the separation in potential between the electrochemical processes of solvent oxidation and reduction, taking place at the particular electrodes used. There may be electro-inactive solutes, however; also, it may be useful to determine the products at the "window frame"—they may explain the electrochemistry of solutes. Generalising among cations can be problematical, such as in the case of the alkali nitrates in which O^{2-}, $[O_2]^{2-}$, and $[O_2]^-$ become involved [3, 45].

Weingärtner and co-workers have obtained dielectric constants for $[EtNH_3][NO_3]$ and some 1,3-dialkylimidazolium salts [46]. The values (14 to 26) suggest that the liquids are neither good ionising solvents (dielectric constant, ε, for water is 80) nor are likely to promote extensive ion pairing of solutes within them (cf. $\varepsilon = 6$ for ethanoic acid). The dielectric constant says little regarding self-ionisation.

14.7 DENSITY

Density measurements are simple. Besides their importance in identifying and separating immiscible phases or mixture compositions, they indicate the molarities of substances of known formulae and, for pure liquid electrolytes, the maximum ionic concentrations. Table 14.2 presents some values for ionic and molecular systems in the form of the content of a 1.66×10^{-24} litre box containing them. Although the space available clearly increases in the sequence solid < liquid < gas, the Li–Cl distance of closest approach decreases in this sequence.

14.8 CONDUCTIVITY: THE CRITICAL MEASURE

If a system contains ions and the ions are free to move, the system will conduct electricity. The procedure for measuring the conductivity of liquid electrolytes is outlined here for convenience.

In practice, the resistance of the liquid between two parallel plate inert electrodes at a fixed separation is measured by means of an applied alternating voltage (the alternating voltage eliminates contributions from the electrode–liquid interface). Comparison with the values for a set of standards such as aqueous KCl solutions [49] is made, and the specific conductance of the liquid reported in siemens per centimeter $(S \cdot cm^{-1})$. From the known or suspected concentrations of the ions present, a molar

TABLE 14.2 A Second Use for Densities—The Content of a 1.66×10^{-24} Litre Box

Medium	Content	Reference
Ideal gas at 298 K and 1 bar	0 (24:1 chance)	
Water at 298 K	55 molecules	
LiCl(l) at 1053 K	35 Li$^+$ + 35 Cl$^-$	47
[S(CH$_3$)$_3$][HBr$_2$](l) at 298 K	<7.3M$^+$, <7.3X$^-$, some MX	48
[emim][Al$_2$Cl$_7$](l) at 298 K	<3.4M$^+$, <3.4X$^-$, some MX	48
LiCl(s) at 883 K	~44Li$^+$, ~44Cl$^-$	27
1 M aq. KCl at 298 K	K$^+$ + Cl$^-$ + 54 H$_2$O	

conductance, in units of S · cm² · mol⁻¹, can be calculated. Conductivity metres, cells, and standard solutions are available commercially.

Specific conductivity, σ, is the practical term—it tells one how well the actual liquid conducts and the corresponding Joule heating from the passage of a given current. The molar (or equivalent) conductivity, Λ, takes into account the concentration of conducting species. For a single liquid salt, the molar value quoted assumes that the salt is fully dissociated into ions; that is, it is a maximum value. For solutions in ionising solvents, the molar or equivalent value is a complex function of the added salt concentration. In particular, good fits of dilute solution data are given by the Debye–Hückel–Onsager equation [28]:

$$\Lambda = \Lambda° - (A + B\Lambda°)c^{1/2}$$

where Λ = conductivity at concentration c, $\Lambda°$ = conductivity at infinite dilution, and A and B include $\varepsilon^{-1/2}$ and $\varepsilon^{-3/2}$, respectively (ε = dielectric constant).

Table 14.3 [15, 28, 48, 51–53] contains a selection of specific and molar conductivities of various liquids, including metals and polar molecules. Specific conductivities, σ, range from 23 S · cm⁻¹ for Li₄SiO₄ through S · cm⁻¹ values for simple salts to mS · cm⁻¹ values for organic salts, H₂SO₄, and aqueous simple salt solutions, down to

TABLE 14.3 A Selection of Electrical Conductivities of Liquids[a]

System	T/K	σ/S · cm⁻¹	Λ/S · cm² · mol⁻¹	M/mol · l⁻¹
HCl	188	1 × 10⁻⁹		
CH₃COOH	298	8 × 10⁻⁹		
H₂O	298	5.7 × 10⁻⁸		
AlCl₃	473	5.6 × 10⁻⁷		
HgCl₂	560	8 × 10⁻⁵		
20% aq. AcOH	298	1.7 × 10⁻³	0.5	(3.414)
[C₄py]BF₄	298	2 × 10⁻³	0.3	5.5
[C₄mim]CF₃CO₂	298	3 × 10⁻³	0.6	5.1
0.5% aq. KHCO₃	298	4.6 × 10⁻³	92	(0.050)
2:1 AcOH/py	298	9 × 10⁻³	1.6	5.5
H₂SO₄	298	1 × 10⁻²		
[C₂mim]Al₂Cl₇	298	1.5 × 10⁻²	4.4	3.4
0.5% aq. KOH	298	2.0 × 10⁻²	298	(0.067)
[S(CH₃)₃][HBr₂]	298	3.4 × 10⁻²	4.7	7.3
2% aq. HCl	298	0.18	325	(0.553)
20% aq. NaCl	298	0.204	52	(3.928)
BiCl₃	520	0.44	37	11.89
LiCl–KCl (059:0.41)	723	1.57	53	29.7
NaCl	1173	3.88	154	25.3
LiCl	1053	7.59	217	35
Li₄SiO₄	2023	23.2	332	7
19% K in KCl	1082	144		
Na	373	1.04 × 10⁵		

[a]Molarities in parentheses refer to ionic solutions. See Reference [50].

$<10^{-7}$ S·cm^{-1} for H$_2$O, CH$_3$COOH, and HCl. Molar conductivities, Λ, in S·cm^2·mol^{-1} are >100 for single alkali halides, close to 50 for other inorganic systems, <10 for organic salts, but >100 for several aqueous solutions.

So far we have omitted the temperature effect(s). Treating conduction as a rate process, conductivities of the inorganic salts show an Arrhenius temperature dependence with activation energies ~ 15 kJ·mol^{-1} [28], while those of the organic salts are best described by empirical VTF (Vogel–Tammann–Fulcher) relations [54], which approximate to activation energies ~ 30 kJ·mol^{-1}. Thus the conductivity of the organic salts is more sensitive to temperature and suggests two processes are involved: molecule dissociation and ion motion.

14.9 EFFECT OF PRESSURE

Our surroundings and the ease of experimentation dictate that most measurements are carried out at constant (atmospheric) pressure. However, measurements at constant volume are enlightening.

The isobaric and isochoric activation energies for molar conductivity are given by

$$(E_\Lambda)_p = -R(\partial \ln \Lambda/\partial(1/T))_p \quad \text{and} \quad (E_\Lambda)_v = -R(\partial \ln \Lambda/\partial(1/T))_v$$

The volume of activation, ΔV_Λ, is given by

$$\Delta V_\Lambda = -RT((\partial \ln \Lambda/\partial P)_T$$

Some values of these parameters for the alkali nitrates at 400 °C and 1 bar are given in Table 14.4 [55]. As the cation size increases, more of the activation energy at constant pressure is required for an increase in volume. Furthermore, the volumes of activation are only 2–12% of the molar volumes, indicating a different charge transport mechanism from that in the solids.

At sufficient pressures the density of the liquid may exceed that of the solid. Measurements at pressures up to 55 kbar showed complex conductivity–pressure behaviour for alkali nitrates, with the activation volume decreasing by 50%.

In the supercritical region, the density can be varied at constant temperature over a wide range by increasing the pressure. Unfortunately, the critical temperatures of alkali halides exceed 3000 K, and those of alkali nitrates exceed their decomposition temperatures. Nevertheless, the moderate conductor BiCl$_3$ [55, 56] and the poorly

TABLE 14.4 Activation Parameters for Alkali Nitrates at 400 °C and 1 bar

Nitrate	$(E_\Lambda)_p$/ kJ mol^{-1}	$(E_\Lambda)_v$/ kJ mol^{-1}	ΔV_Λ/ cm^3 mol^{-1}	V_m/ cm^3 mol^{-1}
LiNO$_3$	15.1	14.6	0.7	40.6
NaNO$_3$	13.4	8.8	3.7	45.9
KNO$_3$	15.5	7.9	6.3	55.4
RbNO$_3$	16.7	7.5	7.2	61.6

conducting mercury(II) halides [11] are amenable to such studies. Using pressures of several kilobar, the molar conductivity of BiCl$_3$ was raised to >100 S cm^2 mol^{-1} at 1200 °C, and that of HgI$_2$ to >10 S·cm^2·mol^{-1} at 300 °C.

One other salt, [NH$_4$]Cl, and two polar liquids were studied, and it was deduced that at the following reduced densities (density/critical density) the following became fully dissociated into ions [11]:

NH$_4$Cl	BiCl$_3$	HgI$_2$	HgBr$_2$	HgCl$_2$	H$_2$O	NH$_3$
1	3.2	3.9	4.2	4.4	6.3	7

What are the possibilities for liquid organic salts? Suppose a nonpolar liquid has a low-lying polar excited state. It appears that sufficient pressure breaks bonds to form ions, not radicals.

14.10 CONCLUSIONS REGARDING THE NATURE OF IONIC SYSTEMS

1. Inorganic salts are highly structured in the solid state, held together by the lattice or Madelung energy, and with electric conduction by defects.
2. Organic salts tend to form glasses (metastable ultraviscous liquids) upon cooling, prior to yielding structured solids.
3. Simple inorganic salts form liquids with conductivities influenced by the electropositive nature of the cation. The liquids have noticeable short-range order, a *pseudo*-Madelung energy, and "holes" permitting transport. While some are fully, or almost fully, dissociated into ions, high P and T regimes can be used to promote ionisation in other liquid salts (and even polar molecules).
4. Liquid organic salts are less concentrated materials and show evidence of incomplete dissociation. Ionic motion in these liquids can be cumbersome as shown by the complex temperature effects on transport properties.
5. Familiar electrolyte solutions usually consist of ions in a sea of molecules. Ions do break the tridymite-like structure of water. Aqueous ionic mobilities often exceed the intrinsic mobilities of ions in liquid organic salts.
6. Some concentrated electrolyte solutions may be considered as ionic if all the molecules are "busy" coordinating ions, for example, constant boiling hydrochloric acid.
7. Salts in the gas phase are molecular, although dimers or even hexamers are possible.

14.11 APPLICATIONS OF LIQUID ELECTROLYTES

Many of the applications of liquid electrolytes are clearly electrochemical in nature. They have tended to involve inorganic salts and aqueous solutions primarily, but

TABLE 14.5 Ion-Dependent Applications of Liquid Electrolytes

1. Electroanalysis: electrogravimetry, coulometry (constant i and v), stripping
2. Electrosynthesis: Kolbe (1843), reductive elimination, adiponitrile (from $CH_2=CHCN$, aqueous Na_2HPO_4, biphasic), PbR_4 (RMgX oxidation at Pb)
3. Electroplating and anodisation
4. Metal winning: Al, Li, Na, Mg (liquid salts); Cu and Zn (hydrometallurgy)
5. Cl_2 and NaOH, F_2, $[ClO_3]^-$, $[ClO_4]^-$, pure MnO_2, $K[MnO_4]$, $K_2[Cr_2O_7]$
6. Batteries: low T, high T (Na/S, Li/FeS)
7. Fuel cells: H_3PO_4 at 200 °C, MCFC (H_2 or CO fuel) at 650 °C, conc. KOH at r.t. (reverse of Stuart electrolyser)
8. Electroflocculation and electrodialysis

liquid organic salts could well see uses in metal winning, electrosynthesis, or fuel cells, for example. Avoiding high-temperature corrosion may need to be balanced against slower desired low-temperature electrode processes. Table 14.5 includes some examples—fuller details of several of them are discussed by Rieger [57].

New processes utilising organic salts include (1) the BASF BASIL preparation of alkoxyphenylphosphine in which the liquid 3-methyl-imidazolium chloride is formed and later decomposed by water [58]; (2) the IFP processes for dimerisation of alkenes [37] in which Lewis acid haloaluminate ions and nickel(II) phosphine complex catalysts are employed; and (3) a developing cellulose conversion process that depends on an organic salt instead of CS_2 as a solvent [58, 59].

Metal-catalysed reactions involving liquid organic salts were discussed recently by Dyson and Geldbach [60], who report the occurrence of over 3000 references through 2004. There is an extensive history of organic chemistry in molten salts, particularly pyridinium chloride [9, 61], which seems to be "on hold."

A wide variety of applications in which ionic liquids are substituted for other materials has been suggested [58]. Many of these have little dependence on the ionicity of the salts. The ionicity is clearly connected to the salt stability and is responsible for unduly low volatility of many systems. Organic cations and anions have their own chemistries—aromatic cations have some aromatic characteristics, for example. Much appropriate background material is available on such items as azolium ions (see Katritzky [62]). The comparison of liquid organic salts with uncharged isomers has been studied, for example, the *PVT* properties of $[NEt_4][BPr_4]$, $[NPr_4][BEt_4]$ and 5,5-dibutylnonane [63].

The ionicity may pose a problem, however; salts, even the supposed hydrophobic ones, tend to be more soluble in water and hence the environment than molecular organic compounds [64]. That suggests that the manufacture (as opposed to preparation) of huge numbers of different liquids should proceed at a cautious rate.

ACKNOWLEDGEMENTS

The author thanks the organisers of EUCHEM 2006 for the opportunity to present this work. Many graduate students and postdoctoral fellows have made sound contributions to various aspects of this work. Recent funding contributions from NSERC (Canada), the Petroleum Technology Research Centre (Regina), and the Saskatchewan Research Council are much appreciated.

REFERENCES

1. MacInnes, D. A., *The Principles of Electrochemistry*, Dover, New York, 1961.
2. Faraday, M., *Experimental Researches in Electricity*, Everyman's Library, London, 1832.
3. Hittorf, W., *Poggendorfs Amalen Phys.* **72**, 481 (1847).
4. Arrhenius, S., *J. Am. Chem. Soc.* **34**, 353 (1912).
5. Gardner, J., *Ber.*, **23**, 1587 (1890).
6. von Zawidzki, J., *Z. Physik. Chem.*, **35**, 129 (1900).
7. Kaufler, F., and Kunz, E., *Ber. Bunsenges. Phys. Chem.* **42**, 385, 2482 (1909).
8. Wilkes, J. S., (a) A short history of ionic liquids—from molten salts to neoteric solvents, *Green Chem.* **4**, 73–80 (2002); (b) Ionic liquids in perspective: the past with an eye toward the industrial future, in *Ionic Liquids: Industrial Applications to Green Chemistry*, R. D. Rogers and K. R. Seddon (Eds.), ACS Symposium Series, Vol. 818, American Chemical Society, Washington, DC, 2002, pp. 214–229; (c) The past, present and future of ionic liquids as battery electrolytes, in *Green Industrial Applications of Ionic Liquids*, R. D. Rogers, K. R. Seddon, and S. Volkov (Eds.), NATO Science Series II: Mathematics, Physics and Chemistry, Vol. 92, Kluwer, Dordrecht, 2002, pp. 295–320.
9. Pagni, R. M., *Adv. Molten Salt Chem.* **6**, 211 (1987).
10. Burwell, R. L., *Chem. Rev.* **54**, 615 (1954).
11. Bardoll, B. and Tödheide, K., *Ber. Bunsenges. Phys. Chem.* **79**, 490 (1975).
12. Porterfield, W. W., *Inorganic Chemistry*, Addison-Wesley, Reading, MA, 1984.
13. Greenwood, N. N. and Martin, R. L., *Q. Rev.* **8**, 1 (1954).
14. (a) Driver, G. and Johnson, K. E., *Green Chem.* **5**, 163 (2003);(b) Johnson, K. E. and Driver, G., in *International Symposium on Ionic Liquids in Honour of Marcelle Gaune-Escard*, H. A. Øye, and A. Jagtøyen (Eds.), Norwegian University of Science and Technology, Trondheim, 2003, p. 233.
15. Greenwood, N. N. and Earnshaw, A., *Chemistry of the Elements*, Pergamon, Oxford, UK, 1984.
16. (a) Kolthoff, I. M. and Willman, A., *J. Am. Chem. Soc.* **56**, 1007 (1934); (b) Kolthoff, I. M. and Bruckenstein, S., *J. Am. Chem. Soc.* **78**, 1 (1956).
17. (a) Owens, B. B. and Argue, G. R., *Science* **157**, 308 (1967); (b) Bradley, J. N. and Greene, P. D., *Trans. Faraday Soc.* **63**, 2516 (1967); (c) Bentie, G. G., *J. Appl. Phys.* **39**, 4936 (1968).
18. Thompson, K. R. and Carlson, K. D., *J. Chem. Phys.* **49**, 4379 (1968).
19. Schäfer, H., Wiese, U., Rinke, K., and Brendel, K., *Angew. Chem.* **6**, 253 (1967).
20. Rankin, K. N., Chandler, W. D., and Johnson, K. E., *Can. J. Chem.* **77**, 1599 (1999).
21. Treble, R. G., Johnson, K. E., and Tosh, E., *Can. J. Chem.* **84**, 915 (2006).
22. (a) Lias, S. G., Liebman, J. F., and Levin, R. D., *J. Phys. Chem. Ref. Data* **13**, 695 (1984); (b) Lias, S. G., Bartness, J. E., Liebman, J. F., Holmes, J. L., Levin, R. D., and Mallard,

W. G., *J. Phys. Chem. Ref. Data* **17**, 1 (1988);(c) Huheey, J. E., Keiter, E. A., and Keiter, R. L., *Inorganic Chemistry*, HarperCollins, New York, 1993;(d) March, J., *Advanced Organic Chemistry*, Wiley, Hoboken, NJ, 1992.
23. Rebelo, L. P. N., Conongia Lopes, J. N., Esperanca, J. M. S. S., and Filipe, E., *J. Phys. Chem. B* **109**, 6040 (2005).
24. Earle, M. J., Esperanca, J. M. S. S., Gilea, M. A., Canongia Lopes, J. N., Rebelo, L. P. N., Magee, J. W., Seddon, K. R., and Widegren, J. A., *Nature* **439**, 831 (2006).
25. Yoshigawa, M., Xu, W., and Angell, C. A., *J. Am. Chem. Soc.* **125**, 15411 (2003).
26. Pitzer, K. S., *J. Am. Chem. Soc.*, **102**, 2902 (1980).
27. Fürth, R., *Proc. Cambridge Philos. Soc.* **37**, 252, 276, 281 (1941).
28. Bockris, J. O. M. and Reddy, A. K. N., *Modern Electrochemistry*, Plenum/Rosetta, New York, 1970.
29. Levy, H. A. P., Argon, P. A., Bredig, M. A., and Danford, M. D., *Ann. N.Y. Acad. Sci.* **79**, 762 (1960).
30. (a) Hardacre, C., Holbrey, J. D., McMath, S. E. J., Brown, D. T., and Soper, A. K., *J. Chem. Phys.* **118**, 273 (2003); (b) Takahashi, S., Suzuya, S., Kohara, S., Koura, N., Curtiss, L. A., and Saboungi, M.-L., *Z. Phys. Chem.* **209**, 209 (1999).
31. (a) Gruen, D. M., in *Fused Salts*, B. R. Sundheim, (Ed.), McGraw Hill, New York, 1964, p. 301; (b) Smith, G. P., in *Molten Salt Chemistry*, M. Blander, (Ed.), Wiley-Interscience, New York, 1964; (c) Johnson, K. E. and Dickinson, J. R., *Adv. Molten Salt Chem.* **2**, 83 (1973).
32. (a) Smith, G. P. and Boston, C. R., *J. Chem. Phys.* **43**, 4051 (1965); (b) *J. Chem. Phys.* **46**, 412 (1967).
33. Hussey, C. L. and Laher, T. M., *Inorg. Chem.* **20**, 4201 (1981).
34. Gable, R. J., Gilbert, B., and Osteryoung, R. A., *Inorg. Chem.* **4**, 1173 (1965).
35. Dent, A. J., Seddon, K. R., and Welton, T., *Chem. Commun.*, 315 (1990).
36. Gilbert, B., Foosn, S. T., and Huglen, R.,U.S. Patent 6639667 (2001).
37. Gilbert, B., Chauvin, Y., and Guibard, I., *Vib. Spectrosc.* **1**, 299 (1991).
38. Hamaguchi, H. and Ozawa, R., *Adv. Chem. Phys.* **131**, 85 (2005).
39. Wassell, D. F., Johnson, K. E., and Mihichuk, L. M., *J. Phys. Chem. B*, **111**, 13578–13582 (2007).
40. Babushkina, O. B., Okresz, S. N., and Nauer, G. E., in *Proceedings of MS7, Toulouse*, 2005, p. 319.
41. Trulove, P. C. and Osteryoung, R. A., (a) *Inorg. Chem.* **31**, 3980 (1992); (b) in *Molten Salts VIII*, Electrochemical Society, Pennington, NJ, 1992, p. 292.
42. Campbell, J. L. E. and Johnson, K. E., *J. Am. Chem. Soc.* **117**, 7791 (1995).
43. Noda, A., Hayamizu, K., and Watanabe, M., *J. Phys. Chem. B* **105**, 4603 (2001).
44. Xiao, L. and Johnson, K. E., *J. Mol. Catal. A Chem.* **214**, 121 (2004).
45. Johnson, K. E., Xiao, L., and Driver, G., in *ACS Symposium 818*, R.D. Rogers, and K. R. Seddon, (Eds.), American Chemical Society, Washington, DC, 2002, p. 230.
46. (a) Wakai, K., Oleinikova, A., Olt, M., and Weingärtner, H., *J. Phys. Chem. B* **109**, 17028 (2005); (b) Weingärtner, H., Knocks, A., Schrader, W., and Kaatze, U., *J. Phys. Chem. A* **105**, 8646 (2001).

47. (a) Plambeck, J. A., *Encyclopedia of Electrochemistry of the Elements, Volume X, Fused Salt Systems*, Marcel Dekker, New York, 1976; (b) Janz, G. J., *Molten Salts Handbook*, Academic Press, New York, 1967. Both give extensive data on high temperature systems.
48. Trulove, P. C. and Mantz, R. C., in *Ionic Liquids in Synthesis*, P. Wasserscheid and T. Welton (Eds.), Wiley-VCH, Weinheim, 2003, p. 112. Data are tabulated, with references for more than 500 systems at 20–35 °C.
49. (a) Lide, D. R.(Ed.), *CRC Handbook of Chemistry and Physics*, 70th ed., CRC Press, Boca Raton, FL, 1989, p. D-221;(b) Lide, D. R.(Ed.), *CRC Handbook of Chemistry and Physics*, 87th ed., CRC Press, Boca Raton, FL, 2006, pp. 5–72, 8–53.
50. This should not be taken as a reference table. The specific conductivities have been gathered from a variety of sources and the individual values, particularly the low ones, could involve 10%, 50% or even 100% uncertainties. The pattern of the data is important and is not in question. The primary sources used are References 15 (molecular liquids), 28 (inorganic salts), 51 (aqueous electrolytes), and 48 (organic salts). Extensive compilations for inorganic [52] and organic[53] salts also exist.
51. Söhnel, O. and Novotny, P., *Densities of Aqueous Solutions of Inorganic Substances*, Elsevier, Amsterdam, 1985.
52. Gaune-Escard, M., private communication.
53. Wasserscheid, P. and Welton, T. (Eds.), *Ionic Liquids in Synthesis*, 2nd ed., Wiley-VCH, Weinheim, 2007.
54. (a) Vogel, H., *Phys. Z.* **22**, 645 (1921); (b) Tammann, G. and Heese, W. Z., *Anorg. Allg. Chem.* **156**, 245 (1926); (c) Fulcher, G. S., *J. Am. Ceram. Soc.* **8**, 339 (1925).
55. Tödheide, K., *Molten Salts I*, Electrochemical Society, Pennington, NJ, 1976, p. 30.
56. Treiber, Von G. and Tödheide, K., *Ber. Bunsenges. Ges.* **77**, 540–548, 1079–1083 (1973).
57. Rieger, P. H., *Electrochemistry*, Prentice Hall, Englewood Cliffs, NJ, 1993.
58. Plechkova, N. V. and Seddon, K. R., Applications of ionic liquids in the chemical industry, *Chem. Soc. Rev.* **37**, 123–150 (2008).
59. (a) Swatloski, R. P., Spear, S. K., Holbrey, J. D., and Rogers, R. D., *J. Am. Chem. Soc.* **124**, 4974 (2002); (b) Barthel, S. and Heinze,T., *Green Chem.* **8**, 301 (2006); (c) Leipner, H., Fischer, S., Brendler, E., and Voigt, W., *Macromol. Chem. Phys.* **201**, 2041 (2000).
60. Dyson, P. and Geldbach, T., *Metal Catalysed Reactions in Ionic Liquids*, Springer, Dordrecht, 2005.
61. Dauzonne, D., Demerseman, P., and Royer, R., *Bull. Chim. Soc. Fr.*, II-601 (1980). This is one of a host of papers by the Royer group on reactions in pyridinium chloride. Many references are to be found in Reference 9.
62. Katritzky, A. R., *Handbook of Heterocyclic Chemistry*, Pergamon, Oxford, UK, 1985.
63. Lind, J. E. Jr., *Adv. Molten Salt Chem.* **2**, 1 (1973).
64. Wells, A. S. and Coombe, V. T., On the freshwater ecotoxicity and biodegradation properties of some common ionic liquids, *Org. Process Res. Dev.* **10**, 794–798 (2006).

15 Protection of a Microstructured Molybdenum Reactor from High Temperature Oxidation by Electrochemical Deposition Coatings in Molten Salts

S. A. KUZNETSOV, A. R. DUBROVSKIY, and
S. V. KUZNETSOVA

Institute of Chemistry, Kola Science Centre RAS, Apatity, Murmansk Region, Russia

E. V. REBROV, M. J. M. MIES, M. H. J. M. DE CROON, and J. C. SCHOUTEN

Laboratory of Chemical Reactor Engineering, Eindhoven University of Technology, Eindhoven, The Netherlands

Abstract

To improve the corrosion resistance of molybdenum, different methods were applied in this study: electrodeposition of hafnium coatings, electrochemical synthesis of hafnium diboride, boronising, siliciding, and boronising of $MoSi_2$. The oxidation resistance of the coatings was investigated by the weight change in an air–water (2.3 vol%) mixture at a temperature of 500 °C for a period up to 700 h. The best results for coatings with a thickness around 5 μm were obtained by boronising of silicided molybdenum substrates ($MoSi_2$/Mo). It was shown that a two-phase microstructure consisting of the $MoSi_2$ matrix phase with 12–15 wt% of the MoB_4 phase greatly improved the oxidation resistance of the molybdenum substrate. The weight gain observed was $6.5 \times 10^{-4}\,\mathrm{mg \cdot cm^{-2} \cdot h^{-1}}$.

Molten Salts and Ionic Liquids: Never the Twain? Edited by Marcelle Gaune-Escard and Kenneth R. Seddon
Copyright © 2010 John Wiley & Sons, Inc.

15.1 INTRODUCTION

Microstructured reactors are the natural platforms for parallel and high-throughput screening of novel catalytic coatings in high exothermic or endothermic reactions at high temperatures [1–3]. A major issue in the development of secondary gas phase screening reactors is to operate all testing sections with different catalyst compositions under the same conditions. Therefore the choice of the basic material for the catalyst support is crucial to provide very fast heat transfer from the reaction zone to the environment. Furthermore, such material should also be stable and catalytically inert in the reaction mixture. The refractory metals can be used as a substrate material due to their high melting points, good corrosion resistances, and relatively low thermal expansion coefficients. Molybdenum was chosen in this study as a basic material due to its high thermal conductivity [4]. However, molybdenum corrodes in contact with aggressive gases and water vapours at elevated temperatures. To achieve protective oxidation of molybdenum, the addition of a third element such as aluminium is often used. Nowadays, coatings are produced mostly by CVD, PVD methods, plasma and detonation spraying. Formation of a protective alumina layer minimises pesting attack at low temperatures. The disadvantage of the method is the low thermal conductivity of alumina, which sets a limit on the thickness of the protective layer within a few microns. At the same time, using ionic melts for coating deposition is highly attractive. Molten salts provide wide possibilities for coatings production by electrodeposition, electrochemical synthesis, and precise surface alloying, by employing noncurrent transfer and reactions of disproportionation [5].

To improve the corrosion resistance of molybdenum, different methods were applied in this study: (1) electrodeposition of hafnium coatings, (2) electrochemical synthesis of hafnium diboride, (3) boronising, (4) siliconising, and (5) boronising of $MoSi_2$.

The electrodeposition of hafnium and hafnium diboride coatings was chosen because they have been successfully used for the protection of niobium from oxidation at temperatures 500–700 °C [6, 7].

Boronising is a thermochemical diffusion surface treatment in which boron atoms are diffused into the surface to form hard borides with the base material [8]. Boronising, being a thermochemical process, is technically well developed and widely used in industry to produce extremely wear-resistant surfaces. Thermal diffusion treatments of boron compounds in molten salt media typically require process temperatures of 700–1000 °C. Boron atoms, owing to their relatively small size and very mobile nature, can diffuse easily into molybdenum, forming borides. A comparative analysis of electrolytic and noncurrent methods of boronising has been performed [9]. Noncurrent transfer of boron from a boron-containing electrode to metallic substrates through molten salts containing boron ions has been reviewed elsewhere [10, 11]. There are a large variety of boron-containing powders, ionic melts, and temperatures applied for noncurrent boronising of different metals [10, 11]. However, there seems to be a unique mechanism of noncurrent boron transfer, including the formation of lower valence boron at the interface of the boron-containing powder and the ionic melt:

$$B + 2B(III) \leftrightarrow 3B(II) \tag{15.1}$$

Boron ions of lower valence reach the metallic substrate by the usual mass-transfer mechanism through the ionic melt and disproportionate at the melt/metal interface, giving energetically more favourable borides:

$$3y\,B(II) + x\,Mo \rightarrow 2y\,B(III) + Mo_xB_y \tag{15.2}$$

Electrochemical boronising has been the subject of a larger number of investigations [10, 11]. The major advantage of electrolytic boriding, as compared with noncurrent boronising, is a faster growth of boride layers. Electrolytic boronising is usually performed at cathodic current densities below $0.2\,A\cdot cm^{-2}$ [10, 11]. Considerably higher cathodic current densities (up to $1.12\,A\cdot cm^{-2}$) were applied in the work of Segers et al. [12] for electrolytic boronising of iron using NaCl–Na[BF$_4$], due to the unusually high diffusion coefficients of boron through the primary Fe$_2$B phase. On the other hand, electrochemical boronising of molybdenum is much more limited by low diffusion coefficients of boron through the growing boride layers.

Molybdenum disilicide is currently one of the potential candidates for oxidation protective coatings. This particular silicide exhibits moderate oxidation resistance, low density, and high electrical and thermal conductivity. Noncurrent deposition of silicide coatings on a molybdenum substrate, for example, of MoSi$_2$ composition, is described by the following equations:

$$2Si + 12F^- + 2SiF_6^{2-} \leftrightarrow 4SiF_6^{4-} \tag{15.3}$$

$$4SiF_6^{4-} + Mo \rightarrow MoSi_2 + 12F^- + 2SiF_6^{2-} \tag{15.4}$$

The total reaction, $2Si + Mo \rightarrow MoSi_2$, is driven by the difference between the thermodynamic activity of pure silicon and that of the siliconised layer. Due to the metallic nature of its bonding, MoSi$_2$ can be electrodischarge machined, thus making it easier for microreactor technology applications. Its deficiencies are low ductility at temperatures below 1000 °C, and the so-called pesting, in which disintegration occurs at temperatures from 400 to 700 °C [13].

Several techniques were developed to deposit the Mo–Si–B coatings on different substrates. A multiphase Mo–Si–B alloy coating, consisting of Mo$_5$Si$_3$, Mo$_3$Si, and Mo$_5$SiB$_2$ phases, was successfully deposited onto Mo–ZrC substrate using a low-pressure plasma spraying method [14]. A high-velocity oxyfuel spraying technique was applied for deposition of Mo–Si–B and Ni–Cr–Mo–Si–B coatings on steel substrates [15, 16]. The effect of coating parameters (spray particle velocity and temperature) on the microstructure and oxidation resistance of the coatings was assessed, and it was found that Mo–Si–B coatings prepared at higher spray velocities are more durable in service [16]. Vacuum-fused Mo–Si–B coatings with addition of Ni and Cr provide a better erosion–corrosion resistance [16]. It is also reported that MoSi$_2$ was previously boronised by molten salt process at 1000 °C for 3 h [17].

The authors reported that oxidation of the MoSi$_2$ phase to Mo$_5$Si$_3$ takes place with simultaneous formation of the Mo$_2$B$_5$ phase. The thickness of the boronised layer was 10–20 mm [17].

Unfortunately, many of the Mo–Si–B multiphase alloys have a poor oxidation resistance in the range of 500–800 °C, due to the competing oxide formation of the molybdenum oxides and the silicate scales [18]. The oxidation kinetics for several systems, Mo$_5$SiB$_2$ (T2 phase)–Mo$_3$Si [19], Mo$_5$Si$_3$B$_x$ (T1 phase)–MoSi$_2$–MoB [20], and boron-doped Mo$_5$Si$_3$ [21], dramatically increase in the 650–760 °C range, where weight loss is observed. Below and above this critical range, the oxidation rate is moderate and depends on the phase composition [21, 22]. It is mainly controlled by diffusion of oxygen through a borosilicate layer and proceeds at a much slower rate [23]. MoO$_2$ oxide-scale formation was observed on Mo–Si–B ternary alloys in dry air above 1000 °C [24]. Water vapours accelerate the growth of the multiphase Mo and MoO$_2$ interlayers in the Mo–Si–B alloys compared to dry air [25]. All of the intermetallics in the Mo–Si–B system (Mo$_5$SiB$_2$, MoSi$_2$, MoB, and Mo$_3$Si) have moderate coefficients of thermal expansion values, ranging from 6.0×10^{-6} K^{-1} for Mo$_5$SiB$_2$, to nearly 14×10^{-6} K^{-1} for the c-axis of Mo$_5$Si$_3$B, and show approximate linear thermal expansion behaviour from room temperature up to 1000 °C [26].

Although these alloys have improved oxidation resistance, as compared to either molybdenum metal or the MoSi$_2$ phase, they are still not appropriate for use in microstructured applications, either due to the relatively large thickness of the protective layers, or due to formation of thick scales that can change the geometry of the channels. Therefore new compositions of protective coatings have to be found that provide sufficient protection even at a thickness of a few microns.

Catalytic coatings can be deposited above the protective layers by different techniques. However, a low metal recession rate is required to guarantee a good stability of catalytic coatings. The engineering goal is estimated to be a mass change of 1 wt% in 100 h, corresponding to a linear mass change rate of 0.0075 mg·cm^{-2}·h^{-1} for the 100 μm thick molybdenum plates designed by Mies et al. [1] for a high-throughput microreactor and tested in this study.

15.2 EXPERIMENTAL

15.2.1 Substrate Materials

Two different types of molybdenum were studied. The samples of the first type (A) had a cylindrical shape and were 6 mm in diameter and 30 mm in length. The samples of the second type (B) were molybdenum sheets (99.99 + wt% Mo; 40 mm × 8 mm), with a thickness of 100 μm. The samples were cleaned by sand jetting before the experiments. Samples A were used in all experiments unless otherwise mentioned. Samples B were designed for application in a high-throughput microreactor [1] and were used in a selected number of experiments to determine the reproducibility of the method.

15.2.2 Electrodeposition of Hafnium and Hafnium Diboride Coatings

Alkali chlorides (NaCl and KCl) were purchased from Prolabo (99.5 % min.). They were dehydrated by continuous and progressive heating just above the melting point under a gaseous HCl atmosphere in quartz ampoules. Excess HCl was removed from the melt by argon. Sodium fluoride (Aldrich 99.5 % min.) was purified by double melt recrystallisation: NaF was dried in a glassy carbon crucible (SU-2000) at 400–500 °C under vacuum, heated up to a temperature 50 °C above its melting point, and finally cooled down 50 °C below the melting point at a rate of $3-4\ °C \cdot h^{-1}$.

Sodium chloride, potassium chloride, and sodium fluoride were mixed in a required ratio, placed in a glassy carbon ampoule (SU-2000 type), and transferred to a sealed stainless steel retort. The latter was evacuated to a residual pressure of 0.67 Pa, first at room temperature and then stepwise at 200, 400, and 600 °C. The cell was heated using a programmable furnace, and temperatures were measured using a Pt-Rh (10 wt %) Pt thermocouple. The retort was then filled with high purity argon (U-grade: <3 ppm H_2O and <2 ppm O_2) and the electrolyte was melted.

Potassium hexafluorohafnate(IV), $K_2[HfF_6]$, was obtained by dissolving type GFI-1 metallic hafnium or "pure-grade" hafnium tetrachloride in a solution of hydrofluoric acid, followed by precipitation with potassium fluoride. After being washed with ethanol and dried in a vacuum, the precipitate was identified with the aid of X-ray powder diffraction, IR spectroscopy, and optical crystallography.

$K_2[HfF_6]$ was added to the NaCl–KCl equimolar mixture, or to NaCl–KCl–NaF melt, for depositing hafnium coatings, and additions of $K_2[HfF_6]$ and $K[BF_4]$ were used for the electrochemical synthesis of hafnium diboride.

Electrodeposition and electrochemical synthesis were carried out in a hermetic electrolyser made of Cr18 N10T stainless steel. A sluice device made it possible to replace the cathodes without disturbing the inert atmosphere over the melt. A molybdenum crucible, whose walls and bottom were lined with metallic hafnium, which served as the anode, was employed as a container for the melt. Electrodeposition of hafnium coatings was carried out mostly from the NaCl–KCl–$K_2[HfF_6]$ (10 wt %)–NaF(5 wt %) melt. The density of anodic current was below $1\ mA \cdot cm^{-2}$, and the cathodic current density was changed between 5 and $10\ mA \cdot cm^{-2}$. Electrochemical synthesis of hafnium diboride was performed from NaCl–KCl–$K_2[HfF_6]$ (5–15 wt %)–$K[BF_4]$(5–10 wt %)–NaF(5 wt %) molten system.

15.2.3 Boronising of Molybdenum

Boronising was carried out using molten salts consisting of NaCl, KCl, $K[BF_4]$, NaF, and B (10 wt %). The preparation of the salts and experimental procedure were similar to that described previously. $K[BF_4]$ was recrystallised from the commercial product in hydrofluoric acid, washed in ethanol, and dried in a vacuum box. Boron (powder, Merck) was used without additional treatment. $K[BF_4]$ and boron were introduced into the solidified melt, the crucible was placed in a retort, which was evacuated at room temperature and then filled by argon, and the salt was heated up to the working temperature. The molybdenum substrates were immersed into the melt through a special tube in the cup of the retort without breach of the inert atmosphere. The samples, after experiments, were washed in distilled water and alcohol and were then weighed. Boronising treatments were performed at 850, 900, 950, and 1050 °C, for 7 h, without current and at cathodic current densities of 5–10 mA·cm^{-2}.

15.2.4 Siliciding of Molybdenum

Siliciding was carried out in a melt consisting of NaCl, KCl, K$_2$[SiF$_6$], NaF, and Si (10 wt %). Siliciding treatments were performed at 850, 900, 950, and 1000 °C, for 7 h, with or without current. In the former case, small cathodic current densities of 5 or 10 mA·cm^{-2} were applied, because the process is limited by silicon diffusion through the growing silicide layer.

15.2.5 Boronising of MoSi$_2$

Boronising was carried out using molten salt systems consisting of NaCl, KCl, K[BF$_4$], NaF, and B (10 wt %). The preparation of the salts and the experimental procedure were similar to that described previously. Boronising of MoSi$_2$ was performed without current for 5 and 7 h, and with a cathodic current density of 5 mA·cm^{-2} for 3 and 5 h at 700, 760, 800, 840, 950, and 1000 °C. Boron was used as the anode for electrochemical boronising. The anodic process is described by the following equation:

$$B - 3e^- \rightarrow B(III) \tag{15.5}$$

Molybdenum does not dissolve in the melt, because its dissolution potential is more positive than that for boron.

15.2.6 Coating Characterisation

All samples will be referred to according with the process type (B—boronising, Si— siliciding), temperature of melts, and the cathodic current density during the synthesis. Identical samples obtained to study the reproducibility of the method will be distinguished by a number in parentheses. In this way, a sample of the siliciding series obtained at 1050 °C at 10 mA·cm^{-2} will be referred to as Si-1050-10 (1) hereafter. Similarly, SiB reperesents boronising of MoSi$_2$/Mo samples, temperature of melts, the cathodic current density, and duration of the synthesis. In this way, the Mo–Si–B/Mo sample obtained at 800 °C at 5 mA·cm^{-2} with a synthesis time of 3 hours will be referred to as SiB-800-5-3 hereafter.

The uniformity of the protective layers was studied by EDS and SEM. To determine distribution of alloying elements from surface to interior, energy dispersive X-ray spectroscopy (EDS) was used. Oxidation experiments were conducted in a tubular oven in which specimens were simultaneously exposed to an oxidising flow of 100 cm^3·min^{-1} for a period of up to 700 h. Two different flows were used: (1) high-purity air containing 10 wt % water vapours, and (2) oxygen containing 10 wt % water vapours. Specimens were periodically retrieved, weighed, and taken for structure analysis. The exposure temperatures were 500 °C for boride coatings, and 500 and 700 °C for silicides. The time between cycles was 16 h (time required for heating up and cooling down the specimens).

15.3 RESULTS AND DISCUSSION

15.3.1 Electrodeposition of Hafnium Coatings

The corrosion resistance of molybdenum was improved by electrochemical deposition of hafnium films with a thickness of ca. 10 μm. Electrodeposition of hafnium coatings was carried out from the NaCl–KCl–$K_2[HfF_6]$(10 wt %)–NaF(5 wt %) melt. The density of anodic current was below $1 \text{ mA} \cdot \text{cm}^{-2}$, and the cathodic current density was changed between 5 and $10 \text{ mA} \cdot \text{cm}^{-2}$. The concentration of the main impurities in the hafnium anode and coatings are given in Table 15.1.

Hafnium coatings with significantly higher purity than that of the hafnium anode were obtained, confirming that metal impurities were removed during electrodeposition. The use of a hafnium getter, placed in the electrolyser around the crucible, made it possible to obtain coatings containing less than 0.01 wt % oxygen.

The coated samples were investigated in a 10 vol % water in oxygen mixture at 500 °C for a period of more than one week. The preliminary results showed the improved behaviour of coated plates in comparison with that of pure molybdenum. The rate of MoO_3 formation was decreased by a factor of 2 on all coated samples. An X-ray photoelectron spectroscopy analysis (XPS), after exposure to the reaction conditions, showed that all the hafnium was oxidised to HfO_2. The uniformity of the HfO_2 layer was studied by energy dispersive X-ray microanalysis (EDS). The results showed a large nonuniformity in the hafnium distribution along the reactor length (Figure 15.1). The SEM data also confirmed a wide distribution in the hafnium particle size in the range of 5–50 μm (Fig. 15.2). In an attempt to improve the deposition procedure, the density of cathodic current was increased stepwise from 10 to $200 \text{ mA} \cdot \text{cm}^{-2}$ and impulse electrolysis was used, but even in these cases a lot of hafnium crystals had a size of greater than 5 μm.

15.3.2 Electrochemical Synthesis of Hafnium Diboride Coatings

The process of electrochemical synthesis of HfB_2 from a NaCl–KCl–$K_2[HfF_6]$ (5–15 wt %)–$K[BF_4]$(5–10 wt %)–NaF(5 wt %) molten system was studied in

TABLE 15.1 Main Impurities in Hafnium Coatings

Material	\multicolumn{11}{c}{Concentration of Impurities/10^3 wt %}
	Fe	Ni	Mn	Si	Al	Cr	Mo	Zr	Ti	Ca	C	N	O
$K_2[HfF_6]$	<0.3	<0.3	<1	3	<1	<1	<0.3	10	<1	<30	—	—	—
Hf anode	20	20	0.1	30	<2	1	50	400	5	10	10	5	10
Coatings	2	2	<0.1	2	≤1	≤0.2	<0.1	400	—[a]	1	2	2	30

[a] Ti was not determined, because its lines were obscured by Zr lines.

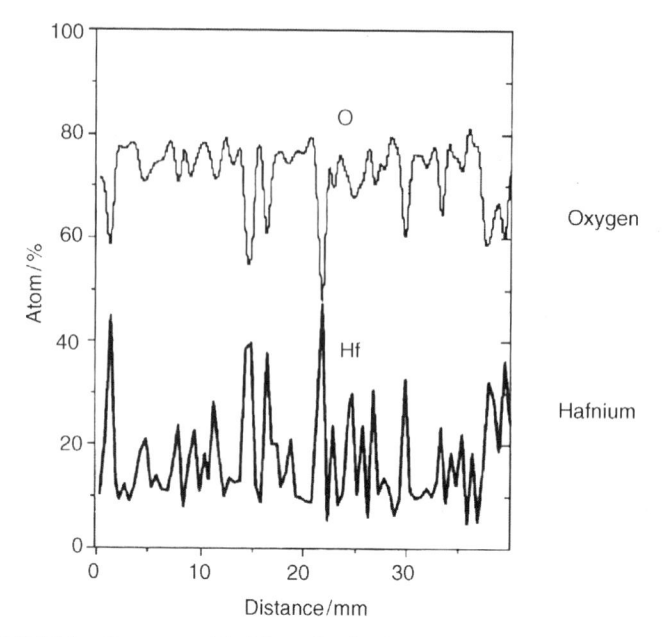

200 PROTECTION OF A MICROSTRUCTURED MOLYBDENUM REACTOR

FIGURE 15.1 Oxygen and hafnium distribution along the reactor compartment.

detail [7]. It was shown that the wave for the electrochemical synthesis of HfB_2 has a more positive potential than the discharge potentials of boron and hafnium. So galvanostatic electrolysis is a convenient regime for deposition of hafnium diboride. The electrochemical synthesis of HfB_2 was performed at a cathodic current density of

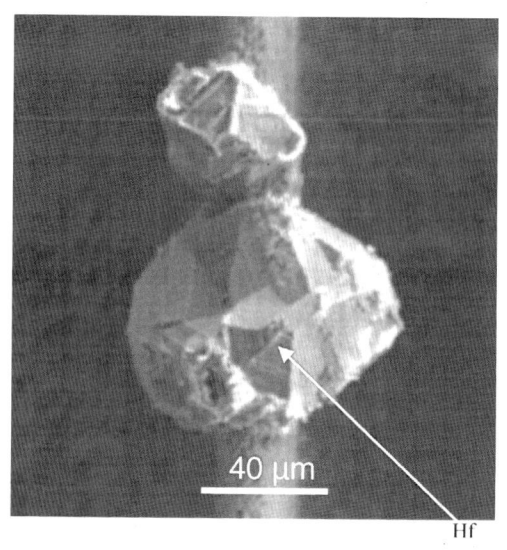

FIGURE 15.2 Scanning electron micrograph of hafnium particles.

FIGURE 15.3 Cross section of hafnium diboride coating on the molybdenum substrate obtained in NaCl–KCl–K$_2$[HfF$_6$](5 wt %)–K[BF$_4$](5 wt %)–NaF(5 wt %) melt. Current density, 30 mA · cm^{-2}; temperature, 850 °C.

5–50 mA · cm^{-2} and a temperature of 700–850 °C. An SEM micrograph of a Cross section of HfB$_2$ on a molybdenum substrate is presented in Figure 15.3.

As can be seen from Figure 15.3 the coating had a column structure, and microhardness determined on the cross section was 3200 ± 250 kg · mm^{-2} [7]. Using a current density ≥70 mA · cm^{-2} in the same molten system led to electrodeposition of two-phase coatings of composition HfB$_2$–Hf. The surface morphology and cross section of the HfB$_2$–Hf coating are presented in Figure 15.4. Hafnium diboride coatings with thickness ≥10 µm showed a good oxidation resistance at 500 °C in an air–water mixture (water content: 2.3 vol %). The coatings of HfB$_2$ are not available for use in microstructured reactors due to the large sizes of the crystals (Figs. 15.3 and 15.4), and consequently to a relatively large thickness of the protective coatings.

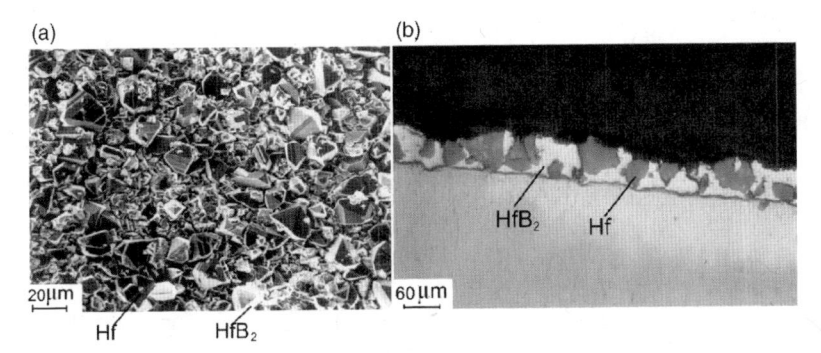

FIGURE 15.4 (a) Morphology and (b) Cross section of two-phase HfB$_2$–Hf coating on the molybdenum substrate obtained in a NaCl–KCl–K$_2$[HfF$_6$](5 wt %)–K[BF$_4$](5 wt %)–NaF (5 wt %) melt. Current density, 100 mA · cm^{-2}; temperature, 850 °C.

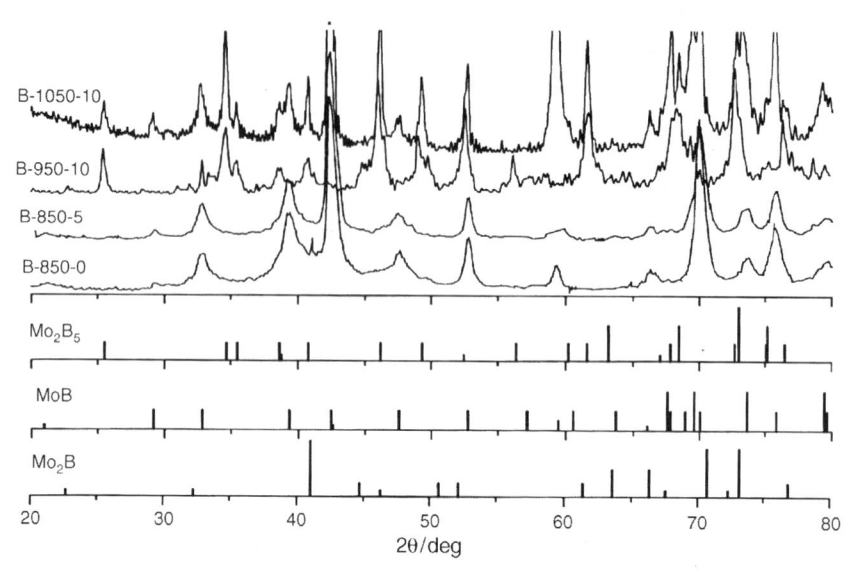

FIGURE 15.5 XRD patterns obtained from the surface of molybdenum samples boronised for 7 h using the molten salt of composition NaCl–KCl, KBF_4, NaF, and B (10 wt %).

15.3.3 Boronising of Molybdenum

Borides are formed by synthesis at 850 °C, as indicated by X-ray diffraction (Fig. 15.5). The thickness and morphology of molybdenum borides are closely related to the process temperature. Representative microstructures are shown in Figure 15.6a. Figure 15.6b shows two EPMA scans across the boronised samples. Based on these data, the composition and thickness of different boride layers were analysed (see Table 15.2). Two boride layers were obtained by diffusing boron atoms into the surface of molybdenum in a temperature range of 850–950 °C: MoB (outermost layer ~1 μm thickness) and Mo_2B intermetallic phase. The formation of MoB can be explained by the fact that Mo_2B acts as a diffusion barrier [27]. Hence the supply of boron for the continued growth of Mo_2B is reduced. Boron atoms accumulating behind the layer of Mo_2B react with it, forming MoB at the outside region.

Similar behaviour was also observed for samples with electrochemically synthesised layers. A thin high-grade boron Mo_2B_5 outermost layer, identified by XRD (see Fig. 15.5), was formed during electrochemical boronising, starting with a melt temperature of 950 °C. However, neither image contrast nor X-ray dot mapping revealed the presence of a boron-rich phase in the cross section of the sample B-950-10; see Figure 15.6b. The surface WDS scan of sample B-1050-10 gave the B/Mo ratio of 2.6.

The following equation describes the synthesis potential of the boride phase Mo_xB_y on the molybdenum cathode (Mo) from the molten salt consisting of

RESULTS AND DISCUSSION   203

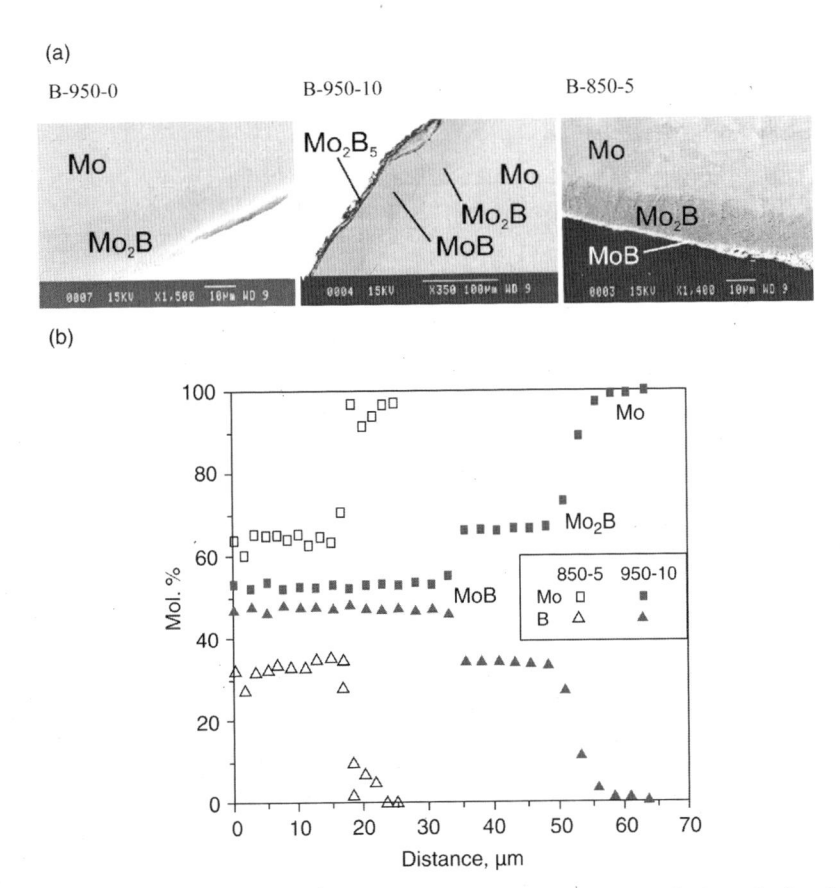

FIGURE 15.6 (a) Scanning electron micrographs of a cross section of a boronised molybdenum sample. (b) WDS scans across the boronised samples shown in (a).

B^{3+} ions [11]:

$$E^{Synthesis}_{Mo_xB_y} = E^{Inert}_B - \frac{\Delta_f G^0_{Mo_xB_y}}{3yF} \qquad (15.6)$$

where E^{inert}_B is the equilibrium deposition potential of boron on the inert cathode without boride formation; $\Delta_f G^0$ is the standard free enthalpy of formation of the given boride; and F is the Faraday constant.

TABLE 15.2 Thickness and Composition of Boride Layers

	Temperature	
Process	850 °C	950 °C
Noncurrent	7 μm (Mo₂B)	16 μm (Mo₂B/MoB)
Electrochemical	14 μm (Mo₂B/MoB)	48 μm (Mo₂B/MoB/Mo₂B₅)

The second term of Eq. (15.6) is the depolarisation due to boride formation. Equation (15.6) shows that the highest depolarisation appears for the boride phase with the lowest boron concentration. Therefore the boride layers with low boron content are formed at the beginning of the boronising experiments. However, due to solid-state diffusion limitations, further boronising will lead to the formation of multilayered phases with increasing boron content from the bulk of the metal to its surface.

From these experiments, it can be concluded that electrochemical boronising gives thicker coatings in comparison with the diffusion process. Furthermore, Mo_2B and MoB phases were obtained in all experiments, while relatively large temperatures exceeding 950 °C and a relatively high cathodic current density were required for formation of the Mo_2B_5 phase. Obviously, the most boron-rich phases form only if there is a high supply of boron atoms to the sample surface.

Figure 15.7 shows the oxidation kinetics data obtained on boronised samples and a sample of pure molybdenum, which were tested in a oxygen–water mixture. Two types of oxidation kinetics can clearly be distinguished. Pure molybdenum and all samples that do not contain the Mo_2B_5 phase showed parabolic weight gain at 500 °C. Several samples were tested also at higher temperatures (550 °C) and they showed similar behaviour. X-ray diffraction of the scales showed that the oxidation scale contained MoO_3.

Furthermore, it is clear that samples B-950-10 (1-3) and B-1050-10, both containing the Mo_2B_5 phase, showed a linear weight gain. The rate of linear weight gain at 500 °C was 0.0053 mg · cm^{-2} · h^{-1}, which satisfies the engineering goal stated earlier.

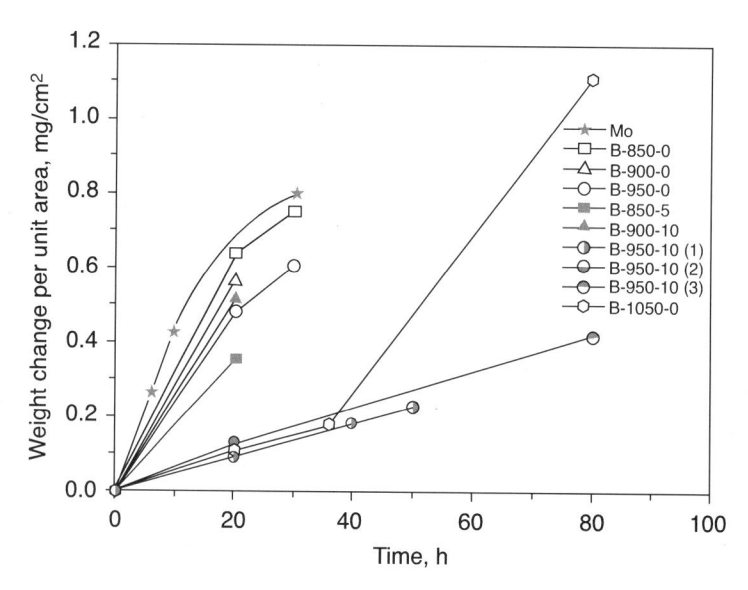

FIGURE 15.7 Oxidation behaviour of boronised samples in air–water mixtures. Temperature, 500 °C; gas flow velocity, 100 cm^3 · min^{-1} (STP); water content, 2.3 vol %. All samples belong under type A, except for sample B-950-10 (3) (type B).

It appears that a Mo_2B_5 layer produces superior oxidation resistance and mechanical properties than a single phase does. However, the specific volume and coefficient of thermal expansion of Mo_2B_5 are quite different from that of the molybdenum substrate, and the presence of several internal layers (Mo_2B, MoB) results in cracking and scaling because of large differences in their inherent stress level. As a result, after three temperature excursions between 20 and 500 °C (heating rate: $10 \, °C \cdot min^{-1}$), sample B-1050-10 started to oxidise in a similar way to pure molybdenum. Another problem is the low mechanical strength of the samples obtained in the temperature range above 950 °C. Because of this, the effect of boronising time should be further investigated.

Summarising, one may conclude that the Mo_2B_5 phase is especially desirable for application as protective coatings. However, an overall thickness of the protective layer has to be further optimised to decrease formation of Mo-rich phases and to improve the mechanical strength of the samples.

15.3.4 Siliciding of Molybdenum

The $MoSi_2$ phase is obtained in the siliciding process. After synthesis at 850 °C, a small amount of the Mo_5Si_3 phase is also observed (see Fig. 15.8). This phase was not observed at higher temperatures. Figure 15.9a shows typical SEM photos and

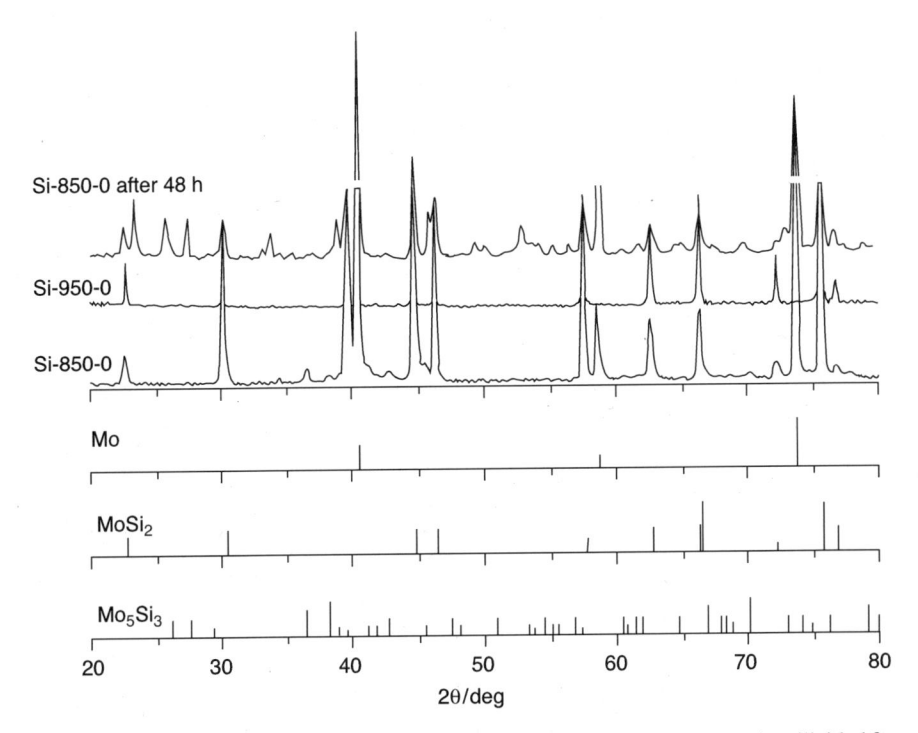

FIGURE 15.8 XRD patterns obtained from the surface of molybdenum samples silicided for 7 h using a molten salt of composition NaCl–KCl, K_2SiF_6, NaF, and Si (10 wt %).

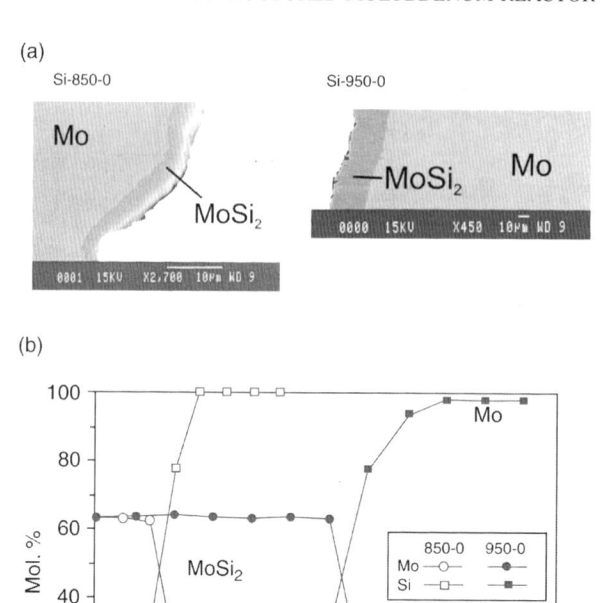

FIGURE 15.9 (a) Scanning electron micrographs of a cross section of silicided molybdenum samples. (b) WDS scans across the silicided samples shown in (a).

Figure 15.9b demonstrates EPMA scans across the silicided samples, where the MoSi$_2$ phases can clearly be identified. The thickness of molybdenum disilicide depends on the process temperature and increases from 6 μm at 850 °C to 20 μm at 950 °C. With increasing coating thickness, the Mo peaks disappear from the XRD pattern.

Figure 15.10 confirms that MoSi$_2$ layers are prone to nonprotective oxidation (a phenomenon known as *pesting*) at low temperatures (500–700 °C) in oxygen–water mixtures, primarily because silica growth rates to form an external continuous scale are extremely low [28]. The XRD pattern of sample Si-850-0 taken after cyclic oxidation for 48 h at 500 °C showed the formation of the Mo$_5$Si$_3$ phase (see Fig. 15.8), which is further oxidised to molybdenum, as was observed by the increasing intensity of the Mo peaks at $2\theta = 40.5°$, $58.7°$, and $73.7°$:

$$Mo_5Si_3 + 3O_2 \rightarrow 5Mo + 3SiO_2 \tag{15.7}$$

It should be noted that the rate of mass gain at 500 °C decreases with increasing MoSi$_2$ coating thickness. The latter, in turn, increases with increasing synthesis temperature. Furthermore, at the same synthesis temperature, the MoSi$_2$ thickness is always higher for coatings obtained by an electrochemical process in comparison with

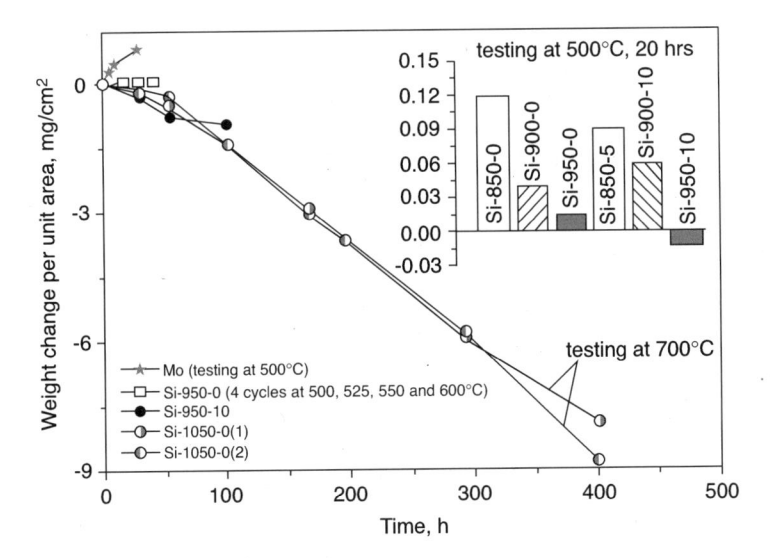

FIGURE 15.10 Oxidation behaviour of silicided samples (type A) in an air–water mixture. Gas flow velocity, 100 cm^3 min^{-1} (STP); water content, 2.3 vol %.

that obtained by a diffusional one. Sample Si-950-0 demonstrated a small weight gain after four oxidising cycles at 500, 525, 550, and 600 °C in an oxygen–water mixture for 40 h (see Fig. 15.10). However, as the testing proceeds, a layer detachment occurs, followed by fast oxidation of molybdenum. The formation of MoO$_3$ at the Mo/MoSi$_2$ interface due to oxygen diffusion is accompanied by volume expansion. As a result, cracks are formed in the coating layer, followed by fast oxidation and coating delamination.

At 700 °C in an air–water mixture, there is a linear weight loss with a rate of 0.0194 mg · cm^{-2} · h^{-1}. Neither the Mo$_5$Si$_3$ nor the MoO$_3$ phase was detected in samples Si-950-0 and Si-1050-10 after testing for 400 h in an air–water mixture. The surface of the silicided samples becomes black after oxidation tests. These results are in line with the literature data reporting that there is a transition from weight gain to weight loss at 650 °C, where the MoO$_3$ phase begins to volatilise [18]. Actually, at 700 °C, the silicided layer provides insufficient protection for the molybdenum toward oxidation. Thus molybdenum (as MoO$_3$) is continually lost, with kinetics of a linear weight loss.

15.3.5 Boronising of MoSi$_2$

According to the results obtained in the preceding section, the duration of the siliciding treatment was fixed at 7 h to obtain a layer of MoSi$_2$ of ca. 5 μm. Figure 15.11a shows a typical XRD pattern of the silicided substrates. MoSi$_2$ is the only phase that is formed on the molybdenum substrate at 900 °C. The appearance of Mo peaks in the XRD pattern of MoSi$_2$/Mo is due to the small thickness of the MoSi$_2$

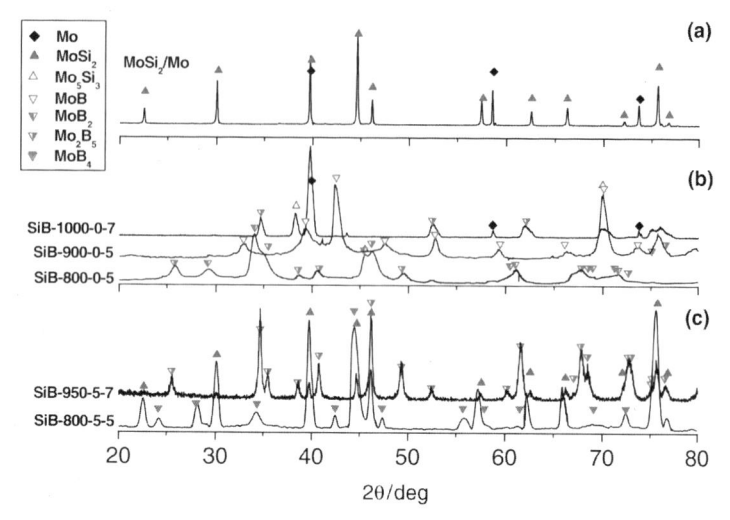

FIGURE 15.11 XRD patterns obtained from the surface of (a) a molybdenum sample silicided for 7 h at 900 °C using a molten salt of composition NaCl–KCl, K$_2$[SiF$_6$], NaF, and Si (10 wt %); (b) MoSi$_2$–Mo samples boronised using a molten salt of composition NaCl–KCl, K[BF$_4$], NaF, and B (10 wt %) in a noncurrent process; and (c) MoSi$_2$–Mo samples boronised in an electrochemical process using the same molten salt of composition as in (b).

layer, which is still transparent for the X-ray beam. Therefore the molybdenum underlayer is also observed in the XRD pattern.

In the following experiments, the influence of the boronising temperature, time, and the nature of the process (electrochemical or noncurrent) was investigated in order to obtain different Mo–Si–B composites, and to study their oxidation behaviour in an air–water mixture.

Noncurrent transfer of boron from a boron-containing electrode to metallic substrates through molten salts containing boron ions has been described in the introduction. Boron ions of lower valence reach the MoSi$_2$ coating by the usual mass transfer mechanism through the ionic melt, and disproportionate at the melt/MoSi$_2$ interface giving the most thermodynamically stable borides:

$$3y\,\mathrm{B(II)} + 6x\,\mathrm{MoSi_2} \rightarrow 2y\,\mathrm{B(III)} + \mathrm{Mo}_x\mathrm{B}_y + x\mathrm{Mo_5Si_3} + 9x\,\mathrm{Si} \qquad (15.8)$$

The thickness of the MoSi$_2$ layer after the boronising step does not change over the whole temperature range. This is because the interdiffusion coefficients of Mo$_x$B$_y$Si are rather small and the major part of the silicon remains on the surface. After boronising, a loss of mass of the samples is observed, because silicon is poorly attached to the surface and is removed during the washing procedure. The elemental analysis of the powder demonstrated that it consisted only of silicon.

Figure 15.11b shows typical XRD patterns obtained after boronising the MoSi$_2$ coatings in the molten salt under diffusional synthesis conditions. Only a few XRD patterns are presented for the sake of clarity. For all samples, the boronised coatings were homogeneously distributed over the surface. The phase array is listed in

Table 15.3. It can be seen that the MoSi$_2$ phase is partially transformed to the Mo$_5$Si$_3$ phase at temperatures equal to or below 950 °C, while it is completely transformed to the Mo$_5$Si$_3$ and boride phases at 1000 °C (see Eq. (15.8)). At the same time, Mo peaks appear in the XRD pattern of MoB-1000-0-7.

The composition of the boride phases depends strongly on the boronising temperature. A noncurrent synthesis at 750 °C for 5 h results in the formation of major amounts of a boride phase, MoB$_2$, which becomes the matrix phase at 800 °C. Increasing the duration of the synthesis at 800 °C to 7 h leads to the formation of the Mo$_2$B$_5$ as the matrix phase. Further increase of the boronising temperature to 900 °C results in the formation of the MoB phase after 5 h, which transforms to the Mo$_2$B$_5$ phase after 7 h. Increasing the molten salt temperature to 1000 °C leads to the formation of the Mo$_2$B$_5$ phase, which is the only boride phase in the coatings under these conditions.

Figure 15.11c shows XRD patterns obtained after boronising of the MoSi$_2$ coatings in the molten salt under electrochemical synthesis conditions. The main advantage of electrolytic boronising compared with noncurrent boronising is a faster growth of borides and a possibility to obtain boron-rich phases at the same composition of an ionic melt. The phase array after electrochemical synthesis is also listed in Table 15.3. The main difference between noncurrent and electrochemical boronising is that there are no peaks corresponding to the Mo$_5$Si$_3$ phase in the XRD patterns over the whole temperature range studied. That means the silicide layer is partially transformed to the boride phases, without its reduction to a molybdenum-rich phase.

$$y\,B(III) + x\,MoSi_2 + 3y\,e^- \rightarrow Mo_xB_y + 2x\,Si \qquad (15.9)$$

The loss of mass of the samples was also observed after electrochemical boronising, because silicon was removed during the washing procedure. The composition of the resulting boride phases depends strongly on the temperature. In contrast to noncurrent boronising, the MoSi$_2$ phase is only slightly transformed to the boride phases at 800 °C. A boron-rich MoB$_4$ phase is obtained after boronising for 5 h. Further increase of the temperature to 900 °C results in the formation of the Mo$_2$B$_5$

TABLE 15.3 Phase Array of Mo–Si–B Coatings[a]

	MoSi$_2$ 6-681	Mo$_5$Si$_3$ 34-371	β-MoB 6-644	MoB$_2$ 6-682	Mo$_2$B$_5$ 38-1460	MoB$_4$ 20-1236
SiB-750-0-5	+	+ +		+ + +		
SiB-800-0-5	+	+ +		+ + +	+	
SiB-800-0-7	+	+ +		+	+ + +	
SiB-900-0-5	+	+ +	+ + +			
SiB-950-0-7	+	+ + +			+	
SiB-1000-0-7		+ + +			+ +	
SiB-800-5-5	+ + +					+ +
SiB-900-5-5	+ +		+ +		+ + +	
SiB-950-5-7	+ +				+ + +	

[a] + + +, matrix phase (>60%); + +, major phase (10–30%); +, minor phase (<10%).

TABLE 15.4 Phase Array of Mo–Si–B Coatings[a]

	MoSi$_2$ 6-681	Mo$_5$Si$_3$ 34-371	Mo$_5$(B,Si)$_3$ 9-292	MoB$_2$ 6-682	Mo$_2$B$_5$ 38-1460	MoB$_4$ 20-1236
SiB-760-0-5	+	+ +		+ + +		
SiB-760-0-7		+ +		+ + +		
SiB-800-0-5	+	+ +		+ + +	+ +	
SiB-800-0-7	+	+ +		+	+ + +	
SiB-840-0-5	+	+ +		+ +	+ + +	
SiB-840-0-7	+	+ +		+	+ + +	
SiB-760-5-3			+ + +	+		+
SiB-760-5-5			+ + +	+ +		+
SiB-800-5-3	+ + +			+		+ +
SiB-800-5-5	+ + +					+ +
SiB-840-5-3	+ + +			+ +	+ +	+
SiB-840-5-5	+ +			+ +	+ + +	+

[a] + + +, matrix phase (>60%); + +, major phase (10–30%); +, minor phase (<10%).

and MoB phases. The latter was also a major phase in the noncurrent boronising under similar conditions (see Table 15.4). Upon further increasing of temperature to 950 °C, the MoSi$_2$ becomes a major phase and the Mo$_2$B$_5$ is the only boride phase formed in the coatings. It should be noted that the Mo$_2$B$_5$ phase was also formed at 950 °C during direct electrochemical boronising of the molybdenum substrates.

The results of cyclic oxidation tests at 500 °C in a flow of wet air containing 2.3 vol % water are shown in Figure 15.12. The sample of pure Mo and the best sample of boronised molybdenum (B-950-10-7) are shown for comparison. The experimental results showed that the MoSi$_2$ samples boronised at 900 °C exhibited linear weight gain that exceeds that observed on the control sample B-950-10-7. The thickness of

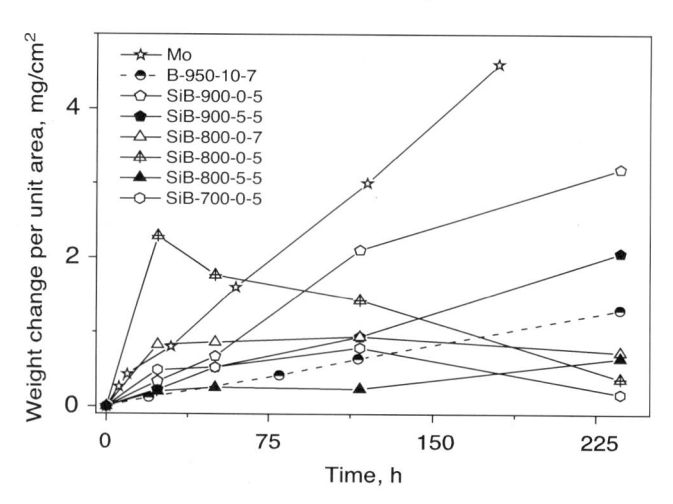

FIGURE 15.12 Oxidation behaviour of Mo–Si–B samples in an air–water mixture. Gas flow velocity, 100 cm$^3 \cdot$ min^{-1} (STP); water content, 2.3 vol %.

the boride layer in the samples boronised at 900 °C and above exceeds 5 µm; therefore boronising also takes place far beyond the MoSi$_2$ layer. On the contrary, weight loss is observed after a short induction period on the samples obtained by noncurrent boronising at 750 and 800 °C for 5 and 7 h. In the latter case, boron diffusion is limited to the MoSi$_2$ layer, which partially oxidises to Mo$_5$Si$_3$. The oxidation resistance of the Mo$_5$Si$_3$ phase is rather low, and it transforms to the SiO$_2$ and MoO$_3$ phases. This process is responsible for initial weight gain and it results in a large volume expansion and development of cracks. In the second stage, the MoO$_3$ evaporates via the cracks from the surface, resulting in the weight loss. However, the oxidation resistance of the specimen boronised at 800 °C (SiB-800-5-5) was double that of the boronised molybdenum (B-950-10-7). The average weight gain observed was 0.0053 mg · cm^{-2} · h^{-1} for B-950-10-7, and 0.0027 mg · cm^{-2} · h^{-1} for SiB-800-5-5. Therefore it was decided to investigate a range of synthesis conditions close to those for sample SiB-800-5-5. In this way, three temperatures of 760, 800, and 840 °C, and three synthesis times of 3, 5, and 7 h were investigated during electrochemical and noncurrent boronising of MoSi$_2$ coatings. In these experiments, the duration of electrochemical boronising was usually shorter compared to noncurrent boronising because the former method results in a faster growth of borides.

Figure 15.13 shows typical XRD patterns of the boronised MoSi$_2$/Mo samples. The phase array is listed in Table 15.4. It can be seen from Figure 15.13 and Table 15.4 that a difference in boronising temperature of 40 °C leads to a different phase composition of the resulting coatings. The MoSi$_2$ phase is almost completely transformed to the Mo$_5$Si$_3$ phase in a noncurrent process. In contrast, the MoSi$_2$ phase is present as the matrix or a major phase after electrochemical synthesis, except the synthesis at 760 °C, when Mo$_5$(B,Si)$_3$ is obtained as the matrix phase. The electrochemical reaction of

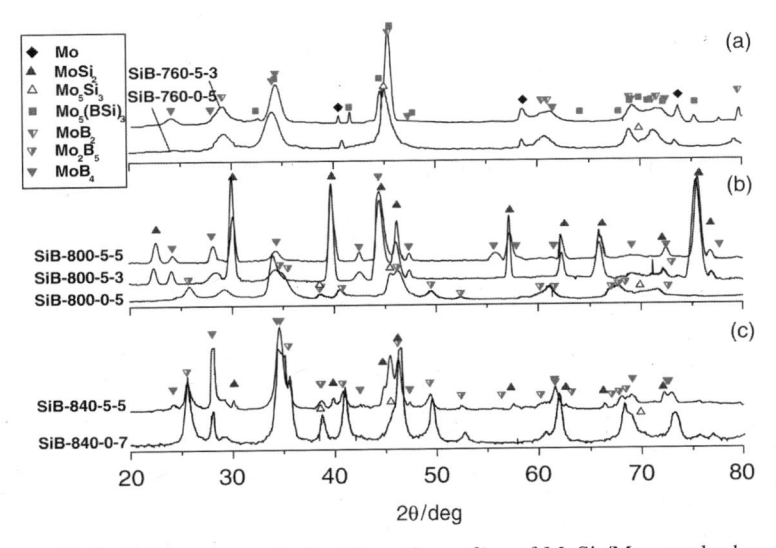

FIGURE 15.13 XRD patterns obtained from the surface of MoSi$_2$/Mo samples boronised (a) at 760 °C, (b) at 800 °C, and (c) at 840 °C, using the molten salt of composition NaCl–KCl, K[BF$_4$], NaF, and B (10 wt%).

discharge of the boron cations on MoSi$_2$ yielding the Mo$_5$(B,Si)$_3$ phase can be expressed by the following equation:

$$5\text{MoSi}_2 + 3\text{B(III)} + 9e^- \rightarrow \text{Mo}_5(\text{B, Si})_3 + 7\text{Si} \quad (15.10)$$

Thus the process of boron discharge and alloy formation is a single electrochemical step. It should be noted that the Mo$_5$(B,Si)$_3$ does not form in noncurrent boronising due to the low solubility of boron atoms in the Mo$_5$Si$_3$ phase [29]. However, this does not prohibit the occurrence of reaction (15.10) in electrochemical boronising.

Molybdenum diboride (MoB$_2$) was the matrix boride phase at 760 °C in noncurrent synthesis and it was a major phase after electrochemical synthesis in the whole range of synthesis times. Minor amounts of the MoB$_4$ phase (<10 wt %) were formed in the latter case. Increasing the temperature to 800 °C, the amount of the MoB$_4$ phase rises to ca. 12–15 wt % in electrochemical boronising, while the Mo$_2$B$_5$ phase becomes a major phase after noncurrent boronising. By increasing the duration of noncurrent boronising from 5 to 7 h, as well as increasing the temperature to 840 °C, the Mo$_2$B$_5$ phase becomes the matrix phase. It also becomes the matrix phase in electrochemical synthesis at 840 °C for 5 h. There is a clear trend in the amount of the MoB$_4$ phase present in the coatings. Comparing the data of Tables 15.3 and 15.4, it can be seen that, first of all, the amount of the MoB$_4$ phase increases when the synthesis temperature rises from 760 to 800 °C, passes through a maximum at 800 °C, and then decreases to zero at 900 °C. Second, the amount of the MoB$_4$ phase does not depend on the synthesis time in the whole temperature range. Finally, it is only the boride phase at 800 °C after 5 h.

The reason for the formation of a single MoB$_4$ phase can be explained by the presence of a rather thick MoSi$_2$ layer on the top of the substrate material. During the electrochemical treatment at the temperature used, active boron atoms are adsorbed on the surface followed by diffusion into the substrate. The MoSi$_2$ layer on top of the substrate acts as a barrier and reduces the diffusion rate of boron atoms. The concentration of active boron atoms at the surface becomes so high that MoB$_4$ is formed, not Mo$_2$B$_5$. By increasing the boron diffusion rate, the Mo$_2$B$_5$ becomes the main phase at boronising above 800 °C.

Figure 15.14 shows typical SEM images taken after boronising of the MoSi$_2$/Mo substrates. All samples have rather uniform thickness of the resulting protective layer of ca. 5–6 μm. There is a single protective layer present in the coatings. The result is

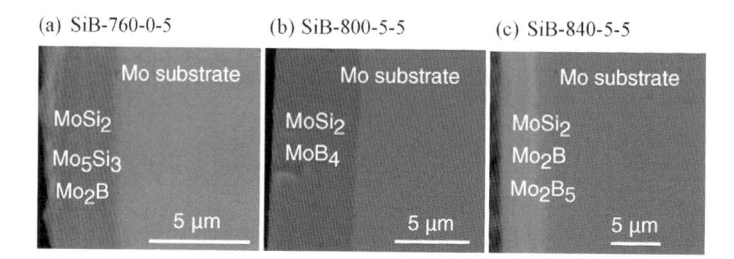

FIGURE 15.14 Scanning electron micrographs of cross section of boronised MoSi$_2$/Mo samples.

RESULTS AND DISCUSSION

TABLE 15.5 Thickness of Protective Layers After Noncurrent Siliciding and Electrochemical Boronising of Molybdenum Samples

	Temperature/°C			
Process	800	850	900	950
Siliciding (noncurrent, 7 h)	<1 µm	3 µm	7 µm	20 µm
Boronising (10 mA · m^{-2}, 7 h)	7 µm	14 µm	n/d	48 µm

opposite to what we obtained during boronising of a pure molybdenum substrate (B-950-10-7) when several boride layers (Mo$_2$B, MoB, and Mo$_2$B$_5$) were observed on top of each other.

Table 15.5 shows the thickness of the resulting coatings after noncurrent siliciding and electrochemical boronising of the original molybdenum substrate. One can see that boronising at 800 °C with a cathodic current density of 10 mA · cm^{-2} results in a similar layer thickness to that obtained after siliciding at 900 °C. At higher temperatures, boronising of the Mo substrate beyond the MoSi$_2$ phase takes place, while at lower temperatures boron dissolves in the silicide layer yielding the Mo$_5$(B,Si)$_3$ phase. Therefore a temperature of 800 °C appears to be the most appropriate for the boronising step. Figure 15.15 shows that the Mo and Si distribution is rather homogeneous both along and across the plate. Therefore one can conclude that no new bulk silicon-containing phases are formed during diffusion of the boron atoms. The thickness of the MoSi$_2$ layer is ca. 5 µm, which is close to the data of Figure 15.14.

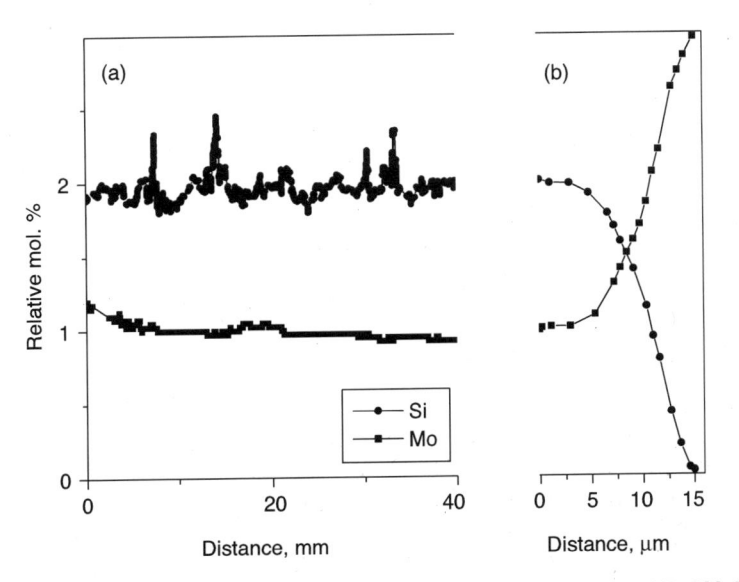

FIGURE 15.15 EDS scans (a) along the length of 4 cm of boronised sample SiB-800-5-5 and (b) at the cross section. The signal of boron is not shown because its lines overlap with those of molybdenum. The relative molybdenum content was taken as 1, and the silicon content is presented relative to the molybdenum content.

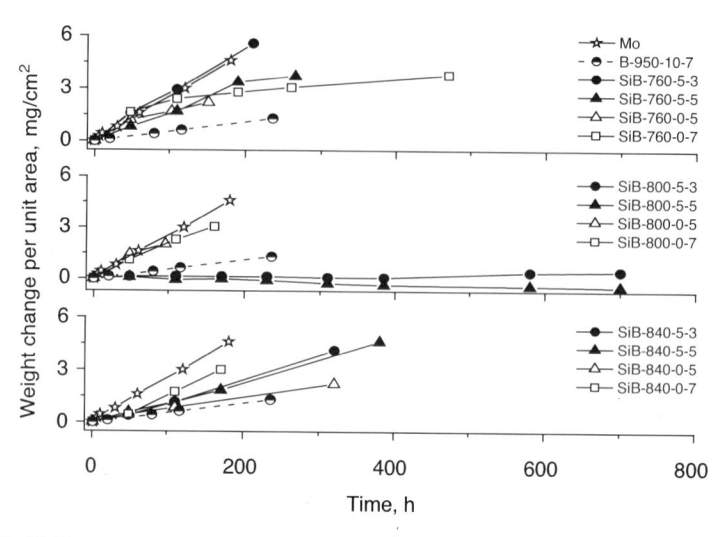

FIGURE 15.16 Oxidation behaviour of Mo–Si–B samples in an air–water mixture. Conditions are similar to those in Figure 15.12.

Figure 15.16 shows the oxidation kinetics data at 500 °C in a flow of wet air containing 2.3 vol % obtained on boronised samples, control sample B-950-10-7, and pure molybdenum. The formation of the matrix $Mo_5(B,Si)_3$ phase does not protect the molybdenum substrate from oxidation. The experimental results showed that the oxidation rate for SiB-760-5-3 is similar to that for pure molybdenum. The oxidation rate for SiB-760-5-5 is 1.5 times lower, due to formation of major amounts of the MoB_2 phase after 5 h. It can be seen that the initial oxidation rate for sample SiB-760-5-5 is close to that for sample SiB-760-0-5, containing MoB_2 as the matrix phase. In contrast, sample SiB-760-0-7 has a higher oxidation rate due to a higher content of the Mo_5Si_3 phase, which is easy to oxidise compared to the $MoSi_2$ phase. All samples obtained at 840 °C demonstrated a higher oxidation rate than that for B-950-10-7. Sample SiB-840-0-5, containing the matrix Mo_2B_5 phase and major amounts of the Mo_5Si_3 phase, as well as sample B-950-10-7, containing the $MoSi_2$, MoB, and Mo_2B_5 phases, demonstrated virtually the same oxidation rates: 5.5×10^{-3} and 5.3×10^{-3} mg·cm^{-2}·h^{-1}, respectively. These values are also very close to that of 5.0×10^{-3} mg·cm^{-2}·h^{-1} for boronised $MoSi_2$ intermetallics obtained in noncurrent boronising [30]. In the latter case, the Mo_2B_5 phase was also the matrix phase.

The oxidation resistance of the boronised specimens (SiB-800-5-3) was considerably higher than that for specimens B-950-10-7 and SiB-840-0-5. The linear weight gain rate observed was 6.5×10^{-4} mg·cm^{-2}·h^{-1}, which is almost eight times lower than that for sample B-950-10-7. Sample SiB-800-5-5 demonstrated a similar oxidation rate over a time interval up to 200 hours. Then, weight loss with a rate of 9×10^{-4} mg·cm^{-2}·h^{-1} is observed. A two-phase microstructure consisting of the $MoSi_2$ matrix phase with 12–15 wt % of the MoB_4 phase is essential for improved oxidation resistance. This microstructure acts as a protective layer for the substrate material against oxidation.

15.4 CONCLUSION

Different methods were used for improving the corrosion resistance of molybdenum, such as electrodeposition of hafnium coatings, electrochemical synthesis of hafnium diboride, boronising, siliciding, and boronising of $MoSi_2$. It was shown that the best protection of molybdenum from oxidation can be achieved by deposition of Mo–Si–B coatings in molten salts. In this process, the molybdenum substrate was first silicided in molten salt at 900 °C for 7 hours to produce a $MoSi_2$ layer, which was boronised in a subsequent step. Results indicated that the presence of small amounts of the MoB_4 phase formed during boronising of $MoSi_2$/Mo coatings in molten salts greatly improved the oxidation resistance of the molybdenum substrates. No pest disintegration of the molybdenum substrate was observed in an air–water mixture at 500 °C after 700 h. To provide maximum protection, the duration of the boronising step was adjusted to boronise an external $MoSi_2$ layer without boronising the molybdenum substrate.

ACKNOWLEDGEMENTS

Financial support by The Netherlands Organisation for Scientific Research (NWO), project No. 047.017.029, and by the Russian Foundation for Basic Research (RFBR), project No. 047.011.2005.016, is gratefully acknowledged.

REFERENCES

1. Mies, M. J. M., Rebrov, E. V., De Croon, M. H. J. M., and Schouten, J. C., Design of a molybdenum high throughput micro-reactor for high temperature screening of catalytic coatings, *Chem. Eng. J.* **101**, 225–235 (2004).
2. Mies, M. J. M., Van den Bosch, J. L. P., Rebrov, E. V., Jansen, J. C., De Croon, M. H. J. M., and Schouten, J. C., Hydrothermal synthesis and characterisation of ZSM-5 coatings on a molybdenum support and scale-up for application in micro reactors, *Catal. Today* **110**, 38–46 (2005).
3. Murphy, V., Volpe, A. F., Weinberg, W. H. Jr., High-throughput approaches to catalyst discovery, *Curr. Opin. Chem. Biol.* **7**, 427–432 (2003).
4. Shields, J. A. Jr., Molybdenum and its alloys, *Adv. Mater. Processes* **142**, 28–31 (1992).
5. Kuznetsov, S. A., Surface coatings for functional materials creation in ionic melts, in *The International Symposium on Ionic Liquids*, H. A. Oye and A. Tagtoyen (Eds.), The Norwegian University of Science and Technology, Trondheim, Norway, 2003, pp. 199–209.
6. Kuznetsov, S. A., Polyakov, E. G., Kuznetsova S. V., and Stangrit P. T., Alloy formation upon the deposition of hafnium on a niobium cathode, *Russ.-J. Appl. Chem.* **61**, 160–161 (1998).
7. Kuznetsov, S. A., Devyatkin, S. V., Glagolevskaya, A. L., Taranenko, V. I., Stangrit, P. T., and Shapoval, V. I., Electrochemical synthesis of hafnium diboride from chloride-fluoride melts, *Rasplavy* **6**, 67–70 (1992) (in Russian).

8. Biddulph, R. H., Boronizing for erosion resistance, *Thin Solid Films* **45**, 341–347 (1977).
9. Makyta, M., Chrenkova, M., Fellner, P., and Matyasovsky, K., Mechanism of the thermochemical boriding process and electrochemical studies in the molten systems based on $Na_2B_4O_7$, *Z. Anorg. Allg. Chem.* **540–541**, 169–176 (1986).
10. Kuznetsov, S. A., Electrochemical synthesis of high temperature borides from molten salts, *Russ. J. Electrochem.* **35**, 1301–1317 (1999).
11. Kaptay, G. and Kuznetsov, S. A., Electrochemical synthesis of refractory borides from molten salts, *Plasmas Ions* **2**, 45–56 (1999).
12. Segers, L., Fontana, A., and Winand, R., Electrochemical boriding of iron in molten salts, *Electrochim. Acta* **36**, 41–47 (1991).
13. Maruyama, T. and Yanagihara, K., High temperature oxidation and pesting of $Mo(Si,Al)_2$, *Mater. Sci. Eng. A* **239–240**, 828–841 (1997).
14. Nomura, N., Suzuki, T., Yoshimi, K., and Hanada, S., Microstructure and oxidation resistance of a plasma sprayed Mo–Si–B multiphase alloy coating, *Intermetallics* **11**, 735–742 (2003).
15. Totemeier, T. C., Wright, R. N., and Swank, W. D., FeAl and Mo–Si–B intermetallic coatings prepared by thermal spraying, *Intermetallics* **12**, 1335–1344 (2004).
16. Shrestha, S., Hodgkiess, T., and Neville, A., Erosion–corrosion behavior of high-velocity oxy-fuel Ni–Cr–Mo–Si–B coatings under high-velocity seawater jet impingement, *Wear* **259**, 208–218 (2005).
17. Yokota, H., Kudoh, T., and Suzuki, T., Oxidation resistance of boronized $MoSi_2$, *Surf. Coat. Technol.* **169–170**, 171 (2003).
18. Supatarawanich, V., Johnson, D. R., and Liu, C. T., Effects of microstructure on the oxidation behavior of multiphase Mo–Si–B alloys, *Mater. Sci. Eng. A* **344**, 328–339 (2003).
19. Parthasarathy, T. A., Mendiratta, M. G., and Dimiduk, D. M., Oxidation mechanisms in Mo-reinforced Mo_5SiB_2 (T2)–Mo_3Si alloys, *Acta Materialia* **50**, 1857–1868 (2002).
20. Thom, A. J., Summers, E., and Akinc, M., Oxidation behavior of extruded $Mo_5Si_3B_x$–$MoSi_2$–MoB intermetallics from 600–1600 °C, *Intermetallics* **10**, 555–570 (2002).
21. Natesan, K. and Deevi, S. C., Oxidation behavior of molybdenum silicides and their composites, *Intermetallics* **8**, 1147–1158 (2000).
22. Akinc, M., Meyer, M. K., Kramer, M. J., Thom, A. J., Huebsch, J. J., and Cook, B., Boron doped molybdenum silicides for structural applications, *Mater. Sci. Eng. A* **261**, 16–23 (1999).
23. Berczik, D. M., Method for Enhancing the Oxidation Resistance of a Molybdenum Alloy, and a Method of Making a Molybdenum Alloy, U.S. Patent. 5, 595, 616 (1997).
24. Park, J. S., Sakidja, R., and Perepezko, J. H., Coating designs for oxidation control of Mo–Si–B alloys, *Scripta Materialia* **46**, 765–770 (2002).
25. Mandal, P., Thom, A. J., Kramer, M. J., Behrani, V., and Akinc, M., Oxidation behavior of Mo–Si–B alloys in wet air, *Mater. Sci. Eng. A* **371**, 335–342 (2004).
26. Yoshimi, K., Nakatani, S., Nomura, N., and Hanada, S., Thermal expansion, strength and oxidation resistance of Mo/Mo_5SiB_2 in-situ composites at elevated temperatures, *Intermetallics* **11**, 787–794 (2003).

27. Brandstötter, J. and Lengauer, W., Multiphase reaction diffusion in transition metal–boron systems, *J. Alloys Compd.* **262–263**, 390–396 (1997).
28. Kuznetsov, S. A., Kuznetsova, S. V., Rebrov, E. V., Mies, M. J. M., De Croon, M. H. J. M., and Schouten, J. C. Synthesis of molybdenum borides and molybdenum silicides in molten salts and their oxidation behavior in an air–water mixture, *Surf. Coat. Technol.* **195**, 182–188 (2003).
29. Huebsch, J. J., Kramer, M., Zhao, H., and Akinc, M., Solubility of boron in $Mo_{5+y}Si_{3-y}$, *Intermetallics* **8**, 143–150 (2000).

16 Molten Salt Synthesis of LaAlO₃ Powder at Low Temperatures

ZUSHU LI

Department of Materials, Imperial College London, London, United Kingdom

SHAOWEI ZHANG

Department of Engineering Materials, University of Sheffield, Sheffield, United Kingdom

WILLIAM EDWARD LEE

Department of Materials, Imperial College London, London, United Kingdom

Abstract

LaAlO$_3$ powder with a rhombohedral crystal structure was synthesised by reacting equimolar La$_2$O$_3$ and Al$_2$O$_3$ in a molten KF–KCl eutectic salt for 3 h between 630 and 800 °C. The lowest synthesis temperature (630 °C) is about 1000 °C lower than that of conventional mixed oxide synthesis, and close to or lower than those used by most wet chemical methods. LaAlO$_3$ particle size increased from <3 µm to 3–7 µm with increasing temperature from 630 to 700 °C, but changed very little on further increasing temperature to 800 °C. On the other hand, it decreased with increasing salt/oxide weight ratio from 1:1 to 6:1. The "dissolution–precipitation" mechanism played a dominant role in the molten salt synthesis of LaAlO$_3$.

16.1 INTRODUCTION

Single crystal lanthanum aluminate (LaAlO$_3$) is widely used as substrates and electrically insulating buffers for depositing high-temperature superconducting films due to its high quality factor and excellent lattice and thermal expansion matching ability [1, 2]. Its use as a gate dielectric material has also been explored [3, 4].

Molten Salts and Ionic Liquids: Never the Twain? Edited by Marcelle Gaune-Escard and Kenneth R. Seddon
Copyright © 2010 John Wiley & Sons, Inc.

In addition, LaAlO$_3$ powder has been studied as a catalyst for oxidative coupling of methane and hydrogenation and hydrogenolysis of hydrocarbons due to its high catalytic activity [5].

LaAlO$_3$ is generally synthesised via a solid-solid reaction route (conventional mixed oxide synthesis (CMOS)) at >1550 °C [2]. Besides the high synthesis temperature, extensive crushing, grinding, and milling are often required to generate sufficient reactivity and chemical homogeneity in the raw materials as well as products. To overcome these drawbacks, several low temperature wet chemical synthesis techniques have been developed. One of these is coprecipitation of lanthanum- and aluminium-bearing precursors followed by calcining the precipitates. Depending on the test conditions, synthesis temperatures for completely converting the coprecipitates to LaAlO$_3$ varied over a large range (from 700 to 1350 °C) [6–9]. A similar low temperature synthesis route is sol-gel processing [10–12]. Sahu et al. [12] prepared nanosized LaAlO$_3$ at as low as 600 °C via a sol-gel process using aluminium and lanthanum chlorides as precursors. In situ polymerisation is another low temperature technique for the synthesis of reactive LaAlO$_3$ powder, which could decrease the synthesis temperature to 700–900 °C [13, 14]. In addition to these wet chemical techniques, several other low temperature routes for synthesis of LaAlO$_3$ have been investigated, including self-igniting combustion [15], PVA (polyvinyl alcohol) evaporation route [16], and mechanochemical alloying (MA) [17].

Although the techniques described above could substantially decrease the synthesis temperature of LaAlO$_3$, they suffer from various drawbacks. For example, the wet chemical synthesis techniques often use expensive and environmentally unfriendly organic/inorganic precursors and solvents, so are applicable only to niche cases and for small scale products. As for the MA technique, besides the requirements of extensive high energy ball milling and postmilling heat treatment, it suffers from a very limited production capacity (only a few grams).

In this work, the molten salt synthesis (MSS) technique has been used to synthesise LaAlO$_3$ powder at low temperatures and using inexpensive oxide raw materials. The synthesised LaAlO$_3$ powders have been characterised using powder X-ray diffraction analysis (XRD) and scanning electron microscopy (SEM/EDS), and the synthesis mechanism discussed is based on the results.

16.2 EXPERIMENTAL

La$_2$O$_3$ powder (Sigma-Aldrich, ~1 µm, ≥99.9% pure), low-soda calcined Bayer-derived (i.e., precipitated) Al$_2$O$_3$ powder (Almatis, D50 = 0.80 µm, 99.8% pure with 0.08% Na$_2$O), ACS reagent KCl (Aldrich, ≥99.0% pure), and reagent grade KF (Aldrich, 98% pure) were used as starting materials. The La$_2$O$_3$ powder used is produced via solvent extraction after the trivalent rare-earth elements are separated from the insoluble rare-earth fraction in the cracking stage of Bastnasite mineral. Equimolar La$_2$O$_3$ and Al$_2$O$_3$ powders were mixed in an agate mortar with various amounts of KF–KCl eutectic salt. The molar ratio of KF to KCl in the eutectic salt was

46.7:53.3, whereas the weight ratio of KF + KCl to $La_2O_3 + Al_2O_3$ (referred to as the salt/oxide ratio) varied from 1:1 to 6:1. The samples were heated to 630, 700, and 800 °C, respectively, and held at the temperature for 3 h. The heating and cooling rates were 3 °C · min^{-1} and 5 °C · min^{-1}, respectively. After cooling to room temperature, the reacted mass was washed with hot distilled water for 2 h, followed by filtration to remove the salts, and this process was repeated five times. The resultant powder was oven-dried at 105 °C for 4 h for further characterisation.

Phases in the resultant powders were identified by XRD analysis (Siemens D500 reflection diffractometer). Patterns were recorded at 30 mA and 40 kV using Ni-filtered Cu Kα radiation ($\lambda = 1.54178$ Å). The scan rate (2θ) was 1° min^{-1} at the step size of 0.02°. Chemical analyses of the synthesised powders were performed using X-ray fluorescence (XRF) (Bruker AXS, Karlsruhe, Germany, SRS 3400, wavelength dispersive) and inductively coupled plasma–atomic emission spectrometry (ICP-AES) (Perkin Elmer 3300 RL, Boston, MA) to check the impurity levels of K, F, and Cl originating from the salts used. Microstructural morphologies of the as-received La_2O_3 and Al_2O_3 and the synthesised $LaAlO_3$ powders were observed using SEM (JEOL6400, Japan). EDS analysis of the elements in the resultant powders was carried out using the attached Turreted Pentafet detector and ISIS 300 processing unit.

16.3 RESULTS

16.3.1 As-Received Raw Materials

The as-received La_2O_3 powder consisted mainly of small (ca. 1 µm) spheroidal particles, most of which were agglomerated (Fig. 16.1a) Compared to the as-received La_2O_3, better dispersion was seen in the as-received Al_2O_3 powder (Fig. 16.1b). The majority of the Al_2O_3 particles were <1 µm, but about 10% were up to 2 µm.

16.3.2 Effect of Heating Temperature on Synthesis

XRD of the powders from samples (with the salt/oxide ratio of 3:1) heated for 3 h at 630, 700, and 800 °C, respectively, revealed that only $LaAlO_3$ with a rhombohedral perovskite crystal structure (ICDD card number 31–22) and no other phases were identified in any samples. Figure 16.2 shows the SEM of the synthesised $LaAlO_3$ powders. At 630 °C, the powder consisted mainly of <3 µm cuboidal particles, although a few noncuboidal particles were also observed. On increasing the temperature to 700 °C, the cuboidal particles became larger (3–7 µm). On further increasing the temperature to 800 °C, the particle size changed very little, but the edges of the cuboidal crystals appeared less sharp. Higher magnification SEM further reveals many submicron particles on some of the large cuboidal particles (Fig. 16.3a). EDS (Fig. 16.3b) revealed that these small particles have an identical (La and Al) composition to that of cuboidal and noncuboidal grains. This along with XRD confirmed that they were all $LaAlO_3$.

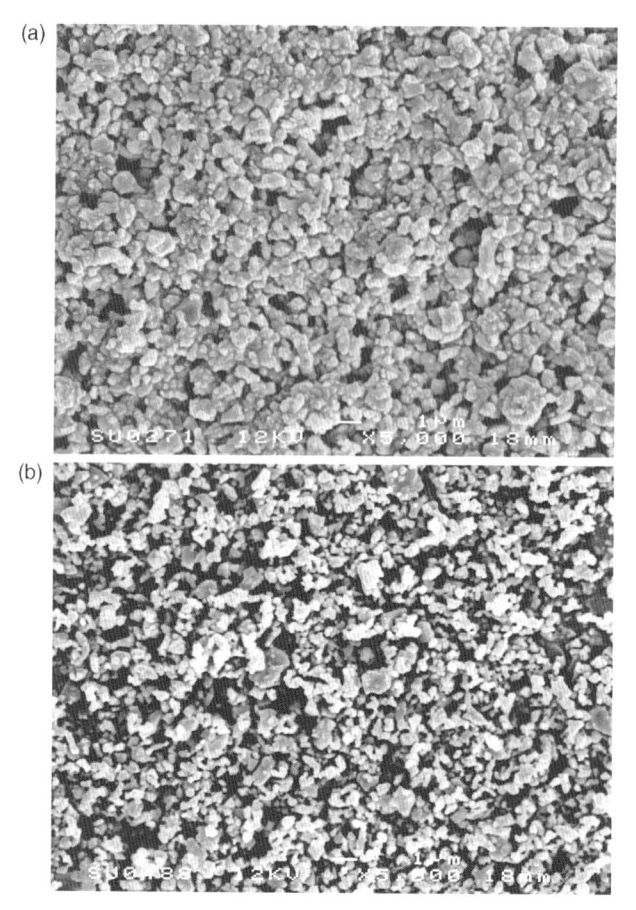

FIGURE 16.1 SEM of as-received raw (a) La_2O_3 and (b) Al_2O_3 powders.

16.3.3 Effect of Salt/Oxide Ratio on Synthesis

XRD of the powders from samples with different salt/oxide ratios heated for 3 h at 700 °C, similar to the case of varying temperature, showed only $LaAlO_3$ in all samples. Nevertheless, SEM (Fig. 16.4 and Fig. 16.2b) reveals that the salt/oxide ratio affected the size of the cuboidal crystals. With increasing salt/oxide ratio from 1:1 to 3:1, the grain size slightly decreased; however, on further increasing the ratio from 3:1 to 6:1, it decreased more significantly from 3–7 µm to $\lesssim 4$ µm.

16.3.4 Impurity Levels in Synthesised Powders

The impurity levels of K, F, and Cl in the raw La_2O_3 and Al_2O_3 powders are considered to be trace (not tested by the supplier) since both are high purity (>99.9% La_2O_3 and 99.8% Al_2O_3 with 0.08% Na_2O, respectively). XRF and ICP-AES analysis results revealed that only 0.016% K, 0.008% F, and 0.01% Cl remained in the $LaAlO_3$

FIGURE 16.2 SEM of LaAlO$_3$ particles synthesised for 3 h using salt/oxide weight ratio of 3:1 at (a) 630 °C, (b) 700 °C, and (c) 800 °C. Cuboidal crystals along with some noncuboidal crystals were observed. Higher magnification micrograph of the rectangular zone in part (b) is shown in Figure 16.3.

224 MOLTEN SALT SYNTHESIS OF LaAlO$_3$ POWDER AT LOW TEMPERATURES

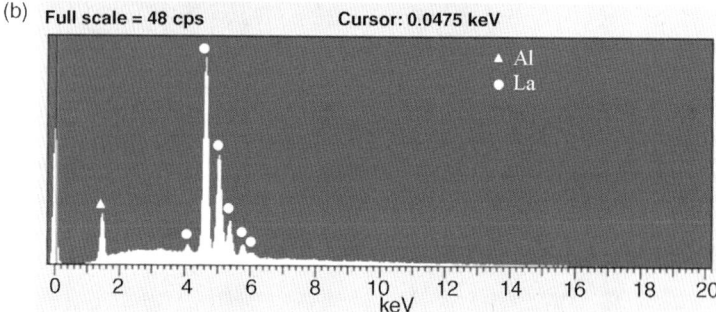

FIGURE 16.3 (a) Higher magnification SEM revealing that some submicron-sized LaAlO$_3$ particles were present on the surfaces of cuboidal LaAlO$_3$ crystals. (b) EDS spectra of cuboidal crystals, submicron particles on the surface of cuboidal grains, and small noncuboidal grains.

powders synthesised in the KF–KCl eutectic salt followed by repeated water-washing. Because the main objective of this chapter is to illustrate the feasibility of MSS for synthesis of high-melting complex oxide (LaAlO$_3$) at low temperatures, no further study has yet been carried out to purify the synthesised LaAlO$_3$ powders and to examine the effects of such levels of salt contamination on the properties of the synthesised LaAlO$_3$ powders.

16.4 DISCUSSION

16.4.1 Synthesis Mechanism

Since the melting temperature of KF–KCl eutectic salts is ~610 °C [18], they are liquid at the test temperatures (630–800 °C). As both La$_2$O$_3$ and Al$_2$O$_3$ are highly soluble in molten potassium fluoride/chloride salts [19–21], they dissolve into the molten eutectic salt according to reactions (16.1) and (16.2) (species in brackets refer to those dissolved in the molten salt):

$$La_2O_{3(s)} \rightarrow (La_2O_3) \quad (16.1)$$

$$Al_2O_{3(s)} \rightarrow (Al_2O_3) \quad (16.2)$$

FIGURE 16.4 SEM of LaAlO$_3$ powders synthesised at 700 °C for 3 h using different salt/oxide ratios: (a) 1:1 and (b) 6:1.

The La$_2$O$_3$ and Al$_2$O$_3$ dissolved in the molten salt are mixed homogeneously at the atomic level, and diffusion is more rapid than in the solid state. Consequently, the dissolved La$_2$O$_3$ and Al$_2$O$_3$ react rapidly to form LaAlO$_3$ in the molten salt according to reaction (16.3). Once the molten salt is oversaturated with LaAlO$_3$, LaAlO$_3$ crystals start to precipitate from the salt according to reaction (16.4) and grow (Figs. 16.2 and 16.4). The precipitation of LaAlO$_3$ from the oversaturated salt leads to further dissolution and reaction of La$_2$O$_3$ and Al$_2$O$_3$, reactions (16.1)–(16.3), and then more precipitation of LaAlO$_3$, reaction (16.4). This cycle, reactions (16.1)–(16.4), is repeated until all of the starting La$_2$O$_3$ and Al$_2$O$_3$ were used up in forming LaAlO$_3$.

$$(La_2O_3) + (Al_2O_3) \rightarrow 2(LaAlO_3) \tag{16.3}$$

$$(LaAlO_3) \rightarrow LaAlO_{3(s)} \tag{16.4}$$

The "dissolution–precipitation mechanism" described here is consistent with the different morphologies of the reactant and product phases. Although spheroidal La_2O_3 and Al_2O_3 starting powders were used (Fig. 16.1), the synthesised $LaAlO_3$ particles showed well-crystallised euhedral shapes (Figs. 16.2–16.4). These results further indicated that the "template formation mechanism" involved in many MSS processes [22, 23] is not significant in the present case.

As shown in Figures 16.2–16.4, small $LaAlO_3$ particles were seen on the surfaces of some cuboidal $LaAlO_3$ crystals. They most likely formed on cooling as some $LaAlO_3$ (theoretical solubility) always remains in the liquid salt. On cooling, with decreasing solubility of $LaAlO_3$, the salt becomes oversaturated with $LaAlO_3$. As a result, small $LaAlO_3$ particles subsequently precipitate from the salt onto the large euhedral crystals already formed.

16.4.2 Effects of Heating Temperature or Salt/Oxide Ratios on Synthesis

XRD shows that $LaAlO_3$ forms at all three test temperatures (between 630 and 800 °C), suggesting that temperature had little effect on the phase composition of the synthesised powder. Nevertheless, it did affect the morphologies/sizes of the $LaAlO_3$ particles synthesised. As shown in Figure 16.2, with increasing temperature from 630 to 700 °C, the particle size increased from $\lesssim 3$ μm to 3–7 μm, attributable to accelerated crystal growth with temperature in the molten salt media. However, on further increasing the temperature from 700 to 800 °C, the particle size changed very little, but the edges of the crystals became less sharp (flatter). The flatter edges might arise from slight "redissolution" of $LaAlO_3$ at 800 °C. The solubility of $LaAlO_3$ in the salt at 800 °C is higher than that at 700 °C, so some edge areas of the already formed $LaAlO_3$ crystals might be redissolving into the salt because of their higher reactivity.

Besides heating temperature, the salt/oxide ratio affected the $LaAlO_3$ particle size. As shown in Figure 16.4, $LaAlO_3$ particle size generally decreased with increasing salt/oxide ratio from 1:1 to 6:1. This might be attributable to the changes of the number of nuclei in the salt from which $LaAlO_3$ can precipitate and grow. With increasing relative content of the salt, the number of nuclei would increase, decreasing the final size.

16.5 CONCLUSION

$LaAlO_3$ powder was synthesised at 630–800 °C using La_2O_3 and Al_2O_3 as reactants and KF–KCl eutectic salts as reaction media. The lowest synthesis temperature (630 °C) is about 1000 °C lower than by CMOS, and close to or lower than those used by most wet chemical techniques. The rhombohedral $LaAlO_3$ crystal size increased from $\lesssim 3$ μm to 3–7 μm with temperature from 630 to 700 °C, but changed very little with further increasing temperature to 800 °C. With increasing salt/oxide weight ratio from 1:1 to 6:1, the $LaAlO_3$ particle size generally decreased. The "dissolution–reaction–precipitation" mechanism played an important role in the MSS process; that is, La_2O_3 and Al_2O_3

initially dissolved and reacted in the molten salt and later LaAlO$_3$ precipitated and grew from the salt oversaturated with it.

ACKNOWLEDGEMENTS

The authors gratefully acknowledge the Engineering and Physical Sciences Research Council (EPSRC), UK for financial support (grant numbers GR/S60037/01 and GR/S99211/01).

REFERENCES

1. Nieminen, M., Sajavaara, T., Rauhala, E., Putkonen, M., and Niinisto, L., Surface-controlled growth of LaAlO$_3$ thin films by atomic layer epitaxy, *J. Mater. Chem.* **11**, 2340–2345 (2001).
2. Sung, G. Y., Kang, K. Y., and Park, S. C., Synthesis and preparation of lanthanum aluminate taget for radio-frequency magnetron sputtering, *J. Am. Ceram. Soc.* **74**(2), 437–439 (1991).
3. Lu, X., Liu, Z., Wang, Y., Yang, Y., Wang, X., Zhou, H., and Nguyen, B., Structural and dielectric properties of amorphous LaAlO$_3$ and LaAlO$_x$N$_y$ films as alternative gate dielectic materials, *J. Appl. Phys.* **94**(2), 1229–1234 (2003).
4. Park, B. and Ishiwara, H., Formation of LaAlO$_3$ films on Si(100) substrates using molecular beam deposition, *Appl. Phys. Lett.* **82**(8), 1197–1199 (2003).
5. Tagawa, T. and Imai, H., Mechanistic aspects of oxidative coupling of methane over LaAlO$_3$, *J. Chem. Soc. Faraday Trans. 1 Phys. Chem. Condensed Phases* **84**(4), 923–929 (1988).
6. Golub, A. M., Maidukova, T. N., and Limar, T. F., Preparation of lanthanum aluminate by coprecipitation, *Izv. Akad. Nauk SSSR Neorg. Mater.* **2**(9), 1608–1611 (1966).
7. Li, W., Zhuo, M. W., and Shi, J. L., Synthesizing nano LaAlO$_3$ powders via co-precipitation method, *Mater. Lett.* **58**, 365–368 (2004).
8. Nair, J., Nair, P., Mizukami, F., Van Ommen, J. G., Doesburg, G. B. M., Ross, J. R. H., and Burggraaf, A., Pore structure evolution of lanthana–alumina systems prepared through coprecipitation, *J. Am. Ceram. Soc.* **83**(8), 1942–1946 (2000).
9. Taspinar, E. and Tas, A. C., Low-temperature chemical synthesis of lanthanum monoaluminate, *J. Am. Ceram. Soc.* **80**(1), 133–141 (1997).
10. Chroma, M., Pinkas, J., Pakutinskiene, I., Beganskiene, A., and Kareiva, A., Processing and characterization of sol-gel fabricated mixed metal aluminates, *Ceram. Int.* **31**, 1123–1130 (2005).
11. Lux, B. C., Clark, R. D., Salazar, A., Sveum, L. K., and Krebs, M. A., Aerosol generation of lanthanum aluminate, *J. Am. Ceram. Soc.* **76**(10), 2669–2772 (1993).
12. Sahu, P. K., Behera, S. K., Pratihar, S. K., and Bhattacharyya, S., Low temperature synthesis of microwave dielectric LaAlO$_3$ nanoparticles: effect of chloride on phase evolution and morphology, *Ceram. Int.* **30**, 1231–1235 (2004).
13. Kakihana, M. and Okubo, T., Low temperature powder synthesis of LaAlO$_3$ through in situ polymerization route utilizing citric acid and ethylene glycol, *J. Alloys Compd.* **266**, 129–133 (1998).

14. Zhou, D., Huang, G., Chen, X., Xu, J., and Gong, S., Synthesis of LaAlO$_3$ via ethylenediaminetetraacetic acid precursor, *Mater. Chem. Phys.* **84**, 33–36 (2004).
15. Shaji Kumar, M. D., Srinivasan, T. M., Ramasamy, P., and Subramanian, C., Synthesis of lanthanum aluminate by a citrate-combustion route, *Mater. Lett.* **25**(11), 171–174 (1995).
16. Adak, A. K. and Pramanik, P., Synthesis and characterization of lanthanum aluminate powder at relatively low temperature, *Mater. Lett.* **30**(3), 269–273 (1997).
17. Zhang, Q. and Saito, F., Mechanochemical synthesis of lanthanum aluminate by grinding lanthanum oxide with transition alumina, *J. Am. Ceram. Soc.* **83**(2), 439–441 (2000).
18. Roth, R. S., Clevinger, M. A., and McKenna, D., *Phase Diagrams for Ceramists*, American Ceramic Society, Columbus, OH, 1984.
19. Balaraju, J. N., Ananth, V., and Sen, U., Studies on low temperature Al electrolysis using composite anodes in NaF–KCl bath electrolyte, *J. Electrochem. Soc.* **142**(2), 439–444 (1995).
20. Belov, S. F., Gladneva, A. F., Matveev, V. A., and Igumnov, M. S., Solubility of lanthanum and neodymium oxides in molten fluorides, *Izv. Akad. Nauk SSSR Neorg. Mater.* **8**(5), 966–967 (1972).
21. Dewing, E. W., The chemistry of the alumina reduction cell, *Can. Metall. Q.* **30**(3), 153–161 (1991).
22. Ebrahimi, M. E., Allahverhi, M., and Safari, A., Synthesis of high aspect ratio platelet SrTiO$_3$, *J. Am. Ceram. Soc.* **88**(8), 2129–2132 (2005).
23. Zhang, S., Jayaseelan, D. D., Bhattacharya, G., and Lee, W. E., Molten salt synthsis of magnesium aluminate (MgAl$_2$O$_4$) spinel powder, *J. Am. Ceram. Soc.* **89**(5), 1724–1726 (2006).

17 Accurate Measurement of Physicochemical Properties on Ionic Liquids and Molten Salts

V. M. B. NUNES

Escola Superior de Tecnologia, Instituto Politécnico de Tomar, Campus da Quinta do Contador, Tomar, Portugal and Centro de Ciências Moleculares e Materiais, Faculdade de Ciências, Universidade de Lisboa, Lisboa, Portugal

M. J. V. LOURENÇO, F. J. V. SANTOS, M. L. S. M. LOPES, and C. A. NIETO DE CASTRO

Centro de Ciências Moleculares e Materiais, Faculdade de Ciências, Universidade de Lisboa, Lisboa, Portugal and Departamento de Química e Bioquímica, Faculdade de Ciências, Universidade de Lisboa, Lisboa, Portugal

Abstract

The measurement of physicochemical properties of liquids has proved a difficult task, especially if the measurements are to be performed in temperature and pressure ranges far from the ambient conditions and if the liquids have properties that make it difficult to mathematically model the instrument to be used. Even for molecular liquids, where the interactions in the liquid state are essentially known and easier to characterise, it is still difficult to measure properties with high accuracy. Ionic liquids, made of isolated or aggregated ions, are electrically conducting, polarisable, and almost nonvolatile, posing several problems for the design and use of standard equipment. They cover a very wide temperature range, from LTILs to HTILs (molten salts). It is the purpose of this chapter to analyse the measurement of several thermophysical properties—namely, viscosity, thermal conductivity, electrical conductivity, electrical permittivity, and binary diffusion of water—using our previous experience in measurements of molecular liquids, electrolyte solutions, and molten salts.

Molten Salts and Ionic Liquids: Never the Twain? Edited by Marcelle Gaune-Escard and Kenneth R. Seddon
Copyright © 2010 John Wiley & Sons, Inc.

17.1 INTRODUCTION

Ionic liquids (or low temperature molten salts) are considered today to constitute a class of liquids with the highest potential to revolutionise the chemical industry, as they constitute innovative fluids for chemical processing, which are generally nonflammable and nonvolatile under ambient conditions and thus perceived as "green" solvents. However, to implement new processes, it is necessary to prove that they are competitive with the traditional processes, not only from the point of view of final products, but for all the technological operations involved in the processes.

The optimal technological design of green processes requires the characterisation of the ionic liquids used. The recent explosion in research and applications of low temperature ionic liquids (LTILs) raises several problems: the need for data is growing exponentially (currently very sparse) and the number of new compounds has been estimated by Earle and Seddon [1] to be on the order of one million.

The accurate measurement of the thermodynamic and transport properties of molten salts, high temperature ionic liquids (HTILs), has proved difficult in the past [2, 3]. The particular intermolecular interactions, localised and nonlocalised, and the difficulties of adapting the existing methods of measurement to this research have created this situation. This is particularly true for the transport properties, where the mechanisms of mass, momentum, heat, and charge transfer in Coulombic liquids are far from being completely understood; this can be illustrated for viscosity, where we know how to measure it at room temperature for normal organic and inorganic liquids with an accuracy of 0.5% or better, while at high temperatures an excellent value is ±2%, most of the times impossible to achieve.

In the following sections we analyse the similarities and differences between molten salts and "ionic liquids", and their relationships to measurement procedures. We restrict our analysis to properties of pure ionic liquid and to the physical/chemical factors that are known to influence these properties. These are the melting points, density, heat capacity, viscosity, thermal conductivity, electrical conductivity, relative permittivity (dielectric constant), and surface tension. Diffusion in ionic liquids + water systems is also discussed, as water is very important in the purity of ionic liquids and in the solubility phenomena of complex processing mixtures.

17.2 MOLTEN SALTS AND IONIC LIQUIDS

Molten salts were the first ionic liquids to be studied experimentally. The naming resulted from the fact that they are usually crystallised salts at room temperature. Table 17.1 shows the main characteristics of molten salts and some of the additional problems for measurement. Figure 17.1 shows results obtained for the viscosity of molten $LiNO_3$, from its melting point to 700 K, taken in our laboratory and by other authors [4]. It can easily be seen that there is no agreement between the different sets of data, within their mutual uncertainty, the differences between different laboratories being sometimes dramatic (10–15%), but that the curves are parallel, which indicates systematic errors in the measurements. Figure 17.2 shows the thermal conductivity

TABLE 17.1 Main Characteristics of Molten Salts and Ionic Liquids

Molten Salts	Ionic Liquids	Additional Problems
• Liquid state in a large temperature range	• Liquid state in a large temperature range	• Medium and high temperatures ($T > 400$ K)
• Ability to dissolve many inorganic and organic compounds, attaining high concentrations	• Ability to dissolve many inorganic and organic compounds, attaining high concentrations	• Purity of compound in the beginning of the measurements
• Low vapour pressure and stability at normal pressures	• Nonflammability	• Change in composition during measurements due to reactions (decomposition, with container and/or environment) or multiple recrystallisations
• Low viscosity, as the ions are mutually independent for most of the cases	• Low vapour pressure and stability at normal pressures	• Materials compatibility between sample, crucibles, and atmosphere
• Chemical inertness (no reaction with air or water)		• Low to high viscosity, as the ions are not mutually independent for most of the cases and can form aggregates
• High heat capacity per unit volume		• Cations and anions have completely different sizes
		• Reaction with water is possible
		• High solubility of water in LTILs, affecting the properties
		• Moderate heat capacity per unit volume

of molten KNO_3 from its melting point up to 730 K. Here the situation is even worse than that found for viscosity, as the differences between different laboratories can be greater than 25% and the temperature slope of the thermal conductivity can be positive, negative, or zero! Several explanations attempt to rationalise this situation:

- The methods of measurement applied were not modeled with the correct physical model and mathematical solution.
- The mechanisms of mass, momentum, heat, and charge transfer in Coulombic liquids are far from being completely understood.
- There is a lack of calibrating fluids for the viscosity range used, namely, for capillary viscometry.

Examples of these items can be capillary viscometers, with corrections for the surface tension based on that of water; not accounting for radiation heat transfer in

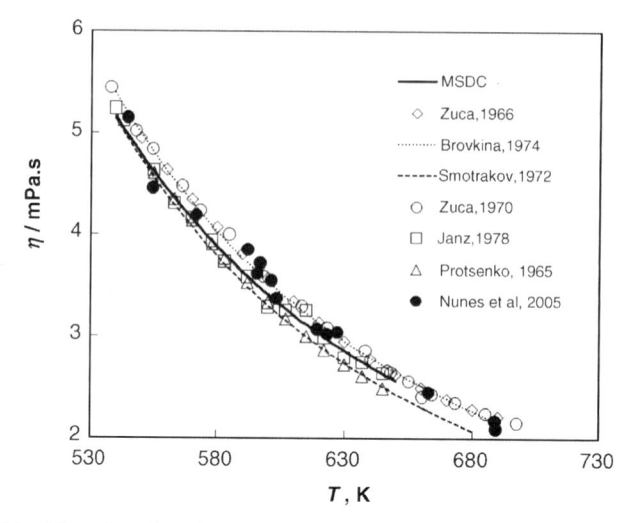

FIGURE 17.1 The viscosity of molten lithium nitrate from its melting point to 700 K, measured in our laboratory and by others. References mentioned can be found in Reference [4].

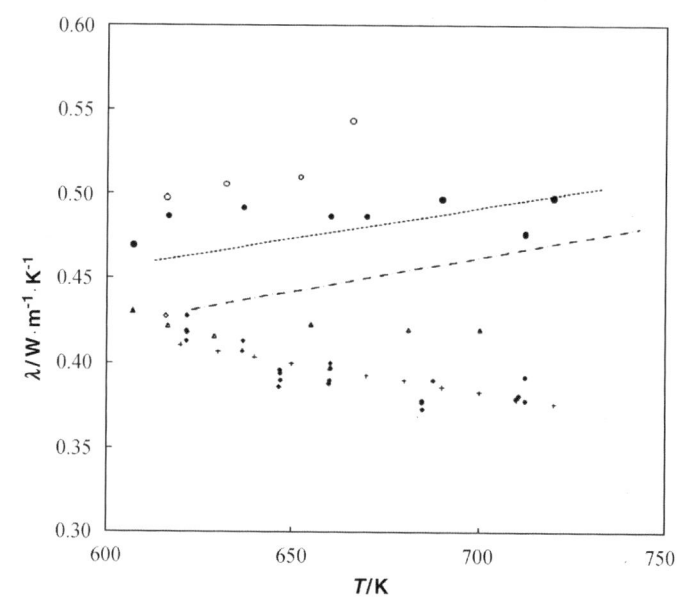

FIGURE 17.2 The thermal conductivity of molten potassium nitrate. ◇, Turnbull [1961]; ○, Bloom et al. [1965]; ▲, White and Davies [1967]; ●, Gustafsson et al. [1968]; •••••, McDonald and Davies [1970]; --•--, Santini et al. [1984]; △, Tufeu et al. [1985]; ◆, Kitade et al. [1989]; +, Nagasaka et al. [1991]. References mentioned can be found in Reference [2].

thermal conductivity measurements; and bad definition of electrical conduction paths between electrodes. From these examples, it can be concluded that the measurement situation in molten salts, especially for transport properties, did not evolve significantly for the last thirty years, that temperatures above 600 K are still a problem, and that there is an urgent need for refinement of methods of measurement.

For the case of LTILs, in spite of the low temperature range of applicability of these liquids, the situation is far from being better, not because one does not know how to make the measurements, but because a great number of problems arise. In Table 17.1, the main properties of ionic liquids are identified, mostly common to molten salts, but with some additional problems that can influence the measurements, if not accounted for. These problems can be systematized in the following manner:

- It is impossible to measure all the properties of all the liquids needed for the design of new processes and plants in a reasonable time frame.
- Some of the liquids have unusual properties, with high or abnormal values, which require reference materials for the measurement of these properties [5].
- Some of the measuring methods need to be modified to account for the ionic character of these salts.
- The nonexistence of reference or calibrating liquids for the properties needed for the chemical process design, namely, for the ranges attained for these fluids, must be resolved.
- The difficulty in having "pure" liquids for the property determination, namely, caused by their extraordinary ability to dissolve water, as well as unreacted molecular compounds from their synthesis, like chloride ions, must be resolved.
- Some of the existing knowledge can change very fast, as some recent publications show, namely, about the volatility of ionic liquids [6].
- The nonapplicability of the available property predictive methods to these compounds must be resolved.

The amount of data obtained so far for the thermophysical properties of ionic liquids does not yet have the quality that allows us to make comparisons similar to those made for the molten salts. The work is more developed in characterising new liquids than in making replicate measurements in different laboratories of a given property or being able to analyse the mutual uncertainty of the data. However, a recent review of the properties of pure ionic liquids—restricted to melting points, density, viscosity, and electrical conductivity [5]—added to our previous experience, can provide some guidance on the actual state of the art in property measurements in ionic liquids.

17.3 PHYSICOCHEMICAL MEASUREMENTS IN IONIC LIQUIDS

17.3.1 The Actual Situation

For the *melting points*, which are known to be influenced by the charge distribution on the ions, hydrogen-bonding ability, symmetry of the ions, and van der Waals interactions, the situation is reasonable, but the impurities in the samples can cause a freezing depression on the melting temperatures, and this has limited the use of much data, where even confusion between glass temperatures and melting points could not be avoided.

For the *density*, it is known that the great majority of LTILs are much denser than water. However, the comparison of data obtained in different laboratories for samples of [C_4mim][PF_6] showed differences of up to 3%. The case is even worse (change of sign of the temperature dependence) for the thermal expansion coefficient and for the isothermal compressibility [5]. The main reason for these discrepancies is probably purity.

Most of the ionic liquids are moderately or highly viscous, with viscosities more comparable to the viscosities of oils, two or three orders of magnitude greater than most organic solvents. This higher *viscosity* has several disadvantages, some related to its negative effect on the heat and mass transfer rates in the processes to which they were applied, others related to the fact that viscosities greater than 150 mPa·s (at 20 °C, 430 mPa·s for [C_4mim][PF_6] and 154 mPa·s for [C_4mim][BF_4]) are difficult to obtain with high accuracy, due to the lack of internationally accepted reference liquids for calibration [7] and the experimental difficulties for its absolute measurement [8, 9]. Moreover, it is well known that viscosity is one of the properties most strongly influenced by impurities, from water to halogen acids. As an example, 1% of water decreases the viscosity of [C_4mim][NTf_2] at 20 °C by 33%, and 0.1% of water decreases the viscosity of [C_4mim][PF_6] at 20 °C by 16.6% [10]. The amount of data existing for the viscosities of ionic liquids, even for less pure samples, is so scarce that a good measurement program is highly recommended [5].

These authors also recommend that the study of *electrical conductivity* is absolutely necessary. A very recent paper [11] shows that 5% of water in [C_2mim][BF_4] increases its conductivity by a factor of 3.2, a value that recommends extreme care of the sample purity during these measurements.

For the *thermal conductivity* of ionic liquids, no direct measurements were found. For *electrical permittivity* (*dielectric constant*) only measurements at high frequency (dielectric relaxation) were recently obtained. For *heat capacity*, the number of results already reported is indeed very small. For the case of *surface tension*, there have been few measurements, but it looks as if the presence of water did not significantly affect the measurements [12]. The surface tensions are lower than that of water, but higher than for linear alkanes.

The facts cited above highlight one of the most important issues to be resolved: before experimental methods of measurement (existing or to be developed) are applied to a given sample or batch of samples of ionic liquids – there must be *the guarantee of purity*. By guarantee, we mean determination by a given analytical method, with a well-known uncertainty/accuracy. Although recent advances in quantitative analytical characterisation have been reported [13], it seems that further developments on the applications of several techniques are strongly needed.

The second point cited above, which has been demonstrated to influence the experimental results obtained for viscosity and electrical conductivity, is the presence of water in the ionic liquid, dissolved or chemically bonded. Both these properties are known to be strongly affected by impurities, and as mentioned before, the presence

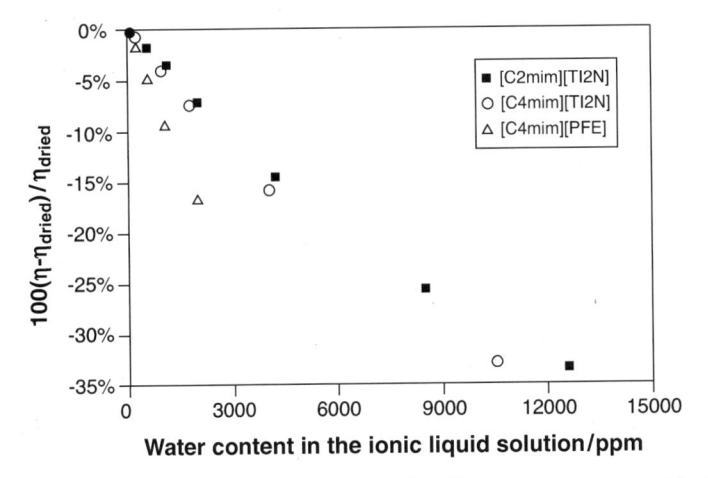

FIGURE 17.3 Percentage change in the absolute viscosity versus water content for [C$_2$mim][NTf$_2$], [C$_4$mim][NTf$_2$], and [C$_4$mim][PF$_6$] [10].

of water and organic solvents has been shown to decrease the viscosity of LTILs [14–16], and the presence of chloride has been shown to increase the viscosity of LTILs [14]. Widegren, et al. [10] noted that little is known about the quantitative effect of low levels of water on physical properties. Water is the most insidious impurity because of its ubiquity. Even "hydrophobic" LTILs, which are not miscible with water, rapidly absorb water from the atmosphere [14, 17–19] or from moist surfaces. Hence, unless an LTIL is carefully dried and handled, it will be contaminated with water. Figure 17.3 shows the percentage change in the absolute viscosity versus water content for three ionic liquids, [C$_2$mim][NTf$_2$], [C$_4$mim][NTf$_2$], and [C$_4$mim][PF$_6$], obtained by these authors [10], using a capillary open to the atmosphere, and with the water content in the sample determined before and after the measurements (a key to these experiments is that the water content of the LTIL was determined before and after each viscosity measurement by coulometric Karl Fischer titration). It can be seen that the addition of 1% (by mass) of water can decrease the viscosity as much as 30%. In addition, time facilitates the diffusion of atmospheric water (with or without chemical reaction), which justifies the measurement of the diffusion coefficient of water at infinite dilution, to better understand the influence of water on the properties of these compounds, viscosity and others. These results were confirmed in our laboratory, and in NIST, for [C$_6$mim][NTf$_2$], where 100 ppm (0.01%) of water was found to decrease the viscosity by 5%, a value much greater than the claimed uncertainty of the data. For the case of electrical conductivity, it is enough to report the data also obtained by the NIST team for four hydrophobic room-temperature ionic liquids, [C$_2$mim][NTf$_2$], [C$_4$mim][NTf$_2$], [C$_6$mim][NTf$_2$], and [C$_4$mim][PF$_6$] [20]. For water + [C$_4$mim][NTf$_2$], the electrical conductivity was found to increase dramatically with increasing w_{H_2O}; for example, in going from $w_{H_2O} = 10^{-5}$ to 10^{-2}, the fractional increase in κ was 0.36 (= $(\kappa_{wet} - \kappa_{dried})/\kappa_{dried}$). The work at NIST [10, 20] illustrates the importance of measuring the water content in LTILs, both

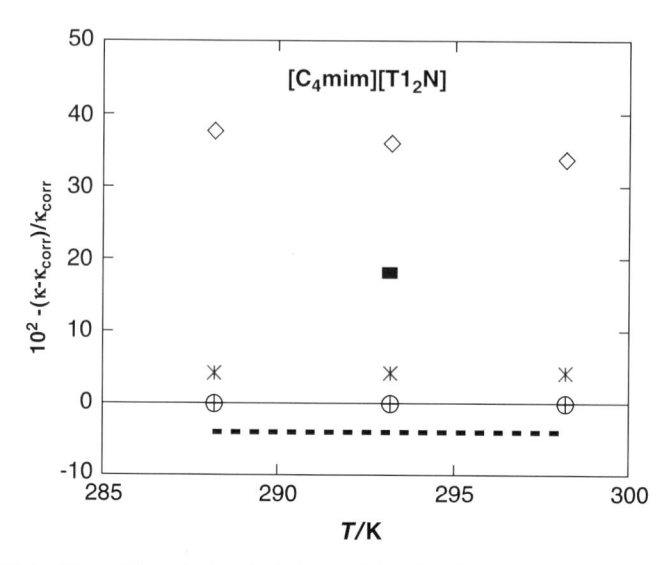

FIGURE 17.4 Plot of the relative deviations of the electrical conductivity of the mixtures [C₄mim][NTf₂] + water from that of pure [C₄mim][NTf₂], for different amounts of water [20]. ◇, $w_{H_2O} = 8.85 \times 10^{-3}$; ✶, $w_{H_2O} = 1.02 \times 10^{-3}$; +, $w_{H_2O} = 8 \times 10^{-5}$; ○, dried ($w_{H_2O} = 10^{-5}$); other symbols [20].

before and after measurements of electrical conductivity. Figure 17.4 shows a sample of the results obtained by these authors for [C₄mim][NTf₂].

From these results, we recommend that to obtain results for these properties with a credible uncertainty, it is fundamental to determine, before and after the measurements, the type and amount of impurities present; and in the case of hydrophobic LTILs,[1] it is necessary to measure the amount of water present and make the measurements under a controlled atmosphere (under flushing dry nitrogen in our laboratory).[2]

17.3.2 The Methods of Measurement

We restrict our analysis of the methods of measurement to properties for which the group has experience: viscosity, thermal conductivity, electrical conductivity, electrical permittivity (dielectric constant), diffusion coefficients, and heat capacity. However, for this last property, no special problems additional to those encountered for nonionic liquids are envisaged. Information about the different methods used to measure the heat capacity in liquids can be found in recent works [21, 22], namely, the measurement of [C₆mim][NTf₂] with adiabatic calorimetry [23].

---

[1] For the case of HTILs, the measurements are usually made above 400 K, which makes the presence of water negligible, except in some systems.

[2] This procedure has proved to be insufficient, as the amount of water in the samples used to measure viscosity increased from 119.3 to 196.4 ppm during the measurements.

Viscosity The viscosity of a fluid is a measure of its tendency to dissipate energy when it is perturbed from equilibrium by a force field, which distorts the fluid at a given rate. The viscosity depends on the thermodynamic state of the fluid, and it is usually specified by pairs of variables (T, P) or (T, ρ) for a pure fluid, to which must be added a composition dependence in the case of mixtures. The dissipative mechanism of shear creates, inevitably, local temperature gradients, which can damage slightly the reference thermodynamic state to which the measurement is assigned, from the initial, unperturbed, equilibrium state. Because it is impossible to measure local shear stresses, the measuring methods must be based on some integral effect, amenable to accurate measurements, from which, by averaging, the reference state is obtained. In addition to the general constraints already referred for primary instruments, namely, the availability of a working equation and a complete set of corrections, we must require that the rates of shear are small enough to maintain a near equilibrium state and that hydrodynamic stability is maintained. Several measurements performed so far are grossly inaccurate due to the disregard of one or several of these conditions. Strictly speaking, there is no primary method[3] at present for the measurement of the viscosity of liquids, as the absolute methods developed so far, in order to achieve high accuracy, involve the use of instrumental constants obtained through experimental calibration. However, the analysis of such methods [7, 8, 24, 25] (designated as relative or secondary) shows completely different levels of departure from the accepted definition of a primary method. Therefore, we chose to designate as "quasi-primary" any method for which a physically sound working equation, relating the viscosity to the experimental–measured parameters, is available, but where some of these parameters must be obtained accurately by an independent calibration with a known standard. Among the existing methods of measurement of viscosity, we can consider as quasi-primary the oscillating body (disk, cup, cylinder, and sphere), the vibrating wire, the torsionally oscillating quartz crystal, and surface light scattering spectroscopy (SLS) methods [8]. The data obtained with these state-of-the-art instruments have smaller uncertainties (or greater accuracies), when compared with data obtained using other secondary methods, namely, the capillary viscometer (in any of its versions), the rolling ball viscometer, or the falling ball viscometer, whose calibration parameters do not possess a clear physical meaning. A complete discussion about the details and application of these methods can be found in previously referenced papers Reference 8 being more detailed about the quality of the measurements performed with the different methods. Results of the application of the oscillating cup to molten alkali nitrates can be found in References 26–28.

The application of such methods to the measurement of viscosity of ionic liquids needs some discussion, because these viscosities are moderate or high on the scale of viscosity (1 mPa s to 1 Pa s). Figure 17.5 shows the availability of calibrating liquids for the different ranges of the viscosity scale (excluding non-Newtonian

---

[3]The definition of a primary method was approved by CCQM (Comité Consultatif pour la Quantité de Matière, BIPM) in 1995: "a primary method of measurement is a method having the highest metrological qualities, the operation of which can be completely described and understood, for which a complete uncertainty statement can be written down in terms of SI units, and whose results are, therefore, accepted without reference to a standard of the quantity being measured."

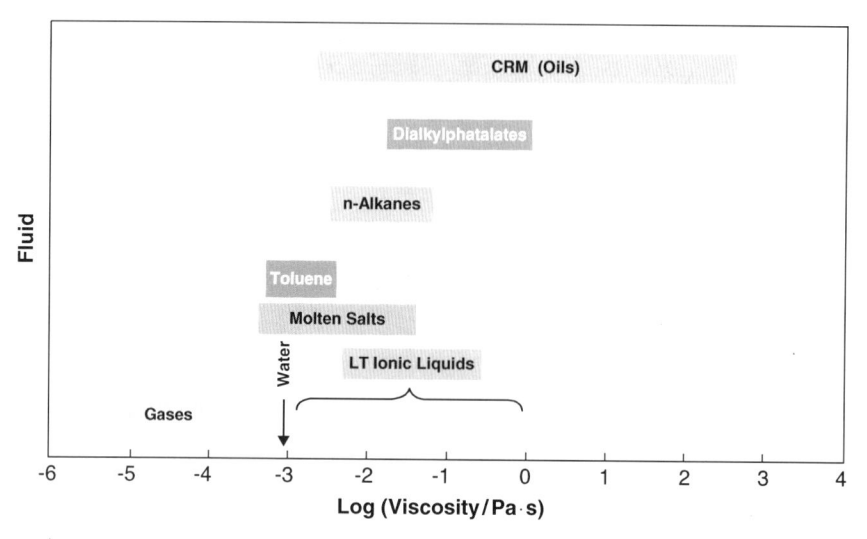

FIGURE 17.5 Viscosity ranges for different reference materials. Also shown are the viscosity domains for molten salts and LTILs.

fluids), some of them not yet used in common calibration procedures for relative viscometers. The ranges for LTIL and HTIL viscosities are also displayed, and it can be seen that they fall in a range with toluene (already accepted as a certified reference material [8]), alkanes, and alkyl phthalate esters, these last not yet approved as international standards,[4] as there is not available a consistent set of fluids and viscosity data to justify their choice, except for diisodecyl phthalate [29]. These fluids were proposed by the authors to resolve the problem of not having primary methods of measurement of viscosity available. The traceability to SI units, always necessary, can therefore be attained in the interim by selecting a set of reference liquids whose viscosity has been determined in different laboratories and by different quasi-primary techniques, whenever possible, covering wide ranges of viscosity. The metrological value of such an approach is much higher than the current route based only on the value of the viscosity of water at 0.1 MPa and 20 °C, and the stepping up procedure. Such a procedure, once fully installed, can avoid the higher uncertainties of the calibrations for high viscosity fluids, and the chemical limitations of pure water, previously discussed. This procedure would resolve most of the existing problems in measuring viscosity of ionic liquids.

Table 17.2 shows the most accurate methods of measurement of viscosity, their type, attainable uncertainty, and their adaptability to measure the viscosity of LTILs. From this analysis we recommend the use of quasi-primary instruments, such as the oscillating body (disk, cup, cylinder, and sphere) and the vibrating wire for high quality work (these viscometers are expensive, as they need a very accurate body machining, and should be used mostly for measuring the viscosity of well chosen

---

[4]From the end-user's point of view, the only viscosity reference point accepted so far is the viscosity of pure water at 20 °C, as recommended by ISO/TR 3666:1998(E), and accepted by IUPAC and OIML.

TABLE 17.2 Existing Methods for the Measurement of Viscosity and Their Applicability to LTILs

Method	Type	Attainable Uncertainty[a]	Adaptability to LTILs — Yes	Adaptability to LTILs — No	Adaptability to LTILs — Maybe if
Oscillating disk	Quasi-primary	1%	For low to moderate viscosity liquids		Disk/plates edge effect corrections became available for high viscosity liquids.
Oscillating cup or cylinder	Quasi-primary	1–2%	For low to moderate viscosity liquids		Meniscus effect must be avoided for high viscosity liquids
Vibrating wire	Quasi-primary	1%	✓		
Torsionally oscillating quartz crystal	Quasi-primary	1%		✓	Correct electrical insulation of gold electrodes deposited on the crystal can be achieved without loss of performance
Surface light scattering spectroscopy	Quasi-primary	<2%	For low to moderate viscosity liquids		Possible to extend to high viscosity liquids, as the vapour pressure of the LTILs is almost negligible
Capillary flow	Secondary/relative	1–3%	For low to moderate viscosity liquids	For high viscosity liquids	The traceability chain is kept with low uncertainty
Falling body	Secondary/relative	2–5%		✓	Not recommended for high quality work

[a] Uncertainty is defined using ISO criteria, using a coverage factor ($k = 2$), that correspond to a 95% confidence level.

reference liquids), and surface light scattering spectroscopy methods, and capillary flow viscometers for current laboratory work. In this last case, the users must be aware of all the problems already mentioned about calibrating liquids and traceability chain to SI units.

Thermal Conductivity The thermal conductivity of a fluid measures its propensity to dissipate energy, when disturbed from equilibrium by imposition of a temperature gradient. However, thermal conductivity is one of the thermophysical properties that is difficult to measure accurately. The main difficulty lies in the isolation of pure conduction from other mechanisms of heat transfer, like convection and radiation, a fact that arises from the contradictory requirement of imposing a temperature gradient on the fluid while preventing its motion. The imposition of a temperature gradient in a fluid in the gravitational field of Earth[5] inevitably creates a state of motion (natural convection). For isotropic fluids, the thermal conductivity is defined by Fourier's law and depends on the thermodynamic state of the fluid prior to the perturbation, and must be related with a reference state, not necessarily equal to the initial one. As it is impossible to measure local fluxes and local gradients [24], the Fourier equation cannot be used directly, and the energy equation adapted to a given geometry is the solution. However, the full energy equation cannot be used, and some approximations have to be introduced. In the first place, we must assure that the perturbation to the state of equilibrium (temperature gradients) is small, so that a near equilibrium state is maintained. Second, fluid movement must be avoided or reduced to a negligible level. It is therefore necessary to make measurements in such a way that the effect of convection is rendered negligible. Finally, it is important to recognise that the transport of energy by radiation is always present and must be corrected for each measuring technique, especially if measurements are performed at high temperatures. The full energy equation can be simplified [24] to give

$$\rho C_P \frac{\partial T}{\partial t} = \lambda \nabla^2 T \tag{17.1}$$

where ρ is the fluid density and C_P is its isobaric heat capacity.

In the last forty years, a variety of experimental methods have been developed, for gaseous, liquid, or supercritical phases, over a wide range of thermodynamic states. These methods are based on the simplified energy equation (17.1) and can be classified into two main categories:

- Unsteady state or transient methods, in which the full equation (17.1) is used, and the principal measurement is the temporal history of the fluid temperature
- Steady-state techniques, for which $\partial T/\partial t = 0$ and the equation reduces to $\nabla^2 T = 0$, which can be integrated for a given geometry

---

[5]Several efforts to measure the thermal conductivity under microgravity conditions showed that we can obtain convection-free measurements. However, the cost of experiments in parabolic flight rockets, in a space shuttle, or in the ISS impedes its routine use. Only experiments designed to prove underlying principles, if financed, can be undertaken [2].

Transient methods are the only ones that can compete for the statute of primary methods, in the sense of CIPM. These methods include the transient hot-wire, the transient hot-strip, and the interferometric technique (adapted to states near the critical point). Details of these methods can be found in References 25, and 30–32. Reference must also be made to three methods for the determination of thermal diffusivity, $\alpha = \lambda/\rho C_P$, which are not direct measurements of thermal conductivity, needing accurate values of density (usually available) and heat capacity (more difficult to obtain): photon-correlation spectroscopy [32], the forced Rayleigh scattering method [33], and the transient grating technique. The first of these is a very versatile (amenable to improvement) method, the second has been applied to molten salts [33], and the third constitutes the only reported measurements so far for the thermal conductivity of LTILs [34]. Details of the forced Rayleigh scattering method and of photon-correlation spectroscopy, both based on light scattering by the material under study, can be found in Reference 25.

From all these methods, only the transient hot-wire can be considered today as a primary method [35]:

- The transient hot-wire technique was identified as the best technique for obtaining standard reference data (Certification of Reference Materials).
- It is an absolute technique, with a working equation and a complete set of corrections reflecting the departure from the ideal model, where the principal variables are measured with a high degree of accuracy. It is accepted by the scientific community as a primary method, the top of the traceability chain for this physical quantity.
- The liquids proposed by IUPAC (toluene, benzene, and water) as primary standards for the measurement of thermal conductivity were measured with this technique with an accuracy of 1% or better.

The theory of the transient hot-wire is well known [36], and a complete uncertainty analysis has been presented previously [36, 37]. The transient hot-wire method is an absolute method, where the thermal conductivity of a fluid is evaluated by monitoring the rate at which the temperature of a thin wire increases with time after a step change of voltage has been applied to it. The constant heat flux per unit length thus generated by a linear source has the effect of producing throughout the fluid a temperature field that increases with time. Details of its applications to nonconducting fluids can be found in References 36, 38, and 39. This method has also been applied to electrically conducting liquids in the past, including polar liquids, molten salts, electrolyte solutions, molten metals, and molten semiconductors. LTILs fall into this group and we shall describe its application in more detail.

When the ordinary transient hot-wire is applied to measure the thermal conductivity of electrical conducting liquids, several problems are encountered [25, 36]:

1. The contact between the bare metallic wire and the conducting liquid provides a secondary path for the flow of current in the cell, and the heat generation in the wire cannot be defined unambiguously.

2. Polarisation of the liquids occurs at the surface of the wire, producing an electrical double layer.
3. The electrical measuring system (an automatic Wheatstone bridge) that detects the changes in the voltage signals in the wire is affected by the combined resistance/capacitance effect, caused by the dual path conduction.

Two main solutions have been developed to overcome these problems. One is the production around the wire of an electrical insulating coating, by direct coating with a polyester insulating layer, as developed by Nagasaka and Nagashima [40], by coating probes with ceramic for high temperature measurements, like Kitade et al. [41], or by anodising a metallic wire at its surface, producing a very stable and thin metallic oxide insulation, like Alloush et al. [42]. When coated wires are applied, one needs to account for the thermal effects of the insulating layer, and this can be consulted in References 36, 38 and 40. The second solution, or polarisation technique, consists in polarising, with a direct current (DC) of high impedance, the wire against the cell wall. Its first application was developed in London [43] with chlorobenzene and applied systematically in Boulder [44, 45] and Lisbon [46–48] to environmentally acceptable refrigerants, all polar liquids, with significant dipole moments. As it was almost impossible to obtain samples of these fluids with high purity, the presence of small amounts of water increases drastically the electrical conductivity of the polar liquid and its capacity to dissolve small amounts of ions, always present in the cells, generating a small voltage difference between the cell walls and the bare metallic wires (usually platinum). The application of a potential between the wires and the cell walls creates electric double layers near the surface of the wires—Stern layers [49]. These double layers contain solvated ions, with a charge opposite to the metallic surface, to maintain the solution's electroneutrality, therefore screening the ions from contact with the surface charges in the wires. Current cannot flow from the wire to the cell wall, through the liquid, because the wires are "electrically" insulated. Figure 17.6 illustrates this phenomenon. Any of these solutions permits the measurement of the thermal conductivity of polar liquids with accuracies comparable to those obtained with nonpolar or nonconducting liquids, and consequently can be applied to LTILs.

The steady-state techniques comprise essentially two types of geometry: the parallel plates and the concentric cylinders. These instruments, although mostly of the absolute type, can be considered as either primary or secondary. Again, details of the different solutions can be found in References 24 and 25. The most popular (and accurate) geometry for liquids is the concentric cylinders method, where the liquid stays inside the annular gap defined by the cylinders. As in all steady-state instruments, the determination of the gap between the cylinders is crucial, and the most acute determinations are made using capacity measurements of the dielectric. An application of the method to the measurement in pure molten salts and salt mixtures has been reported by Tufeu et al. [50], for $Na[NO_3]$, $K[NO_3]$, and $Na[NO_2]$, the equimolar mixture ($KNO_3 + NaNO_3$), and HITEC.[6] In this cell the contribution of

---

[6] HITEC, or HTS, is a ternary mixture (0.07 $NaNO_3$ + 0.4 $NaNO_2$ + 0.53 KNO_3), weight %.

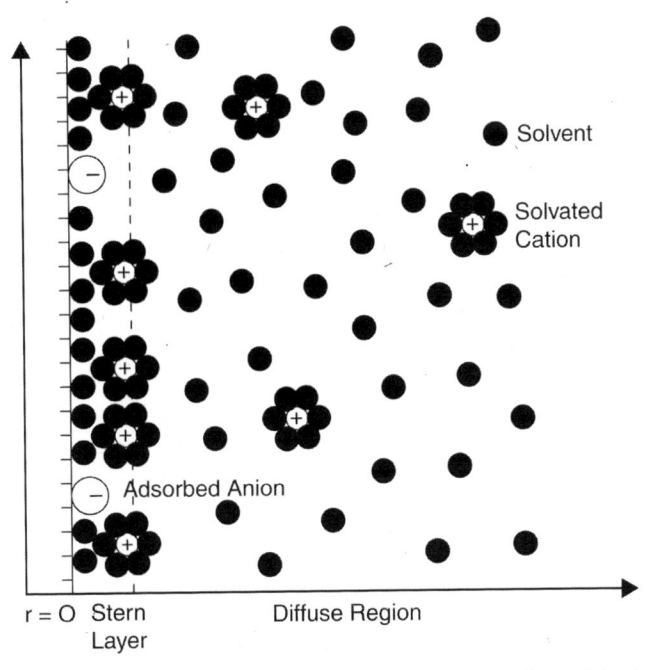

FIGURE 17.6 Model for the electric double layer on the surface of the hot-wires.

radiation to the overall heat transport was minimised by reducing the annular gap to 0.2 mm and using silver as the cylinder's material (low emissivity). The application of the steady-state concentric cylinders method to LTILs is possible, even with greater annular gaps, as the temperature can only go to its degradation temperature, usually on the order of 100–150 °C.

The use of secondary measuring techniques to measure thermal conductivity, those that need calibration with reference liquids of known thermal conductivity, is well supported, except for the high temperature range [35]. The two main works were endorsed by IUPAC and refer to the thermal conductivity of water, at 0.1 MPa, for temperatures between 275 and 370 K [51], and toluene for temperatures between 189 and 553 K [52], along the saturation line. However, for accurate work, the user should follow the discussion presented in Table 17.3, were the principal methods of measurement are analysed, keeping in mind their applicability to the LTILs. From the only results obtained with the transient grating technique [34], the thermal conductivity of [C$_4$mim][BF$_4$], [C$_4$mim][PF$_6$], [C$_4$mim][NTf$_2$], and [C$_2$mim][NTf$_2$] at 298.15 K is about 100–160 mW·m^{-1}·K^{-1}, a value similar to most organic liquids, so no additional problems are to be encountered in using the methods displayed in Table 17.3.

Electrical Conductivity Electrical conductivity (also referred to as electrolytic conductivity) is a measure of a material's ability to conduct an electric current, usually employed as an indicator of the material's purity. The state-of-the-art primary

TABLE 17.3 Existing Methods for the Measurement of Thermal Conductivity (Thermal Diffusivity) and Their Applicability to LTILs

Method	Type	Attainable Uncertainty	Adaptability to LTILs Yes	Adaptability to LTILs No	Adaptability to LTILs Maybe if
Transient hot-wire for conducting liquids	Primary	1%	✓		
Transient hot-strip for conducting liquids	Primary/secondary	1–2%	✓		Can be considered primary if the 3D heat transfer equation is solved for the geometries involved
Steady-state parallel plates	Primary/secondary	1–2%	✓		Recommended for special applications to critical states, critical end points
Steady-state concentric cylinders	Primary/secondary	2%	✓		
Forced rayleigh scattering	Secondary/relative	2–3%	✓		Dyes necessary to enhance the signal are compatible with LTILs
Photon correlation spectroscopy	Secondary/relative	2–3%	✓		Not recommended for high quality work
Transient grating	Secondary/relative	3–5%	✓		Needs a big improvement in the calibration; not recommended for high quality work

measurement method uses cells, of parallel or concentric electrodes with variable spacing, which inhibit electrolytic conduction paths other than the direct way between electrodes, and allows the compensation of fringe effects. To ensure maximum reproducibility, avoiding the consequences of the difference in thermal expansion coefficients, cell materials are currently shifting from the classic pair of glass and platinum to ceramic and platinum thin films on ceramic. Routine measurements (temperatures between 0 and 100 °C at atmospheric pressure) still use standard platinum electrodes, although glass is sometimes replaced by suitable polymers like polypropylene [53, 54]. Primary measurements are usually done at the metrological institutes, like PTB, DFM, and NIST, among others.

Electrical measurements are performed with alternating current (AC) metres or bridges at frequencies of at least 250 Hz, usual values being 1 or 10 kHz. However, at the surface of the electrodes, polarisation phenomena can occur, which influence the measured values. To avoid this error, different measuring techniques can be applied, namely, the use of a higher frequency, the use of four electrode measurements as in platinum resistance thermometry (voltage and current leads), or making inductive or capacitive nonconductive coupling between the electrodes and the measuring instrument.

Electrolytic conductivity is disseminated through the use of primary standards that are aqueous solutions of known potassium chloride concentration [55–57]. The electrolytic conductivity values were originally measured by Jones and Bradshaw, in 1933 [58], on the *demal* scale (equivalents per cubic decimetre), established by Parker and Parker in 1924 [59]. Beginning in 2001, IUPAC made the transition to the more common molality scale based standards, to account for all the effects caused by the improvement in the determination of all the involved physical constants, change of temperature scale, avoidance of the errors associated with volumetric procedures, and so on [60, 61].

The existing practice established to deal with the more common ions (mostly inorganic and usually small) is suitable to deal with most HTILs, but no real experience exists in dealing with the very large ions present in LTILs. Nevertheless, the large size of the ions in LTILs and the expected low ionic mobility might allow the use of lower measurement frequencies without the occurrence of polarisation effects. High accuracy work (primary methods) will usually involve doing measurements at several frequencies and calculating the resistance from

$$R = \lim_{1/f \to 0} \Delta R(f)$$

while routine work (end-user measurements) will just use the values measured at the standard frequency of the measuring instrument, after having calibrated the measurement cell with secondary reference solutions of electrolytic conductivity [54, 57].

For LTILs and HTILs, this property is strongly dependent on temperature and viscosity, and moderately dependent on pressure, with most available studies being on the temperature effect [62, 63]. However, the presence of water is a major concern, as in other properties, because the electrical conductivity of ionic liquids is strongly affected by its presence, even at the low ppm range [20]. Figure 17.4 shows the relative deviations of the electrical conductivity of the mixtures ([C$_4$mim][NTf$_2$] + water)

TABLE 17.4 Existing Methods for the Measurement of Electrical Conductivity and Their Applicability to LTILs

Method	Type	Attainable Uncertainty	Adaptability to LTILs Yes	No	Maybe if
AC impedance method	Primary	0.1–0.5%	✓		
Impedance method, DC or AC, using four electrodes	Primary/ secondary	0.5–1%	✓		
Resistance or conductance method, with cell constant determined by calibration	Secondary/ relative	1–3%	✓		Needs a big improvement in the calibration; not recommended for high quality work

from that of pure [C$_4$mim][NTf$_2$], for different amounts of water, obtained by Widegreen et al. [20], who used an AC impedance bridge technique to measure σ as a function of temperature at about 0.1 MPa, using a commercial cell made of borosilicate glass with two platinum black electrodes, calibrated with standard potassium chloride solutions, with an uncertainty of 2% in σ. The effect of increasing the water content is dramatic and increasing w_{H_2O} by only 2.5×10^{-4} increased the electrical conductivity by 1%. Table 17.4 shows the applicability of the electrical conductivity measurements to LTILs.

Electrical Permittivity The measurement of the electrical permittivity relative to vacuum (dielectric constant) allows the study of fluid molecular behaviour when subjected to an electric field, related to chemical structure and molecular interactions. The measurement of the properties of dielectrics has been the subject of multiple investigations in the past, and the general principles are reasonably established and understood [64–67]. However, when developing new materials for new applications, as in the case of ionic liquids, it is necessary to measure experimentally the dielectric constants as a function of temperature and density in order to understand their electronic structure and behaviour. The response of the different materials to an applied electrical field, of frequency ν, can be characterised by the complex electrical permittivity $\tilde{\varepsilon}(\nu)$, given by [64, 65]

$$\tilde{\varepsilon}(\nu) = \varepsilon'(\nu) - i\varepsilon''(\nu) = \varepsilon^\infty + \Delta\tilde{\varepsilon}(\nu) + \frac{\kappa}{i2\pi\nu\varepsilon_0} \qquad (17.2)$$

The real part $\varepsilon'(\nu)$ reflects the dielectric dispersion and it is usually decomposed into the frequency-dependent contribution $\Delta\tilde{\varepsilon}(\nu)$ and a high frequency contribution ε^∞ due to the electronic and nuclear displacement polarisations in the infrared and optical regimes. The static dielectric constant, ε, is defined as the zero-frequency limit of the dispersion curve:

$$\varepsilon = \varepsilon^\infty + \lim_{\nu \to 0} \Delta\tilde{\varepsilon}(\nu) \qquad (17.3)$$

The imaginary part, $\varepsilon''(\nu)$, reflects dielectric absorption or loss, important for studying relaxation processes in the liquids, amenable, inter alia, by microwave dielectric spectroscopy (or dielectric relaxation spectroscopy) [68, 69] and electrochemical impedance techniques [70]. In addition to the relaxation contribution, the absorption spectrum is superimposed by a diverging low frequency response proportional to the static (DC) electrical conductivity of the sample, κ. This term becomes very important at low frequencies and for highly conducting liquids—see Eq. (17.3)—a fact that has to be very well analysed for measurements in ionic liquids.

Measurements of the dielectric constant have been performed in the past for several liquids, liquid mixtures, and complex materials. In the case of liquids, however, the number of groups that measured dielectric constants in recent years is very limited, and normally restricted to the measurement of the static dielectric constant, measured usually at frequencies[7] between 1 and 100 kHz, well below of the 1 MHz to 20 GHz of the microwave spectroscopy measurements. In this region, the dielectric constant of polar molecular liquids is frequency independent, as shown by Mardolcar et al. [71] for toluene and benzene (1 kHz and 10 kHz), where the values at the same temperature never differ by more than 0.05%. A long series of measurements in environmentally acceptable refrigerants (pure, binary, and ternary mixtures) was recently reviewed by the authors [72], with uncertainties never exceeding 0.16%, including the uncertainty caused by the sample purity (1 part in 10^3). This last work was based on an instrument designed and constructed to operate in an extended thermodynamic range, from 170 K up to 370 K, at pressures up to 30 MPa. The measurements use the direct capacitance method. The description of the cell, based on two concentric cylinders, with a gap of 0.1 mm and 10 cm long, has been presented before by Mardolcar et al. [71] and the sample handling, vacuum, and pressure system are described by Gurova et al. [46]. The measuring process uses fully automated instrumentation, operated from a computer graphics user interface, and described elsewhere [73]. Vacuum capacitance was measured as a function of temperature before filling the cell with the fluid. The technique employed a four-terminal connection to the cell, in order to compensate for parasitic impedances. The mean value of a 10-dimensional sample, taken at a 10 kHz frequency, provides the experimental value of relative permittivity, which proved to be properly suited to the working accuracy. The relative permittivity, ε, of the fluid is determined from the ratio between $C(P, T)$—the geometric capacitance at pressure P and temperature T—and $C_0(T)$—the capacitance under vacuum at a temperature T:

$$\varepsilon = \frac{C(p, T)}{C_0(T)} \qquad (17.4)$$

---

[7] Usually, for normal liquids, a frequency between 1 and 100 kHz is enough to make the frequency high enough to avoid the dielectric relaxation.

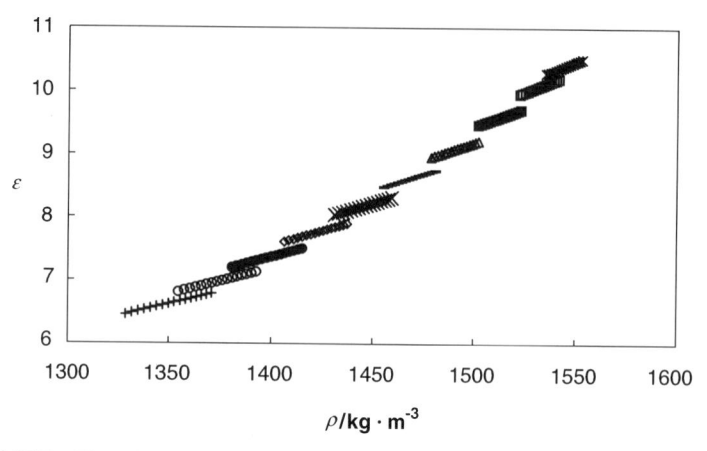

FIGURE 17.7 The relative permittivity ε, of 1,1,1,3,3-pentafluoropropane, as a function of density, ρ, for the different isotherms. +, 303.84 K; ○, 293.23 K; ●, 283.19 K; ◇, 273.12 K; ×, 263.00 K; −, 253.01 K; △, 243.02 K; ■, 233.12 K; □, 224.16 K; ∗, 218.53 K [72].

Figure 17.7 displays the results obtained for 1,1,1,3,3-pentafluoropropane for temperatures between 218.53 and 303.84 K, as a function of density. All the fluids studied showed dielectric constants varying between 5 and 30, covering what are usually considered to be fluids of low to high polarity. In fact, some of these fluids are strongly polar, showing cooperative polarisation effects typical of hydrogen bonding systems, as proved by Cabral et al. [74] for difluoromethane (HFC-32), 1,1,1,2-tetrafluoroethane (HFC-134a), and 1,1,1-trimethylethane (HFC-143a). The method of electrochemical impedance spectroscopy is similar, as it can use an AC stimulus, in the form of a voltage (one measures the current response) or of a current (in this case one measures the voltage response). For dielectric constant measurements, the use of voltage excitation is more common, if the voltage used is well below the decomposition potential of the liquid under study [70, 74]. A special cell was developed by these authors, prepared to measure the electrical properties of highly resistive liquids, using parallel plate electrodes, with a large ratio of electrode area to interelectrode gap, using multiple, closely spaced, parallel plate electrodes, of an air variable capacitor. Measurements with the binary system (butanenitrile + chloroethane) showed values of the dielectric constant varying from 13 (pure chloroethane at 238 K) to 45 (pure butanenitrile at 170 K). The uncertainty of the measurements is on the order of 0.5, which can make the uncertainty on the order of 1–4%, a value that reflects the need for further improvement.

The use of microwave dielectric spectroscopy in the measurement of the dielectric constants of LTILs started recently and has been implemented by Weingärter and co-workers [68, 69]. A very recent paper [71] discusses the values obtained from 14 LTILs, with imidazolium, pyrrolidinium, pyridinium, and ammonium cations, and fluorometallate, triflate, and bis[(trifluoromethyl)sulfonyl]amide anions at 298.15 K, with an uncertainty better than 0.6%. As the reported values of the dielectric constants

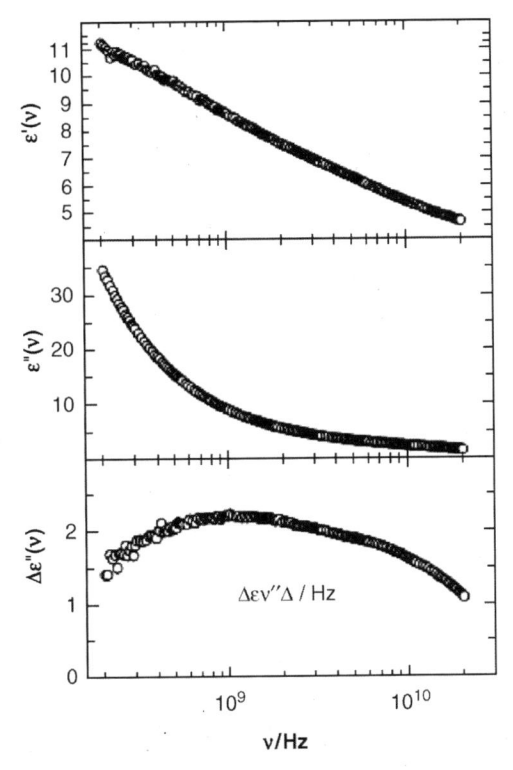

FIGURE 17.8 Real part (negative), imaginary part, and conductance-corrected (negative) imaginary part of the complex permittivity of [C$_4$mim][NTf$_2$] at 298.15 K [69].

of these liquids have values between 10 and 15.1, the uncertainty can amount to 4–6%, a value that is still high and justifies improvement of the method. At this level of uncertainty, the amount of impurities present in the samples and the possible absorption of water cannot play a relevant role. Figure 17.8 displays the real part (negative), imaginary part, and conductance-corrected (negative) imaginary part of the complex permittivity of [C$_4$mim][NTf$_2$] at 298.15 K.

From the discussion above, it is clear that only the direct capacitance method can be considered a primary method for measuring the dielectric constant of polar liquids. It is an absolute method, not needing any type of calibration. Its application to ionic liquids and the complete study of the interaction between the ion mobility and the polarisation in the ionic liquid are under study, but if a careful design of the cell is obtained [75], no special impediments are foreseen. The other two methods—electrochemical impedance spectroscopy and dielectric relaxation spectroscopy—still need some improvement and therefore are considered for now as secondary methods. Table 17.5 resumes this discussion. As a main conclusion, all the methods mentioned can be applied to the measurement of the static dielectric constant of LTILs.

TABLE 17.5 Existing Methods for the Measurement of Electrical Permittivity and Their Applicability to LTILs

Method	Type	Attainable Uncertainty	Adaptability to LTILs Yes	No	Maybe if
Direct capacitance method (concentric cylinder electrodes)	Primary	0.2–1%	✓		
Electrochemical impedance spectroscopy, voltage driven (multiple parallel plates)	Primary/ secondary	1–4%	✓		Can be improved to primary method if a detailed analysis of the current flow between electrodes is made [75]
Dielectric relaxation spectroscopy	Secondary	4–6%	✓		Improvements of the technique are necessary; permits the full frequency spectrum dielectric response

Diffusion Coefficients The diffusion process in liquids is by far one of the most difficult transport phenomena, and can be simply defined as the process whereby an initially nonuniform distribution of species in a mixture proceeds toward a uniform distribution [25, Chap. 9]. The main problems rely on the misinterpretations that affect the discussion about the quality and comparability of the different methods of measurement. Therefore it is necessary to start with some definitions to clarify the issue, namely, those of the self-diffusion coefficient, the interdiffusion (or mutual diffusion) coefficient, and the intradiffusion coefficient. Following, for the sake of consistency, the definitions in Reference 25, the *self-diffusion coefficient* is the coefficient obtained in a system where there is only one component in diffusional motion, and can be studied by techniques like nuclear magnetic resonance. The *interdiffusion coefficient*, the most popular and analysed coefficient because it is the fundamental one in a diffusion process in binary or multicomponent diffusion, appears when there is simultaneous diffusion of several species, where the molar fluxes of the different species are not completely independent but obey the overall continuity equation, in molar quantities [76]. For a binary mixture, the most common case, in the absence of chemical reaction and net movement of the liquid mixture (no mass convection), and when the product $c \mathcal{D}_{AB}$ can be considered constant, the equation of continuity for the mixture reduces to

$$\frac{\partial c_A}{\partial t} = \mathcal{D}_{AB} \nabla^2 c_A$$
$$\frac{\partial c_B}{\partial t} = \mathcal{D}_{BA} \nabla^2 c_B \qquad (17.5)$$
$$\mathcal{D}_{AB} = \mathcal{D}_{BA}$$

Equation (17.5) is called Fick's second law of diffusion, or sometimes simply the diffusion equation. In this equation, c is the total molar concentration, c_i are the molar concentrations of the species under consideration (A, B), and $\mathcal{D}_{AB}$ is the interdiffusion or mutual diffusion coefficient of A in B. Equation (17.5) can easily be generalised for multicomponent diffusion [25, 76]. Finally, the *intradiffusion coefficient* or tracer diffusion characterises the diffusion of each component A or B in a uniform mixture of these components, but where the component under study, A, is chemically identical with component B but can be distinguished by some label such as an isotopic form.

The available methods for the measurement of diffusion coefficients in liquid systems have been comprehensively described [25], although there were some earlier descriptions [77, 78]. Here, we refer only the most adequate, in our judgement, to measurements in LTILs, referring previous applications, if any, to molten salts.

The diffusion coefficients depend parametrically on the thermodynamic state of the liquid and must be assigned a reference state characterised by a triplet of variables (T, P, c). In any measurement of the diffusion coefficients there are several aspects that must be considered. In ideal mixtures, for which there are no volume or enthalpy changes on mixing, when two mixtures of different composition are added together at the same temperature, some very small temperature gradients are created—the Dufour effect. Usually, these temperature heterogeneities are negligible. If the mixture is nonideal, there are volume and enthalpy changes on mixing, creating temperature heterogeneities and convective motion that have to be rendered negligible, to perform isothermal measurements. This is the case in binary mixtures of ionic liquids, mostly nonideal [79]. It follows that a primary instrument for the measurement of diffusion coefficients must satisfy several conditions. First, it is necessary to maintain a near equilibrium state by using small concentration gradients. Second, all the effects resulting from nonideal behaviour must be reduced to a negligible level. Third, the bulk flow motion resulting from free convection must be minimised, as well as the Dufour effect. In addition to these conditions, and as explained in previous sections, it is necessary to have a working equation and a full set of corrections, easily calculable. Not many instruments obey these criteria, and we shall refer to some techniques that are absolute, such as the closed tube method, the Harned conductance cell,[8] the interferometric techniques (Gouy, Rayleigh), and the Taylor dispersion method. All these methods are applicable to LTILs and the only criteria for a given choice are the required uncertainty of the measurements and the average time of duration of an experiment, as most of them, with the exception of the Taylor dispersion technique, last for more than one day for a single experiment. Table 17.6 shows the main characteristics of each of the primary methods mentioned. The close tube methods, namely, the conductance cells developed by Harned and collaborators in the 1950s [80], have been applied to aqueous salt solutions, strong and weak acids, mixed solvents, strong electrolytes, proteins, and ionic dyes; readers can find details in Reference 25. The main disadvantage of these methods, in addition to the duration of a measurement, is the limited temperature range, usually near ambient, a fact that it is not probably important for the LTILs measurements.

---

[8]The Harned conductance cell, used for conductimetric measurements, is in fact a close tube method.

TABLE 17.6 Existing Methods for the Measurement of Mutual Diffusion Coefficients and Their Applicability to LTILs

Method	Type	Attainable Uncertainty	Yes	No	May be if	Average Duration of Experiments
Interferometric techniques	Primary/absolute	0.2%	✓			1–2 days
Closed tube	Primary/absolute	0.4%	✓			Days
Harned conductance	Primary/absolute	0.4–0.8%	✓			Days
Taylor dispersion	Primary/absolute	1–2%	✓			<1 hour

("Adaptability to LTILs" spans the Yes/No/May be if columns.)

However, its application to molten salts has been very restricted, and with uncertainties well above those mentioned in Table 17.5. The same applies to the optical methods that seldom have been applied far from the ambient temperature, but the excellent uncertainties of the methods make them appropriate for standard reference data in a limited range of conditions. The Taylor dispersion technique, although of slightly higher uncertainty, has two great advantages over the other methods. It is very rapid and can be used, without loss of accuracy, in wide ranges of temperature and pressure. Furthermore, it has been applied already to diffusion of water + LTILs [81]. Also, as the authors have experience in the use of this method and intend to use it to measure the diffusion of water in LTILs, at low concentrations of water (in order to understand how to achieve high purity for these compounds), we shall mention it briefly.

In essence, the Taylor dispersion technique uses the velocity profile of the flow of a liquid in a cylindrical tube to enhance the dispersion of a pulse of a mixture in another mixture of slightly different composition, brought about only by molecular diffusion. The coupling of the flow induced dispersion with molecular diffusion results in a significant dispersion in a short time. The method was first conceived by G. I. Taylor in 1953, and its theory, that makes possible its classification as an absolute and primary method, was presented by Alizadeh et al. [82]. Figure 17.9 shows the ideal Taylor diffusion experiment.

A fluid mixture of components A and B flows in laminar flow through a straight, infinitely long, isothermal tube, of uniform circular cross section, of radius a_0, with impermeable walls. The molar concentrations of species A and B in the flowing solution are denoted by c_{Af} and c_{Bf} and the mean velocity of the liquid mixture by $\bar{u}_0$. At time $t=0$, a sample of the liquid mixture of slightly different composition, given by c_{Ai} and c_{Bi}, with a density identical at the flowing stream (to avoid buoyancy forces), is introduced at $z=0$ in the form of a δ pulse. The joint action of convective flow and molecular diffusion results, at a time t down the tube, in a dispersion of the pulse. The governing differential equation for the perturbation of the pulse molar concentration Δc_A is [82]

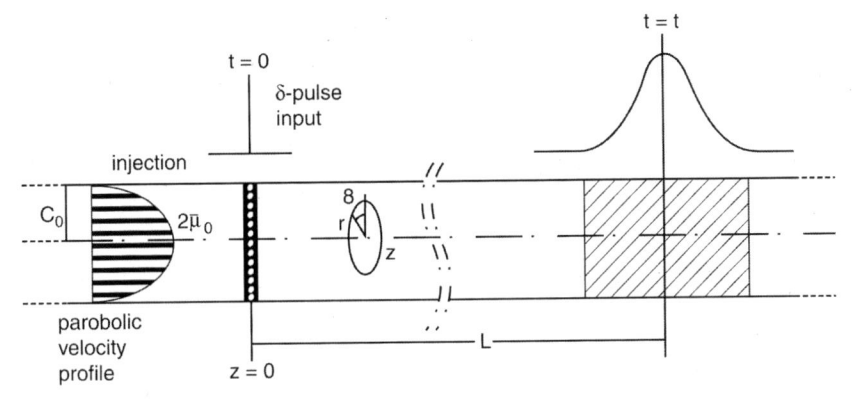

FIGURE 17.9 The ideal Taylor dispersion experiment [81].

$$\frac{1}{\mathcal{D}_{AB}}\frac{\partial(\Delta c_A)}{\partial t} = \nabla^2(\Delta c_A) - \frac{2u_0}{\mathcal{D}_{AB}}\left[1-2\left(\frac{r}{a_0}\right)^2\right]\frac{\partial(\Delta c_A)}{\partial z} \quad (17.6)$$

The initial conditions for its solution are

$$\Delta c_A(r,\theta,0,0) = \delta(z)(c_{Ai}-c_{Af}); \quad \delta(z)\begin{cases} = 0, & z \neq 0 \\ = 1, & z = 0 \end{cases} \quad (17.7)$$

The boundary conditions are

$$\begin{aligned}\frac{\partial(\Delta c_A)}{\partial z} &= 0 \quad \text{at } r = a_0 \text{ for all } t \\ \Delta c_A &\to 0 \quad \text{as } z = \pm\infty \end{aligned} \quad (17.8)$$

This equation cannot be solved directly to obtain $\Delta c_A(z, t)$, but it can be solved for the spatial moments of the concentration distribution at a particular time. In fact, the enhanced dispersion of the pulse results in a broadening of the distribution. If we monitor the cross-sectional averaged concentration distribution at an axial position $z = L$ as a function of time, provided that

$$\frac{\mathcal{D}_{AB}\, t}{a_0^2} > 700 \quad \text{and} \quad u_0 > 700\frac{\mathcal{D}_{AB}}{a_0} \quad (17.9)$$

then, with an error smaller than 0.01%, the first and the second central moments (variance) of the temporal distribution are given by [82]

$$\begin{aligned} \bar{t}_{id} &= \frac{L}{u_0}(1+2\xi) \\ \sigma_{id}^2 &= \left(\frac{L}{u_0}\right)^2(8\xi^2+2\xi) \\ \xi &= \frac{\bar{u}_0 a_0^2}{48\mathcal{D}_{AB}L} \end{aligned} \quad (17.10)$$

If the first and the second central moments (variance) of the temporal distribution are determined at a cross section of the tube at a distance L from the injection point, the interdiffusion coefficient can be obtained from

$$\mathcal{D}_{AB} = \frac{a_0^2}{24\,\bar{t}_{id}} \frac{\left[1 + 4\frac{\sigma_{id}^2}{\bar{t}_{id}^2}\right]^{1/2} + 3}{\left[1 + 4\frac{\sigma_{id}^2}{\bar{t}_{id}^2}\right]^{1/2} + 2\frac{\sigma_{id}^2}{\bar{t}_{id}^2} - 1} \qquad (17.11)$$

Any experimental realisation of the ideal method clearly involves departures from it, and all of them have been considered by Alizadeh et al. [82]. In particular, it was found that the ideal moments have to be corrected by

$$\bar{t}_{id} = \bar{t}_{exp} + \sum_i \delta \bar{t}_i$$
$$\sigma_{id}^2 = \sigma_{exp}^2 + \sum_i \delta \bar{\sigma}_i^2 \qquad (17.12)$$

where $\bar{t}_{exp}$ and σ_{exp}^2 denote the experimentally measured moments and $\delta \bar{t}_i$ and $\delta \bar{\sigma}_i^2$ are the different corrections to the ideal model [82]. They involve the finite volume of the detector, the finite volume of the injected sample, the tube connection to the detection cell, the secondary flow effects induced by the diffusion tube, the nonuniformity and the noncircularity of the tube cross section, as well as the concentration dependence of the interdiffusion coefficient, assumed constant in the ideal model. Following Alizadeh et al. [82], a suitable design can make most of these corrections negligible; the remaining ones do not contribute to more than 0.5% of the moments of the experimental curve.

The reference state of the measured diffusion coefficient is defined by the temperature and pressure of the measurement, and a composition c_{Aref} given by

$$c_{Aref} = c_{A1} + \delta c_A \qquad (17.13)$$

where the correction δc_A depends on the number of moles of the species A present in the injection sample in excess of those in the flowing stream.

Alizadeh and Wakeham [83, 84] and Matos Lopes and Nieto de Castro [85] developed instruments of this type, where the detection was made by using a differential liquid refractometer, normally used in HPLC, for measurements in binary mixtures of alkanes, at atmospheric pressure and temperatures between 290 and 350 K, with an accuracy reported to be better than 1%.

Figure 17.10 shows an eluted concentration profile for an infinite dilution test of heptane in hexane, obtained by Matos Lopes et al. [86] at 24.9 °C and atmospheric pressure. The digital signal, the fitted Gaussian, and the estimates of the errors in the measured variables $\bar{t}_{exp}$ and σ_{exp} all show perfect agreement between the model and the experiment. The later instrument was then applied to study more complex systems like diffusion in homogeneous and micellar solutions [87], and near the consolute point of a binary mixture [88]. This last application was fundamental support for

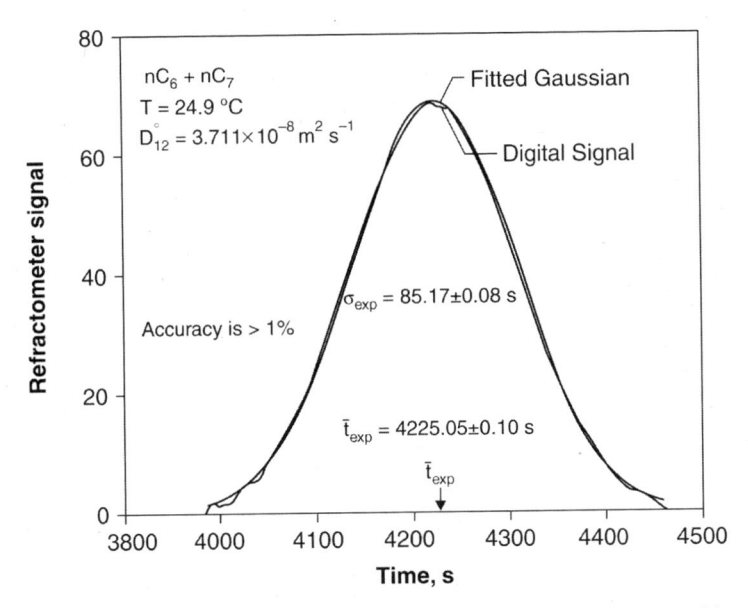

FIGURE 17.10 The eluted concentration profile for an infinite dilution test of heptane in hexane at 24.9 °C. The digital signal is compared with the fitted Gaussian. Estimates of the errors in the principal variables $\bar{t}_{exp}$ and σ_{exp} are also shown [85].

previous results obtained from dynamic light scattering and the theory of the critical enhancement of diffusivity, where the diffusivity decreases upon approaching the critical temperature, eventually vanishing to zero (Fig. 17.11). This result seems very important for ionic liquid applications, namely, when used as a solvent or as a reaction media in supercritical conditions or near immiscibility. Since these measurements, the technique has been applied to several systems, showing its versatility in some applications to wide ranges of temperature and pressure [89]. Many other authors have applied the method since then, and details can be found in Reference 89. It can also be applied to study reference systems, such as the mixture $H_2O + D_2O$ [90]. The results can be applied to study the structure of liquid mixtures, including specific interactions, short-range anisotropy, and packing, as demonstrated by Oliveira et al. [91] for the effect of branching in hexane in the binary diffusion coefficients.

As mentioned before, this technique has been applied with success to alkylimidazolium tetrafluoroborates and hexafluorophosphates in water at room temperature [81] at infinite dilution of the LTILs, the system previously tested with NaCl, with an accuracy of 0.8%. The liquids studied were $[C_n mim][BF_4]$ and $[C_n mim][PF_6]$ ($n = 2, 4, 6$, or 8). Figure 17.12 shows the dependence of the infinite dilution diffusivity as a function of the formula weight for the two series studied. It is interesting to note that the values obtained for the infinite dilution diffusivity in water of $[C_6 mim][BF_4]$ and $[C_4 mim][PF_6]$ are similar to the same quantity for the infinite dilution of heptane in dodecane [85] ($\sim 1.2 \times 10^{-9} \, m^2 \cdot s^{-1}$).

To study the influence of water on the purity of the LTILs, the measurements of the diffusivity have to be performed near infinite dilution of water in the LTILs.

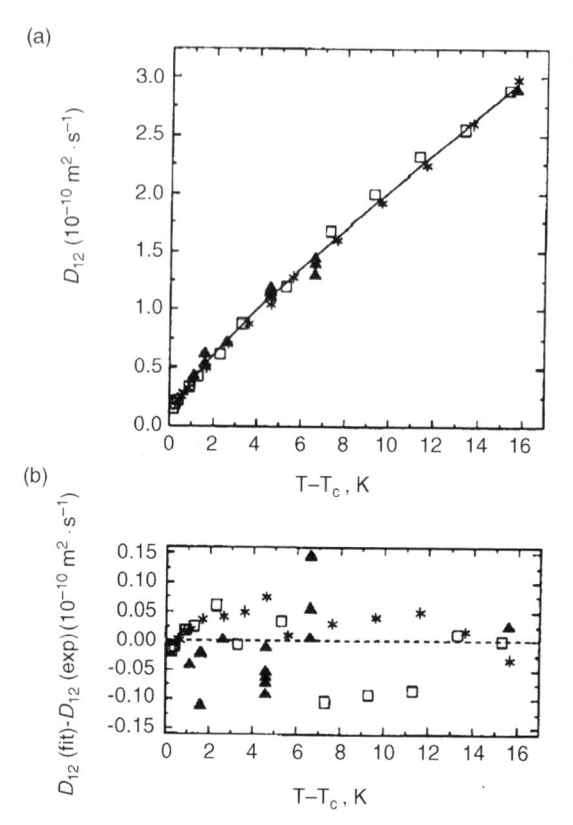

FIGURE 17.11 (a) Mutual diffusivity $\mathcal{D}_{AB}$ of a mixture of hexane and nitrobenzene at the critical composition as a function of $T - T_c$. The curve represents the values calculated using the theory of critical dynamics [87]. (b) Difference between experimental and calculated diffusivities. ▲, cited work; □, Reference 92; ∗, Reference 93.

17.4 CONCLUSION*

Low temperature ionic liquids with strong charge dislocation, polarity, and weak to moderate electrical conduction raise some problems regarding the accurate measurement of thermophysical properties, namely, transport properties. The methods of measurement have to be adapted to these types of liquids. This is indeed true for measuring the thermal conductivity, viscosity, electrical conductivity, and dielectric constant. For these properties, the fact that the liquid is ionic and therefore electrically conducting, and usually highly viscous, when subjected to thermal gradients, stresses, and electrical fields, generates several phenomena like electric double layers, viscous heating dissipation, and the need for sensor electrical isolation that make experiments

---

*The discussion about viscosity measuring methods and uncertainties presented in this chapter, was recently improved in the following paper: C. A. Nieto de Castro, F. J. Vieira dos Santos, J. M. N. A. Fareleira, W. A. Wakeham, "Metrology of viscosity – Have we learned enough?", *J. Chem. Eng. Data*, 54(2), 171–178 (2009), Festschrift for the 90th Birthday of Robin H. Stokes. (Review)

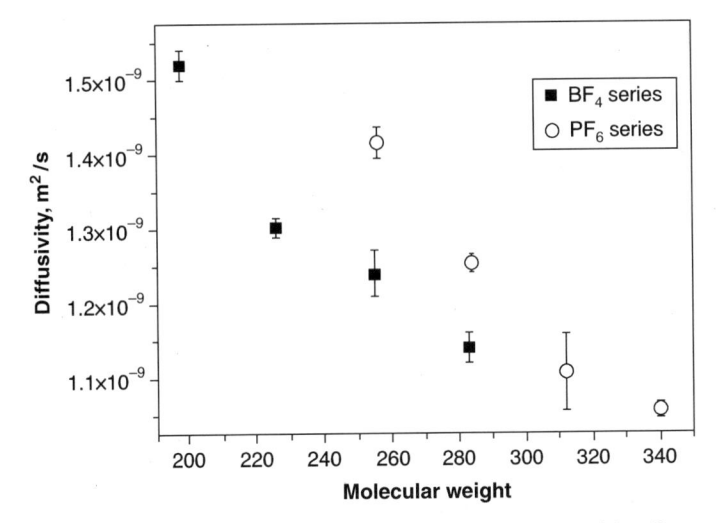

FIGURE 17.12 The variation of the diffusion coefficients of alkylimidazolium tetrafluoroborates and hexafluorophosphates in water with their formula weights [80]. The liquids studied were [C$_n$mim][BF$_4$] and [C$_n$mim][PF$_6$] ($n = 2, 4, 6$, or 8).

difficult to perform with high accuracy, even for temperatures around room temperature. New cells, new cable screenings, new signal analysis, and so on have to be envisaged. New reference liquids for property measurement, necessary for end-user relative methods, have to be devised and internationally accepted.

An international effort, which can strengthen the first efforts to produce new and high quality property data (IUPAC Project, Ken Marsh–coordinator) has to be supported. Among others, the properties to be measured (around room temperature to temperatures up to thermal degradation) are thermal conductivity, heat capacity, viscosity, electrical conductivity, electrical permittivity, surface tension, density, and diffusion of water in ionic liquids. These will support the key data for the necessary changes in existing chemical processes or for the design of new processes, based on green or sustainable chemistry. Our research group is prepared to contribute to this effort by measuring most of these properties in selected ionic liquids.

Finally, the levels of impurities in different LTILs, hydrophilic or not, have to be measured, if the data uncertainty is required to be 3% or better. The fact that the water absorption or dissolution is a time-dominated process at these low concentrations restricts the use of samples in industry for property-sensitive processes, requiring special care in production and quality control.

REFERENCES

1. Earle, M. J. and Seddon, K. R., Ionic liquids. Green solvents for the future, *Pure Appl. Chem.* **72**, 1391–1398 (2000).
2. Nieto de Castro, C. A., Thermal conductivity of molten materials—Is experiment necessary?, in *TEMPMEKO 2004, 9th International Symposium on Temperature*

and Thermal Measurements in Industry and Science, D. Zvizdic (Ed.) Vol. **1**, pp. 49–58, 2005.

3. Nunes, V. M. B., Lourenço, M. J. V., Santos, F. J. V., and Nieto de Castro, C. A., The importance of accurate data on viscosity and thermal conductivity in molten salts applications, *J. Chem. Eng. Data* **48**, 446–450 (2003).

4. Gonçalves, R. J. B., Nunes, V. M. B., Lourenço, M. J. V., Santos, F. J. V., and Nieto de Castro, C. A., Viscosity of molten lithium nitrate, paper presented at the EUCHEM 2006, 16–22 September, Hammamet, Tunisia (2006).

5. Marsh, K. N., Boxall, J. A., and Lichtenthaler, R., Room temperature ionic liquids and their mixtures — a review, *Fluid Phase Eq.* **219**, 93–98 (2004).

6. Earle, M. J., Esperança, J. M. S. S., Gilea, M. A., Canongia Lopes, J. N., Rebelo, L. P. N., Magee, J. W., Seddon, K. R., and Widegren, J. A. The distillation and volatility of ionic liquids, *Nature* **429**, 831–834 (2006).

7. Nieto de Castro, C. A., Santos, F. J. V., Fareleira, J. M. N. A., and Oliveira, C. M. B. P., Metrological references for the measurement of viscosity, in *12th International Metrology Congress*, Electronic Publication in CD-ROM, Collège Français de Métrologie, Lyon, France, 2005.

8. Santos, F. J. V., Nieto de Castro, C. A., Dymond, J. H., Dalaouti, N. K., Assael, M. J., and Nagashima, A., Standard reference data for the viscosity of toluene, *J. Phys. Chem. Ref. Data* **35**, 1–8 (2006).

9. Caetano, F. J. P., Fareleira, J. M. N. A., Oliveira, C. M. B. P., and Wakeham, W. A., Validation of a vibrating wire viscometer: measurements in the range 0.5 to 135 mPa·s, *J. Chem. Eng. Data* **50**, 201–205 (2005).

10. Widegreen, J. A., Laesecke, A., and Magee, J. W., The effect of dissolved water on the viscosities of hydrophobic room-temperature ionic liquids, *Chem. Commun.*, 1610–1612 (2005).

11. Vila, J., Ginés, P., Rilo, E., Cabeza, O., and Varela, L. M., Great increase of the electrical conductivity of ionic liquids in aqueous solutions, *Fluid Phase Equilib.* **247**, 32–39 (2006).

12. Law, G. and Watson, W. P., Surface tension measurements of n-alkylimidazolium ionic liquids, *Langmuir* **17**, 6138–6141 (2001).

13. Rebelo, L. P. N., Najdanovic-Visak, V., Visak, Z. P., Nunes da Ponte, M., Szydlowski, J., Cerdeirina, C. A., Troncoso, J., Romani, L. Esperança, J. M. S. S., Guedes, H. J. R., and Sousa, H. C. de, A detailed thermodynamic analysis of [C$_4$mim] [BF$_4$] + water as a case study to model ionic liquid aqueous solutions, *Green Chem.* **6**, 369 (2004).

14. Seddon, K. R., Stark, A., and Torres, M.-J., Influence of chloride, water, and organic solvents on the physical properties of ionic liquids, *Pure Appl. Chem.* **72**, 2275–2287 (2000).

15. Huddleston, J. G., Visser, A. E., Reichert, W. M., Willauer, H. D., Broker, G. A., and Rogers, R. D, Characterization and comparison of hydrophilic and hydrophobic room temperature ionic liquids incorporating the imidazolium cation, *Green Chem.* **3**, 156–164 (2001).

16. Poole, C. F., Kersten, B. R., Ho, S. S. J., Coddens, M. E., and Furton, K. G., Organic salts, liquid at room temperature, as mobile phases in liquid chromatography, *J. Chromatogr.* **352**, 407–425 (1986).

17. Tran, C. D., De Paoli Lacerda, S. H., and Oliveira, D., Absorption of water by room-temperature ionic liquids: effect of anions on concentration and state of water, *Appl. Spectrosc.* **57**, 152–157 (2003).

18. Cammarata, L., Kazarian, S. G., Salter, P. A., and Welton, T., Molecular states of water in room temperature ionic liquids, *Phys. Chem. Chem. Phys.* **3**, 5192–5200 (2001).
19. Anthony, J. L., Maginn, E. J., and Brennecke, J. F., Solution thermodynamics of imidazolium-based ionic liquids and water, *J. Phys. Chem. B* **105**, 10942–10949 (2001).
20. Widegreen, J. A., Saurer, E. M., Marsh, K. N., and Magee, J. W., Electrolytic conductivity of four imidazolium-based room-temperature ionic liquids and the effect of a water impurity, *J. Chem. Thermodyn.* **37**, 569–575 (2005).
21. Marsh, K. N. and O'Hare, P. A. G. (Eds.), Solution Calorimetry, *Experimental Thermodynamics*, Vol. **IV**, Blackwell Scientific, Oxford, UK, 1994.
22. Lourenço, M. J. V., Santos, F. J. V., Ramires, M. L. V., and Nieto de Castro, C. A., Isobaric specific heat capacity of water and aqueous cesium chloride solutions for temperatures between 298 and 370 K at $p = 0.1$ MPa, *J. Chem. Thermodyn.* **38**, 970–974 (2006).
23. Blokhin, A. V., Paulechka, Y. U., and Kabo, G. J., Thermodynamic properties of [C_6mim] [NTf$_2$] in the condensed state, *J. Chem. Eng. Data* **51**, 1377–1388 (2006).
24. Nieto de Castro, C. A., Absolute measurements of the viscosity and thermal conductivity of fluids, *JSME Int. J. Ser. II* **31**, 387–401 (1988).
25. Wakeham, W. A., Nagashima, A., and Sengers, J. V. (Eds.), *Measurement of the Transport Properties of Fluids, Experimental Thermodynamics*, Vol. **III**, Blackwell Scientific, Oxford, UK, 1991.
26. Nunes, V. M. B., Santos, F. J. V., and Nieto de Castro, C. A., A high-temperature viscometer for molten materials, *Int. J. Thermophys.* **19**, 427–435 (1998).
27. Lança, M. J. C., Lourenço, M. J. V., Santos, F. J. V., Nunes, V. M. B., and Nieto de Castro, C. A., Viscosity of molten potassium nitrate, *High Temp. High Press.* **33**, 427–434 (2001).
28. Nunes, V. M. B., Lourenço, M. J. V., Santos, F. J. V., and Nieto de Castro, C. A., Viscosity of molten sodium nitrate, *Int. J. Thermophys.* **27**, 1638–1649 (2006).
29. Caetano, F. J. P., Fareleira, J. M. N. A., Oliveira, C. M. B. P., and Wakeham, W. A., The viscosity of di-isodecylphthalate: a potential standard of moderate viscosity, *Int. J. Thermophys.* **25**, 1311–1322 (2004).
30. Assael, M. J., Dix, M., Drummond, I., Karagiannidis, L., Lourenço, M. J. V., Nieto de Castro, C. A., Papadaki, M., Ramires, M. L. V., van den Berg, H., and Wakeham, W. A., Towards standard reference values for thermal conductivity of high temperature melts, *Int. J. Thermophys.* **18**, 439–446 (1997).
31. Lourenço, M. J. V., Rosa, S. C. S., Nieto de Castro, C. A., Albuquerque, C., Erdmann, B., Lang, J., and Roitzsch, R., Simulation of the transient heating in an unsymmetrical coated hot-strip sensor with a self-adaptive finite element method (SAFEM), *Int. J. Thermophys.* **21**, 377–384 (2000).
32. Kraft, K., Lopes, M. M., and Leipertz A., Thermal-diffusivity and thermal-conductivity of toluene by photon-correlation spectroscopy—a test of the accuracy of the method, *Int. J. Thermophys.* **16**, 423–432 (1995).
33. Nagasaka, Y., Nakazawa, N., and Nagashima, A., Experimental determination of the thermal conductivity of molten alkali halides by the forced Rayleigh scattering method. I. Molten LiCl, NaCl, KCl, RbCl and CsCl, *Int. J. Thermophys.* **13**, 555–574 (1992).
34. Frez, C., Diebold, G. J., Tran, C. D., and Yu, S., Determination of the thermal diffusivities, thermal conductivities, and sound speeds of room-temperature ionic liquids by the transient grating technique, *J. Chem. Eng. Data* **51**, 1250–1255 (2006).

35. Nieto de Castro, C. A., State of art in liquid thermal conductivity standards, Invited communication to BIPM-CCT-WG9 meeting, Bratislava, Slovak Republic, 2005.
36. Assael, M. J., Nieto de Castro, C. A., Roder, H. M., and Wakeham, W. A., Transient methods for thermal conductivity, in *Experimental Thermodynamics, Volume II, Measurement of the Transport Properties of Fluids*, W. A. Wakeham, A. Nagashima, and J. V. Sengers (Eds.), Blackwell, Oxford, UK, 1991, Chap. 7.
37. Ramires, M. L. V. and Nieto de Castro, C. A., Uncertainty and performance of the transient hot wire method, in *TEMPMEKO 2001, 8th International Symposium on Temperature and Thermal Measurements in Industry and Science*, B. Fellmuth, J. Seidel, and G. Scholz (Eds.), VDE VERLAG GmbH, Berlin, Germany, 2002, Vol. **2**, pp. 1181–1185.
38. Ramires, M. L. V., Nieto de Castro, C. A., Fareleira, J. M. N. A., and Wakeham, W. A., The thermal conductivity of aqueous sodium chloride solutions, *J. Chem. Eng. Data* **39**, 186–190 (1994).
39. Ramires, M. L. V., and Nieto de Castro, C. A., Thermal conductivity of aqueous potassium chloride solutions, *Int. J. Thermophys.* **21**, 671–679 (2000).
40. Nagasaka, Y. and Nagashima, A., Absolute measurement of the thermal conductivity of electrically conducting liquids by the transient hot-wire method, *J. Phys. E Sci. Instrum.* **14**, 1435–1440 (1981).
41. Kitade, S., Kobayashi, Y., Nagasaka, Y., and Nagashima, A., Measurement of the thermal conductivity of molten KNO_3 and $NaNO_3$ by the transient hot-wire method with ceramic-coated probes, *High Temp. High Press.* **21**, 219–224 (1989).
42. Alloush, A., Gosney, W. B., and Wakeham, W. A., A transient hot wire instrument for thermal conductivity measurements in electrically conducting liquids at elevated temperatures, *Int. J. Thermophys.* **3**, 225–235 (1982).
43. Nieto de Castro, C. A., Dix, M., Fareleira, J. M. N. A., Li, S. F. Y., and Wakeham W. A., Thermal conductivity of chlorobenzene at pressures up to 430 MPa, *Physica* **156A**, 534–546 (1989).
44. Laesecke, A., Perkins, R. A., and Nieto de Castro, C. A., Thermal conductivity of R134a, *Fluid Phase Equilib.* **80**, 263–274 (1992).
45. Perkins, R. A., Laesecke, A., and Nieto de Castro, C. A., Polarized transient hot wire thermal conductivity measurements, *Fluid Phase Equilib.* **80**, 275–286 (1992).
46. Gurova, A. N., Barão, M. T., Mardolcar, U. V., and Nieto de Castro, C. A. The thermal conductivity and dielectric constants of HCFC-141b, HCFC-123, HCFC-142b and HFC-134a, *High Temp. High Press.* **26**, 25–34 (1994).
47. Gurova, A. N., Nieto de Castro, C. A., and Mardolcar, U. V., The thermal conductivity of liquid 1,1,1,2-tetrafluoroethane (HFC 134a), *Int. J. Thermophys.* **18**, 1077–1087 (1997).
48. Gurova, A. N., Nieto de Castro, C. A., and Mardolcar, U. V., Thermal conductivity of 1,1-difluoroethane (HFC 152a), *Int. J. Thermophys.* **20**, 63–72 (1999).
49. Bard, A. J. and Faulkner, L. R., *Electrochemical Methods*, Wiley, Hoboken, NJ, 1980.
50. Tufeu, R., Petitet, J. P., Denielou, L., and Le Neindre, B., Experimental determination of the thermal conductivity of molten pure salts and salt mixtures, *Int. J. Thermophys.* **6**, 315–330 (1985).
51. Ramires, M. L. V., Nieto de Castro, C. A., Nagasaka, Y., Nagashima, A., Assael, M. J., and Wakeham, W. A., Standard reference data for the thermal conductivity of water, *J. Phys. Chem. Ref. Data* **24**, 1377–1381 (1995).
52. Ramires, M. L. V., Nieto de Castro, C. A., Perkins, R. A., Nagasaka, Y., Nagashima, A., Assael, M. J., and Wakeham, W. A., Standard reference data for the thermal conductivity

REFERENCES

of toluene over a wide range of temperature, *J. Phys. Chem. Ref. Data* **29**, 133–139 (2000).

53. Schiefelbein, S. L., Fried, N. A., Rhoads, K. G., and Sadoway, D. R., A high-accuracy, calibration-free technique for measuring the electrical conductivity of liquids, *Rev. Sci. Instrum.* **69**, 3308–3313 (1998).
54. Brinkmann, F., Dam, N. E., Deák, E., Durbiano, F., Ferrara, E., Fükö, J., Jensen, H. D., Máriássy, M., Shreiner, R. H., Spitzer, P., Sudmeier, U., Surdu, M., and Vyskočyl, L., Primary methods for the measurement of electrolytical conductivity, *Accred. Qual. Assur.* **8**, 346–353 (2003).
55. OIML R 56 - 1981 (E), Standard Solutions Reproducing the Conductivity of Electrolytes.
56. OIML R 68 - 1985 (E), Calibration Method for Conductivity Cells.
57. Jameel, R. H., Wu, Y. C., and Pratt, K. W., *Primary Standards and Standard Reference Materials for Electrolytic Conductivity*, NIST Special Publication 260–142, 2000.
58. Jones, G. and Bradshaw, B. C., The measurement of the conductance of electrolytes. V. A redetermination of the conductance of standard potassium chloride solutions in absolute units, *J. Am. Chem. Soc.* **55**, 1780–1800 (1933).
59. Parker, H. C. and Parker, E. W., The calibration of cells for conductance measurements. III. Absolute measurements on the specific conductance of certain potassium chloride solutions, *J. Am. Chem. Soc.* **46**, 312–335 (1924).
60. Pratt, K. W., Koch, W. F., Wu, Y. C., and Berezansky, P. A., Molality-based primary standards of electrolytic conductivity (IUPAC Technical Report), *Pure Appl. Chem.* **73**, 1783–1793 (2001).
61. Shreiner, R. H., Preparation and uncertainty calculations for the molality-based primary standards for electrolytic conductivity, *Am. Lab.*, 28–32 (2004).
62. Koronaios, P. and Cleaver, B., Effect of pressure on the conductivity of calcium nitrate tetrahydrate, *J. Chem. Soc. Faraday Trans.*, **94**, 1477–1479 (1998).
63. Songa, Y., Zhua, X., Wang, X., and Wang, M., Characteristics of ionic liquid-based electrolytes for chip type aluminum electrolytic capacitors, *J. Power Sources* **157**, 610–615 (2006).
64. Böttcher, C. J. F., Van Belle, O. O., and Bordewijk, P., *Theory of Electric Polarization*, Vol. **I**, Elsevier, Amsterdam, 1973.
65. Böttcher, C. J. F. and Bordewijk, P., *Theory of Electric Polarization*, Vol. **II**, Elsevier, Amsterdam, 1978.
66. Frölich, H., *Theory of Dielectrics* 2nd ed., Oxford University Press, Oxford, UK, 1990.
67. Hilczer, B. and Malecki, J., Electrets, PWN–Polish Scientific Publishers, Warsaw and Elsevier, Amsterdam, 1986.
68. Wakai, C., Oleinikova, A., Ott, M., and Weingärter, H., How polar are ionic liquids? Determination of the static dielectric constant of an imidazolium-based ionic liquid by microwave dielectric spectroscopy, *J. Phys. Chem. B Lett.* **109**, 17028–17030 (2005).
69. Draguenet, C., Dyson, P., Krossing, I., Oleinikova, A., Slattery, J., Wakai, C., and Weingärter, H., Dielectric response of imidazolium-based room temperatures ionic liquids, *J. Phys. Chem. B* **110**, 12682–12688 (2006).
70. Michnick, R. B., Rhoads, K. G., and Sadoway, D. R., Relative dielectric constant measurements in the butyronitrile–chloroethane system at subambient temperatures, *J. Electrochem. Soc.* **144**, 2392–2398 (1997).

71. Mardolcar, U. V., Nieto de Castro, C. A., and Santos, F. J. V., Dielectric constant measurements of toluene and benzene, *Fluid Phase Equilib.* **79**, 255–264 (1992).
72. Santos, F. J. V., Pai-Panandiker, R. S., Nieto de Castro, C. A., and Mardolcar, U. V., Dielectric properties of alternative refrigerants, *IEEE Trans. Diel. Elect. Insulat.* **13**, 503–511 (2006).
73. Brito, F. E., Gurova, A. N., Mardolcar, U. V., and Nieto de Castro, C. A., Dielectric constant of nearly azeotropic mixture R 410A, *Int. J. Thermophys.* **21**, 415–427 (2000).
74. Costa Cabral, B. J., Guedes, R. C., Pai-Panandiker, R. S., and Nieto de Castro, C. A., Hydrogen bonding and internal rotation of hydrofluorocarbons by density functional theory, *Phys. Chem. Chem. Phys.* **3**, 4200–4207 (2001).
75. Sadoway, D. R., Electrochemical methods for the determination of the thermodynamic and transport properties of molten salts and ionic liquids, Plenary Lecture, in *EUCHEM Conference on Molten Salts and Ionic Liquids,* Hammamet, Tunisia, September 16–22, 2006.
76. Bird, R. B., Stewart, W. E., and Lightfoot, E. N., *Transport Phenomena*, 2nd ed., Wiley, Hoboken NJ, 2002.
77. Tyrrell, H. J. V. and Harris, K. R., *Diffusion in Liquids*, Butterworths, London, 1984.
78. Nieto de Castro, C. A., The measurement of transport properties of fluids — a critical appraisal, in *1987 ASME/JSME Thermal Engineering Joint Conference*, ASME/JSME, New York and Tokyo, 1987.
79. Lopes, J. N. C., Cordeiro, T. C., Esperança, J. M. S. S., Guedes, H. J. R., Huq, S., Rebelo, L. P. N., and Seddon, K. R., Deviations from ideality in mixtures of two ionic liquids containing a common ion, *J. Phys. Chem. B* **109**, 3519–3525 (2005).
80. Harned, H. S. and Blander, M., A glass conductance cell for the measurement of diffusion coefficients, *J. Phys. Chem.* **63**, 2078–2079 (1959) and References 53–75 in Wakeham et al [25].
81. Su, W. C., Chou, C. H., Wong, D. S. H., and Li., M. H., Diffusion coefficients and conductivities of alkylimidazolium tetrafluoroborates and hexafluorophosphates in aqueous solutions, *Fluid Phase Equilib.* **252**, 74–78 (2007).
82. Alizadeh, A., Nieto de Castro, W. A., and Wakeham, W. A., The theory of the Taylor dispersion technique for liquid diffusivity measurements, *Int. J. Thermophys.* **1**, 243–284 (1980).
83. Alizadeh, A., *Binary Diffusion Coefficients for Liquids*, Ph.D. thesis, University of London, London, 1981.
84. Alizadeh, A. and Wakeham, W. A., Mutual diffusion coefficients for binary mixtures of normal alkanes, *Int. J. Thermophys.* **3**, 307–323 (1982).
85. Matos Lopes, M. L. and Nieto de Castro, C. A., Liquid-phase diffusivity measurements in n-alkane binary mixtures, *High Temp. High Press.* **17**, 599–606 (1985).
86. Matos Lopes, M. L., Nieto de Castro, C. A., and Oliveira, C. M. P., Mutual diffusivity in n-heptane + n-hexane isomers, *Fluid Phase Equilib.* **36**, 195–205 (1987).
87. Viseu, M. I., Nieto de Castro, C. A., and Costa, S. M. B., Diffusion coefficients of tetrazolium blue in homogeneous and micellar solutions, *Int. J. Thermophys.* **12**, 323–331 (1991).
88. Matos Lopes, M. L., Nieto de Castro, C. A., and Sengers, J. V., Mutual diffusivity of a mixture of n-hexane and nitrobenzene near its consolute point, *Int. J. Thermophys.* **13**, 283–294 (1992).

89. Erkey, C. and Akgerman, A., Taylor dispersion, in *Experimental Thermodynamics, Volume II, Measurement of the Transport Properties of Fluids*, W. A. Wakeham, A. Nagashima, and J. V. Sengers (Eds.) Blackwell, Oxford, UK, 1991, Chap. 9, Sec. 9.1.4.
90. Matos Lopes, M. L. and Nieto de Castro, C. A., Liquid mutual diffusivities of the H_2O/D_2O system, *Int. J. Thermophys.* **7**, 699–708 (1986).
91. Oliveira, C. M. P., Matos Lopes, M. L., and Nieto de Castro, C. A., Mutual diffusivity in binary mixtures of *n*-heptane with *n*-hexane isomers, *Int. J. Thermophys.* **10**, 973–982 (1989).
92. Haase, R. and Siry, M., Diffusion im kritischen entmischungsgebiet binärer flüssiger systeme, *Z. Phys. Chem., Neue Folge* **57**, 56–73 (1968).
93. Wu, G., Fiebig, M., and Liepertz, A., Messung des binären diffusionskoeffizienten in einem entmischungssystem mit hilfe der photonen-korrelationsspektroskopie, *Int. J. Heat Mass Transfer* **22**, 365–371 (1988).

18 Molten Salt Physics and Chemistry in the Current Development of Spent Nuclear Fuel Management

TORU OGAWA, KAZUO MINATO, and YASUO ARAI

Japan Atomic Energy Agency, Tokai-mura, Naka-gun, Ibaraki-ken, Japan

Abstract

Pyrochemical processing for recovering transuranium (TRU) elements, which is based on molten salt electrorefining with a reactive cathode, is a strong candidate for the reprocessing technology of future nuclear power systems. Recent efforts of international communities filled the gap of our knowledge in the behaviour of TRU elements in these processes. We are now in a position to make a comparison of the integrated tests and detailed computer simulations. Existing thermodynamic data on the electrowinning of TRU elements are discussed.

18.1 INTRODUCTION

Pyrochemical processing of advanced nuclear fuels containing transuranium elements (neptunium, plutonium, americium, and curium) is a major component of future nuclear technology. Since the current power reactors such as light water reactors (LWRs) cannot provide an effective neutron field for fissioning TRU elements, deterioration in the quality of plutonium and accumulation of minor actinides (neptunium, americium, and curium) are inevitable if the recycling has to depend solely on the current reactors. TRU is better managed with a fast neutron field such as fast reactors (FRs) and accelerator-driven systems (ADSs) [1]. The latter is specifically for burning minor actinides (MAS).

For recycling, the commercial reprocessing technology is now based on aqueous chemistry with solvent extraction using TBP (tributyl phosphate) as extractant. The

---

Molten Salts and Ionic Liquids: Never the Twain? Edited by Marcelle Gaune-Escard and Kenneth R. Seddon
Copyright © 2010 John Wiley & Sons, Inc.

aqueous reprocessing technology should predominate in the coming decades, even if FRs partially replace LWRs. On the other hand, the pyroprocesses are favoured for treating the spent fuels from FRs, when it becomes the major components of nuclear power, as well as those from ADSs.

The spent fuels from either FRs or ADSs, even after prolonged cooling over 7–10 years, are characterised by intense radioactivity and decay heat. Decay heat per unit mass of spent fuels, which are mostly due to TRU elements, will increase by tenfold or more compared with those of LWRs, when MAs are recycled along with plutonium in FRs [1]. Accordingly, there will be engineering problems due to the radiolysis of TBP, such as the loss of plutonium into waste streams and a significant increase of the secondary wastes [2]. The radiation stability of molten salts is an important asset of pyroprocesses in this regard.

However, even with the recent developments in research and development (R&D) of pyroprocesses, there still remain technical issues, which basic science can help solve. Since TRU elements are more stable than uranium in molten salts, a cathode matrix is selected so as to preferentially react with TRU elements to form alloys or intermetallic compounds. The reactive cathode matrix also stabilises the trivalent state over the divalent state in the molten salts for some actinide and lanthanide elements. Otherwise, the multivalency in the salts causes significant loss in the current efficiency. The use of a reactive cathode matrix, however, introduces other types of process complexity.

A thermodynamic and electrochemical database has been formed by the efforts of the international community. Closer coupling of experimental demonstration, database, and computer simulation has to be pursued for the pyroprocesses to become reality.

Another important aspect of recent developments is the possible application of room temperature ionic liquids. For extensive application of ionic liquids to TRU element recovery, further studies on α-induced radiolysis is needed, although some promising data appear in the literature [3, 4]. The damage by TRU elements comes not only from α particles, but also from α-recoil heavy nuclei having energies of ~100 keV. Since the latter loses its energy mostly by nuclear scattering, it causes a much larger number of atomic displacements than the former [5].

18.2 RECENT ADVANCES IN PYROPROCESS STUDY AT JAEA

There are many types of pyroprocesses proposed and being studied. Here only the molten salts electrorefining of spent metal and nitride fuels will be discussed. The fundamental scheme of pyroprocess for FR metal fuel, which is based on that proposed by ANL and being studied jointly by CRIEPI and JAEA, is shown in Figure 18.1 [6]. The same scheme is applicable to the ADS nitride fuel, with slight modifications.

In electrorefining the spent fuel, the solid cathode is used for winning uranium and the liquid metal cathode for TRU elements as well as the residual uranium. Currently, cadmium is selected for the liquid metal cathode.

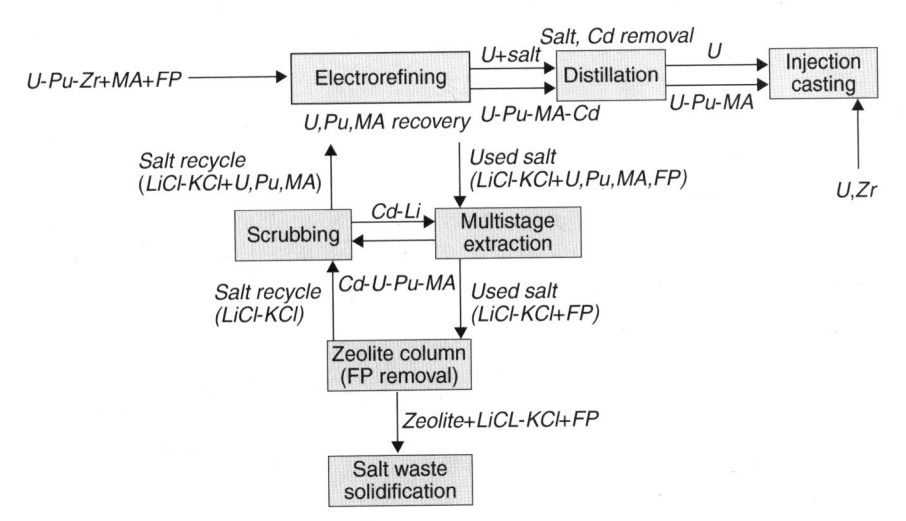

FIGURE 18.1 The fundamental scheme for reprocessing FR metal fuel [6].

Laboratory-scale demonstration has been made on the electrowinning of uranium, plutonium, and americium in a LiCl–KCl eutectic melt with a liquid cadmium cathode (LCC). Recovered actinides have been separated from cadmium and cast into the U–Pu–Zr alloy slug, which will be used in future irradiation experiments. Similarly, nitrides of uranium and plutonium have been prepared from the LCC deposit and used in the fabrication of the nitride pellets [7]. A series of experiments have demonstrated the feasibility of the molten salts electrorefining of irradiated nitrides. Limited data on americium behaviour have also been obtained on both metal and nitride fuels [8, 9].

In the basic studies, XAFS proved to be a useful tool in elucidating the structural features of molten salt mixtures, which give insights into the thermodynamic stability of metal cations in molten salts. It has also been applied to study high-temperature reactions within the molten salts: the chlorination of refractory oxides in molten salts was monitored directly by XAFS [10]. The application of molecular dynamics with structurally optimised potentials to supply the basic parameters for the simulation of pyroprocesses has been discussed elsewhere [11].

18.3 CHEMICAL THERMODYNAMIC APPROACH TO THE ELECTROREFINING OF TRANSURANIUM ELEMENTS

Molten salts electrorefining of actinides from the spent fuel is an essential component of the pyroprocess. Basic data on the electrochemistry, such as the apparent standard potentials of redox systems and the diffusion coefficients of metal cations, have been accumulated. The focus of the technical demonstration is the effective separation of actinides from rare earths, which decides the decontamination of the neutron poisons from the spent fuel, and the limiting current density, which determines the process throughput.

As far as uranium, plutonium, and rare earths are concerned, we are now in the position to make a comparison of the integrated tests and the detailed computer simulations. On the other hand, the database on americium has to be improved; experiments have yet to be made on curium.

A detailed analysis on the separation of TRU elements and rare earth (RE) elements has been made by CRIEPI [12]. In the spent FBR metal fuel, the atomic ratio RE/TRU is about 0.25; RE has to be removed to a level of ~2 wt%/actinides. The analyses have indicated that sufficient decontamination will be obtained by the use of a solid cathode for uranium and a liquid cadmium cathode for TRU elements in the reprocessing of U–TRU–Zr fuels, if the electrolyses are terminated at a certain level of actinide recovery. The remaining actinides in the salts are later partitioned by contacting with liquid Cd–Li alloys after several electrorefining runs.

The following discussions deal with the process for recovering TRU elements. The potential data are referenced to either Cl_2/Cl^- or $Ag/AgCl$. The latter values are converted to the former by adding

$$E(V) = -1.1076 + 0.0002749\, T(K) + 0.03036(1 - X_{AgCl})^2 \\ + (RT/F)\ln X_{AgCl} \pm 0.003 \quad \text{versus} \quad Cl_2/Cl^- \tag{18.1}$$

where X_{AgCl} is the mole fraction of AgCl in the Ag/AgCl reference electrode. Equation (18.1) is based on the free energy of formation of liquid AgCl [17] and the emf data by Yang and Hudson [18].

18.3.1 Relative Stability of Dichloride and Trichloride of Transuranium Elements

Stability of Am^{2+} in the LiCl–KCl eutectic melt has been confirmed by several authors, as shown in Table 18.1. Those data may be compared with the external information as follows.

TABLE 18.1 Apparent Standard Potential of Americium in LiCl–KCl Eutectic Melt

T/K	$E'^\circ\ (Am^{2+/0})$	$E'^\circ\ (Am^{3+}/Am^{2+})$	$E'^\circ\ (Am^{3+/0})$	Reference
673	−2.893	—	—	13
723	−2.852	−2.77	(−2.82)	13
723	−3.00	−2.71	(−2.90)	8
733	−2.911	−2.695	(−2.84)	14
743	−2.945	−2.84	(−2.91)	15
773	−2.838	—	—	13
773	−2.82	−2.74	(−2.79)	16
773	−2.893	−2.676	(−2.82)	14
773	−2.94	−2.68	(−2.85)	9

The free energy of formation of AmCl$_2$ has not been measured, but the relative stability of the divalent and the trivalent chlorides can be estimated by

$$-\Delta H^\circ_{298}(M^{3+}/M^{2+}) = \Delta H^\circ_{f,298}(MCl_2) - \Delta H^\circ_{f,298}(MCl_3)$$
$$= -I_3 - \Delta H^\circ_{f,298}(Cl^-) - (\tfrac{3}{2})RT + \Delta U \quad (18.2)$$

where I_3 is the third ionisation potential of the metal element, $\Delta H^\circ_{f,298}(Cl^-)$ is the standard heat of formation of the chloride ion, and ΔU is the lattice energy difference at 298 K.

$$\Delta U = U_{298}(MCl_3) - U_{298}(MCl_2)$$

$$U_{298}(MCl_3) = \frac{13{,}920\,(kJ\,mol^{-1})}{(r_+ + r_-)}\left(1 - \frac{0.345}{(r_+ + r_-)}\right) \quad \text{for UCl}_3\text{-type structure}$$

$$U_{298}(MCl_2) = U_0(MCl_2) + H_{298} - H_0$$
$$= \frac{7314 \pm 17\,(kJ\,mol^{-1})}{(r_+ + r_-)}\left(1 - \frac{0.345}{(r_+ + r_-)}\right) + H_{298} - H_0$$

$U_{298}(MCl_2)$ and $U_{298}(MCl_3)$ are according to the Kapustinskii type formulations [19], which were fitted to the thermodynamic data of lanthanide chlorides by the present authors. Since the variation of ΔU across the lanthanide and actinide series is relatively small, the relative values of $-\Delta H^\circ_{298}(M^{3+}/M^{2+})$ are largely determined by the variation of I_3 across each series: the larger I_3, the smaller $-\Delta H^\circ_{298}(M^{3+}/M^{2+})$. Equation (18.2) gives $-\Delta H^\circ_{298}(Am^{3+}/Am^{2+}) = 325.1$ kJ·mol^{-1}. With $\Delta H^\circ_{f,298}(AmCl_3) = -977.8$ kJ·mol^{-1} [20], Eq. (18.2) gives $\Delta H^\circ_{f,298}(AmCl_2) = -653$ kJ·mol^{-1}.

The above prediction can be compared with the experiments by Mikheev and Myasoedov [21] on the partition of trace amounts of plutonium, americium, curium, and berkelium between solid LnOCl and the SrCl$_2$–LnCl$_2$–LnCl$_3$ melt. The LnOCl captures M^{3+} but not M^{2+}. The apparent standard potential difference, $\Delta E'^\circ = E'^\circ(Am^{3+}/Am^{2+}) - E'^\circ(Ln^{3+}/Ln^{2+})$, was estimated by changing the rare earth species and the Ln^{2+}/Ln^{3+} ratio. Figure 18.2 shows good agreement between $\Delta E'^\circ = E'^\circ(Am^{3+}/Am^{2+}) - E'^\circ(Pr^{3+}/Pr^{2+})$ and that estimated by Eq. (18.2), $[-\Delta H^\circ_{298}(Am^{3+}/Am^{2+}) + \Delta H^\circ_{298}(Pr^{3+}/Pr^{2+})]/F$. By comparing plutonium and curium in Figure 18.2, we may safely exclude the stability of Cm^{2+} in analysing the pyroprocesses.

Further information on $\Delta H^\circ_{f,298}(AmCl_2)$ comes from the distribution of americium between plutonium metal and NaCl–KCl–PuCl$_3$ melt [22]. The data conformed to the reaction

$$[Am]_{metal} + \tfrac{2}{3}[PuCl_3]_{salt} = \tfrac{2}{3}[Pu]_{metal} + [AmCl_2]_{salt}$$

The enthalpy of reaction was derived,

$$\Delta H^\circ = \Delta H^\circ_f(AmCl_2) - \tfrac{2}{3}\Delta H^\circ_f(PuCl_3) = -14.6\,\text{kJ·mol}^{-1}$$

for temperatures 971–1048 K. With $\Delta H^\circ_{f,298}(PuCl_3) = -959.8$ kJ·mol^{-1} [20], $\Delta H^\circ_{f,298}(AmCl_2) = -654$ kJ·mol^{-1} is obtained.

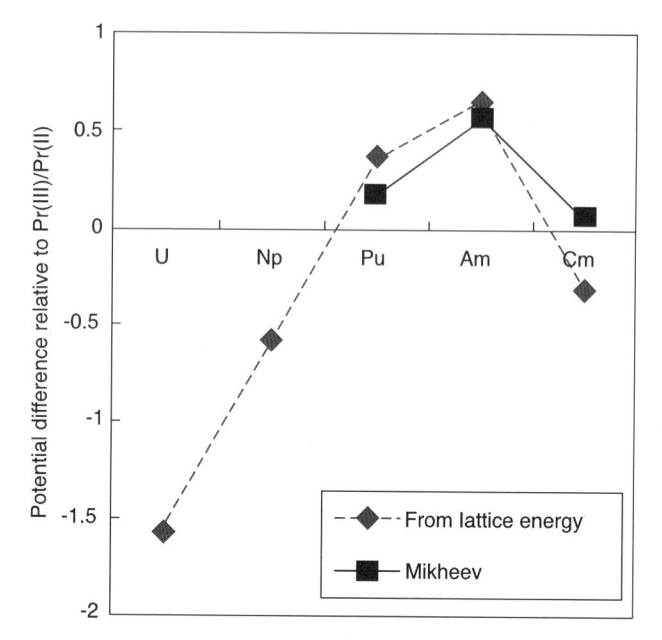

FIGURE 18.2 Comparison of the predicted $[-\Delta H^o_{298}(Am^{3+}/Am^{2+}) + \Delta H^o_{298}(Pr^{3+}/Pr^{2+})]/F$ with the apparent standard potential difference of $\Delta E'^o = E'^o(Am^{3+}/Am^{2+}) - E'^o(Pr^{3+}/Pr^{2+})$ by Mikheev and Myasoedov [21].

Morss [23] has also estimated the heat of formation of $AmCl_2$ to be $-654\,kJ\cdot mol^{-1}$, with the assumption that its enthalpy of solution is similar to that of $SmCl_2$, since the ionic radius of Am^{2+} is close to that of Sm^{2+}.

Thus three sets of approach give the value centered around $-\Delta H^o_{f,298}(AmCl_2) = 653$–$654\,kJ\cdot mol^{-1}$, although each estimation depends on several independent assumptions.

The difference in the entropy of formation between $AmCl_3$ and $AmCl_2$ may be estimated as follows:

$$\begin{aligned}-\Delta S^o_{298}(Am^{3+}/Am^{2+}) &= \Delta S^o_{f,298}(AmCl_2) - \Delta S^o_{f,298}(AmCl_3) \\ &= S^o_{mag}(Am^{2+}) + 2S(Cl^-;2) - 3S(Cl^-;3) + 0.5 S^o_{298}(Cl_2;gas) \\ &= 17.29 + 2\times 32.6 - 3\times 29.3 + 0.5\times 223.12 \\ &= 106.2\,J\cdot mol^{-1}\cdot K^{-1}\end{aligned} \quad (18.3)$$

$S^o_{mag}(Am^{2+}) = R\ln(J+1)$ is the magnetic contribution of Am^{2+} ion, having an f^7 configuration. The magnetic contribution of trivalent americium is zero. $S(Cl^-;n)$ is the contribution of the chloride ion to the standard entropy of solid compound, where the positive ion has the charge n, according to Latimer's method [24]. $S^o_{298}(Cl_2;gas)$ is the standard entropy of chlorine. $S(Cl^-;n)$ values are taken from the review by Puigdomenech et al. [25].

Hence the standard potential for the Am^{3+}/Am^{2+} pair is given by

$$E°(Am^{3+}/Am^{2+}) = [\Delta H°_{298}(Am^{3+}/Am^{2+}) - T\Delta S°_{298}(Am^{3+}/Am^{2+})]/F$$
$$= (-325,100 + 106.2T)/F$$
$$= -2.52 \text{ V at } 773 \text{ K}$$

Adopting $S°_{298}(AmCl_3) = 146.2 \pm 6.0 \text{ J·mol}^{-1}\text{·K}^{-1}$ as estimated by Konings [26], we obtain

$$\Delta G°_f(AmCl_3) = -977.8 + 0.2437T \text{ (kJ·mol}^{-1})$$

and

$$E°(Am^{3+}/Am) = -2.73 \text{ V at } 773 \text{ K}$$

The ΔC_P terms are neglected in these estimations at high temperatures. In estimating the free energy of formation of $LnCl_3$ in the supercooled liquid state around 773 K, the effect of dismissing ΔC_P terms, which amounts to ca. -10 kJ·mol^{-1}, tends to be compensated by that of neglecting the free energy of fusion, which is ca. $+10$ kJ·mol^{-1}.

Then, for Am^{2+}/Am,

$$E°(Am^{2+}/Am) = [3E°(Am^{3+}/Am) - E°(Am^{3+}/Am^{2+})]/2 = -2.83 \text{ V at } 773 \text{ K}$$

Konings estimated the standard entropy of the dichloride as $S°_{f,298}(AmCl_2) = 148.1$ J·mol^{-1}·K^{-1} [26]. This gives a slightly more negative potential of $E°(Am^{2+}/Am) = -2.86$ V.

With decreasing potential, Am^{3+} will first be reduced to Am^{2+}, then Am^{2+} to metal. Direct reduction of Am^{3+} to metal will not be observed. Three redox pairs are estimated to exist within a potential span of about 0.3 V. The difference between $E°(Am^{2+}/Am)$ and $E°(Am^{3+}/Am)$ is a mere 0.1 V.

In the molten salts, both the dichloride and the trichloride are further stabilised by the entropy of mixing and by coordinating to alkali chlorides. Figure 18.3 shows the partial molar excess free energy of mixing of supercooled liquid MCl_3 in a LiCl–KCl eutectic melt: $\Delta \bar{G}_E = RT \ln(\gamma_{MCl_3})$ [27]. With $RT \ln(\gamma_{AmCl_3})$ read at $r(Am^{3+}) = 97.5$ pm in Figure 18.3, the apparent standard potential for Am^{3+}/Am in a LiCl–KCl eutectic melt is estimated: $E'°(Am^{3+}/Am) = -2.89$ V versus Cl_2/Cl^- at 773 K.

As seen in Table 18.1, this value of $E'°(Am^{3+}/Am)$ is consistent with recent data by Hayashi et al. [9], but it is at variance with the apparent standard potential of dichloride, $E'°(Am^{2+}/Am) = -2.84$ V versus Cl_2/Cl^- at 773 K, by emf measurements [13], since $E'°(Am^{3+}/Am)$ should be less negative than $E'°(Am^{2+}/Am)$. The data in Serp et al. [14] appear more consistent with those in Fusselman et al. [13]. Further study is necessary on both chemical thermodynamics and electrochemistry of americium chlorides.

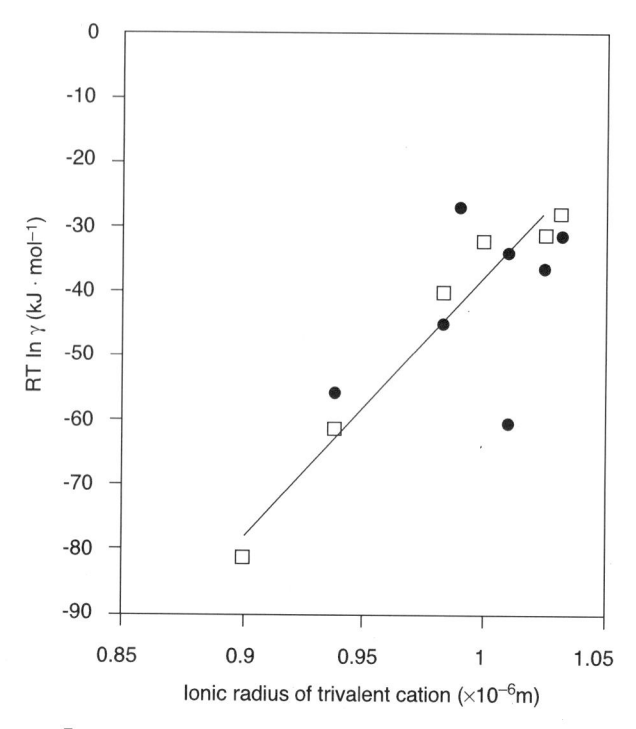

FIGURE 18.3 $\Delta \bar{G}_E = RT \ln(\gamma_{MCl_y})$ in LiCl–KCl eutectic melt as a function of the effective ionic radius of M^{3+} (CN = 6) [27]. The effective ionic radii are from Reference [28].

18.3.2 Electrowinning of Plutonium and Americium with Liquid Cadmium Cathode

Properties of Lithium Over Liquid Cd–Li Alloys It is important to note that lithium forms rather stable alloys with cadmium. Two sets of data on liquid Cd–Li alloys were found in the literature [29, 30]. The activity coefficient of Li, γ_{Li}, in liquid Cd–Li alloys can be modeled by

$$\Delta \bar{G}_E = RT \ln(\gamma_{Li}) = (-41174 - 26983 \times X_{Li})(1 - X_{Li})^2 \quad (18.4)$$

The comparison with the experimental data is shown in Figure 18.4. With the free energy of formation of supercooled liquid LiCl,

$$\Delta G_f^o(\text{liquid LiCl}) = -389.21 + 0.060299T (\text{kJ} \cdot \text{mol}^{-1}) \quad (18.5)$$

and the activity coefficient of 0.646 for LiCl$_{0.584}$KCl$_{0.416}$ [31], one obtains the apparent standard potential of Li$^+$/Li in LiCl–KCl eutectic melt:

$$E'^\circ(\text{Li}^+/\text{Li}) = -3.616 \text{ V versus Cl}_2/\text{Cl}^-$$
$$= -2.380 \text{ V versus Ag/AgCl (1 wt \%) at 773 K}$$

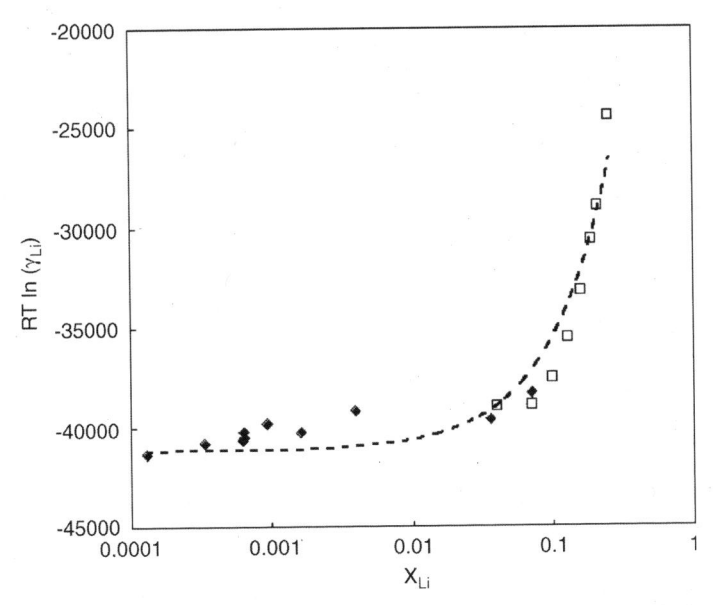

FIGURE 18.4 $\Delta \bar{G}_E = RT \ln(\gamma_{Li})$ in liquid cadmium. Solid symbols are from Reference [29] and open symbols are from Reference [30]. The dashed line is according to Eq. (18.4).

Since lithium in lithium-saturated liquid cadmium (Cd–25.6 mol % Li) is stabilised by $\Delta E = 0.366$ V at 773 K by Eq. (18.4),

$$E'^{\circ}(Li^+/Li;\ Cd\text{-}25.6\ mol\%\ Li) = -3.250\ V\ \text{versus}\ Cl_2/Cl^-$$
$$= -2.014\ V\ \text{versus}\ Ag/AgCl\ (1\ wt\%)\ \text{at}\ 773\ K.$$

Limiting Current Density in Electrowinning TRU Elements The apparent standard potential of Pu^{3+}/Pu is given by

$$E'^{\circ}(Pu^{3+}/Pu) = [\Delta G^{\circ}_{f,773}(\text{liquid PuCl}_3) + RT\ \ln(\gamma_{PuCl_3})]/3F$$
$$= -2.780\ V\ \text{versus}\ Cl_2/Cl^-$$

at 773 K over a solid electrode like tungsten, whose reaction with plutonium can be effectively neglected.

The partial molar enthalpy and excess entropy of mixing, and the solubility limit of plutonium in liquid cadmium are [32]

$$\Delta \bar{H} = (-26570 + 41560 X_{Pu})(1-X_{Pu})^2$$
$$\Delta \bar{S}_E = (-16.72 + 29.54 X_{Pu})(1-X_{Pu})^2$$
$$\ln X_{Pu} = 7.249 - 12151T + (2.662 \times 10^6)T^2$$

From these data, the difference in the apparent standard potential, ΔE, between the solid cathode and the plutonium-saturated liquid-cadmium cathode ($X_{Pu} = 0.018$) is estimated to be 0.275 V at 773 K. Then

$$E'^{\circ}(\text{Pu}^{3+}/\text{Pu}) = -2.505 \text{ V versus Cl}_2/\text{Cl}^-$$
$$= -1.26 \text{ V versus Ag/AgCl(1 wt \%) over liq.}$$

Cd–1.8 mol % Pu alloy at 773 K

In the electrowinning of plutonium, Iizuka et al. [33] observed that there was a limiting current density beyond which the current efficiency significantly deteriorated due to the simultaneous reduction of lithium. The limiting current density depended on the PuCl$_3$ concentration in the salts.

Figure 18.5 shows the result of model calculations, where the diffusion-controlled transportation of Pu^{3+} and the local thermodynamic equilibrium at the salts/cathode interface were assumed. The cathode potential and the equilibrium lithium concentration in the liquid-cadmium cathode are given as functions of the current density. In the electrowinning of plutonium, the competition with the reduction of lithium appears not so important until the current density approaches the theoretical maximum:

$$i_L = -zFD_{\text{Pu}^{3+}}\frac{C_{\text{Pu}^{3+}}}{\delta}$$

where D is the diffusion coefficient in molten salts, and C is the concentration. The diffusion layer thickness, δ, is arbitrarily chosen for the model calculation in Figure 18.5.

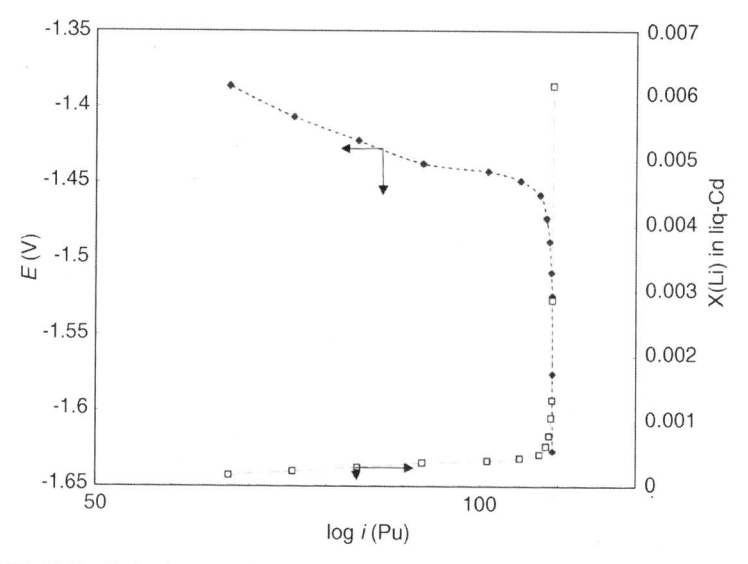

FIGURE 18.5 Cathode potential and equilibrium lithium concentration in a liquid-Cd cathode as functions of the current density in electrowinning plutonium from a LiCl–KCl eutectic melt at 773 K. PuCl$_3$ concentration in the salts is 1.3 mol %.

Since lithium is strongly stabilised in dilute Cd–Li alloys, a small current component due to lithium reduction may exist even at a cathode potential of -1.4 V (versus Ag/AgCl(1 wt %)), but significant reduction of lithium only sets in near i_L, where the cathode potential drops abruptly.

The situation would become more involved in electrowinning of americium. Although thermodynamic data for Cd–Am alloys are lacking, the existence of intermetallic compounds $AmCd_6$ and $AmCd_3$ were inferred in the X-ray diffraction pattern of heated cadmium and americium mixtures [34]. Stabilisation of americium makes the apparent standard potential $E'^\circ(Am^{3+}/Am)$ over liquid cadmium less negative than $E'^\circ(Am^{3+}/Am^{2+})$ in the salts, unless the americium is extremely depleted in the salts. That is, Am^{3+} is directly reduced to metal with a liquid-cadmium cathode. Over the liquid-cadmium electrode, the apparent standard potential of Am^{3+}/Am was measured:

$$E'^\circ(Am^{3+}/Am) = -2.56 \text{ V versus } Cl_2/Cl^- \text{ at about } 773 \text{ K}$$

by CV [9, 16]. Over the americium-saturated liquid cadmium, $E'^\circ(Am^{3+}/Am)$ should be more negative.

The comparison of $E'^\circ(M^{3+}/M)$ of plutonium and americium over liquid cadmium suggests that the massive electrowinning of americium into the liquid-cadmium cathode would be only possible at a limited current density, without causing significant decomposition of LiCl in the LiCl–KCl melt. Figure 18.5 suggests that the difference of 0.1 V in the apparent standard potential of TRU elements may have a significant effect. For the ADS fuel cycle, low throughput might be tolerated due to the limited actinide mass in the system. Still, besides electrowinning, recovery of americium either by the partitioning between molten salts and liquid alloys (such as Cd–Li) or by the metathetical reaction with Li_3N to form AmN precipitates would have to be achieved.

It has also been proposed to use other reactive materials for the cathode, although the incentive was to increase the separation between TRU elements and rare earths. Using an alternative cathode material may also be beneficial in increasing the throughput of americium, depending on the choice of the cathode material.

Aluminum and Gallium Cathodes Bismuth, aluminium, and gallium have been studied as alternative cathode materials. As shown previously for a liquid cadmium cathode, in analysing the electrowinning of TRU elements, thermodynamic data on both TRU elements and lithium in reactive cathode alloys are important.

The emf for the two-phase Al + "LiAl" alloys versus pure lithium at temperatures from 375 to 600 °C has been determined [35]:

$$E^\circ(Al, \text{``LiAl''}) - E^\circ(Li) = 451 - 0.220T(K) \text{ mV} = 281 \text{ mV at } 773 \text{ K}$$

The degree of lithium stabilization in aluminium is slightly less than that in liquid cadmium. Serp et al. [36] measured the CV of plutonium, americium, neodymium, and lanthanum in a LiCl–KCl eutectic melt with an aluminium cathode. The starting potential of the reduction of plutonium and americium were -1.18 and -1.20 V, respectively, versus Ag/AgCl(1 wt %) at 733 K. The difference between tungsten and aluminium electrodes for plutonium was −0.50 V, which is consistent with the partial molar free energy of mixing of plutonium in the Al + PuAl$_4$ diphasic region: $\Delta \bar{G}_{Pu} = -148$ kJ·mol^{-1} at 843 K [32].

In addition to the greater decontamination of rare earths, the larger gap between $E'^{\circ}(M^{3+}/M)$ and $E'^{\circ}(Li^{+}/Li)$ compared with that of the liquid-cadmium cathode is another advantage of using the solid aluminium cathode. However, while cadmium is readily separated from TRU elements by distillation after electrowinning, it is difficult to separate aluminium.

We can make similar arguments for using liquid gallium as the cathode. Lithium stabilization in the cathode is [37].

$$E^{\circ}(\text{Ga, "LiGa"}) - E^{\circ}(\text{Li}) = 929 - 0.532T = 518 \text{ mV at } 773 \text{ K}$$

For plutonium in gallium, Lambertin et al. [38] obtained $\gamma_{Pu} = 5 \times 10^{-8}$ in gallium at 1073 K. From this data, the difference in the apparent standard potential, ΔE, of plutonium between the inert solid cathode and the liquid-gallium cathode would be around 0.6 V at 773 K. Hence, as far as the gap between $E'^{\circ}(M^{3+}/M)$ and $E'^{\circ}(Li^{+}/Li)$ is concerned, the advantage of using the liquid-gallium cathode is not so obvious as for the aluminium cathode. On the other hand, the liquid-gallium cathode is expected to give the greater decontamination of rare earths than bismuth, aluminium, and cadmium, according to the data from Lambertin et al. [38].

As is the case for an aluminium cathode, however, it would also be difficult to separate gallium from TRU elements after electrowinning. Kinetic factors related to the mass transport of TRU elements in the cathode may be favourable for the liquid-gallium cathode compared with the solid aluminium.

ACKNOWLEDGEMENTS

The authors wish to thank Dr. Tadafumi Koyama for updating information on the recent work at CRIEPI.

REFERENCES

1. OECD/NEA, *Accelerator-Driven Systems (ADS) and Fast Reactors (FR) in Advanced Nuclear Fuel Cycles*, 2002.
2. Baetsle, L. H., Application of partitioning/transmutation of radioactive materials in radioactive waste management, *presented at Workshop on Hybrid Nuclear Systems for Energy Production, Utilisation of Actinides and Transmutation of Long-lived Radioactive Waste*, Trieste, September 3–7, 2001.
3. Allen, D., Baston, G., Bradley, A. E., Gorman, T., Haile, A., Hamblett, I., Hatter, J. E., Healey, M. J. F., Hodgson, B., Lewin, R., Lovell, K. V., Newton, B., Pitner, W. R., Rooney, D. W., Sanders, D., Seddon, K. R., Sims, H. E., and Thied, R. C., An investigation of the radiochemical stability of ionic liquids, *Green Chem.* **4**, 152–158 (2002).

4. Berthon, L., Nikitenko, S. I., Bisel, I., Berthon, C., Faucon, M., Saucerotte, B., Zorz, N., and Moisy, P., Influence of gamma irradiation on hydrophobic room-temperature ionic liquids [BuMeIm]PF$_6$ and [BuMeIm](CF$_3$SO$_2$)$_2$N, *Dalton Trans.*, 2526–2534 (2006).

5. Matzke, Hj., Radiation damage in crystalline insulators, oxides and ceramic nuclear fuels, *Radiat. Effects* **64**, 3–33 (1982).

6. Inoue, T. and Yokoo, T., Advanced fuel cycle with electrochemical reduction, in *Proceedings of Global-2003*, New Orleans, Louisiana, November 16–20, 2003.

7. Arai, Y., Iwai, T., Akabori, M., and Minato, K., Nitride fuel cycle technology with pyrochemical processes, presented at *2006 International Pyroprocessing Research Conference*, Idaho Falls, Idaho, August 8–10, 2006.

8. Hayashi, H., Akabori, M., and Minato, K., Electrolysis of americium nitride in a molten salt bath, presented at *2006 Annual Meeting of the Atomic Energy Society of Japan*, Oarai, Japan, 2006.

9. Hayashi, H., Akabori, M., and Minato, K., Electrochemical behaviour of americium at a liquid cadmium electrode in LiCl-KCl eutectic melt, presented at *International Pyroprocessing Research Conference*, Idaho Falls, Idaho, August 8–10, 2006.

10. Okamoto, Y., Yaitam, T., Minato, K., Usami, N., and Kobayashi, K., Chlorination of polyvalent metal oxides in molten salts, *Photon Factory Activity Report* 2004 #22, PartB, 2005, p. 29.

11. Ogawa, T., Minato, T., Okamoto, Y., and Nishihara, K., Nuclear energy and waste management—pyroprocess for system symbiosis, *J. Nucl. Mater.*, **360**, 12–15 (2007).

12. Koyama, T., *private communication*, 2006.

13. Fusselman, S. P., Roy, J. J., Grimmett, D. L., Grantham, L. F., Krueger, C. L., Nabelek, C. R., Storvick, T. S., Inoue, T., Hijikata, T., Kinoshita, K., Sakamura, Y., Uozumi, K., Kawai, T., and Takahashi, N., Thermodynamic properties for rare earths and americium in pyropartitioning process solvents, *J. Electrochem. Soc.* **146**, 2573–2580 (1999).

14. Serp, J., Chamelot, P., Fourcaudot, S., Konings, R. J. M., Malmbeck, R., Pernel, C., Poignet, J. C., Rebizant, J., and Glatz, J. P., Electrochemical behaviour of americium ions in LiCl–KCl eutectic melt, *Electrochim. Acta* **51**, 4024–4032 (2006).

15. Lambertin, D., Lacquement, J., Sanchez, S., and Picard, G., Americium chemical properties in molten LiCl-KCl eutectic at 743 K, *Plasmas Ions* **3**, 65–72 (2000).

16. Laplace, A., Lacquement, J., Maillard, C., and Donnet, L., Electrodeposition of americium on a liquid cadmium cathode from a molten salt bath, in *Proceedings of ATALANTE 2004—Advances for Future Nuclear Fuel Cycles, Nimes, France*, June 21–25, 2004, pp. 1–33.

17. Barin, I., *Thermochemical Data of Pure Substances*, 3rd ed., VCH Verlag, Weinheim, Germany, 1995.

18. Yang, L. and Hudson, R. G., Some investigation of the Ag/AgCl in LiCl–KCl eutectic reference electrode, *J. Electrochem. Soc.* **106**, 986–990 (1959).

19. Johnson, D. A., Stabilities of lanthanide dichlorides, *J. Chem. Soc. A*, 2578–2580 (1969).

20. Fuger, J., Parker, V. B., Hubbard, W. N., and Oetting, F. L., *The Chemical Thermodynamics of Actinide Elements and Compounds—Part 8: The Actinide Binary Halides*, IAEA, Vienna, 1983.

21. Mikheev, N. B. and Myasoedov, B. F., in *Handbook on the Physics and Chemistry of the Actinides*, A. J. Freeman and C. Keller (Eds.), Vol. **3**, North Holland, Amsterdam, 1985, pp. 347–386.

22. Mullins, L. J., Beaumont, A. J., and Leary, J. A., *Distribution of Americium Between Liquid Plutonium and a Fused Salt. Evidence of Divalent Americium*, Los Alamos Scientific Laboratory Report LA-3562, 1966.
23. Morss, L. R. Thermodynamic properties, in *The Chemistry of the Actinide Elements*, 2nd ed., J. J. Katz, G. T. Seaborg, and L. R. Morss (Eds.), Vol. 2, Chapman and Hall, London, 1986, pp. 1278–1360.
24. Latimer, W. M., *The Oxidation States of the Elements and Their Potentials in Aqueous Solutions*, 2nd ed., Prentice Hall, Englewood Cliffs, NJ, 1952, p. 392.
25. Puigdomenech, I., Rard, J. A., Plyasunov, A. V., and Grenthe, I., *Temperature Corrections to Thermodynamic Data And Enthalpy Calculations*, OECD/NEA Report, 1999.
26. Konings, R. J. M., Estimation of the standard entropies of some Am(III) and Cm(III) compounds, *J. Nucl. Mater.* **295**, 57–63 (2001).
27. Ogawa, T., Okamoto, Y., and Konings, R. J., Thermochemical properties of advanced fission fuel materials, in *Innovative Materials in Advanced Energy Technologies*, P. Vincenzini (Ed.), Advances in Science and Technology Series, Vol. 24, Techna Publishers, Faenza, Italy, 1999, pp. 381–392.
28. Shannon, R. D., Revised effective ionic radii and systematic studies of interatomic distances in halides and chalcogenides, *Acta Crystallogr. A* **32**, 751–767 (1976).
29. Lewis, M. A. and Johnson, T. R., A study of the thermodynamic and reducing properties of lithium in cadmium at 773 K, *J. Electrochem. Soc.* **137**, 1414–1418 (1990).
30. Langen, G., Schwitzgebel, G., and Ruppersberg, H., Thermodynamic investigations of liquid (Li,Cd) alloys, *Z. Metallkde.* **74**, 425–429 (1983).
31. Lumsden, J., *Thermodynamics of Molten Salts Mixtures*, Academic Press, New York, 1966.
32. Chiotti, P., Akhachinskij, V. V., Ansara, I., and Rand. M. H., *The Chemical Thermodynamics of Actinide Elements and Compounds—Part 5: The Actinide Binary Alloys*, IAEA, Vienna, 1981.
33. Iizuka, M., Uozumi, K., Inoue, T., Iwai, T., Shirai, O., and Arai, Y., Behaviour of plutonium and americium at liquid cadmium cathode in molten LiCl-KCl electrolyte, *J. Nucl. Mater.* **299**, 32–42 (2001).
34. Hayashi, H., Sato, T., Ogawa, T., and Haire, R. G., Preparation and selected properties of Am–Cd alloys, in *Proceedings of the International Conference on Future Nuclear Systems GLOBAL'99*, August 29–September 3, 1999.
35. Wen, C. J., Baukamp, B. A., Huggins, R. A., and Weppner, W., Thermodynamic and mass transport properties of LiAl, *J. Electrochem. Soc.* **126**, 2258–2266 (1979).
36. Serp, J., Allibert, M., Le Terrier, A., Malmbeck, R., Ougier, M., Rebizant, J., and Glatz, J. P., Electroseparation of actinides from lanthanides on solid aluminum electrode in LiCl-KCl eutectic melts, *J. Electrochem. Soc.* **152**, C167–C172 (2005).
37. Wen, C. J. and Huggins, R. A., Electrochemical investigation of the lithium-gallium system, *J. Electrochem. Soc.* **128**, 1636–1641 (1981).
38. Lambertin, D., Ched'homme, S., Bourges, G., Sanchez, S., and Picard, G. S., Activity coefficients of plutonium and cerium in liquid gallium at 1073 K: application to a molten salt/solvent metal separation concept, *J. Nucl. Mater.* **341**, 131–140 (2005).

19 An Organic Chemist's Perspective on High Temperature Molten Salts and Room Temperature Ionic Liquids

RICHARD M. PAGNI

Department of Chemistry, University of Tennessee, Knoxville, Tennessee

Abstract

Even though a rich organic chemistry has been developed in ionic liquids, these reactions have not been widely used in the synthesis of natural, medical, and unnatural products. Organic reactions in the seemingly more forbidding molten salts have been used even less. This chapter presents an organic chemist's point of view on a select group of organic and organometallic reactions in ionic liquids and molten salts that are as viable as those carried out in traditional solvents. Reactions that are underrepresented, or not represented at all, in ionic liquids and molten salts are also described.

19.1 INTRODUCTION

Ionic liquids have been defined as compounds that are made up of ions and have melting points of less than 100 °C. Molten salts are those ionic compounds having melting points of 100 °C and above. The boundary of 100 °C is completely arbitrary. A more reasonable definition, in my opinion, is to call any ionic compound requiring melting before use a molten salt. Ionic liquids are ready to go, while molten salts must be melted before they can be used for chemistry. *tert*-Butyl alcohol, a covalent compound, has found considerable application in chemistry and photochemistry but,

---

Molten Salts and Ionic Liquids: Never the Twain? Edited by Marcelle Gaune-Escard and Kenneth R. Seddon
Copyright © 2010 John Wiley & Sons, Inc.

with a melting point of 25.8 °C, requires melting on a cold day and cannot be heated under reflux during the winter when the water is cold.

Because heating is required to create a molten salt, the following points should be kept in mind when deciding if a molten salt or an ionic liquid is an appropriate medium in which to run a reaction. Reaction rates generally go up as the temperature of the reaction medium is raised. Faster reactions require less time to run. Selectivities unfortunately go down as reaction temperatures rise. If one desires to maximise the yield of one product compared to another, running the reaction at as low a temperature as possible is important. A chemist thus has to compromise between these opposing trends. Side reactions also become more prominent at higher reaction temperatures, and the desired products themselves may become reactants at sufficiently high temperatures. Solubility is another consideration. Solute solubilities may go up or down as the temperature of the solvent is increased. Unfortunately, there are few data on the solubility of organic compounds in molten salts. One should also remember the maxim: like dissolves like. This can be put in a more quantitative fashion by the use of the Hildebrand solubility parameter [1]. Usually, when a solvent and a solute have solubility parameters within 3 or 4 units of one another, the solute will dissolve in the solvent. One would not expect, however, an organic solute with a solubility parameter of 25 or below to dissolve in a molten salt with a value in the 30s or higher. A chemist must also be concerned with the stability of the reaction medium. Even molten salts may decompose at sufficiently high temperature. Ionic liquids will generally decompose at temperatures well below the decomposition temperatures of molten salts. Surprisingly, many ionic liquids are apparently stable to temperatures well in excess of 300 °C [2].

Most organic chemists have experience in running reactions between roughly −150 and +200 °C. As a result, equipment has been developed to run reactions between these temperature extremes. Running reactions above 200 °C is harder to do and often requires specialised equipment. In my experience, I have run one organic reaction above 200 °C (Eq. (19.1)). [3]. This particular isomerisation, first reported by Meinwald et al. [4], was carried out by passing a solution of the substrate in hexanes over quartz beads heated to roughly 500 °C. There is no reason to believe that this isomerisation could not have been carried out in a molten salt or ionic liquid. Mechanistic, spectroscopic, and physical organic studies become particularly difficult to carry out above ambient temperatures. Ionic liquids have been particularly well characterised [2, 5] because the measurements are relatively easy to make at room temperature. There is a corresponding lack of physical organic data on molten salts.

(19.1)

Even though many ionic liquids [6, 7] and several molten salts [8, 9] have proved very useful in carrying organic and organometallic reactions, these new liquid phases haven't yet attained wide use in the organic community. The major reason for this, I believe, is that most current, practicing chemists have been trained in the use of traditional volatile and flammable solvents. Chemists being, by and large, conservative don't like to try new things. This lack of interest is unfortunate in my opinion because a large number of reactions work as well, if not better, in ionic liquids and molten salts.

I plan to describe in this chapter a select number of reactions that work well in molten salts and ionic liquids. I will also point out a few places where additional research is required before we will know if a particular reaction is viable in an ionic fluid. Because my principal concern is whether a reaction can be run successfully in an ionic fluid, I will not be concerned with the desirable quality of "greenness" of molten salts and ionic liquids. When deciding whether one should use one solvent over another in a reaction, factors other than yield, reaction time, and temperature must also be considered. These factors include cost and availability, ease of synthesis if unavailable commercially, and recyclability. If a solvent is very expensive, one would hope that it can be reused a number of times.

19.2 MOLTEN SALTS

Let us begin by examining the melting points of a representative number of high melting and low melting ionic compounds (Table 19.1). As can be seen from Table 19.1, the melting point range is over 700 °C. Sodium chloride, with a melting point of over 800 °C, has never been used as a medium for an organic reaction. Sodium hydroxide, sodium nitrate, and zinc chloride have much lower melting points, and all have been used as molten media for organic reactions [9]. Eutectic mixtures such as LiI–KI (1:1) with even lower melting points have been widely used in organic chemistry [8, 9]. Of particular interest has been the large number of reactions run in $AlCl_3$-containing molten salts [10]. A large compilation of eutectic mixtures of ionic

TABLE 19.1 Representative List of Ionic Compounds and Their Melting Points

Ionic Compound	Melting Point/°C
NaCl	801
NaOH	323
Na[NO_3]	308
$ZnCl_2$	290
LiI–KI (3:2)	200
NaOH–KOH (1:1)	170
Li[NO_3]–Na[NO_3]–K[NO_3] (30:17:53)	150
Pyridinium chloride	144
KCl–$AlCl_3$ (1:2)	128
[N_{6666}][BF_4]	91
$SbCl_3$	73.4

compounds, many with melting points below 200 °C, may be found in Reference 11. Also found in Table 19.1 are the organic-like pyridinium chloride and tetrahexylammonium tetrafluoroborate, both with relatively low melting points. Antimony(III) chloride, which is nominally ionic, has been used as a medium for a rich and complex organic chemistry [9, 12].

A few comments are in order before examining a number of organic reactions in molten salts. Even though the aluminium(III) chloride-containing molten salts have been the most widely used in organic chemistry, they cannot be recycled because reaction workup entails adding water that destroys the aluminium(III) chloride. Several of these compounds, including aluminium(III) chloride and antimony(III) chloride, are difficult to purify. Although this may not be of concern in carrying out a reaction, it is a serious handicap if one wishes to carry out mechanistic and physical organic studies. A wide range of hard and soft acids and bases are available in molten salts. Changing these characteristics of a solvent may have an effect on the outcome of an organic reaction. Workup of a reaction may be easy to perform when molten salts are involved. If one is not interested in reusing the molten salt, the reaction may be quenched with water and the products extracted with diethyl ether or dichloromethane. If it is important to recycle the molten salt, one may be able to extract the products from the salt using a Soxhlet extractor.

Let us now look at a few interesting organic reactions that occur in molten salts.

19.2.1 Nucleophilic Aromatic Substitution

Nucleophilic aromatic substitution constitutes a broad class of reactions in which a nucleophile attached to an aromatic ring is replaced by another nucleophile [13]. These reactions often occur via intermediates such as benzynes and aryl cations, or by direct attack of a nucleophile on a suitably activated aromatic ring followed by loss of the originally attached nucleophile. Substituents that stabilise negative charge often activate the aromatic ring for attack. Three examples of nucleophilic aromatic substitution occurring in molten salts are shown in Eqs. (19.2)–(19.4). The first example, Eq. (19.2), actually occurs in an ionic liquid at ambient temperature and illustrates the ease with which nucleophilic aromatic substitution can occur when the substrate is activated with electron-withdrawing groups (two nitro groups) and the nucleophile, fluoride, is very potent because it is poorly solvated in the ionic liquid. The next example occurs in Lewis acidic $ZnCl_2$ at 300 °C. What is remarkable about the reaction is that the substrate, phenol, is unactivated for nucleophilic attack. Although its role in the substitution is unclear, $ZnCl_2$ must play a more than passive role in making this extraordinary reaction occur. The last example, Eq. (19.4), also occurs on an unactivated substrate at the relatively low reaction temperature of 252 °C. The KOH is molten at 252 °C, far below the melting point of the pure compound (mp 360 °C), because it contains some water and potassium carbonate. Here, KOH serves both as reactant and solvent. A reasonable mechanism for this reaction is the elimination of $[HSO_3]^-$ from carbons 1 and 2 to form benzyne, followed by addition of hydroxide or water to yield ultimately phenol. What is truly remarkable about the reactions in Eqs. (19.3) and (19.4) is the ease with which unactivated aromatic substrates undergo nucleophilic aromatic substitution in molten salts at relatively low reaction temperatures. As is the case with so many organic reactions carried out in molten salts, no mechanistic details are available.

$$\text{1,3-dinitrobenzene} \xrightarrow[\text{[N}_{4444}\text{]F, room temp}]{\text{almost anhydrous}} \text{3-fluoronitrobenzene} \qquad (19.2)$$

$$\text{PhOH} + \text{4-methylaniline} \xrightarrow[\text{300 °C}]{\text{ZnCl}_2} \text{N-phenyl-4-methylaniline} \qquad (19.3)$$

$$\text{PhSO}_3^- \xrightarrow[\text{252 °C}]{\text{KOH}} \text{PhOH} \qquad (19.4)$$

19.2.2 Aluminium(III) Chloride-Containing Molten Salts

I suspect that more organic reactions have been run in molten salts that contain aluminium(III) chloride than in any other class of molten salts [10]. There are many reasons for this:

1. Aluminium(III) chloride plays a significant role in organic synthesis.
2. Eutectic mixtures of salts with aluminium(III) chloride have sufficiently low melting points to render them useful to organic chemists.
3. Aluminium(III) chloride-containing molten salts are of general interest because aluminium metal is synthesised electrochemically in an aluminium-containing molten salt.

Representative examples of organic reactions occurring in molten salts are shown in Eqs. (19.5)–(19.7). The first reaction, which is an example of the Scholl reaction, occurs at 270 °C in only 4 minutes. The Scholl reaction, which links sp^2-hybridised carbon atoms, occurs frequently in aluminium(III) chloride-containing molten salts, although its mechanism is obscure. The second example, which occurs at 140 °C in 1 h, is in essence an example of Friedel–Crafts acylation and yields a diketone in modest yield. A trace amount of an isomeric diketone is also produced here. Undoubtedly, trace HCl in the molten salt catalyses the isomerisation of the 1-naphthyl ester into the 2-naphthyl ester, followed by Friedel–Crafts acylation. HCl is a potent Brønsted acid in AlCl$_3$-containing molten salts [14, 15]. Lastly, when 1,3-diphenylpropane is treated with a 9:1 mixture of SbCl$_3$ and AlCl$_3$ at 100 °C, benzene and indane are generated in equal amounts. It appears that the reaction is initiated by aluminium(III) chloride abstracting a chloride from antimony chloride to form $[SbCl_2]^+$. The cation in turn abstracts a benzylic hydride from the hydrocarbon to form a carbocation that yields the two products after a few steps.

(19.5)

(19.6)

(19.7)

19.2.3 Isomerisation

The following silver ion-catalysed isomerisation, Eq. (19.8), is a spectacular example of a reaction that works exceedingly well in a molten salt, but not at all in a traditional solvent. The reaction does not occur in boiling benzene, even after 3 days. In the silver ion-containing molten salt, the reaction is quantitative after a much shorter period of time. It is not clear if the difference in reaction temperature is responsible for the vast difference in reactivity. Reactions run at 130 °C are quite easy for organic chemists to carry out.

(19.8)

(a) AgClO$_4$ (10 eq.), refluxing benzene, 3 days — 0% yield
(b) AgNO$_3$/KNO$_3$/AgCl (59:38:17; mp 113 °C), melt at 130 °C for 16 h — 100% yield

19.2.4 Pyridinium Chloride

Organic reactions in molten pyridinium chloride were popularised by the extensive studies of René Royer in the 1960s [16]. Perhaps the most useful reaction he and his co-workers discovered was the efficient dealkylation of aryl alkyl ethers, a still very important synthetic procedure. The pyridinium cation serves as a proton source, which protonates the ether oxygen to form a cation that in turn is dealkylated in an S_N2 reaction, with chloride serving as nucleophile. This procedure still represents one of the best methods to carry out this transformation. A recent example, carried out on a large scale, is shown in Eq. (19.9) [17]. When 2250 g of substrate was treated with 5360 g of pyridinium chloride at 180 °C for 3 h, a 97% yield of dealkylated product was obtained. Numerous other dealkylation procedures were attempted here, but none worked as well as the one in pyridinium chloride.

$$\text{CH}_3\text{O-C}_6\text{H}_4\text{-CH}_2\text{CH}_2\text{-COOH} \xrightarrow{\text{pyridinium chloride}} \text{HO-C}_6\text{H}_4\text{-CH}_2\text{CH}_2\text{-COOH} \quad (19.9)$$

19.3 IONIC LIQUIDS

One of the most significant and amazing developments in chemistry during my long career has been the evolution of the field of ionic liquid chemistry. There are perhaps hundreds of these materials currently known, and they have been used in an incredible number of organic and organometallic reactions [6, 7]. There is every reason to believe that tens of thousands of these liquids could be made with ease. Having new solvents, with little or no vapour pressure, was a great boon to the field of green chemistry. Because these compounds were liquids at room temperature, they were easy to characterise. It is no exaggeration to say that ionic liquids are as well characterised as the traditional, common solvents, many of which have been known for close to two centuries.

I believe the first concern a chemist will have in deciding whether to use an ionic liquid in a chemical process is its viscosity, for which there are many data. Handling viscous liquids, and dissolution of solutes in them, may be problematical. By and large, ionic liquids are more viscous than traditional solvents, but they are not so viscous as to prevent a wide variety of chemical reactions from being carried out in them. We, and others, have easily carried out organic reactions in lithium perchlorate in diethyl ether, a very viscous liquid [18, 19]. Some ionic liquids are hydrophilic, which means they are water soluble, while others are distinctly hydrophobic. The nature of the cation–anion combination dictates this difference. One thus cannot always predict whether a solute will be soluble in an ionic liquid. Comparing the Hildebrand solubility parameter of the solute [1] with that of the ionic liquid may be helpful in this regard. The solubility parameters of several ionic liquids are now known [20, 21]. If a reaction is run in a hydrophilic ionic liquid, reaction workup is easy: just add water and extract with ether or dichloromethane. For reactions run in hydrophobic ionic liquids, one can often extract the product directly with a suitable immiscible solvent. One can clean up and reuse the hydrophobic ionic liquid by dissolution in a miscible solvent, followed by washing with immiscible water, and then removing the volatile organic solvent from the ionic liquid by pumping under vacuum.

Numerous cation types have been used to create ionic liquids but the imidazolium, pyridinium, tetraalkylammonium, tetraalkylphosphonium, and trialkysulfonium cations, all shown below, appear most often in ionic liquids. The type of alkyl group used in these cations is quite varied but long-chain alkyl groups are commonly used. Anions used in ionic liquids are also quite varied, several of which are also shown below. Some anions are basic while others are neutral. $[Al_2Cl_7]^-$, on the other hand, is a potent Lewis acid.

COMMONLY USED CATIONS

$[NR_4]^+ \equiv [N_{nnnn}]^+$ $[PR_4]^+ \equiv [P_{nnnn}]^+$ $[SR_3]^+ \equiv [S_{nnn}]^+$

(where n represents the number of carbon atoms in a linear alkyl chain)

COMMONLY USED ANIONS

$[BF_4]^-$ $[PF_6]^-$ Cl^- Br^- $[Br_3]^-$ $[HF_2]^-$

$[RCOO]^-$ $[N(SO_2CF_3)_2]^-$ $[CF_3SO_3]^-$ $[AlCl_4]^-$ $[Al_2Cl_7]^-$

I believe that the ionic liquids based on the imidazolium and pyridinium cations are currently the most widely used in organic chemistry. It is thus worth making a few comparisons about these two heterocyclic systems. Imidazolium ions are relatively hard to reduce, while pyridinium ions are easier to reduce. This of course will affect the size of electrochemical windows. This difference may also influence the course of redox reactions. This difference also plays a significant role in photochemistry (to be discussed later). N,N'-dialkylimidazolium cations are relatively acidic [22, 23], while N-alkylpyridinium cations are not. When imidazolium ions are deprotonated at the site between the two nitrogen atoms by a strong base, a carbene is generated, which can function as a strong base or as a ligand to a transition metal.

There are far too many reactions to be described in this short space. I have chosen to describe a few that are particularly interesting, or important in the synthesis of natural and unnatural products. I will also mention as I go along some areas that require additional work before their attributes can be assessed. I will begin this discussion with a consideration of protection and deprotection.

19.3.1 Protection–Deprotection

A common procedure in natural product synthesis is to protect a hydroxyl group as a trialkylsilyl ether at one stage of the synthesis and then remove the silyl group at a later stage (Eq. (19.10)) [24]. The silylation is commonly carried out using a trialkylsilyl chloride and base, while the desilylation is usually carried out using fluoride, which is silophilic. These reactions work well either in an ionic liquid or using an ionic liquid as a reagent. A very useful method to silylate alcohols involves the use of hexamethyldisilazine in [C$_4$mim][BF$_4$] with no added base [25]. Benzyl alcohol afforded the silyl ether in 96% yield after 30 min at room temperature (Eq. (19.11)). Contrast this to the 55% yield of product when the reaction was carried out for 24 h in boiling dichloromethane. Desilylation has been carried out successfully using the ionic liquid [C$_2$mim][F(HF)$_{2.3}$] as a reagent in dichloromethane [26]. A beautiful example of this deprotection is shown in Eq. (19.12) Other ionic liquids that contain fluoride or HF will undoubtedly also desilylate silyl ethers [9, 27, 28].

$$ROH \xrightarrow[\text{base}]{R'_3SiCl} ROSiR'_3 \xrightarrow{F^-} ROH + R'_3SiF \qquad (19.10)$$

$$PhOH + [(CH_3)_3Si]_2NH \longrightarrow PhOSi(CH_3)_3 \qquad (19.11)$$

$$\text{R-OSiMe}_3 \xrightarrow[\text{CH}_2\text{Cl}_2,\ 0.1\ \text{eq. MeOH}]{[C_2\text{mim}][F(HF)_{2.3}]} \text{R-OH} \qquad (19.12)$$

19.3.2 Carbon–Carbon Bond Formation

Although there are currently a very large number of reactions used in organic synthesis, I think it fair to say that those that form carbon–carbon bonds are the most important. We already saw an example of this in the Scholl reaction (Eq. (19.5)). Among the oldest of these reactions is the addition of Grignard reagents (RMgX) to compounds containing carbonyl groups. Even older than these are the base-catalysed condensation reactions that proceed through enolate intermediates. Enolate chemistry has proved to be extremely important in the synthesis of natural products. Of recent vintage are the exceptionally useful palladium-catalysed reactions that link sp^2-hybridised carbon atoms. Reactions of phosphorus and sulfur ylids also generate carbon–carbon bonds. All of these disparate reactions have been studied extensively in ionic liquids.

There have been three reports of Grignard reactions involving ionic liquids. In one report a carbonyl compound was added to a mixture of phenylmagnesium bromide in tetrahydrofuran and a tetraalkylphosphonium ion-based ionic liquid, yielding an alcohol in moderate to good yields [29]. It is worth noting that the Grignard reagent did not deprotonate the phosphonium ion to make an ylid.

Alkyltriphenylphosphonium ions, by contrast, are readily deprotonated by Grignard reagents. In the second report, a carbonyl compound in the ionic liquid, 1-butyl-2-isopropyl-3-methylimidazolium bistriflamide, was added to a Grignard reagent in tetrahydrofuran [30]. As was the case in the first report, moderate to good yields of alcohols were produced in the reactions. What is significant here is that the acidic hydrogen at C(2) position of the imidazolium ring has been replaced by an isopropyl group—it is likely that the Grignard reagent would have deprotonated the ionic liquid with a hydrogen atom at this position. It is unclear to me what the advantage is in using ionic liquids in these reactions. Tetrahydrofuran, an ether, is still involved in both procedures.

It is well established that Grignard reagents form in diethyl ether or tetrahydrofuran because the ether molecules help solvate the reagent by complexation to the magnesium in the organometallic reagent. It is thus surprising that a Chinese group has attempted to prepare Grignard reagents in ionic liquids in the absence of ether or tetrahydrofuran [31]. Iodoethane and magnesium did not react to form an organometallic compound in imidazolium ion-based ionic liquids, but did form some type of organometallic reagent in N-butylpyridinium tetrafluoroborate. Can it be that the more reducible pyridinium ion is involved in some type of electron-transfer reaction? Interestingly, when benzaldehyde was added to the organometallic reagent, hydrobenzoin was often produced. When pyridine was added to the organometallic compound before the carbonyl compound was added, poor to excellent yields of alcoholic products were produced. The role of the pyridine in this chemistry is at present unclear. A successful example of this interesting chemistry is shown in Eq. (19.13). If a Grignard reagent is formed in this interesting, but unusual chemistry, it is striking that it did not attack the N-butylpyridinium cation. More work is needed to clarify the nature of the organometallic compound.

$$\text{CH}_2\text{=CHCH}_2\text{I} + \text{Mg} \xrightarrow{[\text{C}_4\text{py}][\text{BF}_4]} ? \xrightarrow{\text{py}} \xrightarrow{\text{PhCHO}} \text{Ph-CH(OH)-CH}_2\text{-CH=CH}_2 \quad (19.13)$$

91% yield

Below is a list of condensation and ylid reactions that have been carried out successfully in ionic liquids:

Aldol	Claisen–Schmidt
Mukaiyama aldol	Conjugate addition
Benzoin	Wittig
Knoevenagel	Baylis–Hillman

Because of space limitations, I will only discuss three of them in a little detail: the Mukaiyama aldol condensation, the Wittig reaction, and the Baylis–Hillman reaction.

The Mukaiyama aldol condensation reaction is the Lewis acid-catalysed addition of an enol silane to an aldehyde to form a β-hydroxylaldehyde. With suitably substituted enol silanes, a diastereomeric mixture of products will be obtained. The bismuth triflate-catalysed version of this reaction has been carried out successfully at room temperature in [C$_4$mim][BF$_4$] [32]. The reaction, which produces good to excellent yields of aldol products, tolerates functional groups such as nitro, methoxy, and carboethoxy. An example of this chemistry is shown in Eq. (19.14). The Mukaiyama procedure is certainly an improvement over the traditional base-catalysed aldol condensation, whose initial aldol product often undergoes another condensation reaction, or eliminates water from the positions alpha and beta to the carbonyl.

$$\text{Ph-CHO} + \text{CH}_2=\text{C(OSiMe}_3)\text{Ph} \xrightarrow[\text{room temp}]{\text{Bi(OTf)}_3 \cdot n\text{H}_2\text{O}, [\text{C}_4\text{mim}][\text{BF}_4]} \text{Ph-CH(OH)-CH}_2\text{-C(O)-Ph} \quad 92\% \text{ yield} \quad (19.14)$$

The Wittig reaction, which entails the addition of a phosphorus ylid to an aldehyde or ketone to form an alkene and a phosphine oxide, has been carried out successfully in [C$_4$mim][BF$_4$] by Bouliare and Grée [33]. The authors studied the reaction of three stabilised ylids with benzaldehyde, affording E,Z mixtures of olefins in excellent yields (Eq. (19.15)). By judicious extraction with diethyl ether, and then with toluene, the researchers were able to separate the desired olefinic products from triphenylphosphine oxide, a problem that has bedevilled chemists ever since Wittig first described the reaction. Bouliare and Grée were able to reuse the ionic liquid five times with no diminution of yield.

$$\text{Ph}_3\text{P}=\text{CHY} + \text{PhCHO} \xrightarrow{[\text{C}_4\text{mim}][\text{BF}_4]} \text{PhCH}=\text{CHY} + \text{Ph}_3\text{PO} \quad (19.15)$$

Y	Olefin Yield (%)	E/Z
COMe	82	97/3
CO$_2$Me	90	96/4
CN	79	50/50

Unlike most other condensation reactions that go back to the 19th century, the Baylis–Hillman reaction, which involves the trialkylamine-catalysed addition of an aldehyde to an alkene with an electron-withdrawing substituent to afford a β-hydroxy-α-methylene product, was first reported in the 1970s. The reaction is initiated by conjugate addition of the amine to the alkene to form an anion that then adds to the carbonyl group of the aldehyde. When catalysed by DABCO (1,4-diazabicyclo[2.2.2]octane) in ionic liquids, the Baylis–Hillman reaction afforded good to excellent yields of product [34]. Interestingly, the reaction gave consistently higher yields of product in [C$_4$C$_1$mim][PF$_6$], where the C(2) position of the imidazolium ring has a methyl substituent, than in [C$_4$mim][PF$_6$], where the C(2) position has a hydrogen (Eq. (19.16)).

$$\text{Cl-C}_6\text{H}_4\text{-CHO} + \text{CH}_2\text{=CH-CO}_2\text{CH}_3 \xrightarrow{\text{ionic liquid}} \text{Cl-C}_6\text{H}_4\text{-CH(OH)-C(=CH}_2\text{)-CO}_2\text{CH}_3 \qquad (19.16)$$

[C$_4$mim][PF$_6$]: 66% yield
[C$_4$C$_1$mim][BF$_4$]: 99% yield

One of the most significant advances in organic synthesis in the last decade and a half has been the development of a wide variety of palladium-catalysed reactions that link sp^2-hybridised carbon atoms to sp^2- and sp-hybridised carbon atoms. These reactions are often carried out in the presence of base, and numerous palladium(0) and palladium(II) compounds have been used successfully in the chemistry. Because heating the reaction mixture is often required as well, molten salts may serve a useful purpose here. The reactions include the Heck reaction (aryl halide with alkene), Suzuki reaction (arylboronic acid with haloarene), Stille reaction (trialkylarylstannane with haloarene), and the Sonogashira reaction (haloarene + terminal alkyne). All of these have been carried out successfully in ionic liquids. In fact, some of these were among the first reactions carried out in ionic liquids. Two examples will suffice to show the utility of the reactions in ionic liquids.

When the commercially available phenylboronic acid and bromobenzene were treated with a palladium(0) complex and base in [C$_4$mim][BF$_4$] at 110 °C, an excellent yield of biphenyl was obtained (93%) (Eq. (19.17)) [35]. The ionic liquid version of the Suzuki reaction works as well as that in traditional solvents or heterogeneous media [7, 36]. As shown in Eq. (19.18), the Sonogashira reaction of iodobenzene and phenylacetylene in [C$_4$mim][BF$_4$] at 60 °C also worked exceedingly well [37]. What is remarkable about this reaction is that no copper salt, a common requirement in this reaction, and phosphine ligand to bind the palladium were needed. The Sonogashira reaction carried out in ionic liquids works as well as that found in other solvents and heterogeneous media [7, 38].

$$\text{PhBr} + \text{PhB(OH)}_2 \xrightarrow[\text{[C}_4\text{mim][BF}_4\text{], 10 min, }\Delta]{\text{Pd(PPh}_3)_4, \text{Na}_2[\text{CO}_3]} \text{Ph-Ph} \qquad (19.17)$$

$$\text{PhI} + \text{H-} \equiv \text{-Ph} \xrightarrow[\text{[C}_4\text{mim][BF}_4\text{], HN(CHMe}_2)_2, \Delta]{\text{[Pd(im)}_2\text{ClMe]}} \text{Ph-} \equiv \text{-Ph} \qquad (19.18)$$

98% yield

An interesting variation of the Suzuki reaction, the reaction of an anhydride or acid chloride with an arylboronic acid, is shown in Eq. (19.19) [39]. In this instance, the reaction was carried out in water with added [C$_4$mim][PF$_6$] or poly(ethylene glycol). The reactions worked about equally well in both solvent mixtures. The role of the addend in the chemistry is unclear. Note that no phosphine ligand was used in this reaction.

$$\text{naphthyl-B(OH)}_2 + \text{Ph-CO-O-CO-Ph} \xrightarrow[\text{H}_2\text{O/[C}_4\text{mim][PF}_6\text{], 60 °C}]{\text{Pd(OAc)}_2, \text{Na}_2[\text{CO}_3]} \text{naphthyl-CO-Ph} \qquad (19.19)$$

95% yield

19.3.3 Pericyclic Reactions

A large number of pericyclic reactions, that is, those that are governed by the Woodward–Hoffmann rules, have been carried out in ionic liquids. The discussion will begin by examining a couple of Diels–Alder reactions, which are undoubtedly the most synthetically useful of the pericyclic reactions.

Jaeger and Tucker were, I believe, the first to study the Diels–Alder reaction in an ionic liquid [40]. They examined the reaction (Eq. (19.20)) of cyclopentadiene with methyl acrylate in ethylammonium nitrate (mp 12 °C) and found a high *endo-* to *exo-* ratio of adducts, which is consistent with the reaction occurring in a highly polar solvent. Ethylammonium nitrate has been known for close to a century, but has only been used as a solvent in recent decades. Lee [41], on the other hand, examined the same reaction in mixtures of [C$_2$mim]Cl and AlCl$_3$ (Eq. (19.20)). Mixtures of [C$_2$mim]Cl and aluminium(III) chloride were probably the first of the "newer" ionic liquids to be prepared and studied. Lee found that, when [C$_2$mim]Cl was in excess (basic medium), a ratio of adducts slightly smaller than that found by Jaeger and Tucker was obtained. When aluminium chloride(III) was in excess (Lewis acidic medium; [Al$_2$Cl$_7$]$^-$ the dominant acid), the ratio increased dramatically, as did the rate of the bimolecular reaction. Aluminium(III) chloride is known to catalyse Diels–Alder reactions in many other solvents, and to increase the *endo-* to *exo-*selectivity of reactions like the one under discussion here. The Diels–Alder reaction in ionic liquids is thus similar in behaviour to that observed in more traditional solvents.

$$\text{cyclopentadiene} + \text{CH}_2=\text{CH-COOCH}_3 \longrightarrow \text{endo-adduct} + \text{exo-adduct} \qquad (19.20)$$

	endo-	exo-
[EtNH$_3$][NO$_3$]	6.7	1
[C$_2$mim]Cl–AlCl$_3$ (basic)	5.25	1
[C$_2$mim]Cl–AlCl$_3$ (acidic)	19	1

Other types of cycloaddition reactions have also been carried out in ionic liquids including the [3 + 2] cycloaddition, shown in Eq. (19.21) [42]. The function of the potassium hydrogencarbonate is to eliminate HCl from the enol ester to form a transient nitrile oxide, which then reacts with the methyl acrylate. The product of the reaction, shown in Eq. (19.21), was obtained in a modest 58% yield.

$$\tag{19.21}$$

The last example of a pericyclic reaction to be described is the Claisen rearrangement, the conversion of an aryl allyl ether to an *ortho*-allyl phenol. This reaction normally requires high reaction temperatures, probably because the initial step of the reaction destroys the aromatic ring. Han and Armstrong [43] synthesised three new dicationic ionic liquids in order to carry out several high temperature Claisen rearrangements, one of which is shown in Eq. (19.22). The reactions afforded very good yields of products (74–82%), which depended on the ionic liquid used, the reaction temperature, and the substituent on the aromatic ring; a regioisomeric mixture of allylphenols was generated in all cases. It would be interesting to know how well these reactions would take place if the ionic liquids were replaced with molten salts.

R = H, OCH$_3$, CF$_3$, CH$_3$

$$\tag{19.22}$$

19.3.4 Reductions and Oxidations

Not surprisingly, a large number of reduction and oxidation reactions have been carried out in ionic liquids. These reactions appear to work as well as comparable reactions carried out in traditional solvents. Below are partial lists of reducing and oxidising agents that have been used successfully in ionic liquids. In the case of the reducing agents, the medium in which the agent was used is shown. In the case of the oxidising agents, the oxidation reaction the agent effected is shown.

Reducing Agents

Borohydride ($R_4PCl + BH_3$ [BH_3Cl]$^-$)
Na[BH_4] (in [PR_4]Cl)
Li[AlH_4] (in [PR_4]Cl; behaviour complex)
BBu$_3$ (in a variety of ILs and MSs)

Oxidising Agents

K[MnO$_4$] (alcohol to aldehyde or ketone)
MnO$_2$ (alcohol to aldehyde)
m-CBA (Baeyer–Villiger oxidation)
Hypervalent iodine (sulfide to sulfoxide)
OsO$_4$ (alkene to vicinal diol)
MnSO$_4$ (alkene to epoxide)
Jacobsen catalyst (asymmetric epoxidation)
MeReO$_3$ (alkene to epoxide)

A remarkable reduction reaction in an ionic liquid is taken from the work of Xiao and Malhotra [44] (Eq. (19.23)). In this study, the researchers asymmetrically reduced a series of aryl ketones with lithium tetrahydroaluminate(III) and an optically active ligand in pyridinium ion-based ionic liquids. The ligands used were the commercially available (R)- and (S)-BINOL (BINOL = 2,2'-dihydroxy-1,1'-binaphthyl). The reduction of acetophenone, which is illustrative of the general procedure, afforded an excellent yield of product, with a very good enantiomeric excess (ee). Interestingly, the yield of product went down slightly as the reaction temperature was lowered, but the enantiomeric excess correspondingly increased. It is interesting to note that the hydride did not reduce the N-ethylpyridinium ion, as one might have expected. This work nicely demonstrates that it is possible to carry out asymmetric reactions in ionic liquids. Asymmetric reactions are certainly at the forefront of current organic synthetic methodology.

$$\text{PhC(O)CH}_3 \xrightarrow[\text{Li[AlH}_4\text{], BINOL}]{[\text{C}_2\text{py}][\text{O}_2\text{CCF}_3]} \text{PhCH(OH)CH}_3 \qquad (19.23)$$

–30 °C; 85% yield; 84% ee

One example will suffice to show the utility of ionic liquids in oxidation reactions. The reaction comes from the work of Farmer and Welton [45]. These researchers used tetrapropylammonium perruthenate to catalyse the oxidation of a series of alcohols to aldehydes and ketones. In the illustrative example, shown in Eq. (19.24), cinnamalcohol was oxidised to cinnamaldehyde, with *N*-methylmorpholine-*N'*-oxide (MNO) serving as oxidising agent. The excellent yield of product (97%) attests to the viability of the ionic liquid procedure.

$$\text{PhCH=CHCH}_2\text{OH} \xrightarrow[\text{room temperature}]{[\text{N}_{3333}][\text{RuO}_4],\ \text{MNO}}_{[\text{C}_4\text{mim}][\text{BF}_4]} \text{PhCH=CHCHO} \quad (19.24)$$

19.3.5 Photochemistry

I think it fair to say that organic photochemistry has been poorly studied in ionic liquids and molten salts. In fact, I don't know of a single example of an organic photochemical reaction that has been studied in molten salts. Although sparse, a number of photochemical and photophysical studies have been carried out in ionic liquids. The literature through roughly 2003 may be found in Reference 46. Some recent, interesting photochemistry in ionic liquids has also been published [47–49]. This lack of effort is surprising because certain transformations are uniquely photochemically induced and new phenomena, as described below, are expected to occur in ionic liquids. The photochemically induced [2 + 2] cycloaddition reaction is perhaps the best manner in which to prepare four-membered rings. E. J. Corey, the Nobel laureate in chemistry, used a [2 + 2] cycloaddition reaction in the syntheses of caryophyllene and isocaryophyllene in the early 1960s [50, 51], and two such reactions in his synthesis of the unusual natural product, pentacycloanammoxic acid, in 2006 [52]. Some examples taken from my research will illustrate the unique qualities of ionic liquids in photochemical reactions.

Anthracene is known to undergo a [4 + 4] dimerisation photochemically in most solvents. When anthracene is photolysed in basic [C$_2$mim]Cl–AlCl$_3$ (excess [C$_2$mim]Cl; Cl$^-$ and [AlCl$_4$]$^-$ present), the same dimerisation reaction is observed (Eq. (19.25)) [53]. When anthracene is photolysed in acidic, [C$_2$mim]Cl–AlCl$_3$ (excess AlCl$_3$; [AlCl$_4$]$^-$ and [Al$_2$Cl$_7$]$^-$ present), a complex series of electron transfer, hydrogen transfer, and dimerisation reactions, too involved to describe here, take place [54]. In acidic medium HCl, a very strong Brønsted acid [14, 15], present in trace amounts, protonates the anthracene at position 9 to form a stable carbocation (Eq. (19.26)), which is a very good electron acceptor in photochemical reactions.

$$\text{anthracene} \xrightarrow[{[\text{C}_2\text{mim}]\text{Cl–AlCl}_3}]{h\nu,\ \text{basic}} \text{[4+4] dimer} \quad (19.25)$$

$$\text{anthracene} + \text{HCl} \xrightarrow[\text{[C}_2\text{mim]Cl–AlCl}_3]{h\nu, \text{ acidic}} \text{9,10-dihydroanthracene cation} \quad (19.26)$$

Isomers *cis*- and *trans*-stilbene are known to interconvert photochemically by a variety of mechanisms (Eq. (19.27)). In basic [C$_2$mim]Cl–AlCl$_3$, the interconversion takes place through the singlet excited states of the isomeric stilbenes [55]. In basic [C$_4$py]Cl–AlCl$_3$, on the other hand, the isomerisation occurs through the radical cations of the substrates, which are formed by electron transfer from the singlet excited states of the stilbenes to the pyridinium ion. The difference in behaviour in the two solvents is due to the fact that the imidazolium ion is hard to reduce, whereas the pyridinium ion is much easier to reduce.

$$\text{Ph–CH=CH–Ph} \xrightleftharpoons[h\nu]{h\nu} \text{Ph, Ph (cis)} \quad (19.27)$$

The photochemistry of 9-methylanthracene (9-MeAn) in basic [C$_2$mim]Cl–AlCl$_3$ is complex, unusual, and distinctive [56]. Electron transfer phenomena rule the day here. At an early stage of the photochemical reaction, the radical cation of 9-MeAn is generated. This radical cation is very acidic and reacts with chloride to generate HCl and the 9-anthrylmethyl radical. At a later stage of the reaction, the radical anion of 9-MeAn is produced, which ultimately affords 9,10-dihydroanthracene. This is one of the rare photochemical reactions known in which both the radical cation and radical anion of the substrate are generated.

$$\text{9-MeAn} \xrightarrow[\text{[C}_2\text{mim]Cl–AlCl}_3]{h\nu, \text{ basic}} [\text{9-MeAn}]^{+\bullet} + [\text{9-MeAn}]^{-\bullet} \quad (19.28)$$

19.4 CONCLUSION

Organic reactions in molten salts have been studied for many decades. The field is rich, useful, and yet vastly underutilised these days. Considering that there are hundreds of molten salts with melting points below 200 °C, this underutilisation is quite surprising. Organic chemists routinely run reactions up to 200 °C. I wonder if anyone will create a new organic chemistry in molten salts, or at least reinvigorate the old molten salt chemistry.

There are now hundreds of ionic liquids and ionic liquid reactions known. I wonder if there will ever be a "universal" ionic liquid, one in which most organic reactions can be run successfully. I suspect this medium will emerge in time. I also wonder if chemistry unique to ionic liquids will be developed. Most of the research to date has dealt with running established reactions in ionic liquids. One area in which unique chemistry may evolve is in the area of electron-transfer phenomena.

Although I doubt that the researchers involved are aware of this, there are many instances in which features of molten salts have been incorporated into ionic liquids. One instance of this unintended merging will illustrate this point. Although not discussed in this chapter, lithium ion-containing molten salts have been used extensively in organic chemistry, especially for the ring opening of epoxides [9]. Yadav and co-workers [57] have shown that lithium salts in [C$_4$mim][PF$_6$] effectively ring open epoxides to form *vic*-halohydrins in excellent yields (Eq. (19.29)). I hope that additional work of this nature will be investigated, as I believe useful synthetic methodologies will result.

(19.29)

REFERENCES

1. Reichardt, C., *Solvents and Solvent Effects in Organic Chemistry*, 3rd ed., Wiley-VCH, Weinheim, 2003.
2. Chiappe, C. and Pieruccini, D., Ionic liquids: solvent properties and organic reactivity, *J. Phys. Org. Chem.* **18**, 275–297 (2005).
3. Pagni, R. M. and Watson, C. R., Reaction of phenalenyl anion with methylene chloride and butyllithium. New method of preparing 1,8-naphtho(C$_4$H$_4$)-hydrocarbons, *Tetrahedron Lett.* **14**, 59–60 (1973).
4. Meinwald, J., Samuelson, G. E., and Ikeda, M., Synthesis and rearrangement to pleiadiene, *J. Am. Chem. Soc.* **92**, 7604–7606 (1970).
5. Rooney, D. W., and Seddon, K. R., Ionic liquids, in *The Handbook of Solvents*, G. Wypych (Ed.), ChemTec Publishing, Toronto, 2001, pp. 1459–1484.
6. Wasserscheid, P. and Welton, T. (Eds.), *Ionic Liquids in Synthesis*, Wiley-VCH, Weinheim, 2003.
7. Dyson, P. J. and Geldbach, T. J., *Metal Catalysed Reactions in Ionic Liquids*, Wiley-VCH, Weinheim, 2005.
8. Gordon, J. E., Applications of fused salts in organic chemistry, in *Techniques and Methods of Organic and Organometallic Chemistry*, D. B. Dewey (Ed.), Vol. 1, Marcel Dekker, New York, 1969, pp. 51–187.
9. Pagni, R. M., Organic and organometallic reactions in molten salts and related melts, in *Advances in Molten Salt Chemistry*, G. Mamantov, C. B. Mamantov, and J. Braunstein (Eds.), Vol. 6, Elsevier, Amsterdam, 1987, pp. 211–346.

10. Jones, H. L. and Osteryoung, R.A., Organic reactions in molten tetrachloroaluminate solvents, in *Advances in Molten Salt Chemistry*, J. Braunstein, G. Mamantov, and G. P. Smith (Eds.), Vol. 3, Plenum, New York, 1975, pp. 121–176.
11. Clark, P. V., *Fused Salt Mixtures*, Sandia Laboratory, Albuquerque, 1968.
12. Smith, G. P. and Pagni, R. M., Homogeneous organic reactions in molten salts. Selected topic, in *Molten Salt Chemistry*, G. Mamantov and R. Marassi (Eds.), Reidel, Dordrecht, 1987, pp. 383–404.
13. Lowry, T. H. and Richardson, K. S., *Mechanism and Theory in Organic Chemistry*, 3rd ed., Harper & Row, New York, 1987.
14. Smith, G. P., Dworkin, A. S., Pagni, R. M., and Zingg, S. P., Brønsted superacidity of HCl in a liquid chloroaluminate $AlCl_3$-1-ethyl-3-methyl-1*H*-imidazolium chloride, *J. Am. Chem. Soc.* **111**, 525–530 (1989).
15. Smith, G. P., Dworkin, A. S., Pagni, R. M., and Zingg, S. P., Quantitative study of the acidity of HCl in a molten chloroaluminate system ($AlCl_3$/1-ethyl-3-methyl-1*H*-imidazolium chloride) as a function of HCl pressure and melt composition (51.0-66.4 mol% $AlCl_3$), *J. Am. Chem. Soc.* **111**, 5075–5077 (1989).
16. Royer, R. and Demerseman, P., Pyridinium chloride induced reactions, *Bull. Chim. Soc. Fr.*, 2633–2648 (1968).
17. Schmid, C. R., Beck, C. A., Cronin, J. S., and Staszak, M. A., Demethylation of 4-methoxybutyric acid using molten pyridinium chloride on a multikilogram scale, *Org. Proc. Res. Dev.* **8**, 670–673 (2004).
18. Pagni, R. M., Kabalka, G. W., Bains, S., Plesco, M., Wilson, J., and Bartmess, J., A chemical, spectroscopic, and theoretical assessment of the Lewis acidity of $LiClO_4$ in ether, *J. Org. Chem.* **58**, 3130–3133 (1993).
19. Springer, G., Elam, C., Edwards, A., Bowe, C., Boyles, D., Bartmess, J., Chandler, M., West, K., Williams, J., Green, J., Pagni, R. M., and Kabalka, G. W., Chemical and spectroscopic studies related to the Lewis acidity of lithium perchlorate in diethyl ether, *J. Org. Chem.* **64**, 2202–2210 (1999).
20. Swiderski, K., McLean, A., Gordon, C. M., and Vaughan, D. J., Estimates of internal energies of vaporisation of some room temperature ionic liquids, *Chem. Commun.*, 2178–2179 (2004).
21. Lee, S. H. and Lee, S. B., The Hildebrand solubility parameters, cohesive energy densities and internal energies of 1-alkyl-3-methylimidazolium-based room temperature ionic liquids, *Chem. Commun.*, 3469–3471 (2005).
22. Alder, R. W., Allen, P. R., and Williams, S. J., Stable carbenes as strong bases, *Chem. Commun.*, 1267–1268 (1995).
23. Kim, Y.-J. and Streitwieser, A. Jr., Basicity of a stable carbene, 1,3-di-*tert*-butylimidazol-2-ylidene, in THF, *J. Am. Chem. Soc.* **124**, 5757–5761 (2002).
24. Smith, M. B., *Organic Synthesis*, 2nd ed., McGraw Hill, New York, 2002.
25. Mojtahedi, M. M., Abbasi, J., and Abaee, M. S., A novel efficient method for the silylation of alcohols using hexamethyldisilazane in an ionic liquid, *Phosphorus, Sulfur, Silicon* **181**, 1541–1544 (2006).
26. Yoshino, H., Matsumoto, K., Hagiwara, R., Ito, Y., Oshima, K., and Matsubara, S., Fluorination with ionic liquid emimF(HF)$_{2.3}$ as mild HF source, *J. Fluor. Chem.* **127**, 29–35 (2006).

27. Bhadhury, P. S., Raza, S. K., and Jaiswal, D. K., A semi-molten mixture of hexadecyltributylphosphonium bromide and potassium fluoride in the synthesis of organofluorine compounds, *J. Fluor. Chem.* **99**, 115–117 (1999).
28. Shodai, Y., Kohara, S., Ohishi, Y., Inaba, M., and Tasaka, A, Anionic species $(FH)_xF^-$ in room-temperature molten fluorides $(CH_3)_4NF \cdot m HF$, *J. Phys. Chem. A* **108**, 1127–1132 (2006).
29. Ramnial, T., Ino, D. D., and Clyburne, J. A. C., Phosphonium ionic liquids as reaction media for strong bases, *Chem. Commun.*, 325–327 (2005).
30. Handy, S. T., Grignard reactions in imidazolium ionic liquids, *J. Org. Chem.* **71**, 4659–4662 (2006).
31. Law, M. C., Wong, K.-Y., and Chan, T. H., Grignard reagents in ionic liquids, *Chem. Commun.*, 2457–2459 (2006).
32. Ollevier, T., Desyroy, V., Debailleul, B., and Vaur, S., Bismuth triflate catalysed Mukaiyama aldol reaction in an ionic liquid, *Eur. J. Org. Chem.*, 4971–4973 (2005).
33. Le Boulaire, V. and Grée, R., Wittig reaction in the ionic liquid [bmim][BF$_4$], *Chem. Commun.*, 2195–2196 (2000).
34. Hsu, J.-C., Yen, Y.-H., and Chu, Y.-H., Baylis–Hillman reaction in [bdmim][PF$_6$] ionic liquid, *Tetrahedron Lett.* **45**, 4673–4676 (2004).
35. Matthew, C. J., Smith, P. J., and Welton, T., Palladium catalysed Suzuki cross-coupling reactions in ambient temperature ionic liquids, *Chem. Commun.*, 1249–1250 (2000).
36. Kabalka, G. W., Wang, L., Pagni, R. M., Hair, M., and Vasudevan, N., Solventless Suzuki coupling reactions on palladium-doped potassium fluoride alumina, *Synthesis*, 217–222 (2003).
37. Park, S. B. and Alper, H., Recyclable Sonogashira coupling reactions in an ionic liquid, effected in the absence of both a copper salt and a phosphine, *Chem. Commun.*, 1306–1307 (2004).
38. Kabalka, G. W., Wang, L., and Pagni, R. M., Sonogashira coupling and cyclisation reactions on alumina: a route to aryl alkynes, 2-substituted-benzo[*b*]furans and 2-substituted-indoles, *Tetrahedron* **57**, 8017–8028 (2001).
39. Xin, B., Zhang, Y., and Cheng, K., Phosphine-free cross-coupling reactions of arylboronic acids with carboxylic anhydrides and acyl chlorides in aqueous media, *J. Org. Chem.* **71**, 5725–5731 (2006).
40. Jaeger, D. A. and Tucker, C. E., Diels–Alder reactions in ethylammonium nitrate, a low-melting fused salt, *Tetrahedron Lett.* **30**, 1785–1788 (1989).
41. Lee, C. W., Diels-Alder reactions in chloroaluminate ionic liquids: accelerations and selectivity enhancement, *Tetrahedron Lett.* **40**, 2461–2464 (1999).
42. Conti, D., Rodriquez, M., Seaga, A., and Taddei, M., 1,3-Cycloaddition of nitrile oxides in ionic liquids. An easier route to 3-carboxy isoxazolines, potential constrained glutamic acid analogues, *Tetrahedron Lett.* **44**, 5327–5330 (2003).
43. Han, X. and Armstrong, D. W., Using geminal dicationic ionic liquids as solvents for high-temperature organic reactions, *Org. Lett.* **7**, 4205–4208 (2005).
44. Xiao, Y. and Malhotra, S. V., Asymmetric reduction of aromatic ketones in pyridinium-based ionic liquids, *Tetrahedron Asymmetry*, **17**, 1062–1065 (2006).
45. Farmer, V. and Welton, T., The oxidation of alcohols in substituted imidazolium ionic liquids using ruthenium catalysts, *Green Chem.* **4**, 97–102 (2002).

46. Pagni, R. M. and Gordon, C. M., Photochemistry in ionic liquids, in *CRC Handbook of Organic Photochemistry and Photobiology*, 2nd ed., W. Horspool and F. Lenci (Eds.), CRC Press, Boca Raton, FL, 2004, pp. 5-1–5-21.
47. Hubbard, S. C. and Jones, P. B., Ionic liquid photosensitisers, *Tetrahedron* **61**, 7425–7430 (2005).
48. Álvaro, M., Carbonell, E., Ferrer, B., Garcia, H., and Herance, J. R., Ionic liquids as a novel medium for photochemical reactions. Ru(bpy)$_3^{2+}$/violegen in imidazolium ionic liquid as a photocatalytic system mimicking the oxido-reductase enzyme, *Photochem. Photobiol.* **82**, 185–190 (2006).
49. Marquis, S., Ferrer, B., Alvaro, M., Garcia, H., and Roth, H. D., Photoinduced electron transfer in ionic liquids: use of 2,4,6-triphenylthiapyrylium as a photosensitiser probe, *J. Phys. Chem. B* **110**, 14956–14960 (22006).
50. Corey, E. J., Mitra, R. B. and Uda, H., Total synthesis of DL-caryophyllene and DL-isocaryophyllene, *J. Am. Chem. Soc.* **85**, 362–363 (1963).
51. Corey, E. J., Mitra, R. B., and Uda, H., Total synthesis of DL-caryophyllene and DL-isocaryophyllene, *J. Am. Chem. Soc.* **86**, 485–492 (1964).
52. Mascitti, M. and Corey, E. J., Enantioselective synthesis of pentacyclanammoxic acid, *J. Am. Chem.* **128**, 3118–3119 (2006).
53. Pagni, R. M., Mamantov, G., Lee, C., and Hondrogiannis, G., The photochemistry of anthracene and its derivatives in room temperature molten salts, *Proc. Electrochem. Soc.* **94–13**, 638–645 (1994).
54. Hondrogiannis, G., Lee, C. W., Pagni, R. M., and Mamantov, G., Novel photochemical behavior of anthracene in a room temperature molten salt, *J. Am. Chem. Soc.* **115**, 9828–9829 (1993).
55. Lee, C., Mamantov, G., and Pagni, R. M., The photoisomerisation of *cis*- and *trans*-stilbene in ionic liquids, *J. Chem. Res. (S)*, 122–123 (2002).
56. Lee, C., Winston, T., Unni, A., Pagni, R. M., and Mamantov, G., Photinduced electron transfer chemistry of 9-methylanthracene. Substrate as both electron donor and acceptor in the presence of the 1-ethyl-3-methylimidazolium ion, *J. Am. Chem. Soc.* **118**, 4919–4924 (1996).
57. Yadav, J. S., Reddy, B. V. S., Reddy, C. S., and Rajeskhar, K., Green protocol for the synthesis of *vicinal*-halohydrins from oxiranes using the [bmim]PF$_6$ reagent system, *Chem. Lett.* **33**, 476–477 (2004).

20 Raman Spectroscopy of High Temperature Melts

G. N. PAPATHEODOROU and A. G. KALAMPOUNIAS
Foundation for Research and Technology Hellas–Institute of Chemical Engineering and High Temperature Chemical Processes, Rio, Greece and Department of Chemical Engineering, University of Patras, Patras, Greece

S. N. YANNOPOULOS
Foundation for Research and Technology Hellas–Institute of Chemical Engineering and High Temperature Chemical Processes, Rio, Greece

20.1 INTRODUCTION

A few years after the discovery by Raman and Krishnan of the new "Raman" effect, one of the first molten salt studies appeared in a new journal addressed to a new, at that time, field [1]. The mercury lamp (Toronto ark) as excitation source and the photographic plate detection system made the use of Raman spectroscopy rather difficult for melt studies; thus up to the early 1960s very limited work on melts was published [2–5]. One of these early authors, T. F. Young [2], was also the first to introduce in the early 1950s the photomultiplier tube (PMT) in the Raman detection system.

The advent of lasers, and the gallant support of molten salt research in the 1960s and early 1970s, resulted in numerous studies on molten salts which were addressed to their structural properties and have been collected in earlier reviews [6–8]. In the same period, different experimental and theoretical approaches were introduced for Raman spectroscopy and presented in books [9–13]. Experiments that were formerly considered difficult and time consuming became practical at that time.

During the 1970s, the replacement of the Ne–He laser red line with the intense green-blue lines from Ar^+ or Kr^+ lasers dramatically improved the quality of the spectra. For colourless melts at temperatures up to ~ 1000 °C, spectra could be measured with no serious interference due to either fluorescing samples or to blackbody radiation from the furnace. An extensive review on the structural properties of melts based on Raman spectroscopy during this period was published in 1983 [14].

Molten Salts and Ionic Liquids: Never the Twain? Edited by Marcelle Gaune-Escard and Kenneth R. Seddon
Copyright © 2010 John Wiley & Sons, Inc.

Starting in the 1980s, considerable and successive improvements in experimental techniques regarding lasers, optical components, instrumentation, detection systems, computer interfaces, and signal processing have made Raman spectroscopy one of the most versatile methods for studying vibrating "molecular" units and obtaining information about their structure and interactions with their environment [15–18]. Furthermore, it became one of the prevalent techniques for both laboratory and remote measurements, as well as for industrial applications [19, 20]. The weak intensity of the Raman effect can be compensated today with modern optical and electronic signal processing techniques, making Raman spectroscopy sufficient for most analytical chemistry/biology requirements [21, 22]. Measurements under the microscope allow micrometer entities to be analysed [16] with a variety of applications ranging from industry [20] to art [23].

All these advances and improvements also had an impact on molten salt research, and a large number of papers have been published during the last two decades. Thus Raman spectra of oxide melts [24] or volatile halide melts [25] have been studied under the microscope; in situ studies of electrode surfaces in melts have been reported [26] and, more recently, containerless techniques for studying oxide melts up to 2200 °C have been advanced [27].

The main purpose of this chapter is to present the experimental methods used to study high temperature melts. The focus is on specific problems faced in handling the melts, and in the construction of the experimental setups for high temperatures using commercially available instrumentation. Furthermore, certain Raman results from representative methods/techniques are presented. A more general background on Raman spectroscopy is included in the books mentioned above. Theoretical aspects can be found in References 9, 10, 15, and 16, while for practical experimental details References 9, 28, and 29 are recommended. Structural studies of melts by Raman spectroscopy have been compiled in Reference 14, as well as in a more recent review [30].

The outline of this chapter is as follows. We first present the basics of the Raman scattering process, with emphasis on the temperature effects on the spectra. Next, we describe the basic experimental design, discuss the role of blackbody radiation, and point out the important role of materials' purity on the quality of the spectra. Following that, we are concerned with well-established macro- and micro-Raman arrangements, and their use for studies of melts in optical furnaces. We devote one section to the containerless methods for measuring Raman spectra at high temperatures, which have been developed primarily in our laboratory. Then we include as subjects the recently advanced UV Raman and FT Raman spectroscopy—the first with potential use at high temperature and the second for studying coloured melts at intermediate temperatures. Finally, We describe methods for measuring in situ Raman spectra in the vicinity of the electrodes during electrolysis in melts.

20.2 RAMAN SCATTERING

20.2.1 Introductory Remarks

Light scattering, in general, has contributed much to our knowledge about the structure and dynamics in solids, liquids, and gases. Photons that have interacted

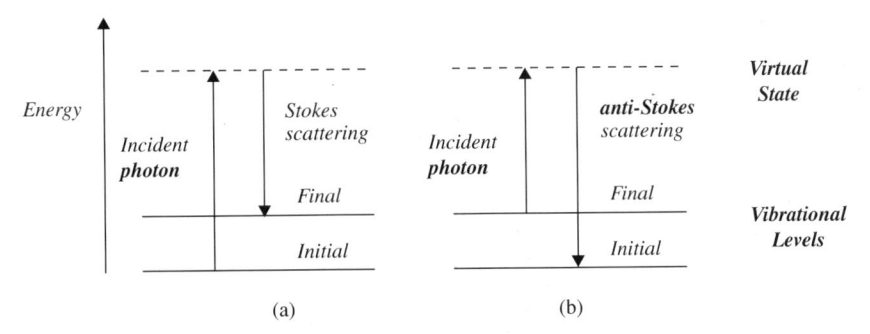

FIGURE 20.1 Energy level diagrams: (a) Stokes Raman scattering and (b) anti-Stokes Raman scattering.

with matter carry an impressive variety of information that, if properly analysed, can provide us with a deep understanding of fundamental phenomena of matter over wide spatial and temporal ranges [15, 31]. Light scattering is complementary, as regards the time and spatial scale probed, to other scattering techniques such as inelastic neutron and X-ray scattering.

The Raman effect arises when a photon impinges on a medium and interacts with the electric dipole of the molecules. In classical terms, the interaction can be viewed as a perturbation of the molecule's electric field. In quantum mechanics, the scattering is described as an excitation to a "virtual" state lower in energy than a real electronic transition, with nearly coincident deexcitation and a change in vibrational energy. The scattering event occurs in 10^{-14} second or less. The virtual state description of scattering is shown in Figure 20.1a, where the energy difference between the incident and scattered photons is represented by the arrows. The energy difference between the initial and final vibrational levels, $\bar{\nu}$, which is the Raman shift in wavenumbers (cm^{-1}), is calculated through

$$\bar{\nu} = \frac{1}{\lambda_{incident}} - \frac{1}{\lambda_{scattered}} \quad (20.1)$$

where $\lambda_{incident}$ and $\lambda_{scattered}$ denote the wavelengths (in cm) of the incident and Raman scattered photons, respectively. At finite temperature, a small fraction of the molecules are in vibrationally excited states. In Raman scattering from vibrationally excited molecules, the scattered photon appears at higher energy, as shown in Figure 20.1b. The anti-Stokes Raman spectrum is always weaker than the Stokes Raman spectrum, but at elevated temperature it is strong enough to be useful for vibrational frequencies $\bar{\nu} < 1000\ cm^{-1}$. The Stokes and anti-Stokes spectra contain the same frequency information. The anti-Stokes spectrum is useful in cases where the Stokes spectrum is problematically recorded, for instance, in cases of limited instrumental sensitivity or in the presence of overlapping fluorescence signals.

20.2.2 Raman Selection Rules and Band Intensities

In a simple classical electromagnetic field description of Raman spectroscopy, an electric field with instantaneous amplitude $\mathbf{E}(t) = \mathbf{E}_0 \cos \omega_L t$, impinging on a molecule, induces a dipole moment $\mathbf{P}(t) = \bar{\alpha}\mathbf{E} + \frac{1}{2}\hat{\beta}\mathbf{E}^2 + \cdots$, where $\bar{\alpha}$ is the molecular polarisability and $\hat{\beta}$ is the hyperpolarisability (responsible for the hyper-Raman effect). $\bar{\alpha}$ is generally a second-rank tensor expressing the scattering anisotropy; namely, the components of $\mathbf{P}$ in some direction in general do not come from the corresponding ones of the electric field vector $\mathbf{E}$ in the same direction. $\bar{\alpha}$ can be written as a symmetric tensor:

$$\bar{\alpha} = \begin{bmatrix} \alpha_{xx} & \alpha_{xy} & \alpha_{xz} \\ \alpha_{yx} & \alpha_{yy} & \alpha_{yz} \\ \alpha_{zx} & \alpha_{zy} & \alpha_{zz} \end{bmatrix} \quad (20.2)$$

The polarisability tensor $\bar{\alpha}$ is a combination of a symmetric part and an asymmetric part. The symmetric part is described by the invariant α,

$$\alpha = \frac{1}{3}(\alpha_{xx} + \alpha_{yy} + \alpha_{zz}) = \frac{1}{3}\mathrm{Tr}\bar{\alpha} \quad (20.3)$$

while the asymmetric part is described by the polarisability anisotropy β,

$$2\beta^2 = [(\alpha_{xx}-\alpha_{yy})^2 + (\alpha_{yy}-\alpha_{zz})^2 + (\alpha_{zz}-\alpha_{xx})^2 + 6(\alpha_{xy}^2 + \alpha_{yz}^2 + \alpha_{zx}^2)] \quad (20.4)$$

Since $\bar{\alpha}$ depends on the distance between the nuclei, it can be written as a function of a generalised vibrational coordinate that quantifies this dependence; then $\bar{\alpha} = \bar{\alpha}(q_i)$. An expansion of $\bar{\alpha}$ in a power series of the vibrational coordinate around its equilibrium value $\bar{\alpha}_0$ at $q_i = 0$ leads to the expression

$$\bar{\alpha}(q_i) = \bar{\alpha}_0 + (\partial\bar{\alpha}/\partial q_i)_0\, q_i + \frac{1}{2}(\partial^2\bar{\alpha}/\partial q_i^2)_0\, q_i^2 + \cdots \quad (20.5)$$

The linear term in q_i determines the *first-order* Raman effect while the quadratic one gives rise to the *second-order* Raman effect. Considering the first-order Raman scattering and assuming a sinusoidal change for q_i, we arrive at

$$\begin{aligned} \mathbf{P}(t) = \bar{\alpha}_0 \mathbf{E}_0 \cos(\bar{\nu}_L t) &+ \frac{1}{2}\left(\frac{\partial\bar{\alpha}}{\partial q_i}\right)_0 \mathbf{E}_0 q_{i0} \cos[(\bar{\nu}_L - \bar{\nu}_S)t] \\ &+ \frac{1}{2}\left(\frac{\partial\bar{\alpha}}{\partial q_i}\right)_0 \mathbf{E}_0 q_{i0} \cos[(\bar{\nu}_L + \bar{\nu}_S)t] \end{aligned} \quad (20.6)$$

It is seen from the above equation that the induced dipole moment will fluctuate not only with the frequency $\bar{\nu}_L$ of the incident light, but also with the combination of the frequencies $\bar{\nu}_L \pm \bar{\nu}_S$ (Stokes Raman frequency, $-\bar{\nu}_S$; anti-Stokes Raman frequency, $+\bar{\nu}_S$). The latter frequencies arise from the modulation of the electronic polarisability $\bar{\alpha}$ by the vibrational motion of atoms that is the prerequisite or *selection rule* in order to observe the Raman effect. The classical treatment predicts correctly the occurrence of the inelastic processes, but leads to an incorrect ratio for their intensities, $I^{\text{Stokes}}/I^{\text{anti-Stokes}} = (\bar{\nu}_L - \bar{\nu}_S)^4/(\bar{\nu}_S + \bar{\nu}_S)^4$, which is less than unity—contradicting the experimental value.

In a quantum mechanical treatment, Raman scattering is the result of an inelastic collision process between the photon and the elementary excitations of the medium.

The photon either loses one or more quanta of vibrational energy (Stokes lines) or gains one or more such quanta (anti-Stokes lines). In *first-order* scattering, only one phonon is involved; this corresponds to the term that is linear in the vibrational coordinate q_i in the expansion of $\bar{\alpha}(q_i)$. In *second-order* scattering, two phonons are involved, corresponding to the term proportional to the term that is quadratic in q_i or to an anharmonic coupling of a phonon that is active in the first-order Raman scattering. The predicted (from quantum mechanics) Stokes and anti-Stokes intensity ratio is

$$I^{\text{Stokes}}/I^{\text{anti-Stokes}} = (\bar{\nu}_L - \bar{\nu}_S)^4 / (\bar{\nu}_S + \bar{\nu}_S)^4 \ \exp(hc\bar{\nu}_S/k_B T) \qquad (20.7)$$

The energy and momentum conservation laws for the first- and second-order scattering processes are summarised as follows:

$$\bar{\nu}_L = \bar{\nu}_S, \quad \mathbf{q}_L = \mathbf{q}_S \quad \text{(Rayleigh scattering)} \qquad (20.8a)$$

$$\left. \begin{array}{l} \bar{\nu}_L = \bar{\nu}_S \pm \bar{\nu}_j(\mathbf{q}) \\ \mathbf{q}_L = \mathbf{q}_S \pm \mathbf{q} \end{array} \right\} \begin{array}{l} \text{(Stokes and anti-Stokes} \\ \text{Raman first-order scattering)} \end{array} \qquad (20.8b)$$

$$\left. \begin{array}{l} \bar{\nu}_L = \bar{\nu}_S \pm \bar{\nu}_{j'}(\mathbf{q}') \pm \bar{\nu}_{j''}(\mathbf{q}'') \\ \mathbf{q}_L = \mathbf{q}_S \pm \mathbf{q}_{j'} \pm \mathbf{q}_{j''} \end{array} \right\} \begin{array}{l} \text{(Stokes and anti-Stokes} \\ \text{Raman second-order scattering)} \end{array} \qquad (20.8c)$$

20.2.3 Experimentally Useful Forms of the Raman Scattered Intensity

Experimentally and theoretically, the scattering cross section may be divided into two parts—polarised and depolarised [15]. On the experimental side, the most common polarisations used in the study of noncrystalline materials (glasses and liquids) are the *polarised* (VV) component, where incident and scattered photons have their electric field vectors parallel, and the *depolarised* (VH) component, where incident and scattered photons have their electric field vectors perpendicular. It can be rather easily shown that the *symmetric* (α) and *asymmetric* (β) parts of the polarisability tensor (cf. Eqs. (20.3) and (20.4)) are related to the *isotropic* and *anisotropic* Raman scattered intensities, respectively, defined as

$$I_{\text{ISO}}(\bar{\nu}) = I_{VV}^{\text{expt}}(\bar{\nu}) - \tfrac{4}{3} I_{VH}^{\text{expt}}(\bar{\nu}) \qquad (20.9a)$$

$$I_{\text{ANISO}}(\bar{\nu}) = I_{VH}^{\text{expt}}(\bar{\nu}) \qquad (20.9b)$$

Based on the fact that I_{ISO} depends on the modulation $\partial \alpha/\partial q_i$ of the symmetric part of the polarisability tensor (α), it is rather straightforward that $I_{\text{ISO}}(\omega)$ depends only on vibrational motion since $\partial \alpha/\partial q_i$ is independent of molecular orientation. On the contrary, I_{ANISO} depends on both vibrational and reorientational motions of the molecule because $\partial \beta/\partial q_i$ depends on molecular reorientation. Equations (20.9a) and (20.9b) are very useful in studies of species equilibria with the aid of Raman scattering when changing either the temperature or the material composition.

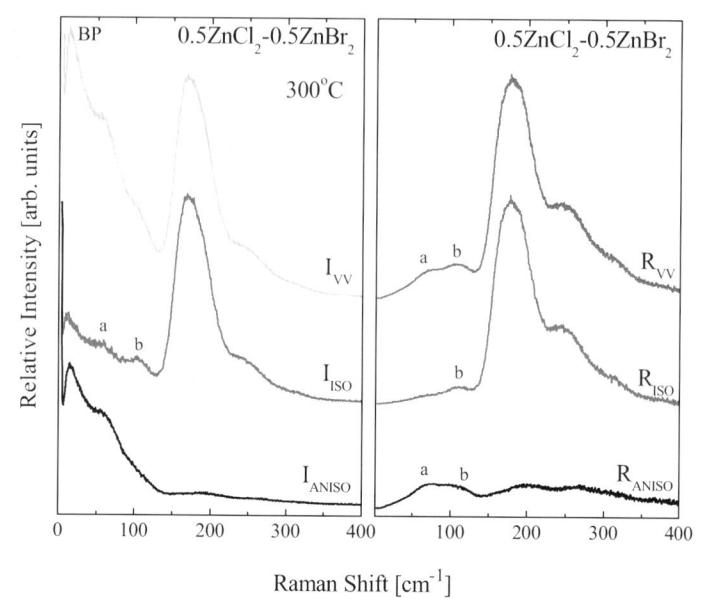

FIGURE 20.2 Raman spectra of the ZnCl$_2$–ZnBr$_2$ (1:1) molten mixtures at 300 °C [32]. The VV, ISO, ANISO (or VH) spectra in the intensity (I, left side) and reduced intensity (R, right side) representation. Note: (i) the use of I_{ISO} resolves the "b" band and indicates that it is polarised; (ii) the R_{ISO} shows that the "a" band is depolarised while the "b" band appears weak in the R_{ANISO}, indicating that it is either partially polarised or that it is composed from a superposition of polarised and depolarised bands. BP is the low frequency boson peak. Spectra conditions: $\lambda_o = 488.0$ nm; laser power = 100 mW; resoloution = 2 cm^{-1}; PMT detection.

In particular, if two bands in a Raman spectrum are assigned to two different "chemical species" or coordination polyhedra, then the relative population of these species can (under some approximations) be determined by comparing the integrated areas of the corresponding Raman peaks. In this case, the use of the I_{ISO} functions, instead of I_{VV}^{expt}, is of crucial importance, because this intensity comes from purely vibrational effects [14, 30]. In the first part of Figure 20.2, the VV, ISO, and ANISO (VH) spectra are shown for a molten mixture of ZnCl$_2$–ZnBr$_2$ [32].

Another important factor that complicates the analysis of Raman spectra is the *thermal population* or *Bose factor* that inherently distorts bandshapes and intensities. Specifically, at any temperature different from absolute zero, there is a finite probability for populating vibrational energy levels other than the true ground state. In this case, the Raman scattering cross section, and therefore the Raman intensities, will depend not only on the vibrational density of states, that is, the number of vibrational modes within a specific energy interval, but also on the particular distribution of the molecules' energies among the vibrational energy levels. Because this distribution is a strongly dependent function of temperature, the comparison of the intensities of a particular Raman peak recorded at various temperatures will not reflect real changes in the population of the species responsible for that peak. It is therefore desirable to be able to disentangle the changes brought about in the vibrational lines by the temperature, and by alterations of local species equilibria or modifications in structure. In order to remove the temperature effect, one can use the *reduced representation* of the Raman spectrum. Due to the boson-like statistical description obeyed by phonons, their mean number at any temperature is given by $n(\bar{\nu}, T) = [\exp(hc\bar{\nu}/k_B T) - 1]^{-1}$, where h and k_B are the Planck and Boltzmann constants, respectively. Therefore the first-order Stokes-side reduced Raman intensity ($I_{\text{Stokes}}^{\text{red-1st}}$) is related to the experimentally measured one via the equation

$$I_{\text{Stokes}}^{\text{red-1st}}(\bar{\nu}) = (\bar{\nu}_0 - \bar{\nu})^{-4} \bar{\nu} \left[n(\bar{\nu}, T) + 1\right]^{-1} I_{\text{Stokes}}^{\text{expt}}(\bar{\nu}) \qquad (20.10)$$

where the term in the fourth power is the usual correction for the wavelength dependence of the scattered intensity; and $\bar{\nu}_0$ denotes the wavenumber of the incident radiation. The use of Eq. (20.10) provides the Raman spectrum that would be measured at "absolute zero." Depending on the polarisation VV, VH (ANISO), and ISO, three different reduced Raman representations can be used, denoted as R_{VV}, R_{ANISO}, and R_{ISO}, respectively (see Fig. 20.2). The advantages for using these representations to analyse the Raman spectra have been discussed in Reference [30]. Finally, it should be stressed that the above equation is valid in the case of *first-order* Raman scattering, where the harmonic approximation holds. In several cases, anharmonic or second-order effects are important. In such circumstances, the second-order reduced Stokes Raman spectrum takes the form [33]

$$I_{\text{Stokes}}^{\text{red-2nd}}(\bar{\nu}) \propto [\bar{\nu}^2/(\bar{\nu}_0 - 2\bar{\nu})] \left[e^{h\bar{\nu}c/k_B T} + e^{-h\bar{\nu}c/k_B T}\right] I_{\text{Stokes}}^{\text{expt}}(\bar{\nu}) \qquad (20.11)$$

Closing this section, it would be instructive to mention that the *depolarisation ratio*, $\rho(\bar{\nu}) = I_{VH}(\bar{\nu})/I_{VV}(\bar{\nu})$, is helpful in identifying the existence of Raman peaks hidden inside complex and broad spectral envelopes. The case where the Raman spectrum contains a peak that is totally depolarised (i.e., $\rho = 3/4$), is very useful to experimentalists, since by using this information one can adjust the polarisation components (polariser and analyser) to achieve the most reliable polarised and depolarised experimental Raman spectra.

20.2.4 Linewidths and Bandshapes of Raman Bands

As pointed out earlier, Raman scattering is mainly a *structural* probe. Actually, the energies of vibrational bands of molecules in the liquid state are the most frequently exploited parameters that are analysed in detail. Vibrational frequencies are related to *static* parameters such as bond lengths, atomic masses, force constants, and short-range order (i.e., coordination numbers, local bonding geometry). On the other hand, much less attention is usually paid to *dynamic* parameters, information on which can be provided by studying linewidths and band profiles. In general, vibrational bands in liquids are much wider than the corresponding bands in the solid state. This broadening can originate from a combination of structural and dynamical effects. Structure is responsible for the *inhomogeneous* broadening. In particular, the vibrational energy involved in the initial and final state of the molecule is modified by intermolecular interaction. Liquids and glasses are locally nonuniform media, and thus intermolecular interactions vary from molecule to molecule. This means that there is no longer a single vibrational transition energy, but a "distribution" of transition energies which gives rise to the observed band. In the homogeneous broadening, all molecules can be considered as being in a *homogeneous* environment; thus all of them have the same vibrational transition energy—in this case fast collisions between molecules are responsible for the band broadening.

Obviously, dynamic parameters are time dependent and hence a Fourier transformation of the spectrum is needed to obtain and analyse the quantity of interest, namely, the corresponding vibrational time correlation function [34]. The term *vibrational relaxation*—which describes changes in the vibrational details of the molecule due to interactions with the environment—has been coined to describe this subject of research. Vibrational relaxation manifests itself in a number of processes [34] such as (1) vibrational energy relaxation (vibrational dissipation), (2) resonant vibrational energy transfer, (3) vibrational dephasing, and (4) intramolecular vibrational relaxation. Representative studies of short-time dynamics in molten salts and glass-forming liquids can be found in References 35 and 36, respectively.

20.2.5 Anharmonicity: Frequency Shifts with Temperature

According to the simple harmonic oscillator model, the frequency of a vibrational band is temperature independent. Because anharmonicity is inherent to condensed matter, vibrational energies experience (in most cases) a red shift with increasing temperature. These shifts can be explained, in a first approximation, by the "quasi-harmonic" model. Temperature effects within this model are taken into account through the volume increase due to thermal expansion, or the Grüneisen parameter [37]:

$$\gamma_i = -\frac{\partial \ln \nu_i}{\partial \ln V} = -\frac{V}{\nu_i}\left(\frac{\partial \nu_i}{\partial V}\right)_P \quad (20.12)$$

Since the thermal expansion coefficient is written as $\alpha = (1/V)(\partial V/\partial T)_P = (\partial \ln V/\partial T)_P$, the Grüneisen parameter transforms to

$$\gamma_{i,T} = -\frac{1}{\alpha}\left(\frac{\partial \ln \nu_i}{\partial T}\right)_P = -\frac{1}{\alpha \nu_i}\left(\frac{\partial \nu_i}{\partial T}\right)_P \quad (20.13)$$

Typical temperature-induced frequency shifts $(\partial \nu_i/\partial T)$ for silicate melts and glasses range between 0.01 and 0.1 cm$^{-1}\cdot$K^{-1} [27, 33].

20.3 BASIC EXPERIMENTAL PROCEDURES AND METHODS

20.3.1 Design of the Experimental Setup

A typical experimental setup for measuring the Raman spectra from molten salts is shown schematically in Figure 20.3. The scattering plane is that of the page. A right

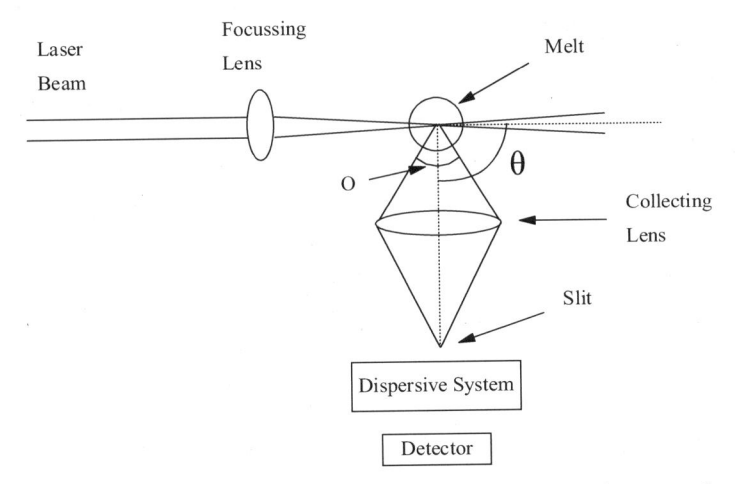

FIGURE 20.3 Schematic diagram of the experimental setup used for measuring Raman spectra.

angle ($\theta = 90°$) scattering geometry is generally used, but backscattering techniques ($\theta = 180°$) have been employed, especially in cases of dark coloured melts. Lines from Ar^+ and Kr^+ ion lasers are usually the sources for the incident light in the visible region. Continuous wave (CW) laser power ranges from a few milliwatts (5–20 mW) up to a few watts. The purpose of the focussing lens is to increase the laser power density at the scattering volume from where the scattered radiation is collected by a set of lenses. The aperture of the collecting lens determines the scattering collection angle, O. Fiber optics as well as microscopes have been also used with the appropriate optical design [16, 19] to measure Raman spectra from melts [26, 38]. The dispersive systems are gratings in a single, double, or triple monochromator. The detector system involves electronic amplification of signals obtained from a PMT or a charged coupled device (CCD) detector. Quasi CW lasers, involving a chopper and lock-in amplifier [39], as well as pulsed lasers and gated techniques, have also been used [40]. Details for the currently used instrumentation will be discussed in a separate section.

However, it should be noted that, for a given melt with a specific scattering cross section and at fixed laser power, there are two main factors improving the intensity of the Raman signal:

1. The spectrometer and optics transmission: this implies careful alignment and matching of all optical components, wide collection angle, and stability of the overall system.
2. The quantum efficiency of the detection systems: for example, the use of high efficiency PMT/or intensified CCD detectors.

High intensity Raman signals minimise the measurement time, which is a very important factor, especially for the study of corrosive melts.

20.3.2 Blackbody Radiation

Until the early 1980s, most Raman setups used scanning techniques for data acquisition. For high temperature Raman measurements, relatively low scanning times are necessary for recording high quality spectra with high resolution as well. Such conditions keep the sample for long times in the furnace, which is not desirable since, in certain cases, the sample could react with the container or slowly volatilise in cooler parts of the cell. Furthermore, stray laser light, bubble formation in the melt, and in some cases particle suspension increase the "Rayleigh" wing and give rise to high background.

An additional important problem is the blackbody radiation coming from the sample and the furnace that is used. The blackbody radiation emitted by any heated material has intensity analogous to the fourth power of temperature for a given wavelength. Figure 20.4 shows the blackbody emission versus wavelength for different temperatures. Above 1000 K, the blackbody radiation is extended up to the visible region of the electromagnetic spectrum and is a few orders of magnitude more intense than the Raman signal of the material.

Recording of high temperature spectra is possible up to certain temperatures, despite the presence of the blackbody radiation, by following an appropriate background subtraction procedure; that is, the spectra of the material superimposed on the blackbody are measured first (Fig. 20.5) and then the laser is blocked and the blackbody spectrum is recorded. The difference between the two spectra corresponds to the pure Raman signal of the material. Near 1800 °C, the method is rather difficult to use while above 2000 °C, the blackbody radiation becomes very intense relatively

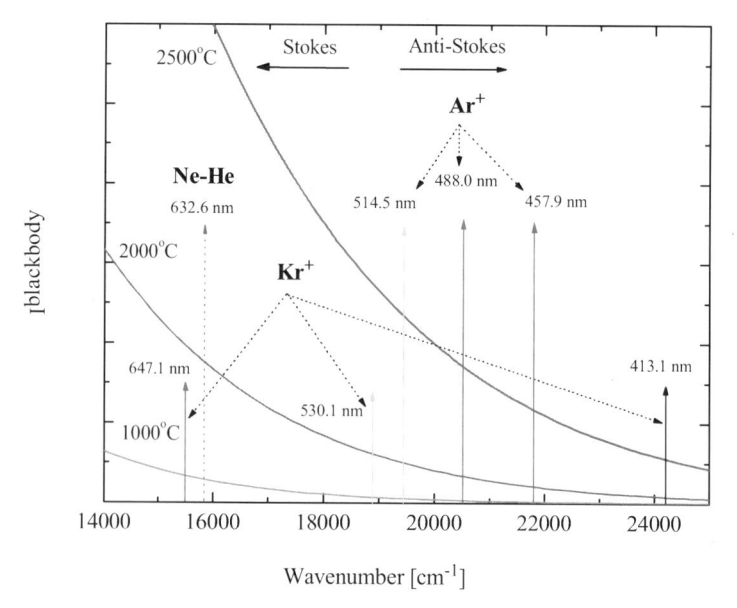

FIGURE 20.4 Relative blackbody thermal emission intensity. The commonly used laser lines are indicated by vertical arrows.

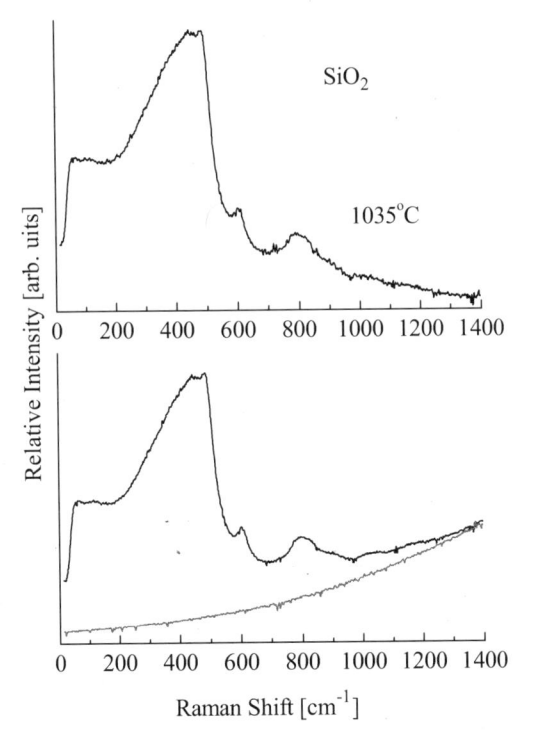

FIGURE 20.5 Procedures for subtracting the blackbody "background" from high temperature spectra. Lower panel: Raman spectra of amorphous silica and blackbody emission at 1035 °C. Upper panel: Silica spectrum after "background" subtraction.

to that of the Raman signal and no spectra can be measured. In such cases, special chopping techniques or pulse excitation methods are used, which permit an "electronic" subtraction of the blackbody radiation background [40, 41].

In some cases, improvement of the high temperature spectra can be achieved by recording the anti-Stokes side of the spectra where the background is lower. This is shown in Figure 20.6a, where the thermal emission is more intense at higher wavenumber. Thus when the blackbody saturates the signal on the Stokes side, the spectra on the anti-Stokes side can still be measured. Furthermore, due to the Boltzmann distribution, the anti-Stokes to Stokes Raman intensity ratio increases with temperature and above 1000 °C the intensities are comparable, as seen in the spectra of HfO_2 [42] at ∼1500 °C in Figure 20.6b.

20.3.3 Materials Purity: Preparation

Sample preparation for Raman studies occasionally present serious difficulties. Even high purity inorganic salts available commercially (e.g., NaCl) may contain traces of organic impurities which upon melting give a slight coloration (e.g., yellowish) and make the melt fluorescent; thus a nonuniform background may appear in the spectra

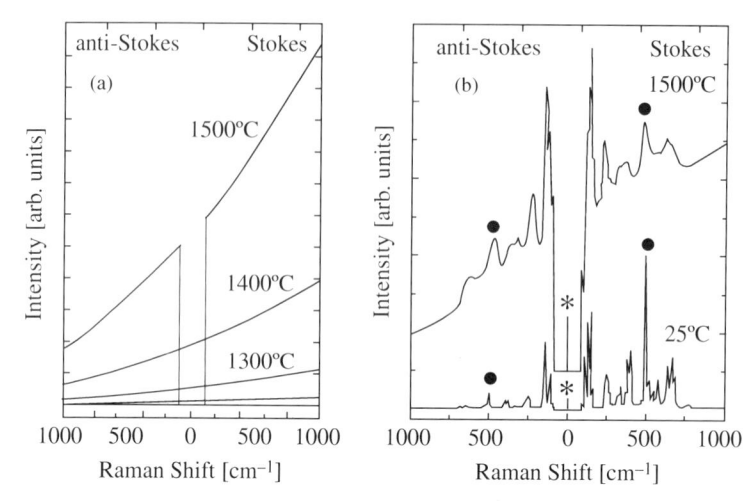

FIGURE 20.6 (a) Thermal emission in the Stokes and anti-Stokes sides for several temperatures; (b) Stokes and anti-Stokes side Raman spectra of HfO_2 at 25 °C and 1500 °C [42].

and, in extreme cases, signal saturation occurs. Oxide formation during the dehydration process of many hygroscopic salts may also create problems. Filtration of fused salts through a sintered frit may be desirable, because the Tyndall scattering from solid particles usually increases the noise and background of the spectra and interferes with data at low Raman shifts. It is often desirable to treat the starting materials (if water soluble) with activated charcoal in water to remove fluorescent and other organic impurities, and then to recrystallise the salt from the aqueous solution. Zone refining by melt crystallisation is also necessary for high melting corrosive melts like metal fluorides. For many salts, however, the ideal means of purification is sublimation under vacuum. Fairly elaborate anhydrous preparation procedures are required for strongly acidic salts like the halides of aluminium, zirconium, and transition metals, while the nitrate, sulfate, and carbonate salts are rather easy to handle.

Dehydration of pure recrystallised salts is not always possible by heating in the air or under dynamic vacuum. Even for obtaining purified alkali halides (e.g., LiCl, LiBr), the process is rather difficult. Such salts are first dried under vacuum by gradually heating to 200–300 °C over a period of at least 4 h. Then hydrogen halide gas is bubbled through the melt for 1–4 h at temperatures 50–100 °C above melting. Small amounts of oxygen are also passed through the melt for 1–2 min to remove (burn) carbonaceous materials or organic contamination (free radicals). Finally, pure dry nitrogen is passed for 15–30 min to remove the hydrogen halide, and the melt is filtered through a quartz frit under vacuum conditions [43] and then kept in sealed containers.

The preparation/purification of each salt has its peculiarities and no general procedures can be outlined. For example, rare-earth halides may be purified through a vapour complexation process [44], while high purity aluminium halides can be prepared by high temperature direct reaction of the metal with a stream of hydrogen halides. In general, it has been found that quenching of the fluorescence of many salts, especially halides, occurs by keeping the melt at 50–100 °C above the melting point for 30–60 min, then fast cooling it on the walls of the container and finally transferring it into the Raman cell.

Finally, metal oxide compounds (e.g., silicates) apart from the classical sol-gel techniques can be prepared by conventional melting in crucibles. The latter can cause contamination and/or heterogeneous nucleation. An alternative method is the CO_2 laser hearth melting technique [45] developed to synthesise high purity samples, which are also used for the levitation experiments described later. A schematic diagram of a modified laser hearth is shown in Figure 20.7. The top of the hearth has an annular groove with a central depression of about 1 cm in diameter and 0.5 cm deep. The compounds are prepared in air or under flowing argon gas and the temperature is measured with an optical two-colour pyrometer and controlled by varying the CO_2 laser power. The diameter of the produced spherical materials varies between 0.1 and 0.5 cm according to the initial amount of the powder due to surface tension. The samples are left to cool freely with gradual decrease of the laser power or fast quench

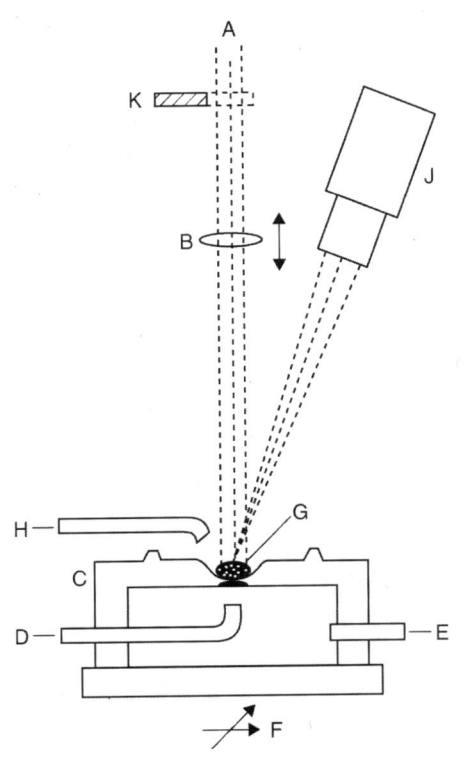

FIGURE 20.7 Schematic illustration of the laser hearth melter [45]: A, CW CO_2 laser beam (max. power: 250 W); B, laser focussing mirror; C, copper hearth; D, cooling water inlet; E, water outlet; F, XYZ positioner; G, molten sample; H, process blanket gas; J, optical pyrometer; K, laser beam block.

by sudden shut down of the laser. The final product is amorphous or crystalline, depending on the cooling rate. The highest temperatures that can be reached are near 2900 °C, causing melting even of very high melting point materials.

All operations concerning the purified and anhydrous salts should be carried out in an inert gas glove box with low water content (<1 ppm) and in torch-sealed fused silica containers. Contact of the salts for long periods with the glove-box atmosphere may be contagious, especially if organic materials (including paper and soft plastics) are present whose "vapours" may be absorbed by the salt, giving rise to fluorescence upon melting. The safest long-term storage of purified chemicals is inside torch-sealed Pyrex or fused silica containers.

20.4 CONVENTIONAL MACRO-RAMAN SPECTROSCOPY

A conventional Raman setup utilises a double or triple monochromator, a laser source, and a detection system with a PMT or a CCD camera detector. A high temperature optical furnace is placed on a heavy XYZ positioner 40–50 cm in front of the entrance slit of the monochromator. Depending on the furnace size and position, the appropriate laser focussing lens and scattered light collecting optics are used. The maximum scattering intensity signal is obtained by matching the collecting optics to those of the monochromator [13]. For measurements above 800 °C, it is useful to introduce a variable aperture pinhole in the path of the scattered light cone in order to partially reject the blackbody radiation. Figure 20.8 shows schematically such a Raman setup.

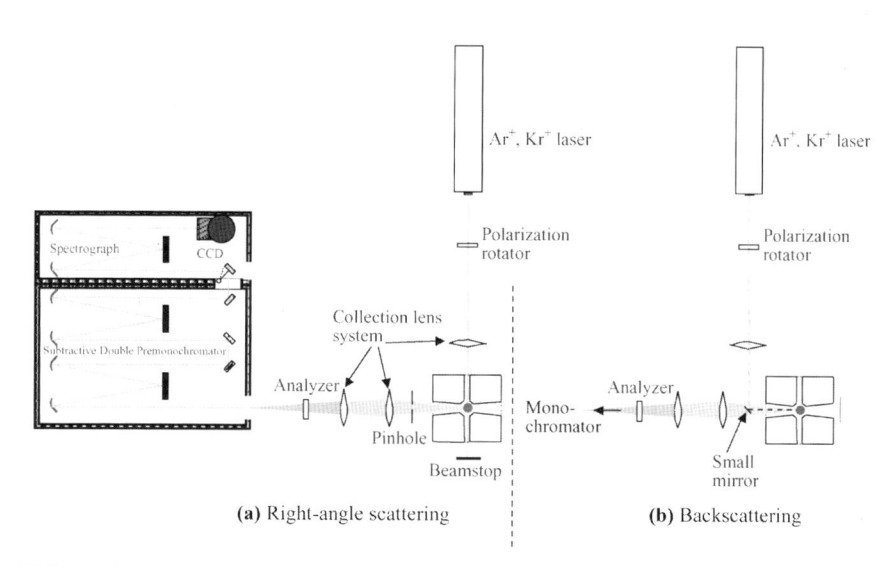

FIGURE 20.8 A conventional macro-Raman system for measurements at high temperatures. A CCD or a PMT detection system can be used: (a) 90° scattering geometry and (b) backscattering arrangement.

20.4.1 Laser Sources

The green-blue lines from an Ar^+ or Kr^+ laser are commonly used, especially for measurements above 500 °C. The red lines of Kr^+ and Ne–He lasers can also be used with PMT or CCD detection up to 500–600 °C, but at higher temperatures due to overlap of the wavelength of these lines with the emission edge of the blackbody radiation, the spectra show a high and asymmetric background. This problem can be quelled by modulation techniques with a chopper–PMT–lock-in amplifier system since "subtraction" of the unwanted background and a rather intense Raman signal can be obtained up to 800 °C [39]. The same chopping techniques can be utilised for measurements at higher temperature, but with the blue-green laser lines as excitation sources.

20.4.2 Furnaces

A typical high temperature furnace for Raman spectroscopy has crossing openings at the centre (Fig. 20.9), where the optical container with the melt is placed. Transmission of the laser beam through the sample and out from the furnace diminishes the stray light, while the opening at the back side of the furnace minimises blackbody radiation. The furnace has two independently controlled heating zones: one at the centre around the opening for the optical paths (windows) and the other at the main furnace body. Using different controller settings, the temperature and temperature gradient of the furnace can be adjusted while the temperature of the sample is measured with an independent thermocouple (J or S). The temperature gradient along the sample can be improved by placing the cell in a metal block with opening for the optical paths.

20.4.3 Containers

Fused silica resists corrosion from a variety of molten salts and has been used for the construction of Raman cells. A simple way for making cylindrical cells is from fused silica tubing with ~0.5–1 mm wall and 2–6 mm ID, as shown in Figure 20.10a. Depending on the cell size, the amount of sample required varies from 50 mg to 1 g. The small ID cells (~2–3 mm) are recommended for handling melts with high vapour pressure (~50 atm). The cylindrical cells are placed in a fixed position inside the optical furnace and the laser beam is directed, with the aid of a micropositioner attached on the focussing lens, so as to pass across the diameter of the cell (Fig. 20.10b). Furthermore, the collecting optics are adjusted to collect light across the laser beam within the cell and to exclude light scattered from the fused silica walls (Fig. 20.10b). This procedure minimises stray light and allows measurements in the low frequency region (<50 cm^{-1}). Figure 20.11 gives an example of spectra measured with such a simple Raman cell at relatively high temperatures and pressures, which show the dimerisation of pure molten $ZrCl_4$ [48].

Spectra from low vapour pressure corrosive melts, like fluorides and oxyhalides, have been measured with a windowless cell designed by Gilbert et al. [49] and shown

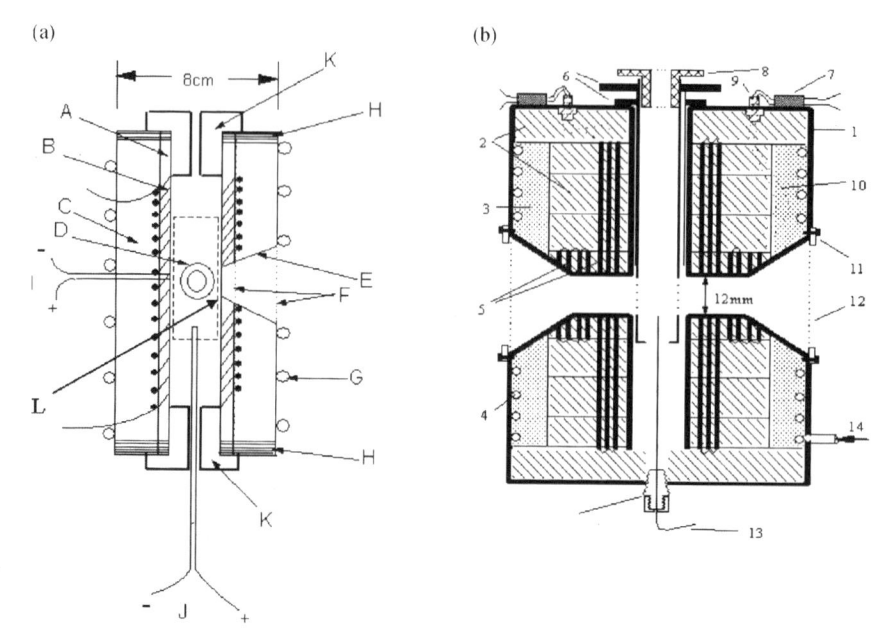

FIGURE 20.9 Schematic diagrams of optical furnaces for Raman measurements at high temperatures. (a) *Furnace with metal block* [46]: A, ceramic rings; B, main heater (coaxial inconel wire); C, insulating material (bubble alumina); D, opening for the incoming laser beam; E, conical opening for collecting the scattered light; F, optical windows from fused silica; G, copper tubing for water cooling; H, bronze metal bases; I, controlling thermocouple; J, measuring thermocouple; K, insulating ceramic covers; L, metal block with central heater (coaxial inconel wire). (b) *Furnace for fluoride melts* [47]: 1, bronze outer jacket (2 mm thick); 2, insulating ceramic; 3, insulating ceramic wool; 4, internal cooling tubes; 5, ceramic shield; 6, bronze supports; 7, electrical connection to heating wire; 8, soft ceramic cover (BN); 9, insulating ceramics; 10, electrical resistance (Kanthal); 11, support of optical window; 12, fused silica optical window; 13, controlling thermocouple; 14, cooling water inlet.

in Figure 20.10c,d. Graphite is the most commonly used material for the construction of the windowless cells, but in cases where the melt creeps on graphite (e.g., CsF), nickel or platinum can be used [50]. From the density of the melt, or by trial and error tests, the appropriate amount of salt is placed into the cell so that upon melting a drop is formed covering the optical openings without creeping out of the cell. The viscosity of the melt and its wettability with the container are determining factors for using this method. For measuring the spectra, the windowless cell is first placed inside a fused silica tube ($\sim$10–12 mm ID), which is sealed under vacuum or $\sim$1 atm inert gas (Ar), and then placed in a fixed position inside the optical furnace. Typical spectra obtained by this method are shown for the ZrF_4–AF [51] binary system in Figure 20.12.

Finally, very viscous melts (e.g., certain silicates) can be melted as large drops on the surface of noncorrosive material and then placed in the furnace within a protective argon atmosphere. The molten drop is aligned at the centre of the optical paths of the

Windowless Graphite Cell

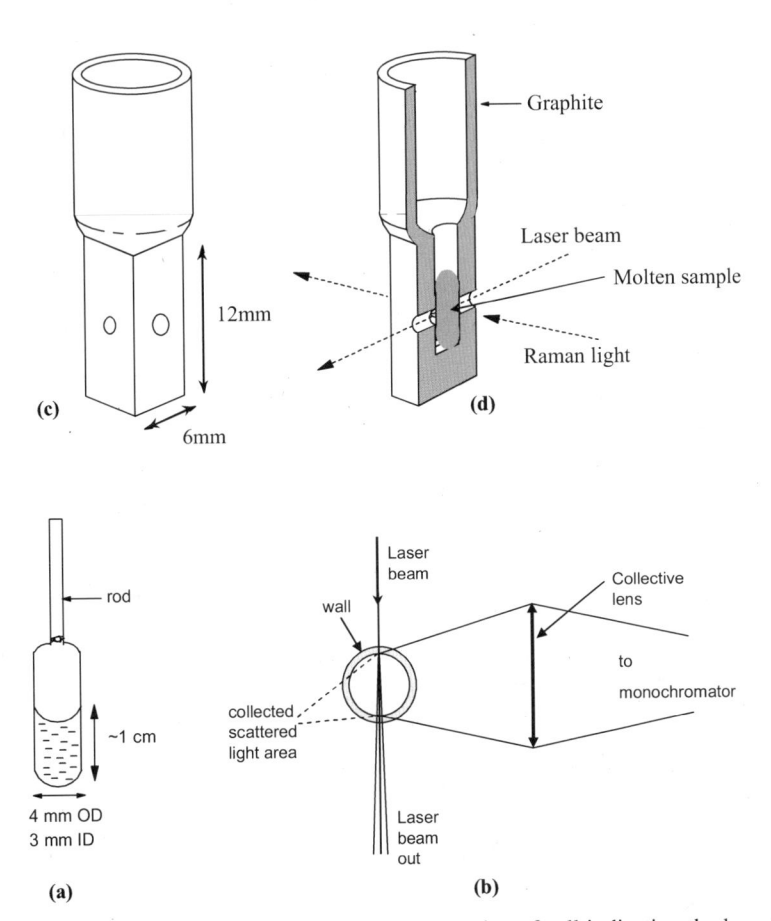

FIGURE 20.10 (a) Fused silica cell and (b) cross section of cell indicating the laser beam passage and the light collection area; the windowless cell (c) empty and (d) with the appropriate melt volume.

experiment, and Raman spectra can be measured in a way similar to that described for the "self-support" techniques.

20.4.4 Dark Coloured Melts

For clear and dust-free melts that do not absorb the laser line, the 90° scattering geometry is the most efficient, giving high intensity spectra with a high S/N ratio. For coloured melts, absorbing the laser beam, the backscattering geometry can be used by placing a small mirror (or prism) in front of the furnace and focussing the beam on the sample through the front furnace window parallel to the direction of the collecting optics (Fig. 20.8b). Alternatively, for melts contained in fused silica containers, the

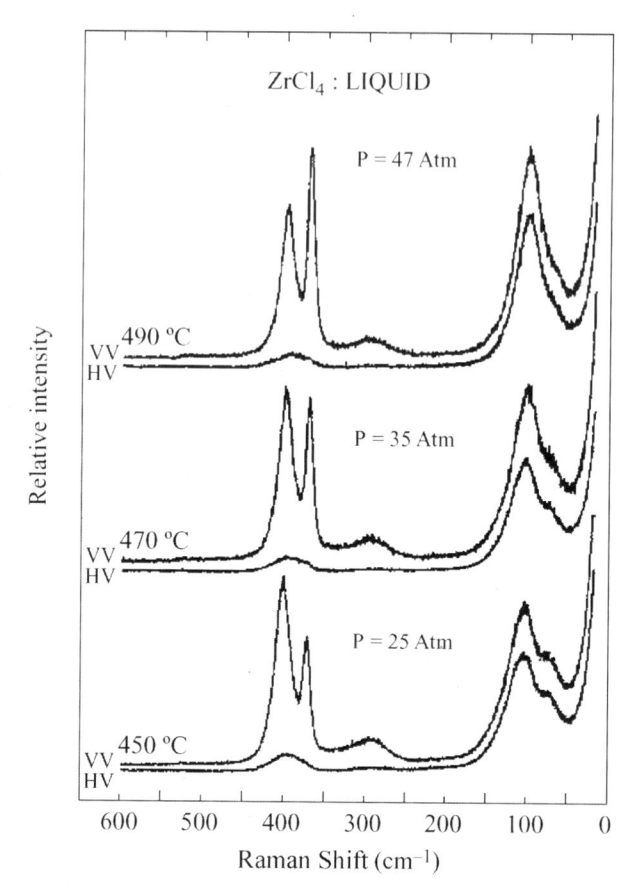

FIGURE 20.11 Temperature dependence of the Raman spectra of molten ZrCl$_4$. Condition $\lambda_o = 488.0$ nm, laser power 100 mW, resolution 3 cm^{-1} PMT/photon counting detection. The marked pressures represent the vapour pressure over the liquid. The M and D bands represent the Zr–Cl stretching frequencies of the monomer and dimer, respectively, which participate in the equilibrium [48]: 2ZrCl$_4$(l) $\leftrightarrows$ Zr$_2$Cl$_6$(l).

90° geometry is kept but the laser beam is directed slightly shifted from the cells' central vertical axis toward the spectrometer; thus the beam is refracted by the cell wall toward the melt creating a pseudo-backscattering geometry as described in Reference 25. At temperatures below 400 °C, the FT Raman spectroscopy can potentially be used for studies of coloured melts.

20.5 MICRO-RAMAN SETUP

The basic optical system, outside the monochromator, for micro-Raman spectroscopy is a metallurgical-like microscope [16, 19–21]. A backscattering configuration is employed, where the laser beam is focused on the sample by means of the microscope

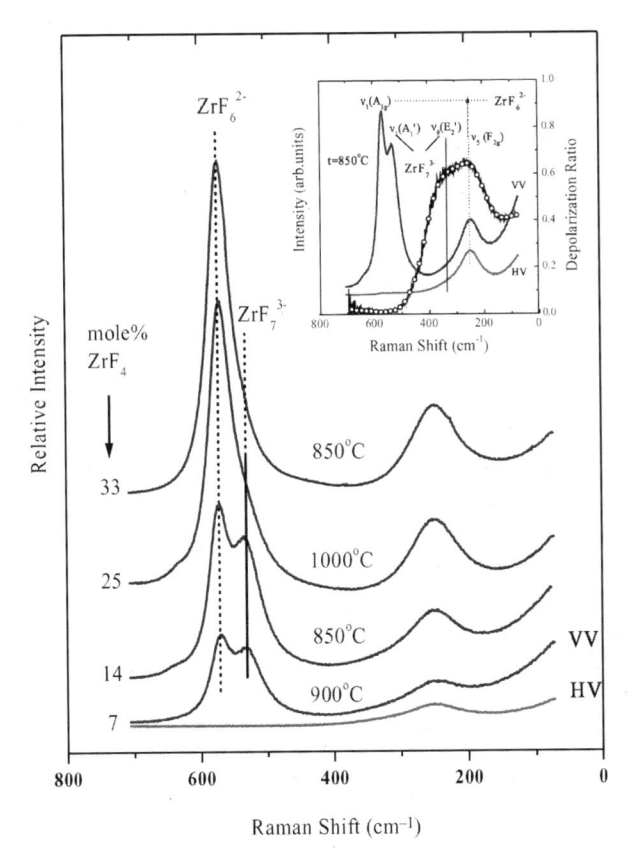

FIGURE 20.12 Raman spectra of molten ZrF$_4$–KF mixtures at different compositions. Inset: Polarised and depolarised spectra of the 14 mol % ZrF$_4$ eutectic mixture at 850 °C and linear depolarisation ratio. Conditions: $\lambda_o = 488.0$ nm; laser power = 500 mW; resolution 5 cm^{-1}; PMT/photon counting detection. The spectral changes with composition indicate that the equilibrium ZrF$_7^{3-}$ ⇌ ZrF$_6^{2-}$ shifts to the right at higher ZrF$_4$ mole fraction and with increasing temperature [51].

objective and the scattered light is collected with the same objective, as shown schematically in Figure 20.13. The sample is placed on the XY positioner of the microscope stage. A video camera attached to the microscope gives a magnified image of the sample and permits an accurate determination of the focussing point of the laser on a sample. The most commonly used laser lines are those of the Ar$^+$ and He–Ne, and spectra can be obtained with a few milliwatts of laser power. The dispersed scattered light from the (final) grating is distributed over a CCD detector and, with relatively small accumulation times (seconds), high intensity Raman signals can be measured.

Microscope objectives with short focal lengths (1–2 mm) define a wider cone solid angle for collecting the scattered light and thus improve the Raman signal. However,

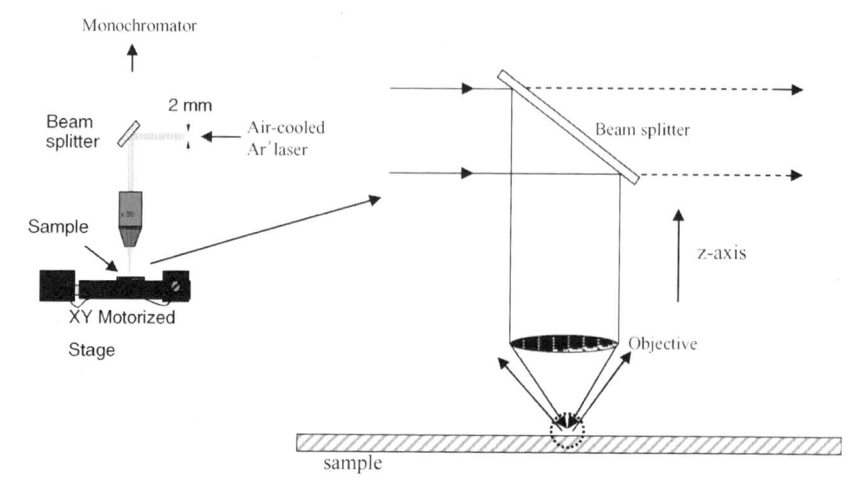

FIGURE 20.13 The micro-Raman optical path toward the sample. The microscope objective serves as focussing and collecting lenses.

for studies at high temperatures, the radiated heat from the sample and heating stage may damage the lower part of the microscope. Thus objectives with longer focal lengths (up to 20 mm) are used with an unavoidable loss of scattered light and Raman intensity. Another difficulty for the study of melts (and liquids in general) is the determination of the polarisation properties and the separation of the isotropic and anisotropic spectra. The terms polarisation direction of the "laser excitation" and direction of "scattering" are difficult to define in such a case, and special procedures should be followed for accurate determination of the polarisation properties [16].

20.5.1 Intermediate Temperatures

An optical microfurnace, presented in Figure 20.14, is placed on the stage, which allows backscattering Raman measurements from melts under the microscope at high

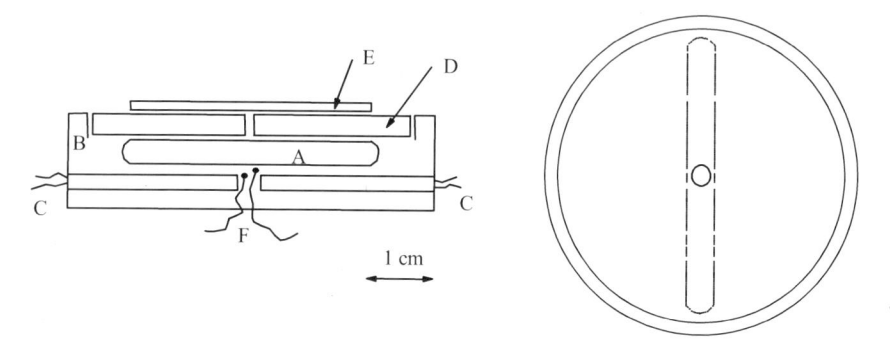

FIGURE 20.14 Schematic diagram of the microscope hot-stage for volatile air-sensitive samples (T_{max} ~800 K); A, Raman cell; B, metal (brass) block; C, heating cartridges; D, insulating disk; E, fused silica cover; and F, controlling and measuring thermocouples [25].

temperatures. Depending on the furnace, material temperatures up to ~800 °C can be achieved. The melt is contained in fused silica cells; these cells, fused at both ends with dimensions OD ≈ 4 mm and ID ≈ 2 mm, can withstand (with a safety factor) approximately 50 atm and thus are appropriate for studies of high vapour pressure melts. The laser beam is focused by a long working distance objective lens on a small diffraction limited spot through the walls of the cell into the melt, and a combined melt/fused silica spectrum is obtained. However, by improving the depth of the field and the axial resolution, and by using two narrow pinholes of 500 and 1000 µm, it is possible to have different degrees of partial rejection of the fused silica Raman signal, and thus measure the spectra of pure molten sample without contributions from the wall of the container [15]. Spectra of dark coloured melts can also be measured with this experimental setup, as shown in Figure 20.15 for the $Cs_2NaFeCl_6$ elpasolite salt.

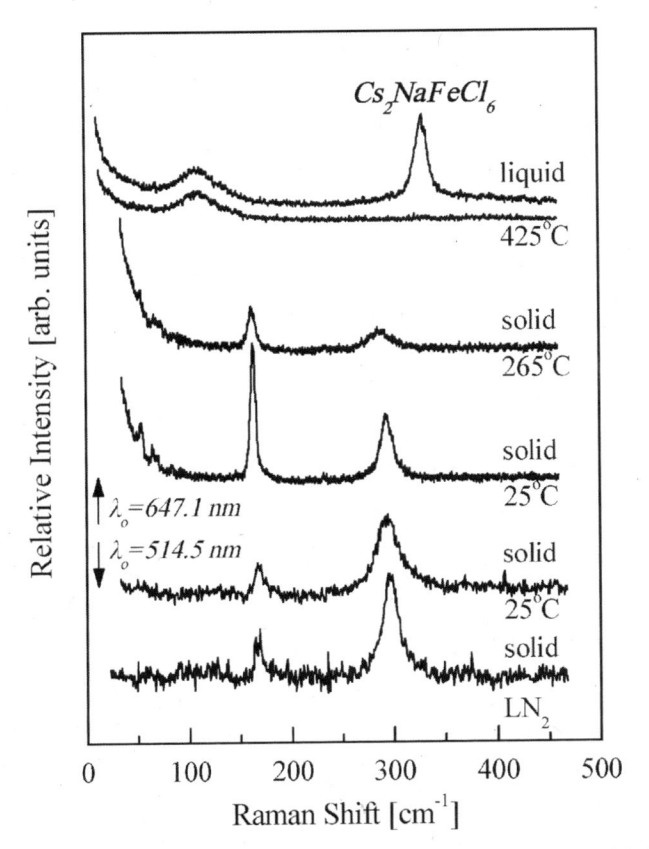

FIGURE 20.15 Raman spectra of solid and molten $Cs_2NaFeCl_6$ recorded with the use of the microscope hot-stage in micro-Raman system [25]. The solid spectra correspond to the vibrational modes of the $[FeCl_6]^{3-}$ octahedra. Upon melting, an octahedral to tetrahedral $[FeCl_4]^-$ coordination change occurs.

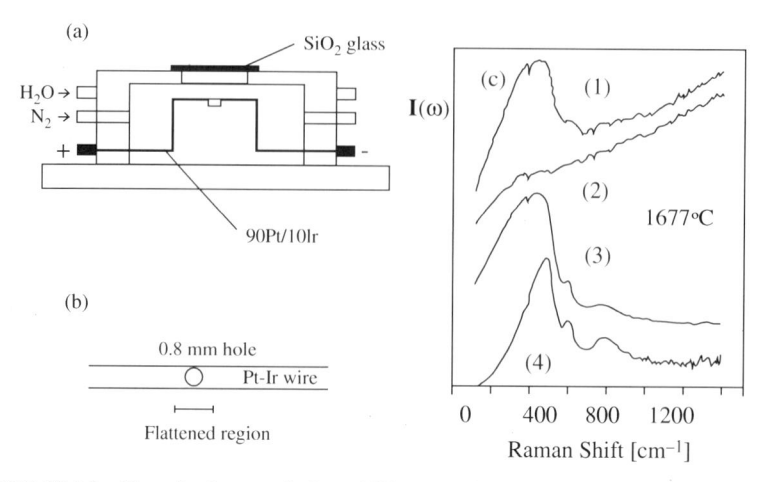

FIGURE 20.16 The wire-loop technique [53]: (a) main furnace, (b) drilled thermocouple holder of melt, and (c) spectra of silica at 1677 °C. The blackbody (furnace) background (2) is subtracted from the measured spectra (1) resulting in the corrected spectra (3); the calculated reduced spectra are also shown (4).

20.5.2 Wire-Loop Technique for Temperatures up to 2000 K

The micro-Raman technique has been combined with the "wire-loop" technique for temperature measurements up to 2000 K [52, 53]. This technique involves a thermocouple (Pt and Pt 10% Rh) with a region flattened near the junction, as shown in Figure 20.16a,b. Through the flattened area at the junction, a hole several hundred micrometers in diameter is drilled. The powdered sample is placed in this hole and melted by applying a heating current to the thermocouple wires. The molten specimen is retained in the hole by surface tension, and the Raman spectra are obtained by focussing the laser beam into the melt. The temperature is controlled by measuring the voltage across the thermocouple wire, which serves as sample holder, and calibrating the system in separate experiments with known melting point samples. Figure 20.16c shows spectra obtained with this method. The uncertainty of the sample temperature and the deposition of the volatile samples at high temperatures on the microscope objective are some of the disadvantages of this method.

20.5.3 Corrosive and Air-Sensitive Melts

Recently, a modified micro-Raman setup consisting of a pulsed copper vapour laser source, a time-resolved detection system, a monochromator, and a high temperature furnace has been used for measurement of corrosive and air-sensitive melts [54]. The furnace shown in Figure 20.17 has been designed so as to be placed under the microscope, and to keep the sample under vacuum or in a controlled gas atmosphere. Sample containers made from noble/refractive metals, amorphous carbon, graphite, or ceramics can be used, and thus the system is suitable for studies of corrosive melts (e.g., fluorides, oxohalides). The heating elements of the furnace

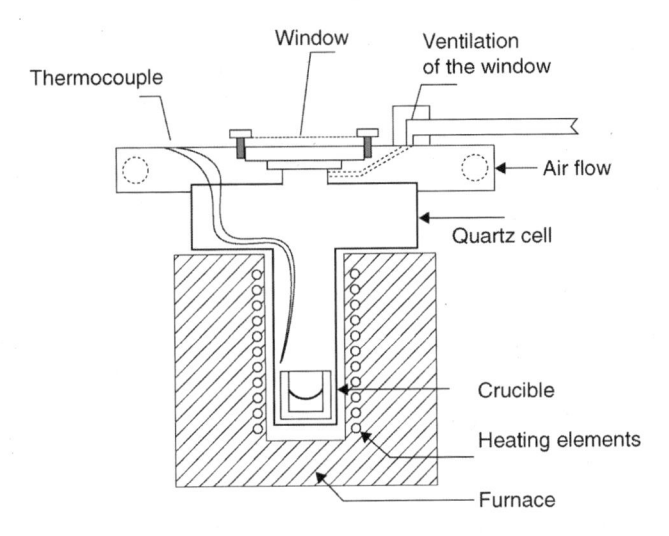

FIGURE 20.17 Schematic diagram of the furnace appropriate for the study of corrosive melts [54].

are lanthanum chromate rods and the temperature can be raised from room temperature to the desired level up to 1000 °C within a reliable time. This design can also be utilised in a backscattering geometry in conjunction with a macro-Raman setup with the appropriate modification of the involved optics. Representative Raman spectra of solid and molten $Na_3Al_3F_{14}$ [54], recorded with this setup under the microscope, are shown in Figure 20.18.

20.6 RAMAN SPECTROSCOPY OF CONTAINER-FREE MELTS

Using the different sample "environments" or "containers" that have already been described, a rich variety of materials were studied in the past at moderately high temperatures. However, Raman studies of materials with very high melting points, and corrosive for any container, were still prevented. This problem was confronted with the adoption of methods that avoid the use of containers and sample holders.

The concept of containerless heating of samples was utilised in the past by metallurgists, who could levitate a metal alloy "spherical" drop inside an induction furnace by applying a magnetic field. In recent years, the advent of infrared lasers, as heating sources, was instrumental in developing a containerless technique that has been used successfully for materials processing [55] and characterisation [56, 57]. The infrared laser is used for melting the material in the form of a "spherical" drop, which is suspended in a noble gas atmosphere with the use of acoustical and/or aerodynamic devices [55, 58].

Very recently, infrared laser heating and containerless techniques have been applied for measuring Raman spectra of molten oxides [27, 59]. Two radiant methods

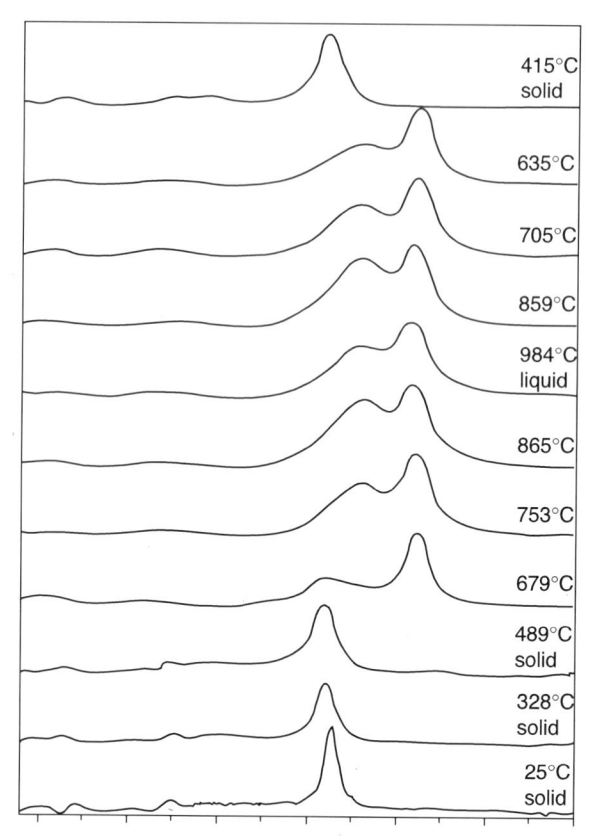

FIGURE 20.18 Raman spectra of chiolite (Na$_5$Al$_3$F$_{14}$) recorded using the furnace of Figure 20.17, appropriate for studies of corrosive melts [54]. The spectra indicate a solid phase transition with increasing temperature. It appears that the vibrational modes of the high temperature solid are transferred into the melt.

are used; levitation and self-support of the melt "drop," which are described in the following subsections.

20.6.1 Levitation Method

A macro-Raman setup, the green-blue Ar$^+$ lines, a CO$_2$ laser, and a levitation device are used for the containerless experiments (Fig. 20.19). The levitation device is a conical jet made of bronze with the appropriate geometry and openings for gas (argon) flow (Fig. 20.20). A protective fused silica tube ($\sim$10 mm ID) is placed around the jet to stabilise gas flow, and the device is mounted on an XYZ stage. Digital mass flow metres are used for adjusting the gas under pressure to equilibrate the sample weight and keep it in a fixed position [58]. The diameter of the droplet is 3–5 mm. Heating of the sample is achieved with the 10.6 µm CO$_2$ laser line with a maximum power of $\sim$240 W. The laser line is directed and

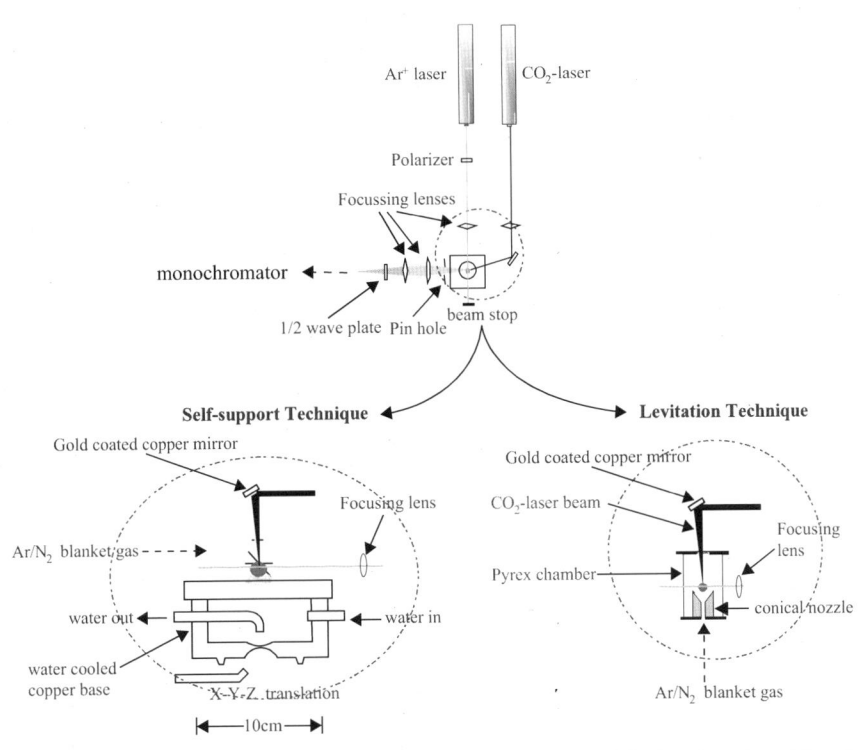

RAMAN SPECTROSCOPY OF CONTAINER-FREE MELTS 325

FIGURE 20.19 The two-laser system for heating the sample (CO_2 laser) and exciting the Raman spectra (Ar^+ laser) [27]. A conventional macro-Raman system is used for recording the spectra. The *self-support* container is similar to that for hearth melting. Details for the *levitation* nozzle are shown in Figure 20.20.

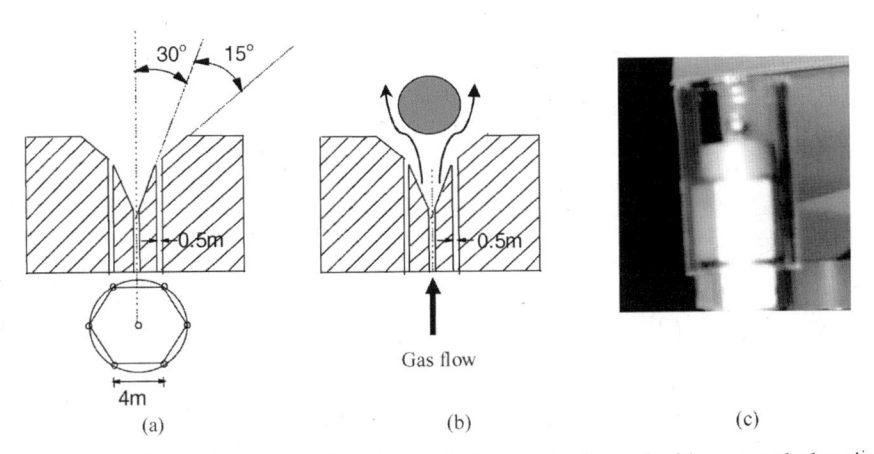

FIGURE 20.20 Levitation nozzle: (a) a conical geometry is used with a central plus six surrounding openings for the flowing gas; (b) the sample position is equilibrated within or slightly above the nozzle cone; and (c) a picture of the nozzle with a protective fused silica tube and the melted "sphere" over the nozzle cone.

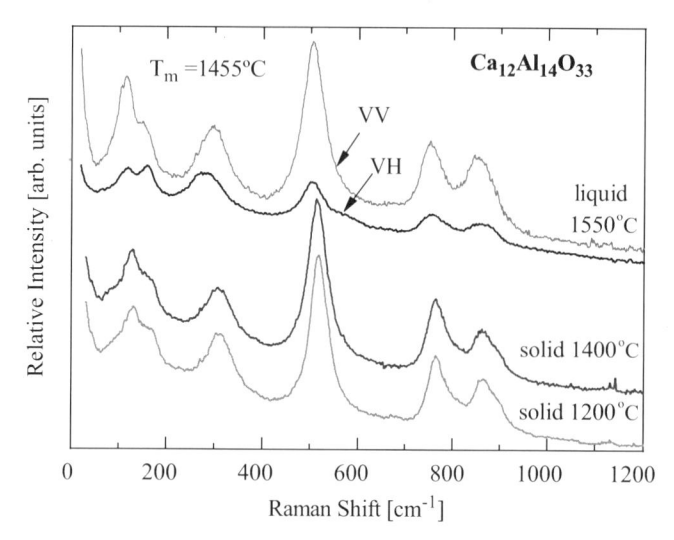

FIGURE 20.21 Representative high temperature Raman spectra of the CaO–Al$_2$O$_3$ binary system recorded with the levitation technique. Conditions: $\lambda_o = 488$ nm; power $= 400$ mW; resolution $= 2$ cm^{-1}; CCD detection.

focused on the sample with the use of water-cooled, concave gold plated mirrors of high reflectivity. The alignment is accomplished with the aid of a red diode laser line collinear with the CO$_2$ laser line. The beam "hits" the sample from the top perpendicularly to the scattering plane, as defined by the Ar$^+$ excitation line and the monochromator slit. For most oxides studied up to 2200 °C, the maximum CO$_2$ laser power used was ~80 W. Figure 20.21 shows the Raman spectra obtained with the described levitation system for the CaO–Al$_2$O$_3$ crystalline and molten phases. One critical and difficult to measure parameter for this experiment is temperature. Above 900 °C, temperature can be measured by the following methods.

Pyrometry A commercially available "two-colour" optical pyrometer is used. The pyrometer is equipped with two detectors measuring (IR intensity) the emissivity of the sample at two different wavelengths and calculating an average temperature of the emitting sample. Due to the small size of the sample, the pyrometer temperature readings are associated with a larger than usual error, which is estimated for the levitation experiments to be 5%.

Spectroscopic Temperature From measurement of the relative intensities between the Stokes and anti-Stokes sides of the spectra (e.g., see Fig. 20.6 and Eq. (20.7)), the temperature of the sample can be estimated. However, at high temperatures where the intensities of the two sides are comparable, the temperature estimation is rather uncertain and the method should be used only to check the validity of the temperature measured by pyrometry.

20.6.2 Self-Support Method

The spectroscopic setup is the same as for the levitation experiment, with the CO_2 and Ar^+ lasers being the heating and the excitation sources, respectively. The overall optical geometry is also the same, having the CO_2 laser line perpendicular to the scattering plane. The sample is partially melted with the CO_2 laser in a way that the liquid is self-supported underneath by the solid phase of the same material. The liquid drop formed has a semispherical shape, due to surface tension, and the Raman excitation line is directed toward the upper part of the drop and through the melt. Depending on the nature of the sample, two types of self-support experiments have been designed.

For solids in powder form, the laser hearth melter (Fig. 20.7) is utilised for supporting the sample. The central depression (cavity) of the copper hearth is overfilled with the powder material, which is then partially melted with the CO_2 laser, forming a drop, resting over the remaining powder, through which the Ar^+ beam passes (as shown in Fig. 20.19). The melter base position and the direction and focussing of the two laser beams are adjusted with micropositioners to facilitate the alignment toward the monochromator. Temperature is measured as described earlier for the levitation method.

For spectra accumulation of solids that can be fabricated in rod form, an alternative "pendant drop" method has been designed and used successfully up to 2300 K [27]. A rod (~5 mm diameter, ~5 cm long) is placed on a micropositioner, perpendicular to the scattering plane and aligned to the incoming CO_2 laser beam (Fig. 20.22). Upon

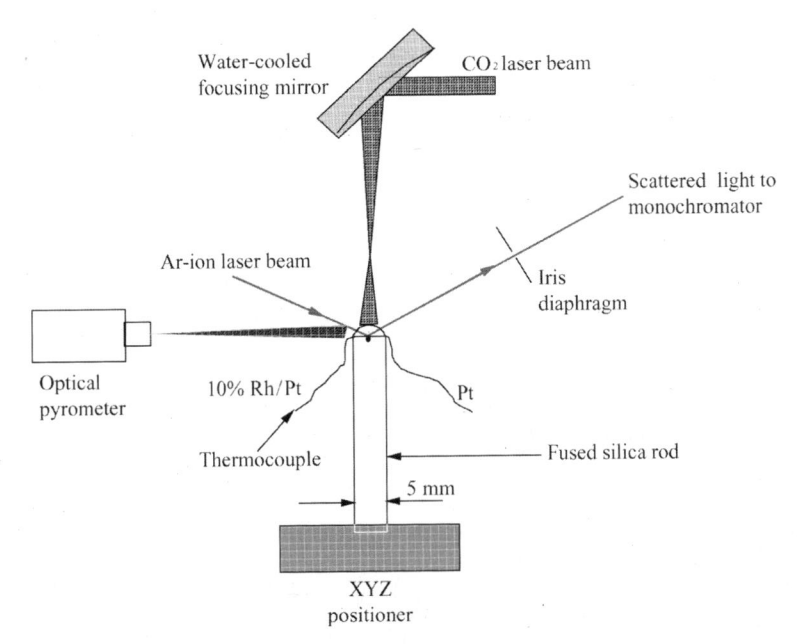

FIGURE 20.22 Schematic diagram of the high temperature (>1300 K) setup for Raman measurements with the CO_2 laser as heating source. A 90° scattering geometry is used and the rod is aligned with an XYZ micropositioner. The iris diaphragm was added at elevated temperatures to minimise the collection of the blackbody radiation.

heating, the top of the rod is melted, forming a pendant drop through which the Raman excitation line is directed. The temperature is measured as before with a pyrometer, but for an additional determination and confirmation, a thermocouple (Pt–Pt 10% Rh) was built—inside the rod through a drilled hole (~500 mm diameter) very near the pendant drop. The thermocouple reading was then calibrated versus the optical pyrometer and the CO_2 laser power used to heat the sample. From a series of temperature cross-checking measurements, an error was estimated of about ±2% up to ~1700 K and ±3–4% up to 2200 K. At temperatures higher than 1400 K, a variable size pinhole was inserted in the collecting optics path to minimise the blackbody radiation coming from the bulk molten/glassy drop, as well as from the emissivity of the nearby thermocouple. Furthermore, a background correction was applied by recording blackbody spectra with the excitation laser off and subtracting it from the overall spectra.

Representative polarised (VV) Raman spectra of SiO_2 [27] at selected temperatures are shown in Figure 20.23. In comparison with previous studies [53] and Figure 20.16c, the present spectra are of higher resolution and can reveal low frequency bands very close to the excitation line.

20.7 RAMAN SPECTROSCOPY IN THE UV AND NEAR-IR REGIONS

All the experimental setups described in the previous section use mainly laser lines in the visible region to excite the spectra. In particular, the 488 nm (blue), 514.5 nm (green), and 647.1 nm (red) lines emerging from the gas lasers (Ar^+ and Kr^+) are the most popular due to the beam quality, stability, high power, availability, and adequate spectral narrowness needed for high resolution Raman measurements. With the advent of lasers (mainly of the solid state type) emitting at the UV and the near-IR region, as well as with the parallel development of optics (mirrors, gratings, polarisers, detectors), Raman spectroscopy extended its limits from the UV to the near-IR regimes [16]. The need for selection of the proper wavelength is driven by the particular application of the user; specific examples are given in the following subsections.

20.7.1 UV Raman Spectroscopy

Development of ultraviolet (UV) laser sources has made possible Raman spectroscopic measurements with excitation sources in the 400–200 nm region [60]. There are certain advantages in using UV Raman spectroscopy:

1. In many samples, there is no interference with fluorescence.
2. There is increased Raman scattering efficiency, due to the ν^4 dependence of the Raman scattering intensity.
3. In many cases, resonance Raman enhancement of the intensity takes place due to the overlap of the (UV) laser line with electronic states of

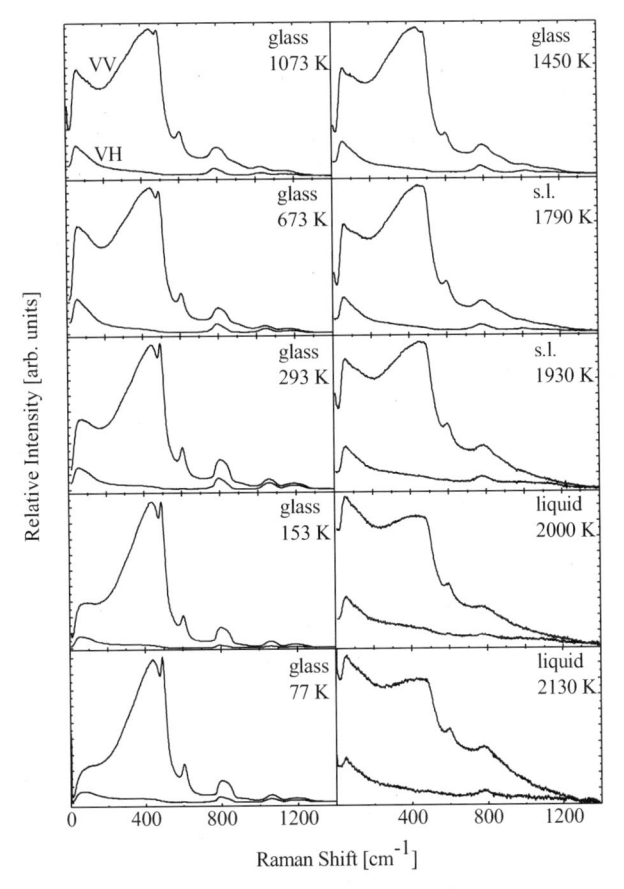

FIGURE 20.23 Representative raw Stokes side Raman spectra of SiO_2 in glassy, supercooled liquid (s.l.) and normal liquid states over a wide temperature range [27]. Lower and upper curves denote the depolarised VH and polarised VV spectra, respectively. Numbers in parentheses indicate the peak intensity of the 450 cm^{-1} band relative to that of the room-temperature glass having an arbitrary value of 100. For the spectra above 1400 K, a correction for blackbody radiation was made. Spectral conditions: excitation line = 488 nm for 293–2150 K and 514.5 nm for 77–293 K; laser power = 400 mW; resolution = 2 cm^{-1}; CCD detector.

the sample. Thus it is possible to observe the Raman bands of substances at low concentrations.

4. Low laser power is required (<100 mW), thus avoiding in many cases overheating or damage of the sample.

Furthermore, UV Raman spectroscopy with CW lasers is more suitable for high temperatures studies, since the overall Raman spectrum is measured at wavelengths away from the intense thermal emission (blackbody).

A schematic description of a UV Raman setup for recording high temperature measurements is shown in Figure 20.24. A backscattering geometry through a

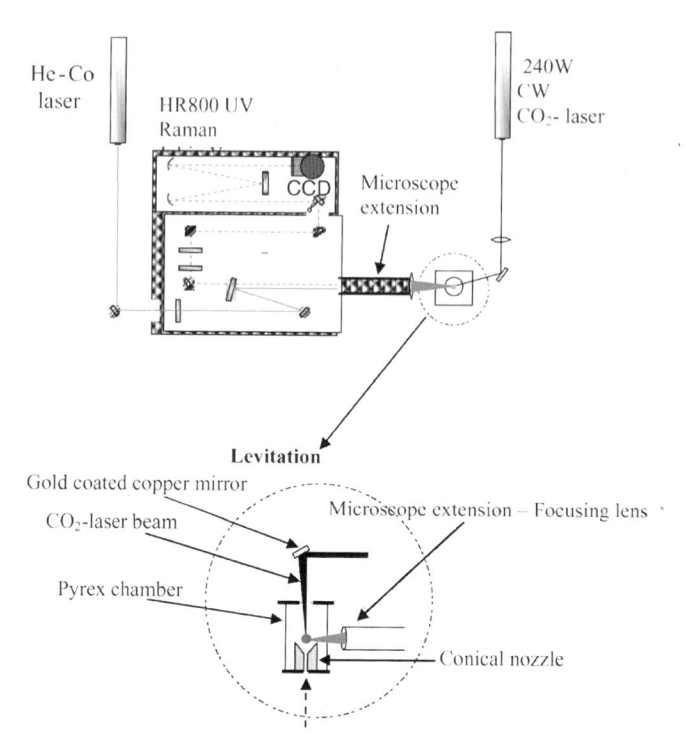

FIGURE 20.24 Schematic diagram of the new high temperature UV Raman setup, which combines the use of "containerless techniques" (levitation or self-support technique) with UV Raman spectroscopy.

microscope objective can be used, permitting measurements up to 2200 °C as shown in the spectra of Figure 20.25. It is noticeable that, below $\sim$350 cm^{-1}, the notch filter strongly attenuates the signal and no spectrum can be obtained. Thus such a system is not useful for low frequency Raman investigations (e.g., studies of low frequency Raman features of amorphous solids and supercooled liquids, as well as low-lying vibrations of crystals). Improved notch filters that provide narrower cutoff of the Rayleigh wing can extend the spectral region to lower frequencies.

20.7.2 Near-IR Raman Spectroscopy

It would be unsafe to define the limits of laser wavelength for near-IR Raman spectroscopy. In the past, limited use of such wavelengths (752.5 and 799.3 nm from Kr$^+$ laser) took place in studies of dark specimens [61]. In recent years, the development of diode lasers and crystal lasers known as diode-pumped solid state (DPSS) lasers at several closely spaced wavelengths in the range 785–1500 nm popularised the use of near-IR Raman spectroscopy [62]. DPSS lasers are characterised by intensity stability, good beam polarisation, and small beam divergence. Germanium detectors or InGaAs detectors are used, especially at wavelengths near

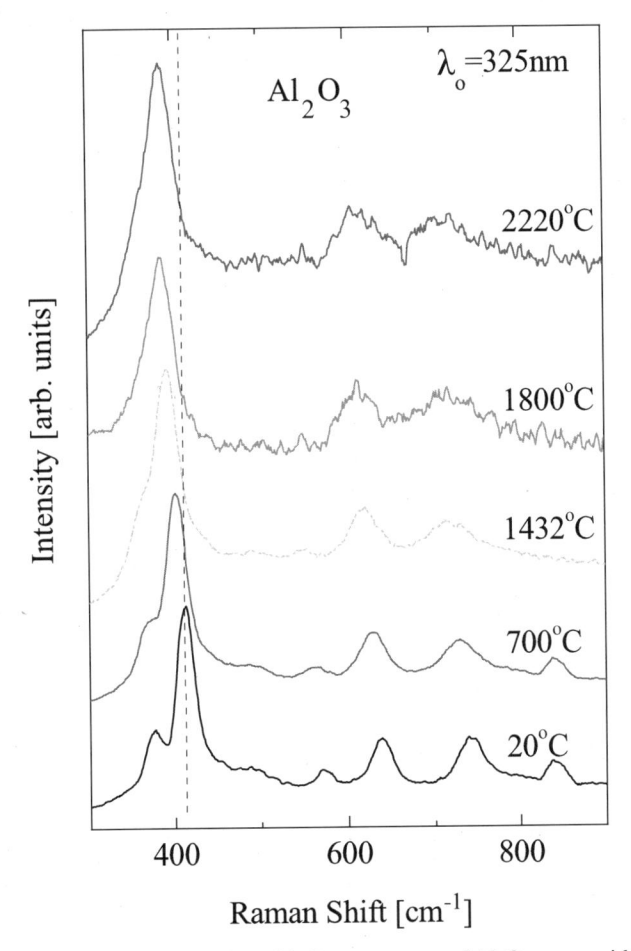

FIGURE 20.25 Representative Stokes-side Raman spectra of Al_2O_3 over a wide temperature range recorded with a UV Raman setup using the levitation technique. Conditions: $\lambda_o = 325$ nm; laser power = 10 mW; resolution = 3 cm^{-1}; CCD detection [unpublished work].

1000 nm. The Ti:sapphire laser is an alternative versatile solution that enables tunable operation over a wide range of wavelengths in the spectral region of 750–950 nm. This laser needs pumping from an intense CW laser; an Ar$^+$ laser operating at the all-line mode or a high power diode laser is usually employed for pumping. The Ti:sapphire as well as other tunable lasers are more susceptible to changes in "cavity length." This results in the undesired mode hopping in which the main resonant mode of the laser jumps from one frequency to another, which could create a problem in obtaining reliable Raman spectra.

Near-IR instruments are commercially available as "FT Raman" setups. In FT Raman spectroscopy [63], a Michelson-type interferometer is used to detect the Fourier transform of the scattered light, which is in turn transformed to provide the vibrational Raman response. The instrument is usually equipped with a Nd:YAG laser

(1064 nm) and operates in both the backscattering and right-angle scattering configurations.

20.7.3 Advantages and Shortcomings of Near-IR Raman Scattering

The use of near-IR Raman scattering has advantages and disadvantages and hence its selection is problem-specific. The major advantages are the following:

1. *Efficient Reduction of the Fluorescence Signal*: The IR laser photon does not have enough energy for the excitation, so that the Raman spectrum can be recorded without fluorescence interference. The reduction or elimination of fluorescence renders possible the detection of very weak Raman signals.
2. *Avoidance of Absorption in Deeply Coloured Samples*: In the case of a medium that absorbs appreciably at the visible wavelengths, the use of near-IR Raman scattering is indispensable in avoiding resonance effects as well as undesired heating that can lead to sample degradation.
3. *Minimisation of Photoinduced Effects in Light-Sensitive Materials*: In some classes of materials, a great variety of photoinduced optical and structural effects takes place upon illumination with bandgap light [64]. The FT Raman technique makes it possible to record reliable Raman spectra from these materials without inducing secondary effects.

Among the shortcomings of the use of the near-IR Raman scattering are the following:

1. *Reduced Scattering Power that Follows the ν^4 Law*: Neglecting resonance and preresonance effects, the scattered intensity associated with a vibrational mode at $500\,cm^{-1}$ will be 25 times stronger when the material is excited with the 488 nm instead of the 1064 nm laser line.
2. *Reduced Accessibility in Low Energy Vibrational Modes*: Due to the use of notch filters, the lowest energy range that can reliably be recorded is at about $100\,cm^{-1}$. This is a major problem in studies of amorphous materials, where the boson peak as well as many bond-bending and lattice modes are located at low energy.
3. *Limitation for High Temperature Measurements*: Due to the fast shift of the blackbody radiation curve toward lower energies with increasing temperature, it is not possible to record Raman spectra at elevated temperatures, especially above 300 °C, with the FT Raman technique.

The last mentioned disadvantage excludes the near-IR Raman technique for studies of melts at high temperatures. On the other hand, the technique is appropriate for studies of "ionic liquids" and their corresponding solids from cryogenic temperature to a few hundred degrees above room temperature. Measurements for such system have recently been reported [65].

20.8 IN SITU RAMAN SPECTRA OF ELECTRODE SURFACES IN MELTS

A knowledge of the species formed on, or in the vicinity of, the electrode surfaces during molten salt electrolysis is important for an understanding of the processes for the production of metals and compounds by molten electrolysis, and gives an insight into the reduction mechanisms occurring in the electrolytic cells. Raman spectroscopy has been used to study in situ, and more extensively ex situ, the species formed at the electrode–solvent interfaces in aqueous room temperature electrolytic cells [66–68], but in situ studies are rather difficult to perform in systems involving melts at high temperatures and only a few successful measurements have been reported [69–71].

For doing these measurements, the laser beam is directed through the molten salt solvent on the surface of the electrode and the scattered light is collected perpendicularly to the surface. Backscattering geometry or a low angle incident excitation laser beam is used, and the scattered light from the electrode surface is superimposed on the strong Raman signal arising from the melt solvent and solutes. Thus rather weak Raman signals are expected to be seen from the species created on or in the vicinity of the electrode. Further difficulties arise from the need of optical furnaces allowing the incident laser beam to enter at different angles, and from the construction of the electrolytic cell offering electrode surfaces accessible to the optical paths needed for Raman scattering. Furthermore, for melts containing high vapour species, the use of vacuum-sealed quartz electrolytic cells is necessary. An electrolytic cell with amorphous graphite electrodes designed for in situ Raman measurements is shown in Figure 20.26a,b. Simple homemade "glass" furnaces made from two coaxial fused silica tubes with the inner tube (20 mm ID, 20 cm long) wound with nichrome wire are appropriate up to 500 °C. For higher temperatures, similar "glass" furnaces with heat reflecting thin gold films are more appropriate [71].

A schematic drawing of the optical configuration is shown in Figure 20.26c. The electrolytic cell is aligned in the furnace so the laser beam hits the working electrode at 30–60°. The incident laser beam is focused by a *cylindrical* lens giving a vertical line ∼2 mm long on the working electrode, which is parallel to the monochromator slit. The scattered light is collected by the appropriate lens(es) matching to the optics of the monochromator.

To measure the Raman spectra in the vicinity of the electrode surface, two different methods have been reported.

20.8.1 Constant Potential Experiments

A potentiostat is used to apply an equilibrium potential on the cell. First, the Raman spectra of the solvent–solute are measured with the laser beam focused on the electrode and no potential on the cell. For example, for the system, KCl/LiCl eutectic (solvent) with 5 mole % $TaCl_5$ (solute), the Raman spectra shown in Figure 20.27c are characteristic of the $[TaCl_6]^-$ species formed in the melt solution, having

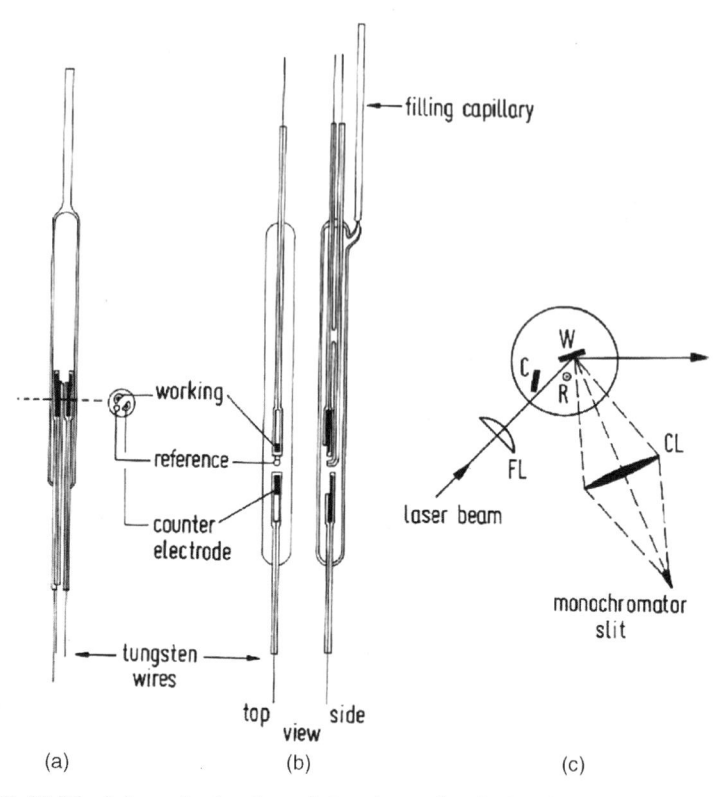

FIGURE 20.26 Schematic drawing of the electrochemical cell with different electrode configurations (a) and (b). A cross section of the optical geometry is shown in (c). (FL, cylindrical focussing lens; CL, collecting lens) [70].

a symmetric stretching frequency $v_1([TaCl_6]^-) \approx 385\,cm^{-1}$. Then, by applying $-200\,mV$ on the electrochemical cell, reduction to Ta(IV) occurs and the Raman spectra, Figure 20.27b, show the presence of another band at $\sim 342\,cm^{-1}$ attributed to $[TaCl_6]^{2-}$. Increasing the equilibrium potential to 500 mV alters the spectra, showing the presence of a band at $\sim 155\,cm^{-1}$ (Fig. 20.27a), which is presumably due to $[TaCl_6]^{3-}$. Simultaneous impedance spectra measurements for the same system in the same cell confirm the stepwise reduction of Ta(V) to Ta(IV) and Ta(III) corresponding to the three octahedral species detected by Raman spectroscopy [70].

Similar constant potential experiments have been conducted in KCl/LiCl–CdCl$_2$ melts, where the reduction process led to the formation of Cd_2^{2+} on the electrode surface (vicinity) [71].

20.8.2 Modulation Experiments

A wave function generator is used to modulate the cell potential. The optical geometry is the same as in Figure 20.26, but the Raman spectra are obtained by connecting the

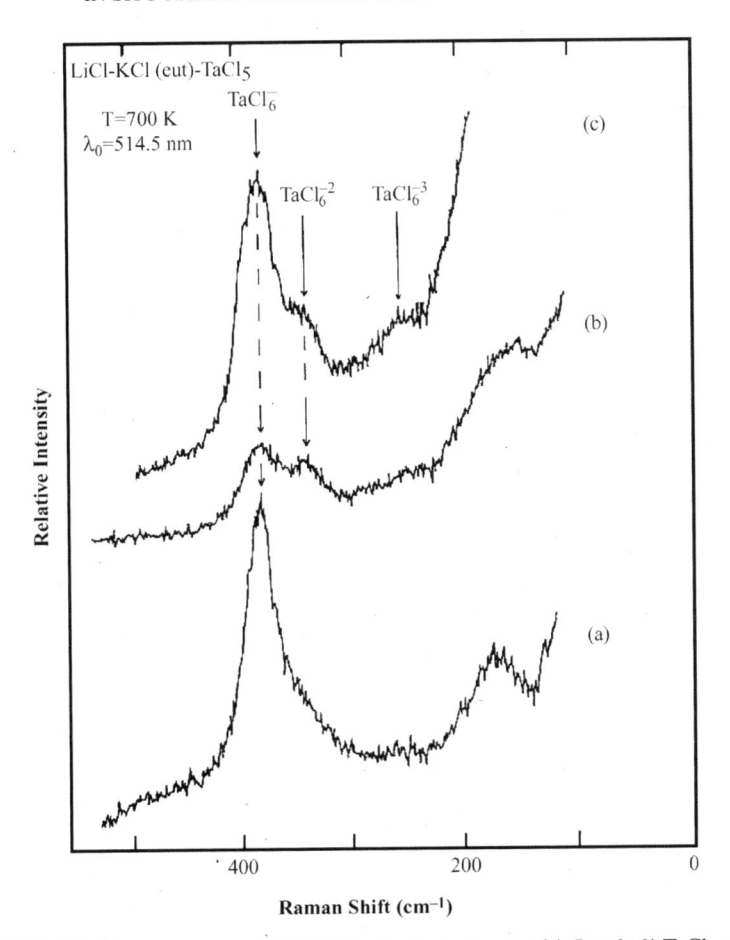

[Figure: Raman spectra plot with three curves (a), (b), (c) showing Raman Shift (cm⁻¹) from 400 to 0 on x-axis and Relative Intensity on y-axis. Labels: LiCl-KCl (eut)-TaCl₅, T=700 K, λ₀=514.5 nm, with peaks marked TaCl₆⁻, TaCl₆²⁻, TaCl₆³⁻.]

FIGURE 20.27 Raman spectra of KCl/LiCl eutectic (solvent) with 5 mole % TaCl₅ recorded in the vicinity of the working electrode and measured simultaneously with impedance spectra. (a) Open circuit potential; the Raman bands are due to the octahedral modes of [TaCl₆]⁻. (b) Cathodic equilibrium potential −500 mV; Raman bands for both [TaCl₆]⁻ and [TaCl₆]²⁻ octahedra are present; (c) With a thin Ta metal layer on the working electrode, an anodic potential of +200 mV is applied; a new species [TaCl₆]³⁻ appears during the anodic dissolution. Spectra conditions: $\lambda_o = 514.5$ nm; laser power = 250 mW; resolution = 8.5 cm⁻¹; PMT detection.

lock-in amplifier with the spectrometer PMT and using as reference the frequency of a square wave from the wave function generator. Thus the Raman signal arises from species formed periodically on the electrode with the frequency of the applied potential. Species not sensitive to the applied potential (i.e., the molten salt solvent and solutes) are not detected.

Figure 20.28 shows modulation Raman spectra obtained from molten KCl/LiCl (solvent)–HgCl₂(solute). The only band present is at ∼150 cm⁻¹, which is due to a subvalent mercury species, Hg₂Cl₂ [69]. The detection of a modulation signal

336 RAMAN SPECTROSCOPY OF HIGH TEMPERATURE MELTS

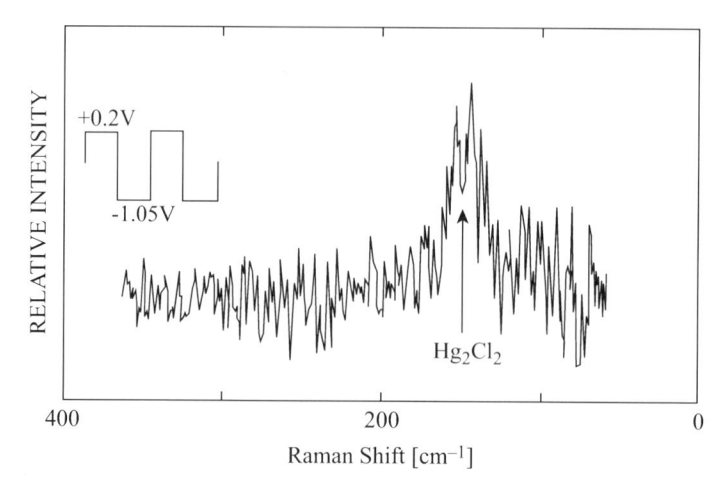

FIGURE 20.28 Modulation Raman spectrum from amorphous carbon electrode in LiCl–KCl eutectic HgCl$_2$ solution at 370 °C. Modulation frequency, 1 Hz; square-wave amplitude, 0.2–1.05 V; lock-in amplifier, $\lambda_o = 514.5$ nm; laser power = 250 mW; SSW = 8.5 cm^{-1} [69].

depends on the "lifetime" of the species formed. For the mercury species, no spectra could be obtained for frequencies higher than 2 Hz, indicating rather slow kinetics for the formation of the Raman active species.

20.9 CONCLUSION

Over the past forty years, developments in conventional macro-Raman spectroscopy have been used successfully for investigating the structural properties of many inorganic melts. Different designs and materials have been introduced for constructing optical furnaces and Raman cells for measuring Raman spectra up to ~1000 °C, including corrosive melts as well as melts with high vapour pressures. The quality of the spectra at high temperature is affected by the purity of the samples; thus special methods have been employed for preparing/purifying the materials used.

Micro-Raman spectroscopy has become a useful and convenient tool for the study of melts, and special microscope "hot" stages have been designed. Containerless techniques with CO$_2$ laser heating have also been introduced, and Raman measurements of melts reaching temperatures above 2200 K have been obtained.

Continued advances in UV lasers, detector systems, and optoelectronics promise a bright future for UV Raman spectroscopic techniques to be used for measurements at "extreme" temperatures of melts, including molten refractive oxides. Similar advances in near-IR lasers and FT Raman spectroscopy provide the possibility for studying, at intermediate temperatures, deeply coloured inorganic melts and room-temperature ionic liquids. Finally, methods for in situ Raman spectroscopy in electrolytic cells, with molten salt solvents, provide important information related to the species formed in the vicinity of the electrode during electrolysis.

REFERENCES

1. Salstrom, E. J. and Harris L., Raman spectra of fused salts, *J. Chem. Phys.* **3**, 241–242 (1935).
2. Irish, D. E. and Young, T. F., Raman spectrum of molten zinc chlorides, *J. Chem. Phys.* **43**, 1765–1768 (1965).
3. Bues, W., Raman-spektren der system $ZnCl_2$-KCl und $CdCl_2$-KCl in der schmelze, *Z. Anorg. Allg. Chem.* **279**, 104–114 (1955).
4. Janz, G. J. and James, D. W., Raman spectra and ionic interactions in molten nitrates, *J. Chem. Phys.* **35**, 739–745 (1961).
5. Janz, G. J. and Wait, S. C., Ionic melts, in *Raman Spectroscopy: Theory and Practice*, H. A. Szymanski (Ed.), Plenum Press, New York, 1967, pp. 139–167.
6. Hester, R. E., Vibrational spectroscopy of molten salts, in *Advances in Molten Salt Chemistry*, J. Braunstein, G. Mamantov, and G. P. Smith (Eds.), Vol. 1, Plenum Press, New York, 1971, pp. 1–57.
7. Mamantov, G. (Ed.), *Molten Salts: Characterisation and Analysis*, Dekker, New York, 1969.
8. Devlin, J. P., The vibrational spectra of ionic vapours, liquids and glasses, in *Advances in Infrared and Raman Spectroscopy*, R. J. H. Clark and R. E. Hester (Eds.), Vol. 2, Heyden, New York, 1976, pp. 153–211.
9. Koningstein, J. A., *Introduction to the Theory of the Raman Effect*, Reidel, New York, 1972.
10. Szymanski, H. A. (Ed.), *Raman Spectroscopy: Theory and Practice*, Plenum Press, New York, 1965.
11. Loader, E. J., *Basic Laser Raman Spectroscopy*, Heyden, London, 1970.
12. Gilson, T. R. and Hendra, P. J., *Laser Raman Spectroscopy*, Wiley Hoboken, NJ, 1970.
13. Tobin, M. C., *Laser Raman Spectroscopy*, Wiley Hoboken, NJ, 1971.
14. Brooker, M. and Papatheodorou, G. N., Vibrational spectroscopy of molten salts and related glasses and vapours, in *Advances in Molten Salt Chemstry*, G. Mamantov and C. B. Mamantov (Eds.), Vol. 5, Elsevier, New York, 1983, pp. 26–183.
15. Long, D. A., *The Raman Effect: A Unified Treatment of the Theory of Raman Scattering by Molecules*, Wiley Hoboken, NJ, 2001.
16. Laserna, J. J. (Ed.), *Modern Techniques in Raman Spectroscopy*, Wiley, Hoboken, NJ, 1996.
17. Chalmers, J. and Griffiths, P. (Eds.), *Handbook of Vibrational Spectroscopy*, Vols. 1–5, Wiley, Hoboken, NJ, 2002.
18. Fredericks, P. M., Frost, R. L., and Rintoul, L. (Eds.) *Proceedings of the XIX International Conference on Raman Spectroscopy: ICORS 2004*, CSIRO Publishing, Collingwood, CO, 2005.
19. Smith, E. and Dent, G., *Modern Raman Spectroscopy: A Practical Approach*, Wiley Hoboken, NJ, 2004.
20. Lewis, I. R. and Edwards, H. G. M. (Eds.), *Handbook of Raman Spectroscopy: From the Research Laboratory to the Process Line*, Dekker, New York, 2001.
21. Pelletier, M. J. (Ed.), *Analytical Applications of Raman Spectroscopy*, Plackwell, New York, 1999.

22. McCreery, R. L., *Raman Spectroscopy for Chemical Analysis*, Wiley Hoboken, NJ, 2000.
23. Edwards, H. G. M. and Chalmers, J. M. (Eds.), *Raman Spectroscopy in Archaeology and Art History*, Royal Society of Chemistry, London, 2005.
24. McMillan P. F., Poe, B. T., Gillet, Ph., and Reynard, B., A study of SiO_2 glass and supercooled liquid to 1950 K via high-temperature Raman spectroscopy, *Geochim. Cosmochim. Acta* **58**, 3653–3664 (1994).
25. Voyiatzis, G. A., Kalampounias, A. G., and Papatheodorou, G. N., The structure of molten mixtures of iron (III) chloride with caesium chloride, *Phys. Chem. Phys.* **1**, 4797–4803 (1999).
26. Børrensen, B., Voyiatzis, G. A., and Papatheodorou, G. N., The Cd_2^{2+} in molten halides at electrode interfaces, *Phys. Chem. Chem. Phys.* **1**, 3309–3314 (1999).
27. Kalampounias, A. G., Yannopoulos, S. N., and Papatheodorou, G. N., Temperature-induced structural changes in glassy, supercooled, and molten silica from 77 to 2150 K, *J. Chem. Phys.* **124**, 014504 (2006).
28. Strommen, D. P. and Nakamoto, K., *Laboratory Raman Spectroscopy*, Wiley, Hoboken, NJ, 1984.
29. Wartewig, S., *IR and Raman Spectroscopy: Fundamental Processing*, Wiley, Hoboken, NJ, 2003.
30. Papatheodorou, G. N. and Yannopoulos, S. N., Light scattering from molten salts: structure and dynamics, in *Molten Salts: From Fundamentals to Applications*, M. Gaune-Escard (Ed.), NATO Science Series, Vol. 52, Kluwer Academic Press, Norwell, MA, 2002, pp. 46–106.
31. Bern, B. J. and Pecora, R., *Dynamic Light Scattering*, Wiley, Hoboken, NJ, 1976.
32. Kalampounias, A. G., Papatheodorou, G. N., and Yannopoulos, S. N., unpublished work.
33. Daniel, I., Gillet, Ph., Poe, B. T., and McMillan, P. F., *In situ* high temperature Raman-spectroscopic studies of aluminosilicate liquids, *Phys Chem. Minerals* **22**, 74 (1995).
34. Rothschild, W. G., *Dynamics of Molecular Liquids*, Wiley, Hoboken, NJ, 1984.
35. Kirillov, S. A., Pavlatou, E. A., and Papatheodorou, G. N., Instantaneous collision complexes in molten alkali halides: picosecond dynamics from low-frequency Raman data, *J. Chem. Phys.* **116**, 9341–9351 (2002).
36. Kirillov, S. A. and Yannopoulos, S. N., Vibrational dynamics as an indicator of short-time interactions in glass-forming liquids and their possible relation to cooperativity, *J. Chem. Phys.* **117**, 1220–1230 (2002).
37. Ashcroft, N. W. and Mermin, N. D., *Solid State Physics*, Saunders College Publishing, New York, 1976.
38. Dai, S., Young, J. P., Begun, G. M., Coffield, G. M., and Mamantov, G., Measurement of molten salt Raman spectra by the use of fiber optics, *Microchim. Acta* **108**, 261–264 (1992).
39. Maroni, V. A., Hathaway, E. J., and Cairns, E. J., Structural studies of magnesium halide-potassium halide by Raman spectroscopy, *J. Phys. Chem.* **71**, 155–162 (1971), and references therein.
40. Iida, Y., Furukawa, M., and Morikawa, H., Comparison of detection schemes for high-temperature Raman spectroscopy, *Appl. Spectrosc.* **51**, 1426–1430 (1997).
41. Exarhos, G. J. and Schaaf, J. W., Raman scattering from boron nitride coatings at high temperatures, *J. Appl. Phys.* **69**, 2543–2548 (1991).

42. Fujimori, H., Kakihana, M., Ioku, K., Goto, S., and Yoshimura, M., Advantage of anti-Stokes Raman scattering for high-temperature measurements, *Appl. Phys. Lett.* **79**, 937–939 (2001).
43. Yannopoulos, S. N., Kalampounias, A. G., Chrissanthopoulos, A., and Papatheodorou, G. N., Temperature induced changes on the structure and the dynamics of the "tetrahedral" glasses and melts $ZnCl_2$ and $ZnBr_2$, *J. Chem. Phys.* **118**, 3197 (2003).
44. Papatheodorou, G. N. and Kucera, G. H., Vapour complexes of samarium(III) and samarium(III) chlorides with aluminium(III) chloride, *Inorg. Chem.* **18**, 385 (1979).
45. Weber, J. K. R., Felten, J. J., and Nordine, P. C., Laser hearth melt processing of ceramic materials, *Rev. Sci. Instrum.* **67**, 522–524 (1996).
46. Kalampounias, A. G., *Development of New High Temperature Techniques Using Infrared Laser (CO_2) for Studying the Network Structure of Inorganic Materials by Means of Raman Spectroscopy*, Ph.D. thesis, University of Patras, 2003.
47. Dracopoulos, V. D., *Raman Spectroscopic Study of Molten Fluoride Mixtures: LnF_3–KF, Ln: La, Nd, Sm, Dy, Yb, Y, ZnF_2–AF, A: K, Cs*, Ph.D. thesis, University of Patras, 2000.
48. Photiadis, G. M. and Papatheodorou, G. N., Vibrational modes and structure of liquid and gaseous zirconium tetrachloride and of molten $ZrCl_4$–CsCl mixtures, *J. Chem. Soc. Dalton Trans.*, 981–989 (1998).
49. Gilbert, B., Mamantov, G., and Begun, G. M., Raman spectra of aluminium fluoride containing melts and ionic equilibrium in molten cryolite type mixtures, *J. Chem. Phys.* **62**, 950–955 (1975).
50. Dracopoulos, V. and Papatheodorou, G. N., Isotropic and anisotropic Raman scattering from molten alkali-metal fluorides, *Phys. Chem. Chem. Phys.* **2**, 2021–2025 (2000).
51. Dracopoulos, V., Vagelatos, J., and Papatheodorou, G. N., Raman spectroscopic studies of molten ZrF_4–KF mixtures and of A_2ZrF_6, A_3ZrF_7 (A= Li, K. or Cs) compounds, *J. Chem. Soc. Dalton Trans.*, 1117–1122 (2001).
52. Mysen, B. O. and Frantz, J. D., Raman spectroscopy of silicate melts at magmatic temperatures: Na_2O–SiO_2, K_2O–SiO_2 and Li_2O–SiO_2 binary compositions in the temperature range 25–1475°C, *Chem. Geol.* **96**, 321–332 (1992).
53. McMillan, P. F., Poe, B. T., Gillet, Ph., and Reynard, B., A study of SiO_2 glass and supercooled liquid to 1950 K via high-temperature Raman spectroscopy, *Geochim. Cosmochim. Acta* **58**, 3653–3654 (1994).
54. Auguste, F., Bessada, C., and Gilbert, B., The structure of fluoroaluminates from room temperature to 1050°C. A Raman and NMR spectroscopy study, in *7th International Symposium on Molten Salts Chemistry & Technology*, P. Taxil (Ed.), Toulouse, August 29–September 2, 2005, pp. 266–272.
55. Kriven, W. M., Jivali, M. H., Zhu, D., Weber, J. K. R., Cho, B., Felten, J. J., and Nordinem, P. C., Synthesis and microstructure of mullite fibers grown from deeply undercooled melts, in *Ceramic Microstructures*, A. P. Tomsia and A. Glaesser (Eds.), Plenum Press, New York, 1996.
56. Weber, J. K. R, Krishnan, S., and Nirdine, P. C., The use of containerless processing in researching reactive materials, *J. Minerals Metals Mater. Soc.* **43** (7), 8–14 (1991).
57. Weber, J. K. R., Rix, J. E., Hiera, K. J., Tangeman, J. A, Benmore, C. J., Hart, R. T., Siewenie, J. E., and Santodonato, L. J., Neutron diffraction from levitated liquids—a technique for measurements under extreme conditions, *Phys. Chem. of Glasses* **46**, 487–491 (2005).

58. Nordine, P. C. and Atkinsm R. M., Aerodynamic levitation of laser-heated solids in gas-jets, *Rev. Sci. Instrum.* **53**, 1456–1464 (1982).
59. Kalampounias, A. G. and Papatheodorou, G. P., Raman spectroscopic measurements of molten ceramic materials at high temperatures, in *Proceedings of the 13th International Symposium of Molten Salts*, P. C. Trulove, H. C. DeLong, R. A., Mantz, G. R. Stafford, and M. Matsunaga (Eds.), The Electrochemical Society, Pennington, USA, 2002, pp. 485–494.
60. Xiong, G., Yu, Y., Geng, Z.-C., Xin, Q., Xiao, F.-S., and Li, C., UV Raman spectroscopic study on the synthesis mechanism of zeolite X, *Microporous Mesoporous Mater.* **42**, 317–323 (2001).
61. Gorman, M. and Solin, S. A., Transmission Raman and depolarisation spectra of bulk α-Se from 13 to 300 cm^{-1}, *Solid State Commun.* **18**, 1401–1404 (1976).
62. Yannopoulos, S. N., Andrikopoulos, K. S., Voyiatzis, G. A., Kolobov, A. V., and Tominaga, J., Raman scattering study of the a-GeTe structure and possible mechanism for the amorphous-to-crystal transition, *J. Phys. Condens. Matter* **18**, 965–979 (2006).
63. Hendra, P. J., Fourier transform Raman spectroscopy, in *Modern Techniques in Raman Spectroscopy*, J. J. Laserna (Ed.), Wiley, London, 1996, pp. 73–108.
64. Kastrissios, D. T., Papatheodorou, G. N., and Yannopoulos, S. N., Study of vibrational modes in the athermal photoinduced fluidity regime of glassy As_2S_3, *Phys. Rev. B* **64**, 2142031–2142039 (2001).
65. Fox, D. M., Fylstra, P. A., Henderson, W. A., Gilman, J. W., Trulove, P. C., and DeLong, H. C., The preparation and characterisation of bombyx mori silk nanocomposites using ionic liquids, in *Proceedings of the 15th International Symposium of Molten Salts XV*, R. A. Mantz (Ed.), Electrochemical Society, 2007.
66. Fleischmann, M., Graves, P., Hill, I., Oliver, A., and Robinson, J., Simultaneous Raman spectroscopic and differential double-layer capacitance measurements of pyridine absorbed on roughened silver electrodes, *Electrochim. Acta* **150**, 33–42 (1983) and references therein.
67. Inaba, M., Yoshida, H., Ogumi, Z., Abe, T., Mizatani, Y., and Asano, M., In situ Raman study on electrochemical Li—intercalation into graphite, *J. Electrochem. Soc.* **142**, 20–26 (1995).
68. Kovac, M., Milicev, S., Kovac, A., and Pejovnik, S., In situ Raman and electrochemical characterisation of the role of electrolyte additives in Li/SOCl$_2$ batteries, *J. Electrochem. Soc.* **142**, 1390–1395 (1995) and references therein.
69. Papatheodorou, G. N., Boviatsis, I. V., and Voyiatzis, G. A., In situ Raman spectra of electrode products during electrolysis of HgCl$_2$ in molten LiCl–KCl eutectic, *J. Appl. Electrochem.* **22**, 517–521 (1992).
70. Bachtler, M., Freyland, W., Voyiatzis, G. A., and Papatheodorou, G. N., Electrochemical and simultaneous spectroscopic study of reduction mechanism and electron conduction during electrodeposition of tantalum in molten alkali chlorides, *Ber. Bunsenges. Phys. Chem.* **99**, 21–31 (1995).
71. Børrensen, B., Voyiatzis, G. A., and Papatheodorou, G. N., The Cd_2^{2+} in molten halides at electrode interfaces, *Phys. Chem. Chem. Phys.* **1**, 3309–3314 (1999).

21 Thermodynamic Properties of LnI$_3$–MI Binary Systems (Ln = La or Nd; M = K, Rb, or Cs)

LESZEK RYCERZ

Chemical Metallurgy Group, Faculty of Chemistry, Wroclaw University of Technology, Wroclaw, Poland and Ecole Polytechnique, Mecanique Energetique, Technopôle de Château-Gombert, Marseille, France

M. F. BUTMAN

Ivanovo State University of Chemical Technology, Ivanovo, Russia and Ecole Polytechnique, Mecanique Energetique, Technopôle de Château-Gombert, Marseille, France

MARCELLE GAUNE-ESCARD

Ecole Polytechnique, Mecanique Energetique, Technopôle de Château-Gombert, Marseille, France

Abstract

Temperatures and molar enthalpies of the phase transitions of the M$_3$LnI$_6$ congruently melting compounds (Ln = La or Nd; M = K, Rb, or Cs) were determined and compared with data obtained for analogous chloride and bromide compounds. This comparison showed that M$_3$LnI$_6$ compounds could be divided into two groups: compounds that are formed at higher temperatures from M$_2$LnI$_5$ and RbI, and compounds that are stable or metastable at ambient temperature. Moreover, compounds formed at higher temperatures can exist at ambient temperature as metastable phases when rapidly cooled. On subsequent heating, thermograms exhibit an exothermic effect related to the abrupt decomposition of "supercooled" phases. The heat capacities of M$_3$LnI$_6$ compounds have been obtained for the first time. They were fitted by equations, which provide a satisfactory representation up to temperature of

---

Molten Salts and Ionic Liquids: Never the Twain? Edited by Marcelle Gaune-Escard and Kenneth R. Seddon
Copyright © 2010 John Wiley & Sons, Inc.

the C_p discontinuity. Heat capacity measurements also confirmed the existence of "supercooled" metastable phases of these compounds.

21.1 INTRODUCTION

The rare earth bromides and iodides are attractive components for doses in high-intensity discharge lamps [1–3] and new highly efficient light sources with energy saving features [4]. They offer the opportunity, when combined with other metal halides, to design light sources with high efficiency and good colour rendition. The properties of many of the rare earth halides, however, are poorly characterised. Comprehensive thermochemical parameters for the lanthanide triiodides are not available in any of the standard compilations of thermodynamic properties. These properties are required for the calculation of the chemical composition in the lamp, to assist in the understanding and hence prediction of tungsten corrosion, silica corrosion, and spectral output. The data are also needed for predicting the behaviour of doses by modelling of multicomponent metal halide systems [5].

Thermodynamic properties (enthalpy of fusion, heat capacity) of pure LaI_3 and NdI_3 were measured with a copper block drop calorimeter by Dworkin and Bredig [6]. Phase diagrams of LnI_3–MI (Ln = La or Nd) were investigated by means of differential thermal analysis (DTA) (M = Na, K, or Cs) [7] and differential scanning calorimetry (DSC) (M = Li or Rb) [8, 9]. These systems with LiI and NaI are of a simple eutectic type, while those including KI, RbI, and CsI exhibit several compounds of different stoicheiometry. The existence of congruently melting M_3LnI_6 compounds is the common feature of these systems. The only exception is the LaI_3–KI system, characterised by a congruently melting K_2LaI_5 compound (K_3LaI_6 compound does not exist in this system).

As far as the thermodynamics is involved, we have paid much attention to the M_3LnI_6 stoicheiometric compounds, which have a more extended stability range than those of different stoicheiometry. No information is available in the literature in this aspect, and present work reports thermodynamic investigations of some M_3LnI_6 compounds (Ln = La, Nd; M = K, Rb, Cs). These investigations are part of a large program focused on thermodynamic properties, structure, and electrical conductivity of lanthanide halides and lanthanide halide–alkali metal halide systems.

21.2 EXPERIMENTAL

21.2.1 Sample Preparation

Lanthanide triiodides (LaI_3 and NdI_3) were prepared by direct reaction of metal with iodine vapour by the Drudding and Corbett [10] modified method. Lanthanide metal (min. 99.9%) was obtained by metallothermic reduction of $LnCl_3$ (99.99%, Department of Chemistry, Wroclaw University of Technology) with calcium (99.999%) [11]. Prior to use, this metal was cut into smaller pieces to facilitate loading into the reaction vessel and to increase the reaction surface with iodine vapour.

EXPERIMENTAL

Iodine (POCH Gliwice, Poland, reagent grade) was sublimed in a Pyrex tube at about 350 K prior to use. A quartz ampoule with a side arm was used as a reaction vessel in the synthesis of lanthanide(III) iodide. Lanthanide metal and iodine (10 mol % excess with respect to stoicheiometry) were loaded into a tantalum crucible, which was placed inside the quartz ampoule. The reaction ampoule was then evacuated, with the evacuation time kept short in order to minimise the loss of iodine by sublimation. The final evacuation to a pressure of 0.1 Pa was carried out with simultaneous cooling of the part of the ampoule containing the reagents with liquid nitrogen. When the pressure inside the ampoule had reached a value of ca. 0.1 Pa, the ampoule was flame-sealed above the connection with the side arm. The section of ampoule containing the metal and iodine was then placed in a vertically oriented furnace and heated up to 350 K. At the same time, the side arm located outside the furnace was cooled with liquid nitrogen and all the iodine was sublimed to this side arm. The temperature of the ampoule section with lanthanide metal was gradually increased up to 1070 K, whereas the temperature of the side arm with iodine was increased up to 450 K. These conditions were maintained for 20 hours to ensure complete reaction. This procedure, which proved satisfactory, was developed empirically, and no attempt was made to reduce the duration of the reaction time, although this may be possible. Finally, the tantalum crucible containing crude LnI_3 was maintained at 500 K for 10 hours while the side arm was cooled with liquid nitrogen to condense unreacted iodine. The ampoule was opened in a glove box, and crude LnI_3 was subsequently purified by sublimation. The sublimation process was carried out in a quartz ampoule containing the tantalum crucible. LnI_3 was loaded into this crucible, and the ampoule was evacuated to below 0.1 Pa and then sealed off. A vacuum sublimation was performed by placing the ampoule in a vertical furnace with about 15 cm of the ampoule exposed above the furnace. The temperature of the hot part of the ampoule was maintained about 40 K lower than the melting temperature of LnI_3 for 60–70 h. Pure LnI_3 was obtained in the cold part of the ampoule. The conditions of sublimation were developed empirically, and no attempt was made for their optimisation. Because of their hygroscopicity, all reaction products were handled in a glove box and stored in sealed glass ampoules.

Alkali metal iodides (KI, RbI, and CsI) were Merck Suprapur reagents (min. 99.9%). Prior to use, they were progressively heated up to fusion under a gaseous HI atmosphere. The excess of HI was then removed from the melt by argon bubbling.

The mixtures of LnI_3 and MI (in proportions corresponding to the M_3LnI_6 stoicheiometry) were melted in vacuum-sealed quartz ampoules in an electric furnace. Melts were homogenised by shaking and solidified. These samples were ground with a pestle and an agate mortar in a glove box. Although only a small amount of sample (300–500 mg) was used for differential scanning calorimetry (DSC) experiments, about 15 g of each compound was synthesised in order to avoid deviation from stoicheiometry

All chemicals were handled in an argon glove box with a measured volume fraction of water of about 2 ppm and continuous gas purification by forced recirculation through external molecular sieves.

21.2.2 Measurements

The temperatures and enthalpies of phase transitions of M_3LnI_6 compounds were measured with a Setaram DSC 121 differential scanning calorimeter. The apparatus and the measurement procedure were described in detail previously [12, 13]. Samples of 300–500 mg were contained in quartz ampoules (about 6 mm diameter, 15 mm length) sealed under a reduced pressure of argon. The sidewalls of the ampoules were ground in order to fit the cells snugly into the heat flow detector. Experiments were conducted at heating and cooling rates ranging between 5 and $0.1 \, K \cdot min^{-1}$.

Heat capacity was measured with the same Setaram DSC 121 operated in a stepwise mode. In this so-called step method, which has already been described [13–16], the sample temperature is alternated in a series of heating and isothermal steps, the length of the isothermal steps being adjusted to allow sufficient time for reaching steady state conditions. In the experimental heat flow plot against time, the area under each individual heating peak represents the heat required to increase the temperature at each step. The heat capacity of the sample was determined over an extended temperature range from two experimental series run in an identical stepwise mode: the first one, with two empty cells (containers) of identical mass, and the second one, with one of these cells loaded with the sample. For each heating step, the difference of heat flux between the two series is proportional to the amount of heat (Q_i) necessary to increase the sample temperature by a small temperature increment ΔT_i. Therefore, in the absence of any phase transition, the heat capacity of the sample is equal to

$$C_{p,m} = (Q_i \cdot M_s)/(\Delta T_i \cdot m_s) \tag{21.1}$$

where m_s is the mass of the sample, M_s is the molar mass of the sample.

The same operating conditions (i.e., initial and final temperatures, temperature increment, isothermal delay, and heating rate) were used in the two experimental series. Experimental monitoring and data acquisition and processing were performed with the Setaram Setsys software.

The DSC 121 apparatus was calibrated by the Joule effect and some test measurements were performed separately with a NIST 720 α-Al_2O_3 Standard Reference material prior to investigation of the terbium bromide mixtures. These tests resulted in $C_{p,m}$ values consistent with standard data for Al_2O_3 (difference in the whole temperature range less than 1.5%), which validated the step method used for heat capacity measurements. In the present heat capacity experiments, each 5 K heating step was followed by a 400 s isothermal delay. The heating rate was $1.5 \, K \cdot min^{-1}$. All experiments were performed in the 300–1100 K temperature range. The mass difference of the quartz cells in any individual experiment did not exceed 1 mg (cell mass: 400–500 mg).

21.3 RESULTS AND DISCUSSION

21.3.1 Enthalpy of Phase Transitions

The difference in enthalpies determined from heating and cooling runs did not exceed 2%. However, because supercooling effects were observed, only temperatures and

TABLE 21.1 Molar Enthalpies of Phase Transitions of M_3LnI_6 Compounds

Compound	T_{form}/ K	T_{trs}/ K	T_{fus}/ K	$\Delta_{form}H_m$/ kJ·mol^{-1}	$\Delta_{trs}H_m$/ kJ·mol^{-1}	$\Delta_{fus}H_m$/ kJ·mol^{-1}	Reference
K_3NdI_6	628	696	805	27.5	9.7	31.2	This work
	—	629, 683	804	—	—	—	7
Rb_3LaI_6	686	706	856	31.6	8.4	32.0	This work
Rb_3NdI_6	—	718	899	—	8,8	39.2	This work
Cs_3LaI_6	—	763	941	—	10.4	45.1	This work
	—	744, 751	941	—	—	—	7
Cs_3NdI_6	—	755	962	—	10.1	47.7	This work
	—	744, 756	966	—	—	—	7

enthalpies determined from heating cycles were taken into account in this work. They are presented in Table 21.1.

Since no structural information was available in the literature regarding M_3LnI_6 compounds, the nature of the solid–solid phase transition was deduced on the basis of comparative studies for analogous chloride and bromide compounds.

M_3LnX_6 congruently melting compounds (M = K, Rb, or Cs; Ln = lanthanide; X = Cl, Br, or I) exist in almost all lanthanide halide–alkali metal halide systems [9, 17–27]. Systematic thermal, structural, and electrochemical measurements performed on chloride and $LaBr_3$–MBr systems [9, 17–27] indicated that all the M_3LnX_6 compounds can be divided into two groups: (1) compounds that are formed at high temperatures from M_2LnX_5 and MX, and (2) compounds that exist at low temperatures as stable or metastable phases. Those compounds that are formed at high temperatures have only one (cubic, elpasolite type) or two (low temperature–monoclinic, Cs_3BiCl_6 type; high temperature–cubic, elpasolite type) crystal structures. Their formation from M_2LnX_5 and MX is a reconstructive phase transition [28]. The other group of compounds existing at low temperatures have both low temperature (monoclinic, Cs_3BiCl_6 type) and high temperature (cubic, elpasolite type) crystal structures. The transition from low to high temperature modification is a non reconstructive phase transition [28]. We have ourselves measured formation (or transition) enthalpies of a number of these compounds [9, 12, 29–31]. These results provide evidence that, in the M_3LnX_6 compounds, the large enthalpy effects (44–55 kJ·mol^{-1}) are related to their formation from M_2LnX_5 and MX, while the modest enthalpy changes (6–9 kJ·mol^{-1}) correspond to solid–solid phase transitions (nonreconstructive phase transitions).

Taking into account the large molar enthalpy values (Table 21.1) corresponding to the thermal effect in K_3NdI_6 at 628 K, and in Rb_3LaI_6 at 686 K (27.5 and 31.6 kJ·mol^{-1}, respectively), it is very likely that it may result from the compound formation from M_2LnI_5 and MI. The second thermal effect at 696 K (K_3NdI_6) and 706 K (Rb_3LaI_6) can be ascribed to the solid–solid phase transition in these compounds. The smaller values of enthalpy related to this effect (9.7 and 8.4 kJ·mol^{-1}, respectively) suggest that it is a nonreconstructive phase transition [28]. Intriguing and interesting phenomena that confirm this conclusion about compound formation at

TABLE 21.2 Thermal Effects Related to the Formation and Decomposition of K₃NdI₆ as a Function of Heating and Cooling Rate

Heating Rate/ K·min⁻¹	T_{exo}/ K	$\Delta_{exo}H_m$/ kJ·mol⁻¹	T_{form}/ K	$\Delta_{form}H_m$/ kJ·mol⁻¹	Cooling Rate/ K·min⁻¹	T_{dec}/ K	$\Delta_{dec}H_m$/ kJ·mol⁻¹
15	504	24.3	630	27.0	15	—	—
10	502	24.5	628	27.5	10	—	—
5	502	24.5	628	27.5	5	—	—
1	502	24.6	628	27.3	1	—	—
0.1	502	24.9	628	27.7	0.1	584	27.6
5[a]	—	—	628	27.6			

[a]Measurement performed directly after cooling with a cooling rate of 0.1 K·min⁻¹.

higher temperatures were observed during heating and cooling runs of K₃NdI₆ and Rb₃LaI₆ compounds with different heating and cooling rates. The results of these experiments are presented in Tables 21.2 and 21.3 (measurements were performed one after another in the order presented in the tables).

The heating rate has no influence on the temperature and enthalpy of the thermal effect related to the K₃NdI₆ formation (Table 21.2). However, faster cooling rates (1–15 K·min⁻¹) induced a lack of compound decomposition on the cooling curves. On subsequent heating, a well-shaped exothermic effect was observed at 502 K. The enthalpy change related to this effect (about 24.5 kJ·mol⁻¹) is comparable to the enthalpy of formation (27.5 kJ·mol⁻¹). When the cooling rate was decreased to 0.1 K·min⁻¹, compound decomposition could be observed at 584 K with an enthalpy change of 27.6 kJ·mol⁻¹, and on subsequent heating (heating rate 5 K·min⁻¹), an exo-effect was not observed.

Very often, solid state reactions are a special kind of reconstructive phase transition. These are transitions in which the arrangement of ions is drastically changed. Ions have to move from one site to another, passing strong potential walls of other ions. The resulting "kinetic hindrance" can cause great differences between reaction temperatures measured in DSC heating and cooling curves (thermal hysteresis) [32]. In extreme cases, in cooling experiments, the "supercooling" can become so strong that the reaction does not occur on the time scale of DSC. The reactions themselves need time—they can be fast or retarded. Hence it is of fundamental

TABLE 21.3 Thermal Effects Related to the Formation and Decomposition of Rb₃LaI₆ as a Function of Heating and Cooling Rate

Heating Rate/ K·min⁻¹	T_{exo}/ K	$\Delta_{exo}H_m$/ kJ·mol⁻¹	T_{form}/ K	$\Delta_{form}H_m$/ kJ·mol⁻¹	Cooling Rate/ K·min⁻¹	T_{dec}/ K	$\Delta_{dec}H_m$/ kJ·mol⁻¹
5	408	20.5	686	31.6	10	623	10.9
5	410	20.9	686	31.7	5	626	11.3
5	410	20.0	687	31.5	2	642	32.3
5[a]	—	—	686	31.4			

[a]Measurement performed directly after cooling with a cooling rate of 2 K·min⁻¹.

importance whether the reaction time is smaller or greater than the time of measurement. If, in a cooling experiment, the reaction time is too large, it may occur that the sample is cooled to a temperature at which the reaction is frozen. If it does not occur at all, one can maintain the high temperature system in a metastable state [33].

Such a situation takes place in the case of K_3NdI_6. During fast cooling, the decomposition did not occur for kinetic reasons, and this compound exists as a metastable phase at ambient temperature. On subsequent heating, "supercooled" decomposition occurred abruptly and an exothermic effect is observed on the DSC curve. When the cooling rate is slow enough ($0.1\,\text{K}\cdot\text{min}^{-1}$), the time of measurement is greater than the reaction time and compound decomposition can be completed.

In the case of Rb_3LaI_6, the heating rate has no influence on temperature and enthalpy of the thermal effect at 686 K, related to its formation (Table 21.3). However, when the cooling rate is greater than $2\,\text{K}\cdot\text{min}^{-1}$, the decomposition enthalpy is significantly smaller than the formation enthalpy (ca. 11 and $31.6\,\text{kJ}\cdot\text{mol}^{-1}$, respectively). On subsequent heating, the DSC curve exhibits an exo-effect at about 410 K, with a related enthalpy change of about $20\,\text{kJ}\cdot\text{mol}^{-1}$). The sum of the enthalpy of this exothermic effect and the enthalpy of the decomposition during cooling gives ideally the value of the enthalpy of formation. So it is evident that, for kinetic reasons, the decomposition of Rb_3LaI_6 is incomplete, and part of this compound exists at ambient temperatures as a metastable phase. This metastable phase decomposes abruptly during subsequent heating experiments, and an exothermic effect is observed. When the cooling rate is smaller ($2\,\text{K}\cdot\text{min}^{-1}$), the time of measurement is long enough in comparison to the time of the decomposition reaction, and decomposition of Rb_3LaI_6 can be completed (the enthalpy of decomposition is ideally equal to the enthalpy of formation: 32.3 and $31.6\,\text{kJ}\cdot\text{mol}^{-1}$, respectively). Any exothermic effect is thus observed on a subsequent heating DSC curve.

In the case of other M_3LnI_6 compounds (Rb_3NdI_6, Cs_3LaI_6, and Cs_3NdI_6), the small enthalpy values (8.4–$10.1\,\text{kJ}\cdot\text{mol}^{-1}$) suggest that the first thermal effect corresponds to a structural transition (the transition from low to high temperature modifications of M_3LnI_6 compound). No formation/decomposition effect was evidenced, so one can assume that these compounds are stable or metastable at low temperatures. We did not observe the second thermal effect in the solid state reported by Kutscher and Schneider [7] for Cs_3LaI_6 and Cs_3NdI_6.

21.3.2 Heat Capacity

The compounds investigated exhibit some features that are uncommon for the binary stoicheiometric compounds of halides. They have, firstly, a limited temperature range of existence and, secondly, a tendency to form metastable phases at temperatures below decomposition. These factors, together with the occurrence of solid–solid phase transitions, give a rather complicated heat capacity dependence on temperature. The polynomial equation

$$C_{p,m} = A + BT + CT^2 \qquad (21.2)$$

TABLE 21.4 Molar Heat Capacity of M_3LnI_6 Compounds:
$C_{p,m}$ (J · mol^{-1} · K^{-1}) = A + BT + CT^2

Compound	Temperature Range/K	A/ J · mol^{-1} · K^{-1}	B/ 10^{-2} J · mol^{-1} · K^{-2}	C/ 10^{-4} J · mol^{-1} · K^{-3}
"$K_3NdI_{6(s)}$"	298–618	257.01	9.0028	—
$K_3NdI_{6(s)}$	642–682	351.92	—	—
$K_3NdI_{6(s)}$	696–752	269.58	15.821	—
$K_3NdI_{6(c)}$	816–911	450.54	—	—
"$Rb_3LaI_{6(s)}$"	298–682	274.84	−3.0748	0.2956
$Rb_3LaI_{6(s)}$	692–707	313.10	—	—
$Rb_3LaI_{6(s)}$	717–807	282.23	3.6201	—
$Rb_3LaI_{6(c)}$	871–950	350.24	—	—
$Rb_3NdI_{6(s)}$	298–712	333.80	−31.1637	3.842
$Rb_3NdI_{6(s)}$	722–812	282.03	4.517	—
$Rb_3NdI_{6(c)}$	906–1006	356.71	—	—
$Cs_3LaI_{6(s)}$	298–757	331.29	−32.789	3.557
$Cs_3LaI_{6(s)}$	772–866	24.972	50.4743	−2.213
$Cs_3LaI_{6(c)}$	951–1035	332.99	—	—
$Cs_3NdI_{6(s)}$	298–747	237.95	3.8738	—
$Cs_3NdI_{6(c)}$	967–1080	352.04	—	—

where $C_{p,m}$ is molar heat capacity (in J · mol^{-1} · K^{-1}), was used to fit the experimental results. Polynomial equation coefficients A, B, and C are presented in Table 21.4. The notation "M_3LnI_6" is used for mixtures of M_2LnI_5 and MI with a stoicheiometry corresponding to M_3LnI_6 compounds, which exist below formation temperature.

Heat capacity measurements performed on K_3NdI_6 and Rb_3LaI_6 (Figs. 21.1 and 21.2) confirmed the conclusion about formation of these compounds at higher temperatures according to the reaction $M_2LnI_5 + MI = M_3LnI_6$, and about the possibility of their forming metastable phases at ambient temperatures. Figures 21.1 and 21.2 present results of the heat capacity measurements on K_3NdI_6 and Rb_3LaI_6, prepared by melting and homogenisation of stoicheiometric amounts of LnI_3 and MI, and subsequent cooling of the liquid mixtures with different cooling rates. Open circles represent results obtained for compounds prepared by cooling at 0.1 K · min^{-1}, whereas the black circles represent results for compounds prepared by cooling at 10 K · min^{-1}. When cooled at a rate of 0.1 K · min^{-1}, both K_3NdI_6 and Rb_3LaI_6 fully decompose into M_2LnI_5 and MI. Heat capacity values of compounds prepared in this way increase monotonically with temperature up to the formation temperature (T_{form}). Discontinuities in the heat capacity dependence on temperature are observed at temperatures of compound formation (T_{form}), solid–solid phase transition (T_{trs}), and fusion (T_{fus}).

Cooling of K_3NdI_6 and Rb_3LaI_6 during preparation at a rate of 10 K · min^{-1} leads to complete blocking of K_3NdI_6 decomposition, and only partial decomposition of Rb_3LaI_6. So these compounds exist at ambient temperature as metastable phases and influence the results of heat capacity measurements. In both cases, K_3NdI_6 (Fig. 21.1)

RESULTS AND DISCUSSION 349

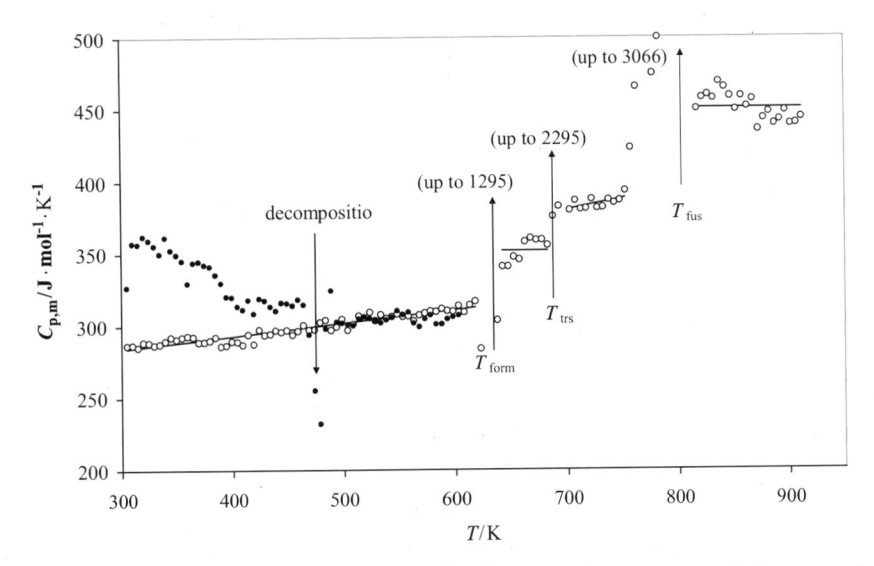

FIGURE 21.1 Molar heat capacity of K_3NdI_6 versus temperature: open circles—experimental values for the compound prepared by cooling at $0.1\,K\cdot min^{-1}$; solid line—polynomial fitting of experimental results; black circles—experimental values for the compound prepared by cooling at $10\,K\cdot min^{-1}$.

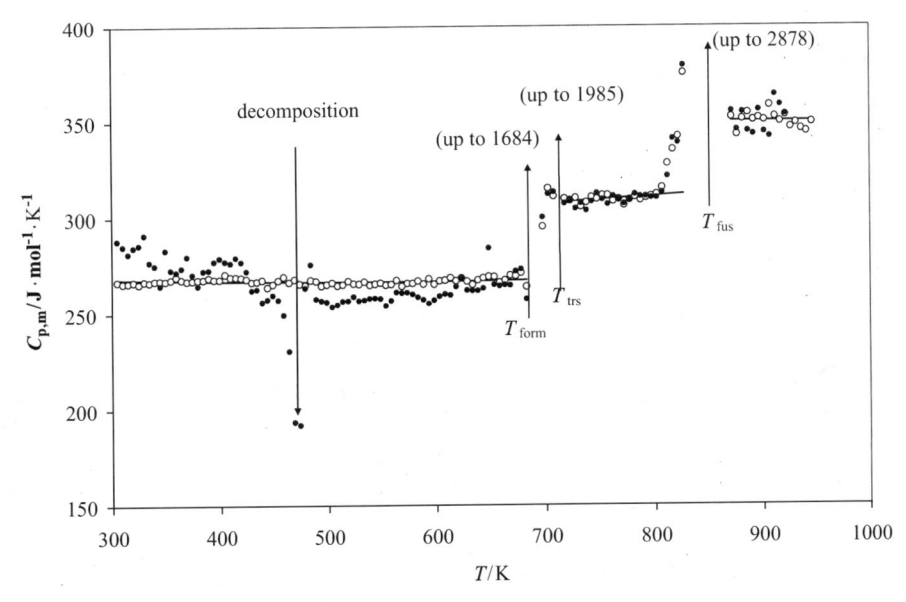

FIGURE 21.2 Molar heat capacity of Rb_3LaI_6 versus temperature: open circles—xperimental values for the compound prepared by cooling at $0.1\,K\cdot min^{-1}$; solid line—polynomial fitting of experimental results; black circles—experimental values for the compound prepared by cooling at $10\,K\cdot min^{-1}$.

and Rb$_3$LaI$_6$ (Fig. 21.2), the value of the heat capacity at 300 K is significantly larger than the value for compounds prepared by cooling at 0.1 K · min^{-1}, and decreases with temperature (exo-effect). The temperature of the heat capacity minimum corresponds well to the temperature of the exo-effect on the DSC curves, that is, the temperature of decomposition of the "supercooled" metastable phase. Starting from this temperature, the heat capacity dependence on temperature is identical with dependence observed for compounds prepared by cooling at 10 K · min^{-1}.

The second group of M$_3$LnI$_6$ compounds (Rb$_3$NdI$_6$, Cs$_3$LaI$_6$, and Cs$_3$NdI$_6$) exhibit an unusual $C_{p,m}$ increase with temperature below the temperature of phase transition (T_{trs}), as exemplified in Figure 21.3 for Rb$_3$NdI$_6$. It is very likely that, as in the case of M$_3$LnCl$_6$ and M$_3$LnBr$_6$, which are stable or metastable at ambient temperature [9, 30, 31], this unusual heat capacity increase is the result of disordering of the cationic sublattice formed by alkali metal cations. This disordering takes place in a continuous way. It starts with the low temperature modification at temperatures significantly lower than the temperature of nonreconstructive phase transition (T_{trs}). The unusual increase of heat capacity at increasing temperature (Fig. 21.3) is the result of this disordering. The state of complete "structural disorder" is obtained in the high temperature modification. Thus the heat capacity dependence on temperature exhibits a λ shape. Different from the compounds of the first group, in which the first-order transition (compound formation) initiates the order–disorder transition, this second group of M$_3$LnCl$_6$ compounds includes the superimposition of order–disorder and first-order (low temperature–high temperature) transitions.

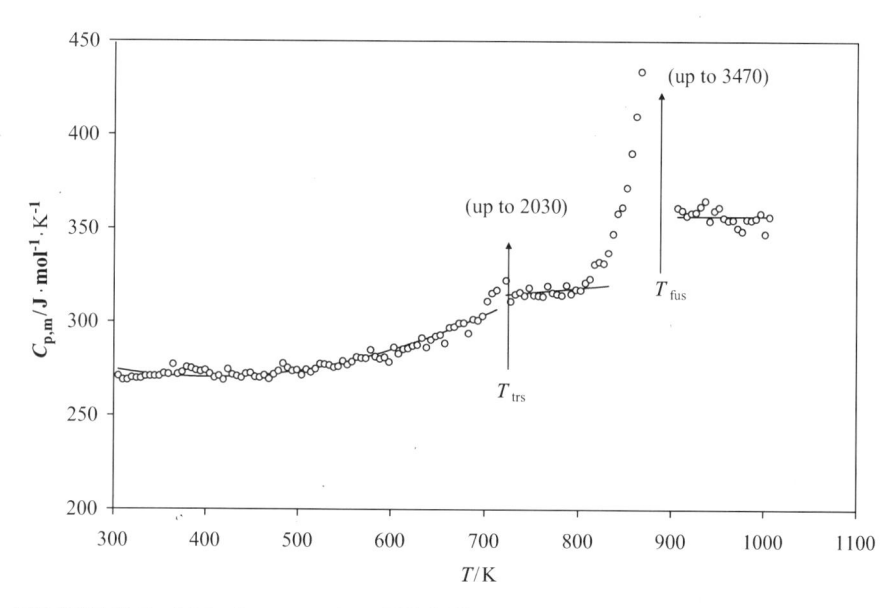

FIGURE 21.3 Molar heat capacity of Rb$_3$NdI$_6$ versus temperature: open circles—experimental values; solid line—polynomial fitting of experimental results.

21.4 CONCLUSION

This work focused on the LnI_3–MI lanthanide alkali metal iodide systems (Ln = La or Nd; M = K, Rb, or Cs) and showed that 3:1 compounds of stoicheiometry M_3LnI_6 exist in all systems. Enthalpy and heat capacity experimental data made it clear that complex phase equilibria exist in the LnI_3–MI mixtures.

M_3LnI_6 compounds split into two groups, the first containing compounds at higher temperatures than the other M_2LnI_5 and RbI compounds, and the second with compounds stable or metastable at high temperature. Enthalpy and heat capacity data obtained on these M_3LnI_6 compounds revealed intriguing and uncommon features (limited temperature range of existence and a tendency to form metastable phases at temperatures below decomposition), behaviour very different from that observed for the analogous chloride (bromide) series of M_3LnCl_6 (M_3LnBr_6) compounds.

In conclusion, one can emphasise that the heat capacity dependence on temperature in the solid–liquid temperature range for M_3LnI_6 serves as the basis for further determination of the standard enthalpies of formation of these compounds. Such a global determination, including the vapour phase, would require knowledge of the thermodynamic activities of MI and LnI_3 in binary systems, which may be obtained by Knudsen effusion mass spectrometry (KEMS) [34]. However, in contrast to intermetallic systems, for which this technique applies successfully [35], accurate activity determinations by KEMS are difficult for ionic salt systems like MI–LnI_3. It stems from the complexity of mass spectral interpretation, for which novel interpretation techniques are being developed for this purpose [36].

REFERENCES

1. Markus, T., Ohnesorge, M., Mollenhoff, M., and Hilpert, K., Aspects of high temperature chemistry in metal halide lamps, in *Proceedings of the 7th International Symposium on Molten Salts Chemistry and Technology*, P. Taxil, C. Bessada, M. Cassir, and M. Gaune-Escard (Eds.), Toulouse, France, August 29–September 2, 2005, pp. 743–746.
2. Mucklejohn, S. A. and Dinsdale, A. T., The behaviour of metal halide systems in high intensity discharge lamps, in *Proceedings of the 7th International Symposium on Molten Salts Chemistry and Technology*, P. Taxil, C. Bessada, M. Cassir, and M. Gaune-Escard (Eds.), Toulouse, France, August 29–September 2, 2005, pp. 761–764.
3. Van Erk, W., Transport processes in metal halide discharge lamps, *Pure Appl. Chem.* **72**, 2159–2166 (2000).
4. Guest, E. C., Mucklejohn, S. A., Preston, B., Rouffet, J. B., and Zissis, G., "NumeLiTe:" an energy effective lighting system for roadways and an industrial application of molten salts, in *Proceedings of International Symposium on Ionic Liquids in Honour of M. Gaune-Escard*, H. A. Oye and A. Jagtoyen (Eds.), Carry le Rouet, France, June 26–28, 2003, pp. 37–45.
5. Mucklejohn, S. A., Trindell, D. L., and Devonshire, R., An assessment of the thermochemical parameters for the lanthanide triiodides, in *6th International Symposium on the Science & Technology of Light Sources*, Budapest, Hungary, 1992.

6. Dworkin, A. S. and Bredig, M. A., Enthalpy of lanthanide chlorides, bromides and iodides from 298–1300 K: enthalpies of fusion and transition, *High Temp. Sci.* **3**, 81–90 (1971).

7. Kutscher, J. and Schneider, A., Untersuchung der Zustandsdiagramme von Lanthaniden (III)-jodiden im Gemisch mit Alkalijodiden, *Z. Anorg. Allg. Chem.* **386**, 38–46 (1971).

8. Rycerz, L. and Gaune-Escard, M., Phase diagrams of LnI_3–LiI binary systems (Ln = La, Nd), unpublished results.

9. Rycerz, L., Thermochemistry of lanthanide halides and compounds formed in lanthanide halide–alkali metal halide systems, in *Scientific Papers of the Institute of Inorganic Chemistry and Metallurgy of Rare Elements of the Wroclaw University of Technology*, Vol. 68, Wroclaw University, Wroclaw, 2004.

10. Drudding, L. F. and Corbett, J. D., Lower oxidation states of the lanthanides. Neodymium(II) chloride and iodide, *J. Am. Chem. Soc.* **83**, 2462–2467 (1961).

11. Bogacz, A., Rycerz, L., Rumianowski, S., Szymanski, W., and Szklarski, W., Technology of light lanthanide metal production, *High Temp. Mater. Proc.* **3**, 461–474 (1999).

12. Gaune-Escard, M., Rycerz, L., Szczepaniak, W., and Bogacz, A., Enthalpies of phase transition in the lanthanide chlorides $LaCl_3$, $CeCl_3$, $PrCl_3$, $NdCl_3$, $GdCl_3$, $DyCl_3$, $ErCl_3$ and $TmCl_3$, *J. Alloys Comp.* **204**, 193–196 (1994).

13. Rycerz, L., Ingier-Stocka, E., Cieslak-Golonka, M., and Gaune-Escard, M., Thermal and conductometric studies of $NdBr_3$ and $NdBr_3$–LiBr binary system, *J. Therm. Anal. Cal.* **72**, 241–251 (2003).

14. Rycerz, L. and Gaune-Escard, M., Heat capacity of K_3LnCl_6 compounds with Ln = La, Ce, Pr, Nd, *Z. Naturforsch.* **54a**, 229–235 (1999).

15. Gaune-Escard, M., Bogacz, A., Rycerz, L., and Szczepaniak, W., Heat capacity of $LaCl_3$, $CeCl_3$, $PrCl_3$, $NdCl_3$, $GdCl_3$, $DyCl_3$, *J. Alloys Comp.* **235**, 176–181 (1996).

16. Rycerz, L. and Gaune-Escard, M., Thermodynamics of $SmCl_3$ and $TmCl_3$: experimental enthalpy of fusion and heat capacity. Estimation of thermodynamic functions up to 1300 K, *Z. Naturforsch.* **57a**, 79–84 (2002).

17. Seifert, H. J., Fink, H., and Thiel, G., Thermodynamic properties of double chlorides in the systems $ACl/LaCl_3$ (A = Na, K, Rb, Cs), *J. Less-Common Metals* **110**, 139–147 (1985).

18. Seifert, H. J., Sandrock, J., and Thiel, G., Thermochemical studies on the systems $ACl/CeCl_3$ (A = Na–Cs), *J. Therm. Anal.* **31**, 1309–1318 (1986).

19. Seifert, H. J., Sandrock, J., and Uebach, J., Zur Stabilitat von Doppelchloriden in den Systemen $ACl/PrCl_3$ (A = Na–Cs), *Z. Anorg. Allg. Chem.* **555**, 143–153 (1987).

20. Seifert, H. J., Fink, H., and Uebach, J., Properties of double chlorides in the systems $ACl/NdCl_3$ (A = Na–Cs), *J. Therm. Anal.* **33**, 625–632 (1988).

21. Thiel, G. and Seifert, H. J., Properties of double chlorides in the systems $ACl/SmCl_3$ (A = Na–Cs), *Thermochim. Acta* **133**, 275–282 (1988).

22. Seifert, H. J. and Sandrock, J., Ternare Chloride in den Systemen $ACl/EuCl_3$ (A = Na–Cs), *Z. Anorg. Allg. Chem.* **587**, 110–118 (1990).

23. Seifert, H. J., Sandrock, J., and Thiel, G., Ternare Chloride in den Systemen $ACl/GdCl_3$ (A = Na–Cs), *Z. Anorg. Allg. Chem.* **598/599**, 307–318 (1991).

24. Mitra, S., Uebach, J., and Seifert, H. J., Ternary chlorides in the systems $ACl/TbCl_3$ (A = K, Rb, Cs), *J. Solid State Chem.* **115**, 484–489 (1995).

25. Zheng, Ch. and Seifert, H. J., Ternary chlorides in the systems $ACl/TmCl_3$ (A = K, Rb, Cs), *J. Solid State Chem.* **135**, 127–131 (1998).

26. Sebastian, J. and Seifert, H. J., Ternary chlorides in the systems ACl/TmCl$_3$ (A = K, Rb, Cs), *Thermochim. Acta* **318**, 29–37 (1998).
27. Seifert, H. J. and Yuan, Y., Thermochemical studies on the systems ABr–LaBr$_3$ (A = Na, K, Rb, Cs), *J. Less-Common Metals* **170**, 135–143 (1991).
28. Seifert, H. J., Ternary chlorides of the trivalent early lanthanides. Phase diagrams, crystal structures and thermodynamic properties, *J. Therm. Anal. Cal.* **67**, 789–826 (2002).
29. Gaune-Escard, M. and Rycerz, L., Solubility of lanthanide chlorides (LnCl$_3$) in alkali metal chlorides (MCl): thermodynamics and electrical conductivity of the M$_3$LnCl$_6$ compounds, *Monatsh. Chem.* **134**, 777–786 (2003).
30. Rycerz, L. and Gaune-Escard, M., Thermodynamic and transport properties of K$_3$TbX$_6$ congruently melting compounds formed in TbX$_3$–KX binary systems (X = Cl, Br), *J. Nucl. Mater.* **344**, 124–127 (2005).
31. Gaune-Escard, M. and Rycerz, L., Unusual nature of lanthanide chloride–alkali metal chloride M$_3$LnCl$_6$ compounds in the solid state, *J. Alloys Comp.* **408–412**, 76–79 (2006).
32. Seifert, H. J., Ternary chlorides of the trivalent late lanthanides. Phase diagrams, crystal structures and thermodynamic properties, *J. Therm. Anal. Cal.* **83**, 479–505 (2006).
33. Seifert, H. J., Retarded solid state reactions, in *Proceedings of the 10th International Symposium on Thermal Analysis.*, THERMANS 95, Konpur, 1995, pp. 1–6.
34. Lisek, I., Kapała, J., and Miller M., Thermodynamic study of the CsCl–NdCl$_3$ system by Knudsen effusion mass spectrometry, *J. Alloys Comp.* **278**, 113–122 (1998).
35. Balducci, G., Brutti, S., Ciccioli, A., Guido, G., Manfrinetti, P., Palenzona, A., Butman, M. F., and Kudin, L., Thermodynamics of the intermediate phases in the Mg–B system, *J. Phys. Chem. Solids* **66**, 292–297 (2005).
36. Kapała, J., The interpretation of mass spectra of inorganic molecules, *Int. J. Mass Spectrom.* **251**(1), 59–65 (2006).

22 Materials Informatics for Molten Salts Chemistry

CHANGWON SUH

Combinatorial Sciences and Materials Informatics Collaboratory (CoSMIC), NSF International Materials Institute, Department of Materials Science and Engineering, Iowa State University, Ames, Iowa

SLOBODAN GADZURIC

École Polytechnique, Mecanique Energetique, Technopôle de Château-Gombert, Marseille, France and Faculty of Science, Department of Chemistry, University of Novi Sad, Novi Sad, Serbia

MARCELLE GAUNE-ESCARD

École Polytechnique, Mecanique Energetique, Technopôle de Château-Gombert, Marseille, France

KRISHNA RAJAN

Combinatorial Sciences and Materials Informatics Collaboratory (CoSMIC), NSF International Materials Institute, Department of Materials Science and Engineering, Iowa State University, Ames, Iowa

Abstract

In this chapter, we demonstrate how one of the prediction data mining techniques, partial least squares (PLS) analysis, can be used to combine descriptors to trace final properties (enthalpy and Gibbs free energy of formation) of lanthanide halide systems. Robustness of the analysis model was examined by experiments and the PLS model gives good agreement with experimental results for two tested compounds. The developed PLS model was also extended to examine the dependence of the enthalpy of formation on the temperature. Detailed mathematical explanations of PLS are given as well.

Molten Salts and Ionic Liquids: Never the Twain? Edited by Marcelle Gaune-Escard and Kenneth R. Seddon
Copyright © 2010 John Wiley & Sons, Inc.

22.1 INTRODUCTION

New materials usually have been found through fortuitous discovery because of the number of experiments and related variables. In the last few years, material design processes have been significantly reduced by high throughput experimentation that handles many parallel experiments based on combinatorial synthesis and analysis techniques. However, due to the enormous number of materials tested in parallel in combinatorial experiments, the resulting data sets are massive and need to be analysed in an accelerated way. In other words, a challenging problem is that huge amounts of samples are created from high throughput experimentation and there exist many variables from lots of different process conditions and from measurements with various analysis tools. Therefore we face the problem of the size and shape of data sets generated from high throughput experimentation. For instance, when the regression techniques are to be used for modelling, traditional regression tools cannot be used if the data set has a much different number of variables than samples or if there is collinearity in the data. It is also difficult to deal with the multivariate features of data sets, since we need to consider all possible combinations of variables to trace some trends in the data.

In previous work, we proposed the concept of virtual combinatorial experiments [1, 2] to deal with the materials design problems described above. The work was to identify the key combinations of multivariables required to have the desired functionality (objective function) in the class of materials being studied by applying data mining techniques. In this work, we will present this discussion in detail in the context of a study on prediction of thermodynamic properties.

There are many techniques dealing with the multivariate problems such as principal component analysis (PCA), canonical correlation analysis (CCA), or partial least squares (PLS) [1–4]. The common feature in most of the high dimensional data analyses is that they deal with an eigenvalue problem to find optimal solutions corresponding to the eigenvectors, which leads to the solution formed by a spectral decomposition of the data [5]. In this chapter, we will focus on a PLS approach to extract information from a large data set, such as an estimation of the influence of each variable (or descriptor) on the final properties. PLS is an effective tool to handle "short and fat" matrices (many more columns than rows), or "long and lean" data (with many more rows than columns) [6]. Therefore a "long and lean" shape of molten salts data used in this chapter is a good example to describe the usefulness of PLS. The shape of the data set is due to the arrays arranged by temperature. Consequently, prediction of thermodynamic properties will serve to show how one may extract thermodynamic reference data by integrating the materials informatics technique.

22.2 MATHEMATICAL METHODOLOGY FOR MATERIALS MODELLING

Partial least squares (PLS) is a multivariate predictive technique of latent variables. It performs linear regression by projecting the data to a low dimensional space while simultaneously constructing a linear regression function in the reduced space minimising the least square loss function. In PLS, the collinear data matrix is transformed into a latent structural matrix with orthogonal vectors. PLS also goes beyond traditional regression in that it models by maximising the covariance between the predictor set (X) and response variables (Y).

According to the literature [7, 8], PLS analysis finds weight vectors w and c such that

$$w_k = \underset{w^T w = 1}{\mathrm{argmax}} \left[\mathrm{cov}(Xw, Yc) \right]^2 \qquad (22.1)$$

subject to the orthogonality constraint

$$w_i^T X^T X w_j = 0 \quad \text{for all } 1 \leq j < k \qquad (22.2)$$

For PLS in this chapter, all descriptors are autoscaled because they have different ranges and scales. An autoscaled data matrix has zero mean and unit variance. Cross-validation is used to choose the optimum number of latent variables (LVs) of the calibration model. In this process, several model parameters are calculated: *RMSECV* (root-mean-square error of cross-validation), *RMSEC* (root-mean-square error of calibration), R^2 (coefficient of determination), and Q^2 (the cross-validated coefficient of determination). The procedures for PLS analysis are shown in Figure 22.1, and detailed mathematical explanations can be found in the Appendix.

22.3 PREDICTION OF THERMODYNAMIC PROPERTIES

Molten salts have been studied for decades due to their wide range of industrial applications. For example, they have provided a unique opportunity to process and produce metals. They are also used to replace pyroreduction, metallothermic reduction, hydrometallurgical methods, or aqueous electrochemical techniques when they were not feasible due to thermodynamic or kinetic constraints.

In previous work, we showed how one found intercorrelations of properties from the molten salt database by PCA, while the physical properties could be estimated by combining the influence of proper parameters by PLS [9]. In this chapter, we extend our previous work for the thermodynamic investigation of lanthanide halide systems that are not available in the Janz database [10–14], and the large experimental data sets that were obtained during systematic investigations of different properties. As the first step of this work, we identify a set of descriptors associated with lanthanide halide systems to build the input for PLS analysis. These descriptors describe some physical characteristic that may be relevant to our thermodynamic objective functions (ΔH_{form} and ΔG_{form}). Having extended experiences on molten salt chemistry, the relative ionic potential and the difference of electronegativity between cation and anion were considered as good descriptors for categorisation of melts, especially those including ions with different charges. Considered descriptors in this work are as follows: atomic number, equimolar weight, electronegativity difference between cation and anion

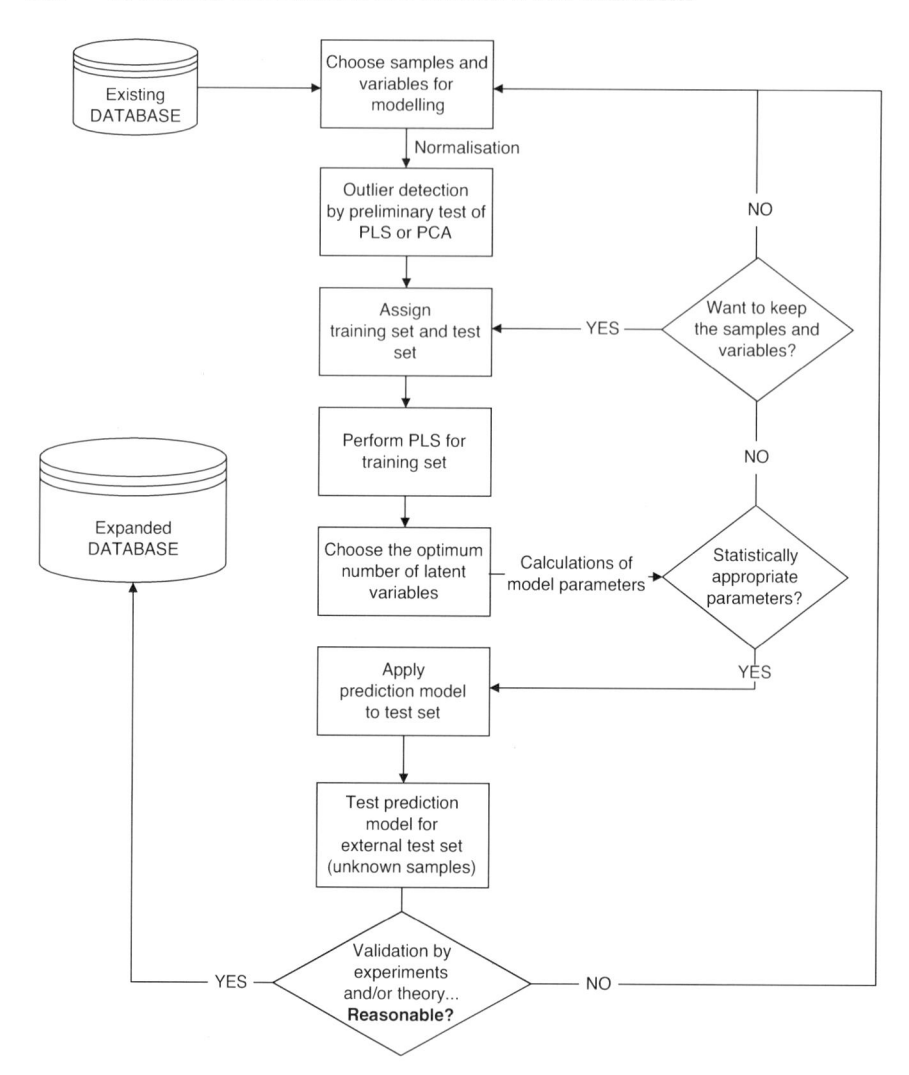

FIGURE 22.1 A flowchart of PLS modelling. Please note that some steps can be omitted depending on the size of the data set or data distribution. For example, outlier detection can be excluded from PLS steps when all data points are critical for modelling. If sample size is small, we can use whole data sets with cross-validation techniques instead of assigning training (calibration) and test sets as well.

(ΔE), ratio of cationic charge and radius (Z_c/r_c), temperature of melting (T_{melt}), and enthalpy and Gibbs free energy of formation at the temperature (ΔH_{form} and ΔG_{form}). Compounds as a training set are given in Table 22.1.

In order to assess the value of which LV and how many LVs may be appropriate to avoid overfitting the data, we need to test different models based on the selection of descriptors and LVs. For the PLS calculation, the SIMPLS algorithm was used

TABLE 22.1 Training Set of Rare Earth Halides and Seven Descriptors

ID	Compound	Equimolar Weight	Atomic Number	ΔE	Z_c/r_c	T_{melt}/K	ΔH_{form}/kJ·mol^{-1}	ΔG_{form}/kJ·mol^{-1}
1	LaCl$_3$	81.76	57	2.06	0.02560	1127	−987	−771
2	CeCl$_3$	82.17	58	2.04	0.02609	1086	−990	−770
3	PrCl$_3$	82.42	59	2.03	0.02655	1061	−982	−761
4	NdCl$_3$	83.55	60	2.02	0.02671	1032	−964	−751
5	SmCl$_3$	85.58	62	1.99	0.02732	950	−955	−728
6	EuCl$_3$	86.10	63	2.04	0.02760	894	−862	−627
7	GdCl$_3$	87.87	64	1.99	0.02783	873	−944	−732
8	TbCl$_3$	88.43	65	1.95	0.02822	854	−941	−723
9	DyCl$_3$	89.62	66	1.96	0.02852	919	−929	−686
10	TmCl$_3$	91.76	69	1.91	0.02852	1092	−932	−688
11	YbCl$_3$	93.13	70	1.95	0.02976	1148	−914	−640
12	LaBr$_3$	126.22	57	1.86	0.02560	1058	−848	−720
13	NdBr$_3$	128.01	60	1.82	0.02671	956	−834	−673
14	TbBr$_3$	132.88	65	1.75	0.02822	1103	−838	−596
15	LaI$_3$	173.21	57	1.56	0.02560	1047	−704	−489
16	NdI$_3$	174.98	60	1.52	0.02671	1058	−665	−446
17	EuCl$_2$	111.43	63	2.04	0.01527	1125	−806	−634
18	YbCl$_2$	121.97	70	1.95	0.01724	994	−771	−605
19	EuBr$_2$	155.89	63	1.84	0.01527	941	−754	−560

and the appropriate number of LVs were chosen using the leave-one-out (LOO) cross-validation. The procedure of LOO is done by fitting a PLS model to $n-1$ samples and making a prediction of the y value for the omitted sample. This calculation is repeated for every sample in the data matrix [15]. Since the value of *RMSECV* at two LVs for all models is minimum, two LVs were chosen and they explained 62.3% of the variance in the descriptors in model I for ΔH_{form}. While atomic number was included in model I, it was excluded in model II. The results are summarised in Table 22.2. After considering the balance of R^2 and Q^2, we chose model I as our final calibration model (Figs. 22.2 and 22.3). Having this final calibration model, we used it to predict ΔH_{form} and ΔG_{form} of lanthanide halides in the test set (Table 22.3), consisting of divalent and trivalent rare earth halides. The predicted ΔH_{form} and ΔG_{form} of the test sets are given in Table 22.4. Once the values of ΔH_{form} and ΔG_{form} in the test sets were predicted, we checked predicted values from experimental thermodynamic investigations on compounds. As examples,

TABLE 22.2 Statistical Parameters[a] of Each Model for ΔH_{form}

Model	R^2	Q^2	RMSECV/kJ·mol^{-1}	RMSEC/kJ·mol^{-1}
I (5 descriptors)	0.92 (0.85)	0.84 (0.75)	38.07 (45.23)	27.65 (34.99)
II (4 descriptors)	0.90 (0.78)	0.82 (0.67)	41.10 (52.81)	30.02 (42.67)

[a] Note that parentheses are used for statistical parameters of each model for ΔG_{form}.

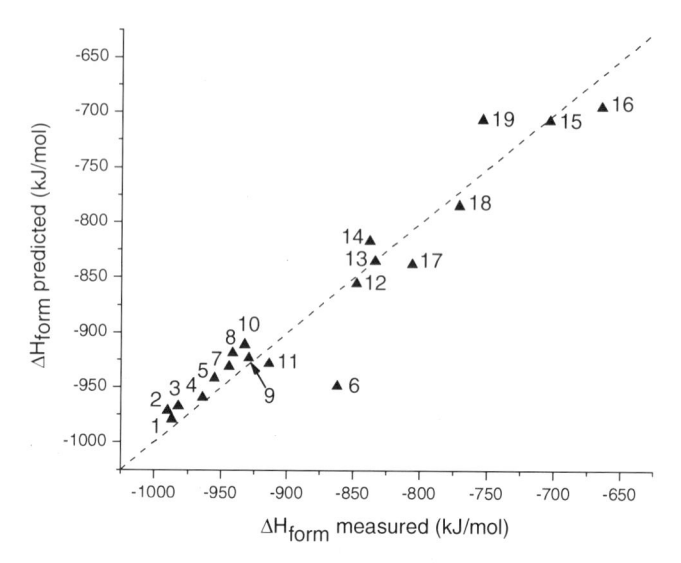

FIGURE 22.2 Predicted versus experimental ΔH_{form} values for the training set for PLS model I. All numbers in Figures 22.2–22.5 correspond to the compounds in Table 22.1.

experimental values of ΔH_{form} for $CeBr_3$ and $GdBr_3$ were -856 and $-811 \, kJ \cdot mol^{-1}$, respectively. The results are in good agreement with the values from the PLS model as shown in Table 22.4 (-846.38 and $-824. \, kJ \cdot mol^{-1}$, respectively).

To check the dependence of ΔH_{form} and ΔG_{form} on the temperature, the same procedures of PLS were applied to the whole temperature range in the liquid phase to

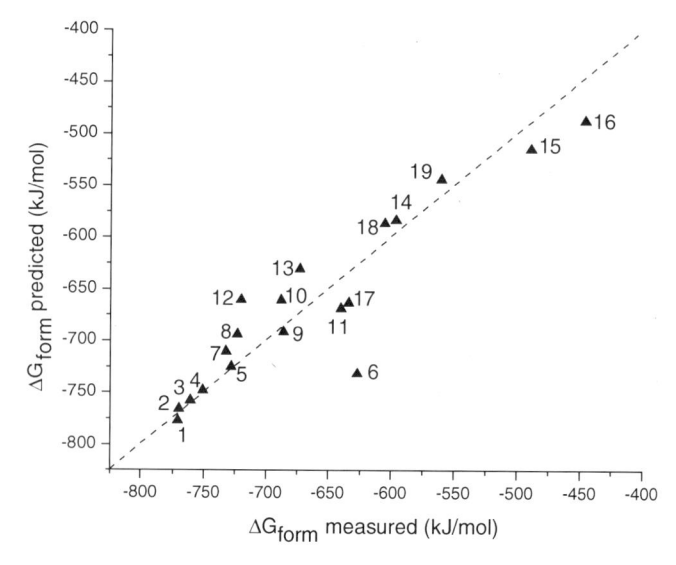

FIGURE 22.3 Predicted versus experimental ΔG_{form} values for the training set for PLS model I.

PREDICTION OF THERMODYNAMIC PROPERTIES

TABLE 22.3 Test Set Used for PLS Models and Predicted Values of ΔH_{form} and ΔG_{form} by PLS Model I

Compound	Equimolar Weight	Atomic Number	ΔE	Z_c/r_c	T_{melt}/K	Temperature Range/K
YbBr$_2$	166	70	1.75	0.0172	1050	1100–1300
YbI$_2$	213	70	1.49	0.0172	1053	1100–1300
EuI$_2$	203	63	1.58	0.0153	853	900–1300
SmCl$_2$	111	62	1.99	0.0142	1132	1150–1300
SmBr$_2$	155	62	1.79	0.0142	973	1000–1300
SmI$_2$	202	62	1.53	0.0142	793	900–1300
CeBr$_3$	127	58	1.84	0.0261	995	1000–1300
PrBr$_3$	127	59	1.83	0.0267	966	1000–1300
SmBr$_3$	130	62	1.79	0.0273	937	1000–1300
EuBr$_3$	131	63	1.84	0.0276	975	1000–1300
GdBr$_3$	132	64	1.79	0.0278	1058	1100–1300
DyBr$_3$	134	66	1.76	0.0285	1152	1180–1300
TmBr$_3$	136	69	1.71	0.0285	1227	1240–1300
YbBr$_3$	138	70	1.75	0.0298	1229	1240–1300

all compounds in the training set (Table 22.1). The temperature range is from the melting point to 1300 K. In this case, R^2 with the 4-LV PLS model was 0.88 and 0.80 for ΔH_{form} and ΔG_{form}, respectively. The results are shown in Figures 22.4 and 22.5. Based on the prediction model over the temperature range, the ΔH_{form} equations as a function of temperature are given in Table 22.4.

TABLE 22.4 Predicted Values of ΔH_{form} and ΔG_{form} for Test Set by PLS Model I[a]

Compound	ΔH_{form}/ kJ·mol^{-1}	ΔG_{form}/ kJ·mol^{-1}	Temperature Dependence of ΔH_{form}	Temperature Range/K
YbBr$_2$	−669.37	−472.27	$-8.1726\text{E}{-}10T^2 + 0.0746T - 775.30$	1100–1300
YbI$_2$	−531.88	−337.38	$-8.1726\text{E}{-}10T^2 + 0.0746T - 642.29$	1100–1300
EuI$_2$	−562.60	−407.14	$-1.0908\text{E}{-}10T^2 + 0.0746T - 649.47$	900–1300
SmCl$_2$	−822.48	−653.56	$-5.4210\text{E}{-}20T^2 + 0.0746T - 950.15$	1150–1300
SmBr$_2$	−692.66	−534.90	$2.5281\text{E}{-}10T^2 + 0.0746T - 800.49$	1000–1300
SmI$_2$	−542.67	−396.25	$1.0908\text{E}{-}10T^2 + 0.0746T - 638.58$	900–1300
CeBr$_3$	−846.38	−648.76	$-2.5281\text{E}{-}10T^2 + 0.0746T - 929.83$	1000–1300
PrBr$_3$	−842.95	−641.09	$-2.5281\text{E}{-}10T^2 + 0.0746T - 924.27$	1000–1300
SmBr$_3$	−822.01	−609.60	$-2.5281\text{E}{-}10T^2 + 0.0746T - 905.80$	1000–1300
EuBr$_3$	−835.34	−617.99	$2.5281\text{E}{-}10T^2 + 0.0746T - 911.14$	1000–1300
GdBr$_3$	−824.71	−598.76	$3.3433\text{E}{-}9T^2 + 0.0746T - 916.71$	1100–1300
DyBr$_3$	−820.20	−580.49	$-3.4965\text{E}{-}9T^2 + 0.0746T - 923.84$	1180–1300
TmBr$_3$	−801.25	−548.02	$2.3810\text{E}{-}8T^2 + 0.0746T - 921.97$	1240–1300
YbBr$_3$	−815.42	−554.95	$2.3810\text{E}{-}8T^2 + 0.0746T - 922.36$	1240–1300

[a] From the calibration model including temperature range (Fig. 22.3), the temperature dependence for the test set was calculated.

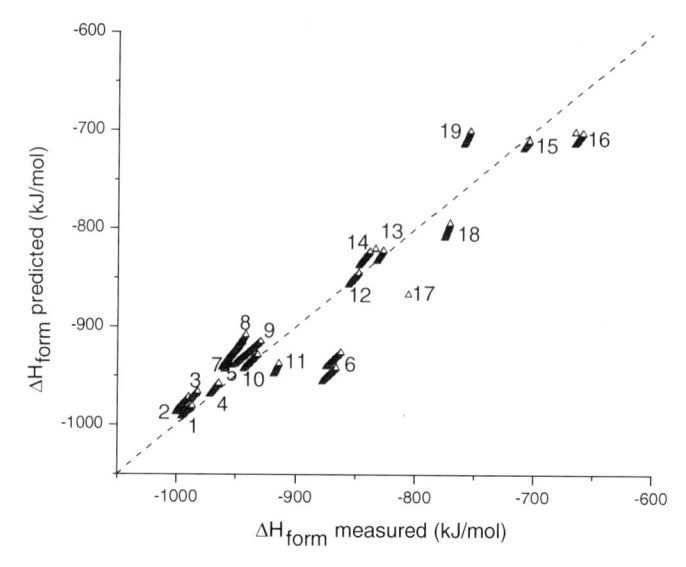

FIGURE 22.4 Predicted versus experimental ΔH_{form} values in the temperature range.

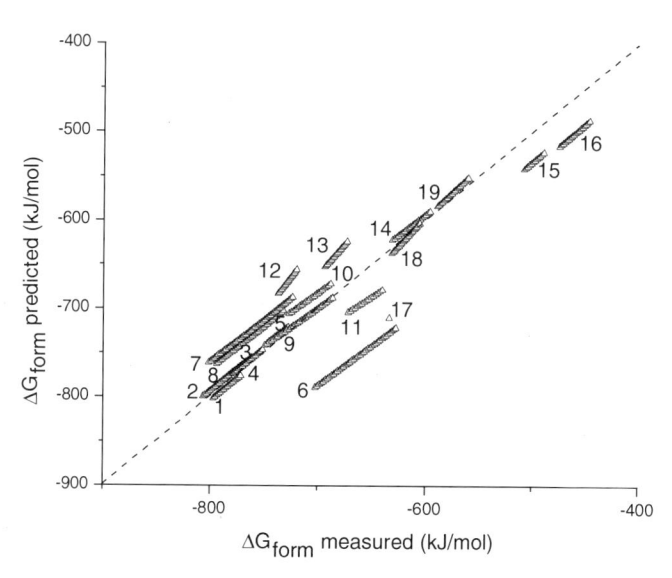

FIGURE 22.5 Predicted versus experimental ΔG_{form} values in the temperature range.

22.4 CONCLUSION

This example with lanthanide halide data serves to demonstrate how carefully selected basic properties (e.g., electronegativity, equimolar weight) appear to influence the collective properties (ΔH_{form} and ΔG_{form}) by using PLS. We also investigated these properties over a wide temperature range of the liquid phase with high accuracy. The use of such data mining techniques needs to become an essential part for establishing and calculating thermodynamic reference data.

22.5 APPENDIX: MATHEMATICAL FOUNDATION OF PARTIAL LEAST SQUARES

Suppose we predict M response variables Y_1, K, Y_M using K predictor variables X_1, K, X_K. With N observations of K variables, X is then a $N \times K$ observation mean-centered data matrix and Y is $N \times M$. Mean centering is expressed as

$$x_i \leftarrow x_i - \bar{x} \quad \text{and} \quad y_i \leftarrow y_i - \bar{y} \tag{22.3}$$

where x_i is the i th element of the vector x and $\bar{x}$ is the mean of its elements [16].

When $N < K$, ordinary least squares (OLS) cannot be applied since the $K \times K$ covariance matrix $X^T X$ is singular. PLS finds the correlation factors between X and Y that have maximum variance as shown in Eq. (22.1). The following mathematical descriptions are based mainly on the treatment of Wold et al. [17].

The predictor variables from the latent component decomposition are expressed as

$$x_{ik} = t_{i1} p^T_{1k} + t_{i2} p^T_{2k} + \mathsf{L}, + t_{iA} p^T_{Ak} + e_{ik} = \sum_a t_{ia} p_{ak} + e_{ik} \quad \text{or} \quad X = TP^T + E \tag{22.4}$$

where e_{ik} are the X residuals. The columns of T are called the scores (or latent variables). Similarly, for predicted variables Y, if the scores of Y are u_a and the weights c_{qm}, then

$$y_{im} = \sum_a u_{ia} c_{am} + g_{im} \quad \text{or} \quad y = UC^T + G \tag{22.5}$$

Since scores of X are good predictors of Y in PLS, then

$$y_{im} = \sum_a t_{ia} c_{am} + f_{im} \quad \text{or} \quad Y = TC^T + F \tag{22.6}$$

where F represents the error between observed values and the predicted response. P and C are loadings of X and Y, respectively. The scores of X, the latent components t_a ($a = 1, 2, \mathsf{K}, A$ where A is the number of PLS components), are expressed as linear combinations of the original variables with the weights w^*_{ka}. The mathematical expression is

$$t_{ia} = \sum_k w^*_{ka} x_{ik} \quad \text{or} \quad T = XW^* \tag{22.7}$$

where $k = (1,\ldots, K)$. Then

$$y_{im} = \sum_a c_{am} \sum_k w^*_{ka} x_{ik} + f_{im} = \sum_k b_{mk} x_{ik} + f_{im} \quad \text{or} \quad Y = XW^*C^T + F = XB + F \tag{22.8}$$

After A dimensions have been extracted, the following equations are available [14]:

$$T_A = XW^*_A, \quad P_A = X^T T_A (T_A^T T_A)^{-1}, \quad W^* = W_A (P_A^T W_A)^{-1} \tag{22.9}$$

The PLS regression coefficients β_{mk} and matrix B of regression coefficients for the model will be

$$\beta_{mk} = \sum_a c_{am} w^*_{ka} \quad \text{or} \quad B = W^* C^T \tag{22.10}$$

Once T is constructed, C can be calculated from the least squares solution

$$C^T = (T^T T)^{-1} T^T Y \tag{22.11}$$

The prediction of y then is written

$$y_{PLS} = T_n (T_n^T T_n)^{-1} T_n^T y \tag{22.12}$$

The original algorithm PLS is called noniterative partial least squares (NIPALS). Other algorithms have also been studied by various researchers. One common algorithm is the SIMPLS algorithm whose PLS1 (univariate Y variable) is identical to that of NIPALS but slightly different for the multivariate Y case [18]. SIMPLS is described by de Jong as well [18]. SIMPLS can directly calculate the PLS factors as linear combinations of the original variables by maximising a covariance with orthogonality and normalisation restrictions. Computationally, the SIMPLS algorithm is faster than NIPALS. Detailed descriptions of this algorithm can be found in the literature [16, 18].

All of the above equations should be implemented in the algorithms. Since NIPALS is the most common algorithm for PLS, it will be explained by following the treatment of Wold et al. [17] and Geladi and Kowalski [19]. After the data are scaled and centered, the steps of NIPALS algorithm are as follows.

1. Take $u_{start} = $ some y_j (for a single y, $u = y$). (Usually the column with greatest variance in Y is chosen.)

In the X block:

2. The weights $w = X^T u / u^T u$. (The weights are calculated from the covariance between X and u.)
3. Normalise, $w_{new} = w_{old} / \|w_{old}\|$.
4. Calculate X scores, $t = Xw/w^T w$. (Score vectors of X are a linear combination of the columns of X with weights in step 2.)

In the Y block:

5. The weights, $c = Y^T t / t^T t$. (The weights are calculated from the covariance between Y and t.)
6. Normalise, $c_{new} = c_{old}/||c_{old}||$.
7. Calculate Y scores, $u = Yc/c^T c$. (Score vectors of Y are a linear combination of the columns of Y with weights in step 5.)
8. Compare t in step 4 with the one in the preceding iteration step. If they are equal, go to step 9, otherwise go to step 2 and use u calculated in step 7. If the Y has only one variable (so called PLS1), the procedure converges in a single iteration and go to the step 9.
9. Calculate X loadings, $p = X^T t / t^T t$.
10. Rescale the scores and weight:

$$p_{new} = \frac{p_{old}}{||p_{old}||}, \quad t_{new} = \frac{t_{old}}{||p_{old}||}, \quad w_{new} = \frac{w_{old}}{||p_{old}||}$$

11. Find the regression coefficient b: $b = u^T t / t^T t$.
12. Remove the present component from X and Y and use deflated matrices as new X and Y in the next component:

$$X = X - tp^T \quad \text{and} \quad Y = Y - tc^T$$

(This step represents the residuals after extracting each component.)

13. Continue with the next component (step 1).

From this algorithm, the first weight vector w is the first eigenvector of the following matrix:

$$X^T Y Y^T X w = \lambda w \qquad (22.13)$$

The following weight vectors are eigenvectors from a deflated matrix. Thus $X_{deflated}^T Y Y^T X_{deflated} w = \lambda_{new} w$, where $X_{deflated} = X_{a-1} - T_{a-1} P_{a-1}^T$. We can similarly show the score vector (t_i) of X is an eigenvector of the matrix $XX^T YY^T$, while the score vector (u_i) of Y is an eigenvector of the matrix $YY^T XX^T$. From these relationships, the weight vectors form an orthonormal set and the score vectors are orthogonal to each other.

ACKNOWLEDGEMENT

We gratefully acknowledge support from the National Science Foundation International Materials Institute program for Combinatorial Sciences and Materials Informatics Collaboratory (CoSMIC-IMI)—grant no. DMR-0603644.

REFERENCES

1. Suh, C. and Rajan, K., Combinatorial design of semiconductor chemistry for bandgap engineering: "virtual" combinatorial experimentation, *Appl. Surf. Sci.* **223**, 148–158 (2004).
2. Suh, C. and Rajan, K., Virtual screening and QSAR formulation for crystal chemistry, *QSAR Comb. Sci.* **24**, 114–119 (2005).
3. Rajagopalan, A., Suh, C., Li, X., and Rajan, K., "Secondary" descriptor development for zeolite framework design: an informatics approach, *Appl. Catal. A.* **254**, 147–160 (2003).
4. Hardoon, D. R., Szedmak, S., and Shawe-Taylor, J., Canonical correlation analysis: an overview with application to learning methods, *Neural Comput.* **16**, 2639–2664 (2004).
5. Shawe-Taylor, J. and Cristianini, N., *Kernel Methods for Pattern Analysis*, Cambridge University Press, Cambridge, UK, 2004.
6. Eriksson, L., Johansson, E., Kettaneh-Wold, N., and Wold, S., *Multi- and Megavariate Data Analysis, Principles and Applications*, Umetrics AB, Umeå, Sweden, 2001.
7. Nguyen, D. V. and Rocke, D. M., Tumor classification by partial least squares using microarray gene expression data, *Bioinformatics* **18**(1), 39–50 (2002).
8. Rosipal, R. and Krämer, N., Overview and recent advances in partial least squares, in: *Subspace, Latent Structure and Feature Selection Techniques*, C. Saunders, M. Grobelnik, S. Gunn, and J. Shawe-Taylor (Eds.) Springer, Berlin, 2006, pp. 34–51.
9. Suh, C., Rajagopalan, A., Li, X., and Rajan, K., in *Proceedings of the International Symposium on Ionic Liquids in Honour of Marcelle Gaune-Escard*, H. Øye and A. Jagtøyen (Eds.), Carry le Rouet, France, June 26–28, 2003, pp. 587–597.
10. Gaune-Escard, M., Bogacz, A., Rycerz, L., and Szczepaniak, W., Heat capacity of $LaCl_3$, $CeCl_3$, $PrCl_3$, $NdCl_3$, $GdCl_3$, $DyCl_3$, *J. Alloys Comp.* **235**(2), 176–181 (1996).
11. da Silva, F., Rycerz, L., and Gaune-Escard, M., Thermodynamic properties of $EuCl_2$ and the $NaCl$-$EuCl_2$ system, *Z. Naturforsch.* **56a**, 647–652 (2001).
12. Rycerz, L. and Gaune-Escard, M., Thermodynamics of $SmCl_3$ and $TmCl_3$: experimental enthalpy of fusion and heat capacity. Estimation of thermodynamic functions up to 1300 K, *Z. Naturforsch.* **57a**, 79–84 (2002).
13. Rycerz, L. and Gaune-Escard, M., Thermodynamics of $EuCl_3$: experimental enthalpy of fusion and heat capacity and estimation of thermodynamic functions up to 1300 K, *Z. Naturforsch.* **57a**, 215–220 (2002).
14. Rycerz, L. and Gaune-Escard, M., Enthalpies of phase transitions and heat capacity of $TbCl_3$ and compounds formed in $TbCl_3$–MCl systems (M = K, Rb, Cs), *J. Therm. Anal. Cal.* **68** (3), 973–981 (2002).
15. Livingstone, D., *Data Analysis for Chemists: Applications to QSAR and Chemical Product Design*, Oxford University Press, Oxford, UK, 1995.
16. Phatak, A. and de Jong, S., The geometry of partial least squares, *J. Chemom.* **11**, 311–338 (1997).
17. Wold, S., Sjöström, M., and Eriksson, L., PLS-regression: a basic tool of chemometrics, *Chemom. Intell. Lab.* **58**, 109–130 (2001).
18. de Jong, S., SIMPLS: an alternative approach to partial least squares regression, *Chemom. Intell. Lab.* **18**, 251–263 (1993).
19. Geladi, P. and Kowalski, B. R., Partial least-squares regression: a tutorial, *Anal. Chim. Acta* **185**, 1–17 (1986).

23 A Novel Ionic Liquid-Polymer Electrolyte for the Advanced Lithium Ion Polymer Battery

DAISUKE TERAMOTO

Pionics Co., Ltd., Kusatsu, Shiga, Japan

RYO YOKOYAMA

Trekion Co., Ltd., Otsu, Shiga, Japan

HIROSHI KAGAWA

Pionics Co., Ltd., Kusatsu, Shiga, Japan

TSUTOMU SADA

Pionics Co., Ltd., Kusatsu, Shiga, Japan and Trekion Co., Ltd., Otsu, Shiga, Japan

NAOYA OGATA

Tokyo and Chitose Institute of Science and Technology, Chitose, Japan

Abstract

We have studied an ionic liquid/poly(vinylidene fluoride) (PVdF or modified PVdF) composite electrolyte (or ICE electrolyte) having a microphase separated structure as a gel-form polymer in a lithium battery, which has excellent properties such as high ionic conductivity and strong mechanical strength (4–6 MPa on 5 cm · min^{-1} at 300 K). Thus Trekion Co., Ltd. has successfully developed gel-form polymer electrolytes of several kinds, having properties of high ionic conductivity (>3 mS · cm^{-1} at 243–353 K), nonflammability, and superior low temperature performance (1 mS · cm^{-1} at 243 K) by altering a grafting ratio reacting with PVdF, for use in a lithium gel polymer battery, which Pionics Co., Ltd. has commercialised. This is a first step in the development of a lithium metal polymer solid state battery for operation at ambient temperature.

Molten Salts and Ionic Liquids: Never the Twain? Edited by Marcelle Gaune-Escard and Kenneth R. Seddon
Copyright © 2010 John Wiley & Sons, Inc.

23.1 INTRODUCTION

The development of a lithium polymer rechargeable battery has been executed actively since the mid-1990s by many battery makers and battery materials manufacturers, and these research reports have been published as bulletins, scientific treatises, and so on. The authors have contributed to developing a polyoxyalkylene (PAO) structure of polymer electrolyte in a lithium metal battery development program of the USABC (United States Advanced Battery Consortium) project, conducted mainly by Hydro Quebec, an electricity company located in Montreal, Canada, since 1989. It has created a film with conductivity of $10^{-3}\,S\cdot cm^{-1}$ by coating a PAO electrolyte on a lithium metal. The PAO structure of the polymer electrolyte, however, has a critical issue at the crystallisation point (T_g) during operation at ambient temperature, causing deterioration of the diffusion coefficient of the lithium ion at less than 40 °C.

To solve this issue, Trekion proposed a polymer lithium battery electrolyte operated at ambient temperature, as well as a polymer composite electrolyte (having flame retardant to nonflammable properties) with a fluoride-type host conjugated polymer, poly(vinylidene fluoride) (PVdF) or poly(tetrafluoroethylene) (PTFE), and the nonflammable, superior low temperature properties of a quaternary ammonium- or pyridinium-based ionic liquids. From an economics point of view, PVdF is a more reasonable polymer for commercialisation than ionic liquids and polymer composite electrolytes. In this development, imidazolium- and pyridinium-based ionic liquids have been studied for lithium gel polymer applications but have several disadvantages, including the existence of a composition of imidazolium ionic liquids in a reduced phase, and high viscosity, making for difficult injection into a separator. Despite these disadvantages, they have been utilised for making a microphase separated and conjugated polymer film having a strong enough tensile strength (TS), because of the following advantageous properties of the ionic liquid:

- High ionic conductivity
- Extremely low volatility
- Nonflammability
- Wide potential window
- Excellent solvent power
- Excellent thermal stability

23.2 CHALLENGES IN MAKING AN IONIC LIQUID AND CONJUGATED POLYMER COMPOSITE ELECTROLYTE (ICE) FILM

Our development work is not based on a brand new idea, since the polymer PVdF used is classified as one of the conventional battery materials and has been modified in use to enhance ion conductivity properties at low temperature and nonflammability. This PVdF is a replacement for poly(tetrafluoroethylene) (PTFE).

23.2.1 Technical Guidance for Composite Materials

As a gel or solid polymer electrolyte having safety properties of flame retardant to self-extinguishing, new ionic liquids (for cations, see Fig. 23.1) have been selected as three reactive and six nonreactive grades and four tetrafluoroborate salt grades (refer to Table 23.1) for making an ionic liquid and conjugated polymer composite electrolyte with grafted PVdF as a host polymer.

PVdF has been selected as being generally stable to general chemicals, although it might be swollen in certain polar solvents, having relatively good thermal stability, and possessing enough mechanical strength to maintain a film formation as an electrolyte and 30–50% crystallisation of crystal polymer having electron receptor on $-CF_2-$ radicals. The electrochemical properties of PVdF are summarised in Table 23.2.

The following electrochemical and physical properties of ionic liquids are desirable for use in making ICE films:

- Nonaqueous and nonvolatile properties at ambient temperature
- Nonflammable materials
- Low bulk resistance for electrochemistry
- High chemical stability
- High thermal stability (greater than 300 °C)
- Wide electric window (-0.5 to ca. $+5.7$ V)
- High decomposition voltage (property stabilised against decomposition)

This development work consisted of simple blend type (HOMO) and grafted polymer cross-linking with the ionic liquids shown in Table 23.1, with anions $[BF_4]^-$ or $[NTf_2]^-$, which gave almost the same ion conductivity performance. As a result of clarifying film formation conditions, film structure, and physical properties, the highest ion conductivity was achieved by vacuum drying at 110 °C for 30 min. In addition, such grafted PVdF is performed differently, dependent on alternation of the grafting ratio.

TABLE 23.1 Ionic Liquids and Their Properties Used as the Ion Conductive Electrolyte

Ionic Liquid	Melting Point/ °C	P^a
$[N_{111}(C_2Omac)][NTf_2]$	34	⊙
$[N_{111}(C_3NHmac)][NTf_2]$		⊙
$[N_{111}(vbz)][NTf_2]$	76	△
$[N_{112}(C_2Oac)][NTf_2]$	30	⊙
$[N_{112}(C_3NHac)][NTf_2]$		×
$[N_{11aa}][NTf_2]$		○
$[C_4mpip][NTf_2]$	10	×
$[N_{122}(C_2OMe)][NTf_2]$	<-50	×
$[N_{222}(C_2OMe)][NTf_2]$	<-50	×
$[A][BF_4]$ (A = N_{11aa}, C_4mpip, $N_{122}(C_2OMe)$, or $N_{222}(C_2OMe)$)		×

[a] P = ease of polymerisation.

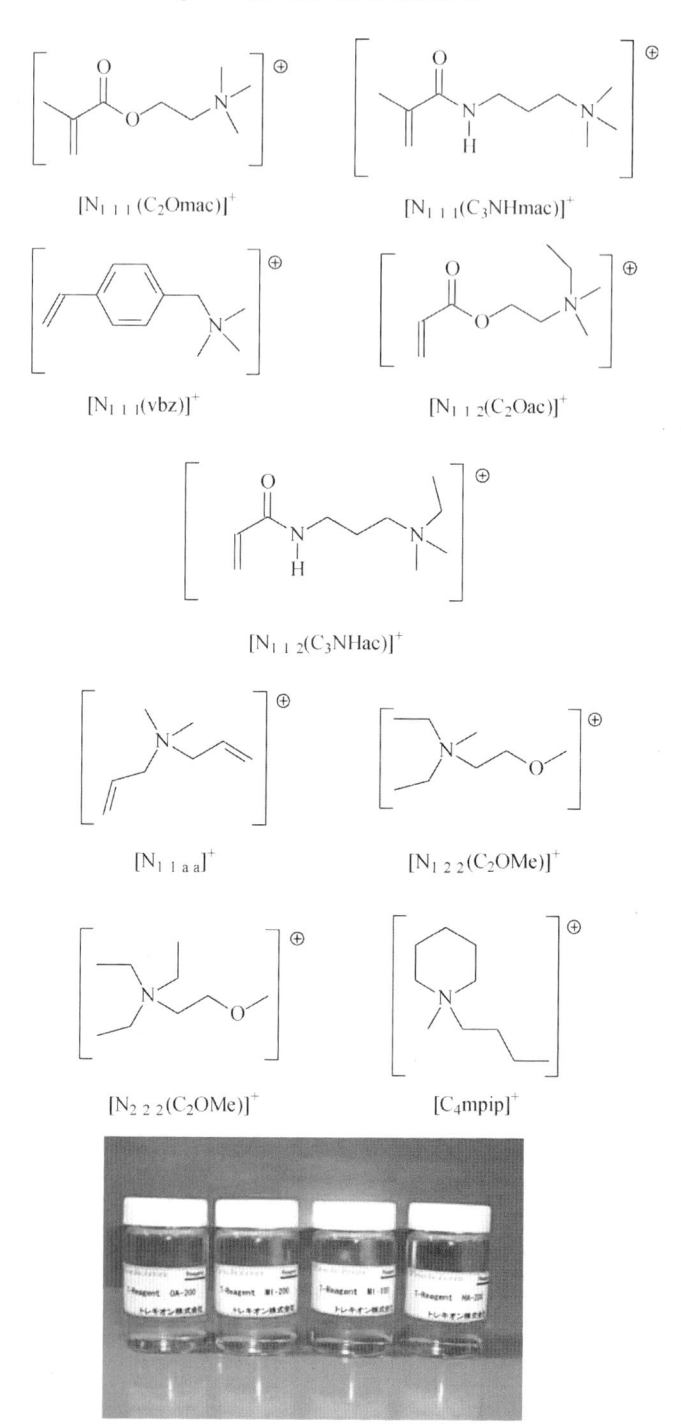

FIGURE 23.1 Cations used in this study and (below) Trekion's ionic liquids (nonflammable).

TABLE 23.2 Electrochemical Properties of PVdF[a]

Dielectric Constant	Loss tan(δ)	Volumetric Resistivity	Surface Resistivity	Dielectric Breakdown Strength
ASTM D150		ASTM D257		ASTM D149
1 kHz		DC 1 kV/ $\Omega \cdot$cm	DC 500V/ $\Omega \cdot$cm^{-2}	DC Short Time Method/MV$\cdot$m^{-1}
10	0.020	10^{15-16}	10^{16-17}	300

[a] Refers to PVdF from Kureha Kagaku Kogyo Kabusikikaisya catalog.

In conclusion, several kinds of ICE film electrolyte have produced an ion conductivity of >3 mS$\cdot$cm^{-1}, and good enough physical stability at 4–6 MPa at 5 cm$\cdot$min^{-1} at 300 K with a thickness of <10 μm. A low temperature, the ion conductivity is maintained excellently at above 1 mS$\cdot$cm^{-1} at −30 °C.

23.3 EXPERIMENTAL

Composite polymer electrolytes were prepared by two methods, one is a *simple blend method* and the other is a *blend method accompanied with graft polymerisation*. PVdF, the ionic liquid, and Li[NTf$_2$] were dissolved in *N,N*-dimethylethanamide (DMA). In the graft blend method, denatured PVdF was used as the polymer. As ionic liquid, the quaternary ammonium salts ([C$_4$mpip][NTf$_2$], [N$_{122}$(C$_2$OMe)][NTf$_2$], and [N$_{222}$(C$_2$OMe)][NTf$_2$]), which possess wide redox windows, were selected. As a polymerised ionic liquid, [N$_{11aa}$][NTf$_2$] was used. Then the solution was coated on a glass plate and dried in vacuum at 383 K for half an hour.

For evaluation of the ICE gel electrolyte for a lithium ion rechargeable battery application, the cathode electrode is lithium manganese oxide and the anode electrode is graphite carbon, together with KB carbon conductive agent and vapour grown carbon fibre. A half cell device was prepared by installing this ICE gel electrolyte, by one of three different procedures: (1) direct mixing with electrode materials, (2) injection of the electrolyte into a porous electrode, or (3) component injection of the ionic liquid as an electrolyte element. In addition, an initial test had been done with a CR2032 coin cell (Fig. 23.2) to evaluate the possibility of utilising this ICE film electrolyte in the solid-state electrolyte type of lithium metal secondary battery, by using a lithium metal foil as the anode electrode. The cycling at ambient temperature for a lithium ion battery with an ICE gel electrolyte is shown in Figure 23.2.

23.3.1 Film Formation Method of ICE Gel Electrolyte

An 11 vol% solution was prepared by dissolving PVdF or DBF in DMA. For formulating with an ionic liquid or salt, a solution was made by adding 2 wt% of cross-linking agent (e.g., benzoyl peroxide) to the monomer in DMA. After mixing

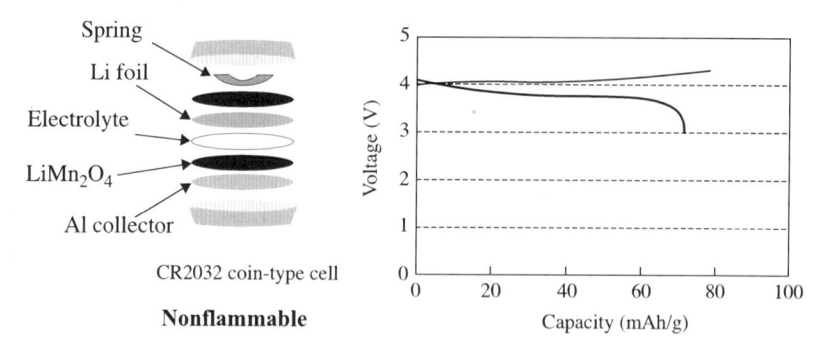

FIGURE 23.2 Typical cycle behaviour, at ambient temperature, using an ICE gel electrolyte.

those two solutions until homogeneous, the ICE film electrolyte was made by coating on a glass plate with an applicator and drying in vacuo at 110 °C for 30 min.

23.3.2 Selection of Host Polymer

PVdF has been selected for a combination of its tensile strength (MPa) and tear strength (%) with KF polymer (Kureha Chemical Industry Co., Ltd.) and Kynar Polymer (ARKEMA). The polymer grade KF-4820 has 48 MPa as tensile strength and 20% as tear strength. Test results for a selection of these polymers are shown below:

KF-4820 > KF-4802 > Kynar-3616 > Kynar-3420
(48 MPa; 20%) (48 MPa; 2%) (36 MPa; 16%) (34 MPa; 20%)

23.3.3 Selection of Ionic Liquid Formulation

KF-4820 was the preliminary selection as PVdF homopolymer and was used with $[N_{111}(C_2Omac)][NTf_2]$ and $[N_{11aa}][NTf_2]$ as polymerisable ionic liquid (PIL) from both the polymerisation and ion conductivity perspectives, as well as with $[C_4mpip][NTf_2]$, $[N_{222}(C_2OMe)][NTf_2]$, and $[N_{122}(C_2OMe)][NTf_2]$ as nonreactive ionic liquids with wide electrochemical windows (low reduction limit potential). Two trials of making ICE electric films have been carried out by simple blending and also grafted-blending with homo type and cross-linking in the graft reaction.

A HOMO-gel type ICE electrolyte film with PVdF/IL = 35/65 wt ratio and 15 wt % or more (>1.4 M) of Li[NTf$_2$] has been made to give an ion conductivity of 10^{-3} S·cm^{-1} or better at 20 °C and 4–6 MPa of tensile strength (TS). In general, dopant Li[NTf$_2$] is limited at 1.5 M for a conventional organic electrolyte in a lithium ion rechargeable battery application, but this formulation makes it possible to dope up to 2.5 M, giving higher ion conductivity but lower tensile strength. On the other hand, a graft cross-linking type ICE electrolyte has a lower ion conductivity and larger tensile strength. There needs to be improved diffusion coefficients of Li$^+$ by supplementary formation of electron networks on interface with the anode active materials for a commercial use, and Pionics has successfully optimised this network for a commercial application.

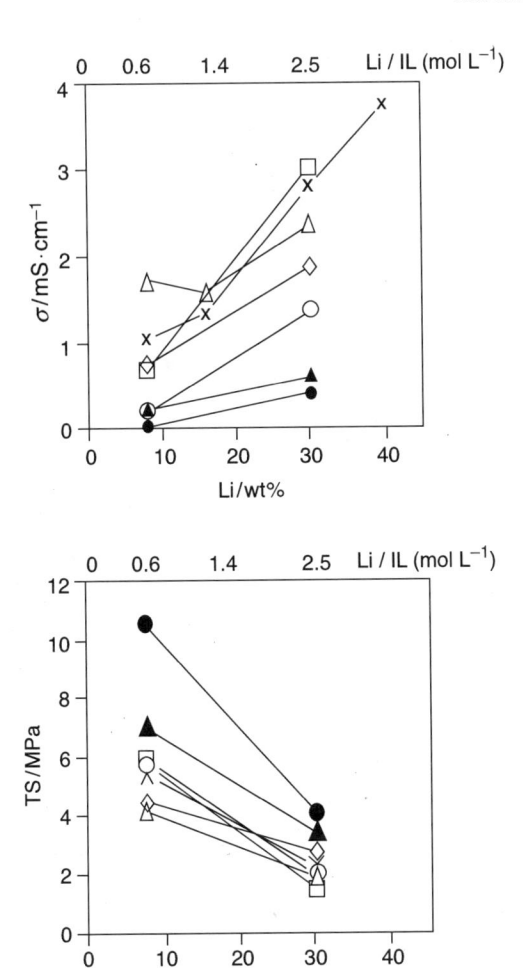

FIGURE 23.3 Ionic conductivity (upper) and tensile strength (lower) of two kinds of HOMO-type ICE polyelectrolyte (gel and solid) as a function of lithium salt concentration (PVdF: IL = 35:65). □ = [C$_4$mpip][NTf$_2$]; × = [N$_{222}$(C$_2$OMe)][NTf$_2$]; ◇ = [N$_{122}$(C$_2$OMe)][NTf$_2$]; ○ = [N$_{111}$(C$_2$Omac)][NTf$_2$]; △ = [N$_{11aa}$][NTf$_2$]; ● = [N$_{111}$(C$_2$Omac)][NTf$_2$] (solid); ▲ = [N$_{11aa}$][NTf$_2$] (solid).

This resulted in showing a symmetrical phenomenon with conventional understanding as higher concentrations of lithium salt in ionic liquid resulted in higher ion conductivity (see Fig. 23.3).

Two formulations of gel ICE electrolyte as PVdF/reactive IL/IL/Li[NTf$_2$] = 28/21/21/30 and solid ICE electrolyte as PVdF/reactive IL/IL/Li[NTf$_2$] = 28/42/0/30 are shown in Figure 23.4, which performed with some efficiency with polymerised ionic liquids (PIL), and this has already shown enhanced ion conductivity in further development work.

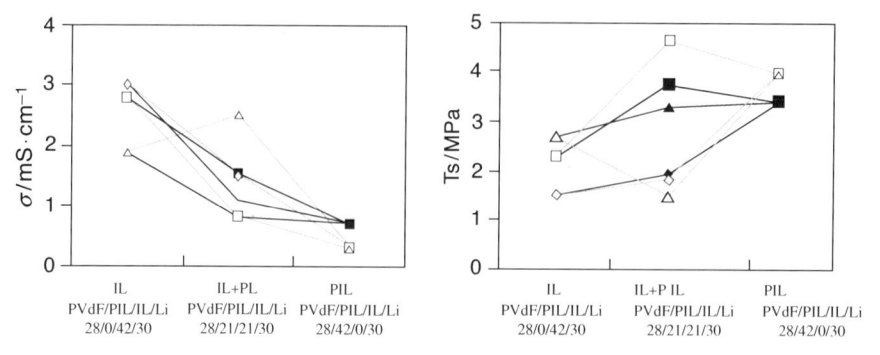

FIGURE 23.4 Ionic conductivity (left) and tensile strength (right) for IL–PVdF compound electrolytes (HOMO-type). ♦ = [N$_{11aa}$][NTf$_2$]-[C$_4$mpip][NTf$_2$]; ■ = [N$_{11aa}$][NTf$_2$]–[N$_{222}$(C$_2$OMe)][NTf$_2$]; ▲ = [N$_{11aa}$][NTf$_2$]–[N$_{122}$(C$_2$OMe)][NTf$_2$]; ◇ = [N$_{111}$(C$_2$Omac)][NTf$_2$]-[C$_4$mpip][NTf$_2$]; □ = [N$_{111}$(C$_2$Omac)][NTf$_2$]–[N$_{222}$(C$_2$OMe)][NTf$_2$]; △ = [N$_{111}$(C$_2$Omac)][NTf$_2$]–[N$_{122}$(C$_2$OMe)][NTf$_2$].

By comparison of Figures 23.4 and 23.5, the ionic conductivity of GRAFT-type is superior to that of HOMO-type, although results of the tensile strengths are similar.

Figures 23.6 and 23.7 show ionic conductivity and temperature charts of PVdF/IL = 35/65 with lithium dopant 30 wt% (HOMO-type) and DBF/[N$_{11aa}$][NTf$_2$]/[N$_{222}$(C$_2$OMe)][NTf$_2$] or [N$_{122}$(C$_2$OMe)][NTf$_2$] = 40/30/30, lithium dopant 30 wt% gel-form, DBF/[N$_{11aa}$][NTf$_2$] = 40/60, Li salt 30 wt% solid form (GRAFT-type), respectively. They show ion conductivity at 10^{-3} S·cm^{-1} at −30 to +80 °C (HOMO-type) and 10^{-3} S·cm^{-1} at −15 to ca. +80 °C with [N$_{11aa}$][NTf$_2$]/[N$_{222}$(C$_2$OMe)][NTf$_2$] as a graft-blending type (GRAFT-type), respectively.

Through altering the grafting rate of the grafted polymer concerned, properties such as flame retardancy and physical strength could be controlled, as well as conditions of gel formation, such as viscose gel to pudding gel, which leads to a decreasing internal resistance to ion conductivity at the interface of electrodes.

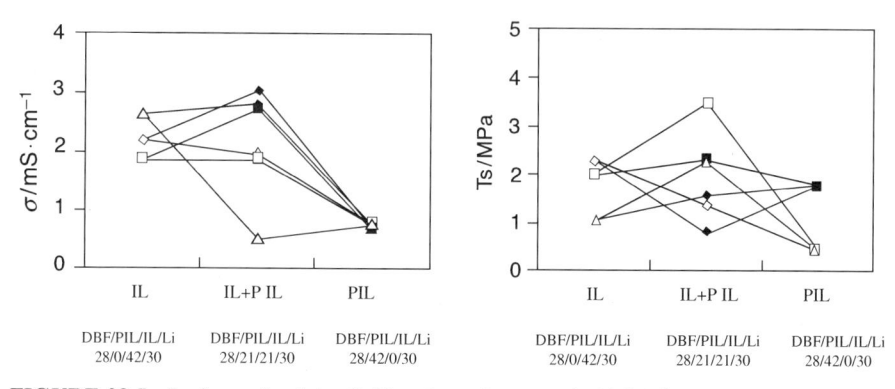

FIGURE 23.5 Ionic conductivity (left) and tensile strength (right) for IL–DBF compound electrolytes (GRAFT-type). Key as for Figure 23.4.

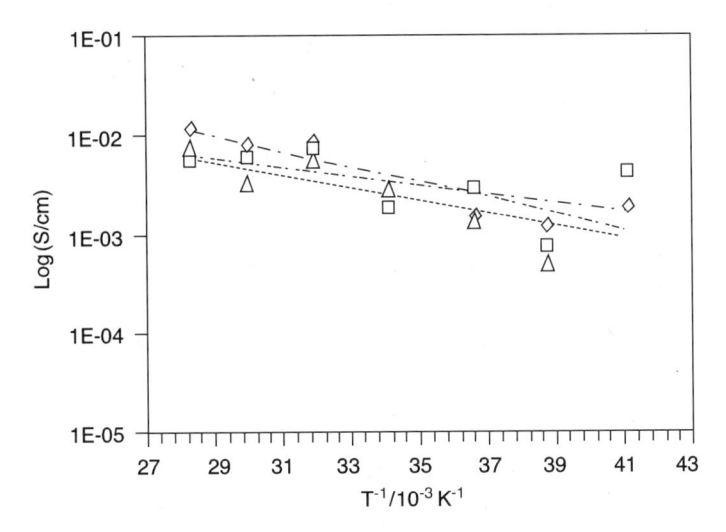

FIGURE 23.6 Temperature dependence of ionic conductivity of HOMO-type gel electrolytes. ◇ = [C$_4$mpip][NTf$_2$]; □ = [N$_{122}$(C$_2$OMe)][NTf$_2$]; △ = [N$_{222}$(C$_2$OMe)][NTf$_2$].

Besides, a composite electrolyte could be achieved with a performance good enough to stream on a commercial plant.

Figure 23.8 shows that PVdF (HOMO-type) performs better than grafted PVdF on ion conductivity as a function of higher grafting rate; the better cycle property is dependent on the viscose binding at the interface of the electrodes.

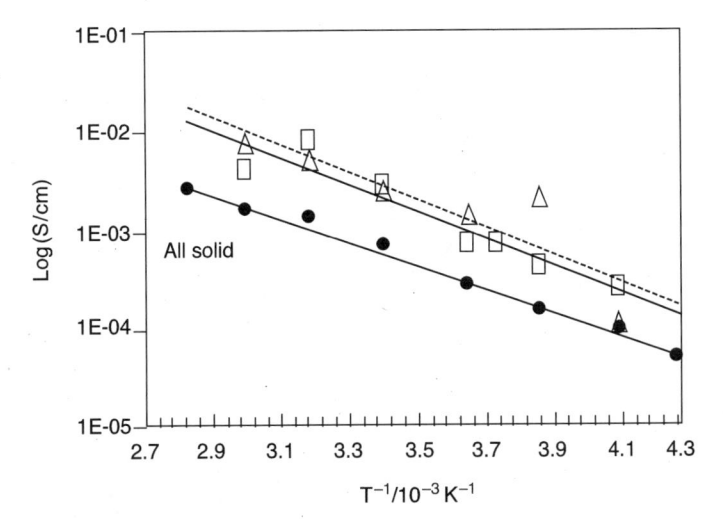

FIGURE 23.7 Temperature dependence of ionic conductivity of GRAFT-type gel electrolyte; DBF/[N$_{11aa}$][NTf$_2$]/IL = 40/30/30, Li[NTf$_2$] = 30 wt%, DBF/[N$_{11aa}$][NTf$_2$] = 40/60, Li[NTf$_2$] = 30 wt% as all solid. ◆ = [N$_{11aa}$][NTf$_2$]; △ = [N$_{11aa}$][NTf$_2$]–[N$_{222}$(C$_2$OMe)][NTf$_2$]; □ = [N$_{11aa}$][NTf$_2$]–[N$_{122}$(C$_2$OMe)][NTf$_2$].

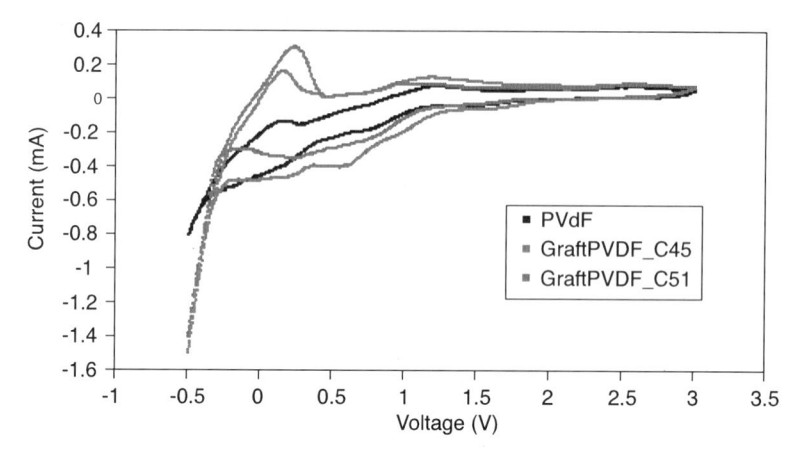

FIGURE 23.8 CV characteristics of PVdF HOMO-type versus grafted PVdF GRAFT-type (50 mV · s^{-1}, 23 °C). Grafted rate: C45, 26.8%; C51, 15.0%.

Furthermore, a microphase separated structure of ICE electrolyte, by compositing grafted PVdF with the ionic liquid, shows better ion conductivity and stable physical properties, adaptable for commercial use. (See Table 23.3).

In Figure 23.9, an ionic liquid (with a quaternary ammonium cation) performed well on formation of a surface electrode interface (SEI) membrane on a tin–metal alloy anode active material, and suppressed a decomposition in the reduction phase of the ionic liquid itself. Imidazolium-based ionic liquids can also be used in this application.

23.3.4 Best ICE Electrolyte Formulation

As a result of evaluating each component, such as (1) type of polymer host, (2) type of ionic liquid, (3) effect of anion in the lithium salt, and (4) difference between HOMO-type and GRAFT-type, ionic conductivity at each temperature and tensile strength of the ICE electrolyte film have been affirmed as a basic technology to determine that the optimum ICE electrolyte recipe is PVdF/PIL/IL/Li = (20–30)/(10–20)/(30–40)/(20–30) on wt ratio.

TABLE 23.3 Result of Ion Conductivity (mS · cm^{-1}) with the Optimum Recipe of Electrolyte (PVdF/PIL/IL/Li = 28/21/21/30 Type)

Type	PIL/IL	[N$_{11aa}$][NTf$_2$] 20 °C	−15 °C	[N$_{111}$(C$_2$Omac)][NTf$_2$] 20 °C	−15 °C	[N$_{111}$(C$_2$Omac)][BF$_4$] 20 °C	−15 °C
HOMO-type	[C$_4$mpip]	**1.1**	0.1	**1.5**	0.04	**5.1**	0.3
	[N$_{222}$(C$_2$OMe)]	1.5	0.2	0.8	0.06	**3.6**	**1.3**
	[N$_{122}$(C$_2$OMe)]	0.8	0.1	**2.5**	0.09	1.6	0.1
Microphase separation grafting type	[C$_4$mpip]	**3.1**	**1.7**	**1.9**	0.3	0.7	—
	[N$_{222}$(C$_2$OMe)]	**2.7**	**2.2**	**1.9**	0.2	**2.3**	**1.0**
	[N$_{122}$(C$_2$OMe)]	**2.8**	0.4	0.5	0.09	**3.1**	—

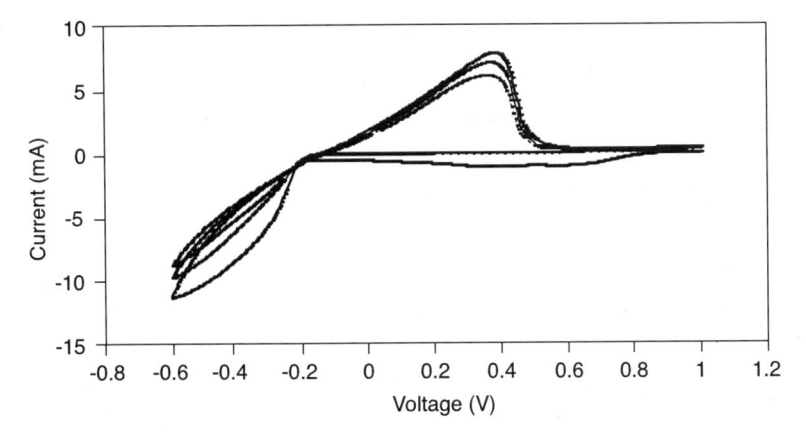

FIGURE 23.9 CV characteristics of a grafted PVdF/IL composite electrolyte (50 mV · s^{-1}, 23 °C).

23.4 APPLICATION OF ICE ELECTROLYTE FOR LITHIUM GEL POLYMER BATTERY

23.4.1 Film Formation and Its Structure

ICE electrolyte films formed by the dry film method were used as described; composite polymer electrolytes were prepared by two methods, one a *simple blend method* and the other a *blend method accompanied with graft-polymerisation*. PVdF, ionic liquid, and Li[NTf$_2$] were dissolved in DMA. In the graft blend method, denatured PVdF was also used as the polymer. As the ionic liquid, the quaternary ammonium salts ([C$_4$mpip][NTf$_2$], [N$_{122}$(C$_2$OMe)][NTf$_2$], and [N$_{222}$(C$_2$OMe)][NTf$_2$]), which possess wide redox windows, were selected. For the polymerised ionic liquid, [N$_{11aa}$][NTf$_2$] was used. Then the solution was coated on a glass plate and dried in vacuo at 383 K for half an hour.

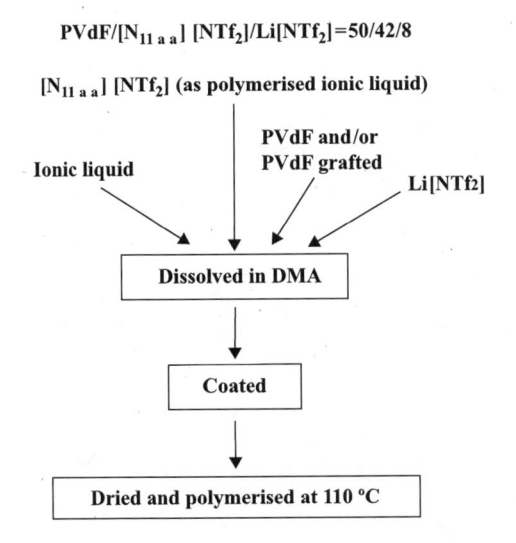

PVdF/[N$_{11\,a\,a}$][NTf$_2$]/Li[NTf$_2$]=50/42/8

[N$_{11\,a\,a}$][NTf$_2$] (as polymerised ionic liquid)

From SEM and polarisation microscope observations, the ICE electrolyte film consists of micron scale (μm) particles (Photo 1) and nano scale (~15 nm) particles inside micron-scale particles (Photo 2), which is recognised as a "microphase separated" structure. In the space between larger particles, there seem to be many vacant spaces and the Li/IL are partially packed in those spaces (from polarisation microscope photo; Photo 3). This phenomenon had been reported as easier film formation from a low melting solvent, as scanty solvent solution and higher crystallised PVdF, according to Michot et al. [35].

The next question is where the ionic liquid and lithium salt are located in the microphase separated structure. A preliminary examination of the photos at larger than 100 K eV SEM indicates the presence of nanoscale particles (<15 nm) on the micron-scale particles (1–2 μm). According to the mapping observation of sulfur atom concentration by energy dispersive X-ray (EDX) analysis, the ionic liquid and lithium salt are concentrated in between nanoscale particles (Photo 4 S). Also, the melting point (mp) of the ICE electrolyte film has been measured and the mp of the PVdF in the film is lower by about 10 °C than that of pure PVdF. Some ionic liquid may be inserted in that ICE electrolyte film, and made incomplete (Photo 4).

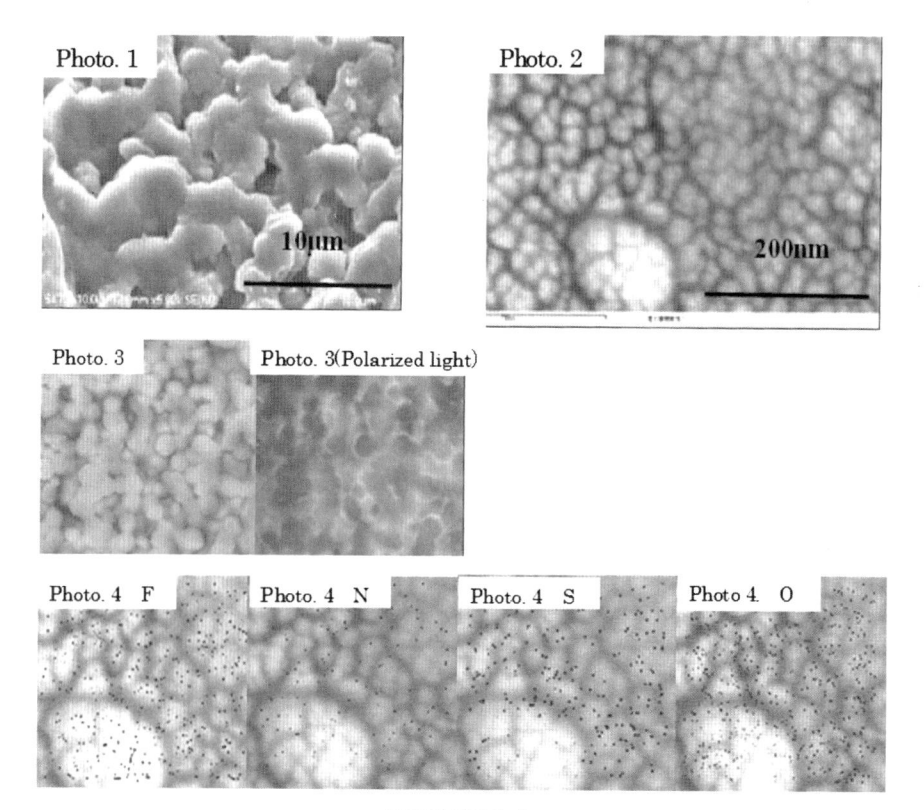

PHOTOS 1–4.

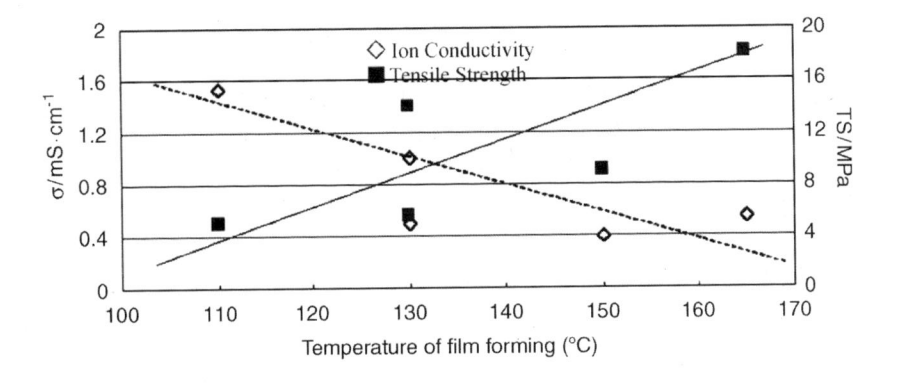

APPLICATION OF ICE ELECTROLYTE FOR LITHIUM GEL POLYMER BATTERY 379

Appearance	White	White to Clear	Clear
Particle size/μm	4-	1-	1-5
Ion conductivity/mS·cm^{-1}	1.5	1.2~0.5	0.5
Strength/MPa	5	6~14	18
Melting point/°C	157	160	161
Polarity/%	29	→8	0

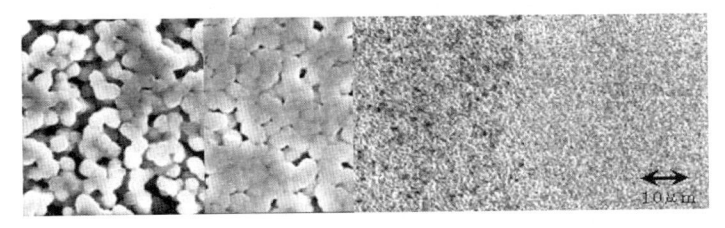

FIGURE 23.10 Some properties of ICE films.

The matrix of ionic conductivity, temperature of film forming, and tensile strength is freely controlled in different microphase separation layers and also lamellar structures. Fig. 23.10 shows the structural conditions exhibiting different ICE functions dependent on the required factors suitable for a lithium battery electrolyte specification and other application needs.

To summarise, characteristics of ICE electrolyte films can be optimised by the materials' formulation and the film formation conditions. The composite formulation study has been performed at 110 °C for 30 minutes at low pressure drying conditions.

23.4.2 Microphase Separated Condition of ICE Electrolyte Film by TEM Analysis

Three kinds of ICE electrolyte films have been made by dissolving PVdF in DMA to make a 30 wt % PVdF solution, to make additionally 21 wt % of ionic liquid, and to

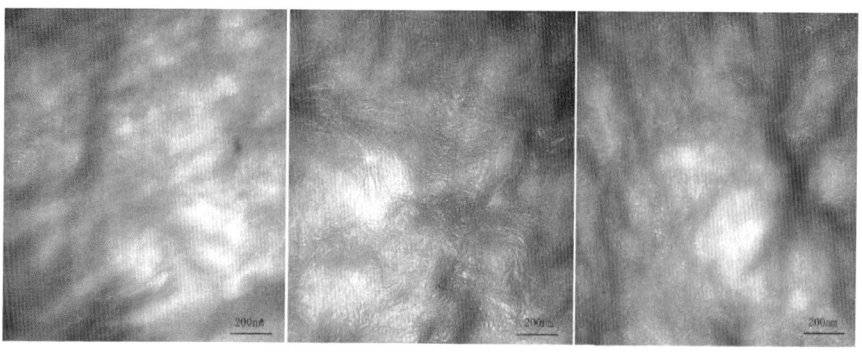

Photo 5 TEM picture of PVdF homomembrane, × 100,000.

Photo 6 TEM picture of PVdF/Li[NTf$_2$] membrane, × 100,000.

Photo 7 TEM picture of ICE gel membrane, × 100,000.

make 30 wt % of lithium salt, Li[NTf$_2$], and then vacuum drying the three different solutions at 80 °C for 30 min. Those films have been observed by a TEM (Hitachi H-800) and revealed a lamellar structure, as shown in Photos 5–7.

In particular, Photo 7 shows the recognisable microphase separated structure of an ICE electrolyte film, the ionic liquid existing in between PVdF molecules. This test has been conducted with grafted PVdF to replace natural PVdF, and also revealed the same lamellar structure on the ICE grafted PVdF composite electrolyte film.

23.4.3 Combustion Test of ICE Electrolyte Film and ICE Gel Polymer

A combustion test has been conducted by heating an ICE composite electrolyte film with the open flame from a gas burner: self-extinguishing was recognised within 1 second, but the film had slightly shrunk (Photo 8).

Dell, Mac (Apple), and others are now recalling their sulfur–lithium batteries installed in notebook computers after the fire accident of such a notebook in Osaka in June 2006 (Photos 9 and 10). Safety issues on lithium rechargeable batteries are critical in the notebook PC industry. Pionics' lithium battery (PLB) can ease consumers' minds by introducing a PLB pack that solves the safety issue in the global market.

This was a sensational event, since they had relied on the safety of operation of notebook PCs. A short circuit on the lithium battery had occurred from time to time, but two protection devices are organised: a shutdown of the separator by heating up, as well as a protection circuit with a thermal sensor. This fire incident resulted in their not operating in a timely manner.

The most risky issue with conventional lithium ion batteries is to utilise flammable organic solvents as the electrolyte liquid, and a polyethylene or polypropylene separator that may burn when a short circuit occurs (Fig. 23.11). Many organic solvent electrolyte manufacturers are developing additive type flame retardant chemicals, like bromine and/or phosphorite, but they are not guaranteed to extinguish a fire in a lithium battery.

PHOTO 8 Combustion test of an ICE film.

PHOTOS 9–10 Recall issue of PC from Dell, Mac, and others at a hotel in Osaka, June, 2006.

A self-extinguishing electrolyte is definitely required for ultimate protection in a lithium ion battery in the case of a short circuit fire accident: the ICE formulation is one of the best choices to prevent the risk of a short circuit fire.

An innovative lithium ion rechargeable battery, known as PLB **PelLicle**®, has been newly developed by Pionics and it has achieved the high energy density of 615 $W \cdot h \cdot L^{-1}$ at present (800 $W \cdot h \cdot L^{-1}$ in the near future) on 513665 of aluminium laminating package. This PLB consists of a tin alloy anode and nickel-type cathode active material, together with the ICE self-extinguishing electrolyte. The higher the energy density, the more risk of fire there is with conventional organic solvent electrolytes in a short circuit fire accident. To minimise the high risk of fire in a lithium ion battery, nonflammable battery materials must be used, especially in the electrolyte

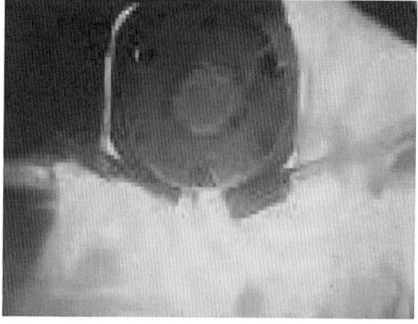

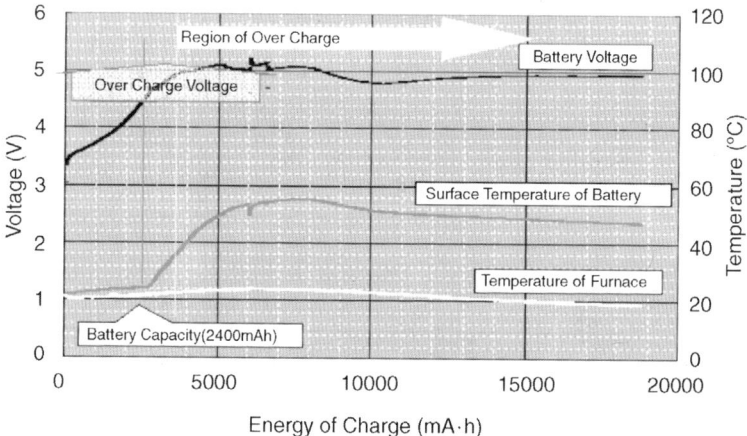

FIGURE 23.11 No gas generation and no fire in the Neil test (upper left). Instant fire upon inserting a nail on a conventional cell installed in a notebook PC (upper right). Charging test for the Pionics' lithium battery (PLB) (lower).

layer, including the modified conventional separator. Thus even if a short circuit happened inside the battery cells, the resulting fire should be extinguished instantly by itself, the so-called self-extinguishing system.

It is feasible for an ICE electrolyte to be installed as a liquid to gel-form polymer at present, or as a solid state polymer in the future.

Under the protection of these ICE self-extinguishing systems, a further enhancement program for increased energy capacity for LIBs should be promoted as safety mode battery systems. Pionics' PLB fulfills such requirements with nonflammable electrolyte materials based on ionic liquids, PVdF, grafted PVdF, PTFE, grafted PTFE, ethylene carbonate, and so on.

23.4.4 Applicable Use for Lithium Gel Polymer Rechargeable Battery

According to a preliminary attempt to make CR2032 coin cells, ICE electrolyte battery coin cells performed at $104 \text{ mA} \cdot \text{h} \cdot \text{g}^{-1}$, with a theoretical discharge capacity,

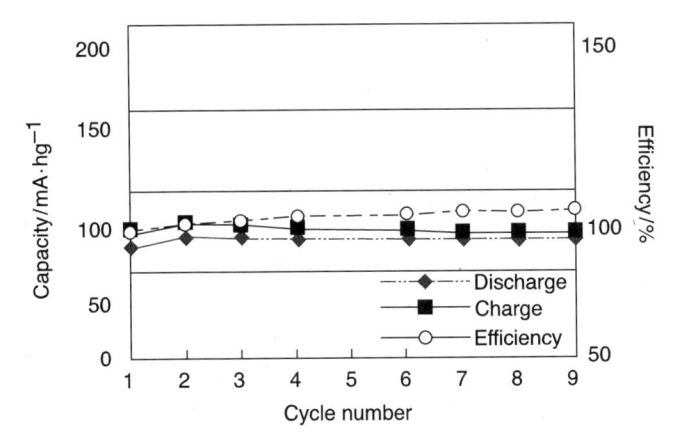

FIGURE 23.12 Charge–discharge capacities as a function of cycle number for an ICE gel polymer lithium battery.

and maintained a rate of initial capacity (after operating for ten cycles) of 97%, being stable at 93% Coulomb efficiency (see Fig. 23.12). After ten cycles, one of these coin cells was dismantled and checked—there was no odour of amine and no colour change, indicating good electrochemical stability. This formulation could be certified to make lithium gel polymer batteries operating at ambient temperature.

23.5 PLB CELL PERFORMANCE

The charge–discharge profile shows stable cycle properties and thermal stability at both 23 °C and 0 °C (see Fig. 23.13).

Over 300 cycles, this battery shows efficient utilisation rates of 78% in 300 cycles and 61% in 500 cycles, respectively (see Fig. 23.14). Improvement is still required,

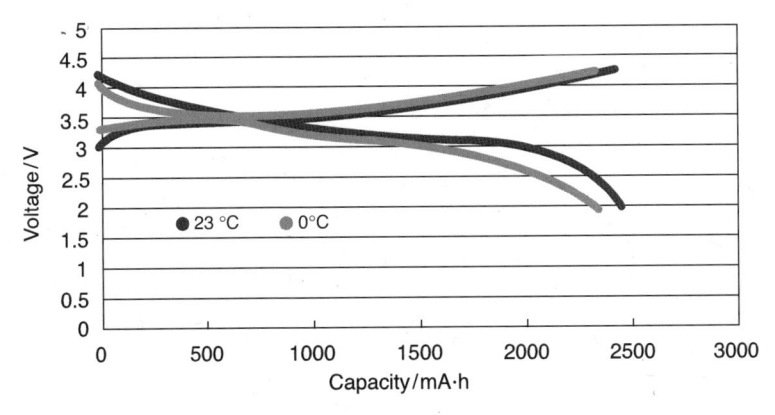

FIGURE 23.13 Charge–discharge profile of a Pionics LIB cell—conventional grade—CC = 0.2 C; discharging at 23 °C and 0 °C.

384 A NOVEL IONIC LIQUID-POLYMER ELECTROLYTE

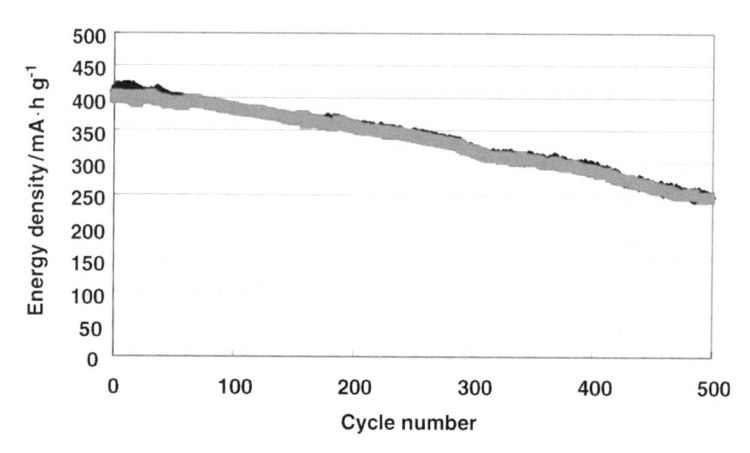

FIGURE 23.14 Cycle properties of a Pionics LIB cell (CC = 0.2 C); the charging cycle (black) and discharge (grey) are almost superimposable.

although it is unnecessary to achieve a target of over 85% in 300 cycles—the most necessary improvement is in the initial property, to maintain the cycle property at a flat level until at least 100 cycles.

At a 1.0 C rate, it maintains a satisfactory profile, but at over 2.0 C, it deteriorates (see Fig. 23.15); further work is required to improve such properties.

23.6 SUMMARY OF THE PIONICS LIB CELL'S FUNDAMENTAL CHARACTERISTICS

The new cell contains specific characteristics:

- A tin alloy anode electrode system to restrain swelling and shrinkage of the cell itself

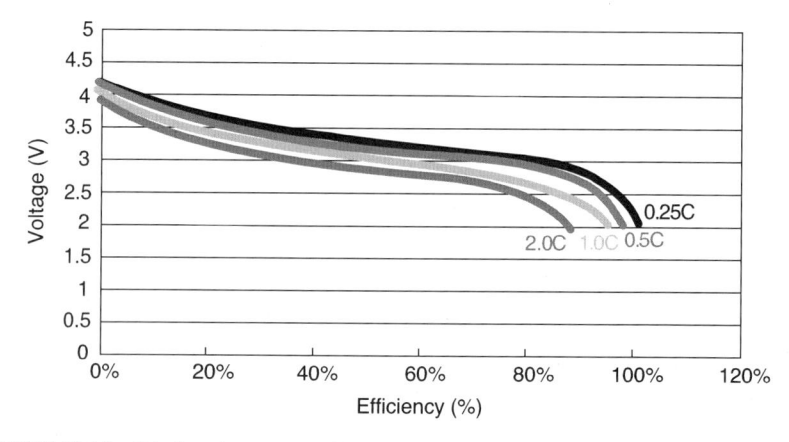

FIGURE 23.15 Discharging rate profile of Pionics LIB cell—conventional capacity grade (capacity = 2400 mA·h at 23 °C).

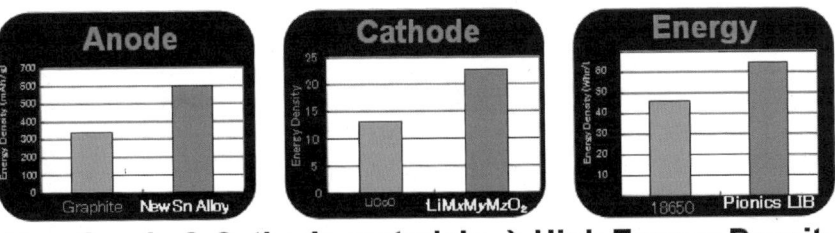

FIGURE 23.16 An illustration of the advantages of the Pionics LIB cell.

- A new formulation of anode–cathode cell design
- Utilisation of an ionic liquid–polymer compound electrolyte that is nonflammable, developed by Trekion Co., Ltd.

These are compared with conventional systems in Figure 23.16.

23.7 CONCLUSION

To sum up, this chapter introduces the concept of ionic liquid–conjugated polymer composite electrolytes, and a way to approach commercial production through an industrialisation program of PLB. As a result, Pionics is now constructing a mass-production commercial plant in Shiga, Japan, to produce ICE self-extinguishing lithium ion rechargeable batteries, as well as tin alloy type lithium ion rechargeable batteries. And then a final target might be realised—to produce a lithium metal solid-state rechargeable battery, operated at ambient temperature, with nonflammable battery materials, which would minimise the risk of fire in a short circuit accident in the cells for mobile devices, such as mobile phones, notebook PCs, and ultramobile PCs (UMPCs) (see Fig. 23.17).

Our research and development work should help elucidate the oxidation and reduction reactions in the charge–discharge operation, and the material science of the metal itself, metal oxides, and chemicals. In particular, we are using conjugated polymer selection together with a nonflammable ion-conductive electrolyte in the matrix formation of these composites (so as to achieve safety with ionic liquids over a

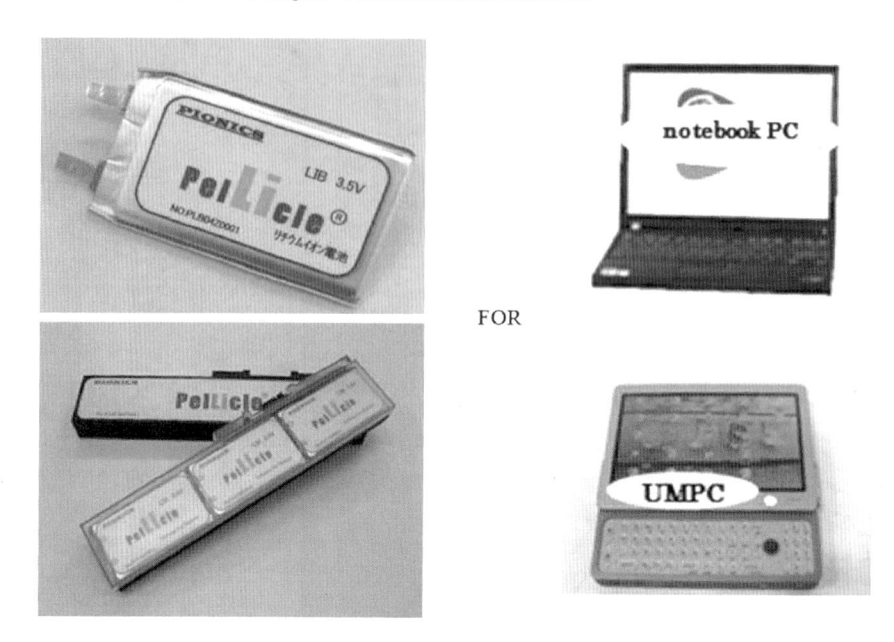

FIGURE 23.17 ICE system PLB cell and battery packs for notebook PC and UMPC.

PLB, the highest energy capacity, cell and battery pack and ICE, self-extinguishing, system installed

wide range of operating temperatures). By maintaining good diffusion coefficients for lithium ions for higher energy capacity, we gain the high rating performance of the lithium battery required for various applications, such as mobile phones, UMPCs, notebook PCs, and other devices in both consumer and industrial markets.

So far, all the lithium battery manufacturers are putting their best efforts to enhance the energy capacity of lithium battery cells from 4.2 to 4.385 V with nickel cathode materials, but Pionics has a different concept of how to achieve effective energy generation range by utilising not only higher voltages like 4.385 V with nickel type cathode, but also lower voltages like 2.25 V with tin alloy anode electrodes. This gives additional advantages, such as low temperature operation and low voltage consumption of AC–DC adaptors on notebook PC applications.

Today, the market requires safe operation of mobile devices with a reliable lithium ion rechargeable battery, without any chance of fire accident while the product is in use. The ICE self-extinguishing system PLB utilising ionic liquids is confidently recommended for the universal operation of mobile devices with installed lithium ion rechargeable batteries. Moreover, this ICE system is applicable for many other applications, such as electrolytes in solar cell batteries, coatings on various plastic film and fabric materials, and blending paints and adhesives, as well as rubber materials, in consumer and industrial markets. Trekion is now developing many kinds of ionic liquid for a wide range of applications for innovative concept materials.

Further information is available from the website, http://www.trekion.co.jp.

BIBLIOGRAPHY

1. Kono, M., Hayashi, E., and Watanabe, M., Preparation, mechanical properties, and electrochemical characterization of polymer gel electrolytes prepared from poly (alkylene oxide) macromonomers, *J. Electrochem. Soc.* **146** (5), 1626–1632 (1999).
2. Asamov, M. K., Yul'chibayev, A. A., Yakubov, I. Yu., and Usmanov, Kh. U., Synthesis and properties of the graft copolymers of polyethylene and fluorine-containing polymers, *Polym. Sci.* **28** (8), 1760–1765 (1986).
3. Percolla, R. and Calgagno, L., Grafting of styrene in polyvinylidene fluoride by high energy ion irradiation, *Ionizing Radiat. Polym.* **105** (1–4), 181–185 (1995).
4. Flint, S. D. and Slade, R. C., Investigation of radiation-grafted PVDF-*g*-polystyrene-sulfonic-acid ion exchange membranes for use in hydrogen oxygen fuel cells, *Solid State Ionics* **97** (1–4), 299–307 (1997).
5. Liu, Y., Lee, J. Y., Kang, E. T., Wang, P., and Tan, K. L., Synthesis, characterization and electrochemical transport properties of the poly(ethyleneglycol)–grafted poly(vinylidenefluoride) nanoporous membranes, *Reactive Functional Polym.* **47** (3), 201–203 (2001).
6. Kang, E. T., Neoh, K. G., Tan, K. L., and Wang, P., Plasma-induced immobilization of poly (ethylene glycol) onto poly(vinylidene fluoride) microporous membrane, *J. Membr. Sci.* **195** (1), 103–114 (2002).
7. Zhai, G., Kang, E. T., and Neoh, K. G., Poly(2-vinylpyridine)- and poly(4-vinylpyridine)-graft-poly(vinylidene fluoride) copolymers and their pH-sensitive microfiltration membranes, *J. Membr. Sci.* **217** (1–2), 243–259 (2003).
8. Kang, E. T., Neoh, K. G., Winata, A. Y., Ying, L., and Zhai, G., pH effect of coagulation bath on the characteristics of poly(acrylic acid)-grafted and poly(4-vinylpyridine)-grafted poly(vinylidene fluoride) microfiltration membranes, *J. Colloid Interface Sci.* **265** (2), 396–403 (2003).
9. Kang, E. T., Neoh, K. G., and Ying, L., Characterization of membranes prepared from blends of poly(acrylic acid)-graft-poly(vinylidene fluoride) with poly(*N*-isopropylacrylamide) and their temperature- and pH-sensitive microfiltration, *J. Membr. Sci.* **224** (1–2), 93–106 (2003).
10. Dahlan, K. Z. M., Nasef, M. M., and Suppiah, R. R., Preparation of polymer electrolyte membranes for lithium batteries by radiation-induced graft copolymerization, *Solid State Ionics* **171** (3–4), 243–249 (2004).
11. Armand, M. B., U.S. Patent 4,578,326 (1986).
12. Armand, M. B., U.S. Patent 4,758,483 (1988).
13. Akashi, H., Takahashi, K., and Tanaka, K., Research on gel-form electrolyte, characteristics of 4V type lithium battery using Tan form gel, *Prep. Polym. J. Jpn.* **44**, 381 (1995).
14. Aksahi, H., Sekai, K., and Tanaka, K., A novel fire-retardant polyacrylonitrile-based gel electrolyte for lithium batteries, *Electrochim. Acta* **43** (10–11), 1193 (1998).
15. Akashi, H. and Sekai, K., Function characteristics of polyacrylonitrile(PAN) type gel-form electrolyte, in *Advanced Technology of Polymer Battery*, CMC Co., Ltd., 1999, pp. 114–127.
16. Wright, P.V., *Br. Polym. J.* **7**, 319 (1975).
17. Armand, M., Chabagno, J. M., and Duclot, M., in *Fast Ion Transport Solids*, P. Vashishta, J. N. Mundy, and G. K. Shenoy (Eds.), North Holland, New York, 1979.

18. Shirakawa, H., Polyacetylene, in *Advanced Technologies for Polymer Battery, Volume 3, Polymer Cathode Materials*, pp. 68–74 (1998).
19. Gauthier, M., Armand, M. B., and Muller, D., Aprotic polymer electrolytes and their applications, in *Electroresponsive Molecular and Polymeric Systems*, T. Scotheim (Ed.), Vol. I, Marcel Dekker, New York, 1988, pp. 41–81.
20. Watanabe, M. and Ogata, N., Ionic conductivity of polymer electrolytes and future applications, *Br. Polym. J.* **20**, 181 (1988).
21. Vallee, A., Duval, M., Brochu, F., Kono, M., Hayashi, E., and Sada, T., U.S. Patent 5,755,985 (1998).
22. Armand, M. B. and Gauthier, M., U.S. Patent 4,303,748 (1981); U.S. Patent 4,357,401 (1982).
23. Bockris, J. O'M., Amulya K., and Reddy, N., *Modern Electrochemistry*, Vol 2, Plenum Press, New York, 1998.
24. Aoki, T., Ogata, N., Sanui, K., Takeoka, Y., and Watanabe, M., Electrochemical studies of a redox-active surfactant correlation between electrochemical responses and dissolved states, *Langmuir* **12** (2), 487–493 (1996).
25. Nakagawa, H., Development of lithium ion rechargeable battery using ambient temperature molten salt, *Battery Conf.* **41**, 364 (2000).
26. Yata, S., Polymer anode material, polyacene type material, in *Advanced Technologies for Polymer Battery*, p. 26 (1998).
27. Uetani, Y., *Industrial Mater.* **47** (2), pp. 48–49 (1999).
28. Hayashi, E., Kono, M., and Watanabe, M., Preparation, mechanical properties, and electrochemical characterization of polymer gel electrolytes prepared from poly (alkylene oxide) macromonomers, *J. Electrochem. Soc.* **146** (5), 1626–1632 (1999).
29. Watanabe, M., Characteristics of polyether type ion gel as internal polymer network (IPN) of ionic liquid, *Electrochemistry Conf.* **68**, 200 (2001).
30. Ohno, H., Synthesis of triple ion type imidazolium salt and evaluation of ion conductive property, *Electrochemistry Conf.* **68**, 203 (2001).
31. Yoneyama, H., Application of lithium battery with conductive polymer composite, *Battery Theory Meeting* **33**, 73 (1997).
32. Maruyama, T., Research of redox reaction with solid battery, *Monthly Mag. Nihon Kinzoku Gakkai* **19**, 11 (1980).
33. Kishi, T., Possible application toward lithium battery cathode with graphite phase materials including highly oxidized transition metals, *Electric Chem.* **11**, 990 (1995).
34. Matsumoto, H., Now and prospect of application technology for ionic liquid, *NTN Conf.*, p. 55 (2003).
35. Michot, T., Nishimoto, A., and Watanabe, M., Electrochemical properties of polymer gel electrolytes based on poly(vinylidene fluoride) copolymer and homopolymer, *Electrochim. Acta* **45** (8–9), 1347–1360 (2000).

24 Solubility of Al_2O_3 in NaCl–KCl Based Molten Salt System

Y. XIAO, C. R. MAMBOTE, G. A. WIERINK, and A. VAN SANDWIJK

Metals Production, Refining and Recycling, Department of Materials Science and Engineering, Delft University of Technology, Delft, The Netherlands

Abstract

Metal recovery is a crucial factor in aluminium scrap recycling, which depends not only on the recyclability of the charging scrap, but also to a certain extent on the properties of molten salt flux, for example, the impurity intake capacity and the solubility of the oxides. In the present study, the solubility of Al_2O_3 in various salt mixtures has been determined experimentally. The effects of temperature and various additives on the solubility of Al_2O_3 in NaCl–KCl based salt flux were investigated. The results prove that the temperature and cryolite addition have a positive influence on the solubility in the chloride melt. Other additives of AlF_3, KF, and NaF also have positive effects on the solubility of alumina, ranked by AlF_3, Na_3AlF_6, KF, and NaF.

24.1 INTRODUCTION

Molten salts, and particularly molten chlorides, are well known as good reaction media for performing selective solubilisation and have already been proposed as a promising route for the treatment of raw materials and subsequent recovery of valuable metals [1]. In the Hall–Héroult process for primary aluminium production, cryolite (Na_3AlF_6) is the main constituent of the electrolyte because of the high solubility of Al_2O_3. Therefore the solubility of alumina in cryolite-based salt mixtures has been studied extensively [2–5].

In recent years, secondary aluminium production has constantly been growing worldwide and has recently increased to over 40% in western Europe. Secondary aluminium

---

Molten Salts and Ionic Liquids: Never the Twain? Edited by Marcelle Gaune-Escard and Kenneth R. Seddon
Copyright © 2010 John Wiley & Sons, Inc.

has been widely used in the field of transportation, building, and packaging industries. Aluminium scrap is normally processed in a rotary furnace or a hearth furnace. The whole recycling process includes scrap collection, scrap pretreatment, melting in the furnace, refining and alloying of liquid aluminium in holding furnaces, and casting or transportation to the industrial end users. The operation temperature is normally around 800 °C. Salt flux is used to protect metal from burning and to absorb and remove contaminants and oxides.

In the secondary aluminium industry, due to the complexity of compositions and contaminants in the various types of aluminium scraps, an understanding of the behaviour of oxide impurities from the scraps during melting is important in the recycling process. High slag viscosity will lead to more fine aluminium metal entrapped in the salt slag, and thus increase the load of salt slag recycling. It was found that the viscosity of the salt flux is increased with the amount of entrapped nonmetallic components, which affects the settling of heavier materials [6]. In aluminium recycling, metal loss is a crucial factor that depends also on the salt properties. The salt flux absorbs the oxides and contaminants from the scrap and protects the aluminium melt from oxidation loss (burn-off). It consists mainly of NaCl and KCl, and some additional cryolite or CaF_2. After melting, aluminium metal and salt slag are tapped from the furnace. Depending on the scrap type, usually a large amount of salt slag is generated, which contains mainly oxides, carbides, nitrides, chlorides, and some residual aluminium metal or alloys. Usually a high salt flux factor is used to reach a higher metal recovery for melting lower grade scrap, which leads to an increased quantity of the salt slags. Due to the high consumption of the salt flux and thus high generation of salt slags, it has to be cleaned and recycled. The salt slag is actually not slag but rather slurry, which contains the salt flux and the products from the chemical reactions that occur between the scrap and the salt flux during melting. Since the properties of the salt flux affect the separation efficiency of the metal from salt slags and the metal loss, research was carried out to understand the dissolution reactions between aluminium oxide and molten salt.

Reviewing the information in the literature, no solubility data for aluminium oxide in NaCl–KCl based salt systems were found. The information on the solubility of Al_2O_3 in the molten salt system is important to understand the impurity intake of the molten salt during aluminium scrap recycling. In the present work, to provide the basic knowledge for aluminium scrap recycling, the solubility of Al_2O_3 in NaCl–KCl based molten salt systems was determined experimentally. The effects of temperature and various additives in the salt system on the solubility are investigated.

24.2 EXPERIMENTAL

24.2.1 Raw Materials

The γ-Al_2O_3 powder with 99.98% purity was used for the solubility experiment. The chemicals (NaCl, KCl, Na_3AlF_6, NaF, AlF_4, KF, $AlCl_3$, and $MgCl_2$) were of laboratory grade, and were stored and handled in a glove box in an argon atmosphere.

24.2.2 Solubility Measurements

In the experiment, about 50 g of the salt mixture with additional 6 wt % alumina was homogeneously mixed and charged in the working crucible, and placed into a high temperature gas-tight chamber furnace. The furnace was closed and flushed with 2 L-min^{-1} Ar. After equilibrium, the melt was quenched with compressed air and sampled, and the samples were ground into powder and leached in 100 cm^3 of 1 M HCl for 24 h. The leachate was analysed by atomic absorption spectrophotometry (AAS) for the dissolved amount of aluminium ions in the molten salt. The solubility of Al_2O_3 in the system was calculated based on the AAS results and the input compositions. Two types of crucibles were used: a graphite crucible for melting cryolite and fluorides in the salt system, and an alsint crucible for melting the salt mixtures without cryolite and fluoride. A crucible lid was placed on the top of the working crucible. To obtain a representative result for each salt system, some repeated melting experiments have been conducted. In addition, during each melting test, three samples were analysed from different locations in the crucible. The average of the analysis results was determined as the solubility of the reaction system. X-ray diffraction (XRD) was applied to identify the formed equilibrium compounds during melting.

24.3 RESULTS AND DISCUSSION

In order to obtain a better understanding of the molten chemistry of alumina in various salt fluxes, the solubility of Al_2O_3 over a wide range of additives was measured in NaCl–KCl based salt systems at 750, 850, and 950 °C, respectively. The alumina equilibrated molten salt was sampled, dissolved, and analysed, and the solubility of Al_2O_3 was calculated. The measured solubility results are listed in Table 24.1.

24.3.1 The System of NaCl–KCl–Al_2O_3

The thermal physical properties of the binary system NaCl–KCl have been well studied. The salt flux on equimolar composition (i.e., 44 wt % NaCl–56 wt % KCl) has the lowest melting point, which corresponds to the eutectic temperature of about 650 °C. In the present study, the salt mixtures have the compositions of 50 wt % NaCl and 50 wt % KCl, with an additional 6 wt % Al_2O_3. The mixtures were homogeneously mixed and melted at 850 °C; after equilibrium the samples were quenched with air. The solidified melt of NaCl + KCl + Al_2O_3 has two separated layers, the bottom layer with precipitated white alumina and the bulk melted salt flux with a pale pink colour. It was found that there is almost no solubility of Al_2O_3 in the NaCl–KCl binary system.

24.3.2 The System of NaCl–KCl–Na_3AlF_6–Al_2O_3

In order to increase the solubility of Al_2O_3 in the molten salt, cryolite was added into the NaCl–KCl salt system. The determination of aluminium ions was calibrated

TABLE 24.1 Measured Solubility of Al$_2$O$_3$ in Various Molten Salt Flux

Temperature/ °C	Salt Compositions/ wt%	Additions/ 6 wt%	Solubility/ wt%
850	NaCl (50) + KCl (50)	γ-Al$_2$O$_3$	0.00
850	NaCl (45) + KCl (45) + Na$_3$AlF$_6$ (10)	γ-Al$_2$O$_3$	1.20
850	NaCl (42.5) + KCl (42.5) + Na$_3$AlF$_6$ (15)	γ-Al$_2$O$_3$	1.60
850	NaCl (40) + KCl (40) + Na$_3$AlF$_6$ (20)	γ-Al$_2$O$_3$	2.14
750	NaCl (45) + KCl (45) + Na$_3$AlF$_6$ (10)	γ-Al$_2$O$_3$	1.10
750	NaCl (42.5) + KCl (42.5) + Na$_3$AlF$_6$ (15)	γ-Al$_2$O$_3$	1.81
750	NaCl (40) + KCl (40) + Na$_3$AlF$_6$ (20)	γ-Al$_2$O$_3$	2.27
950	NaCl (45) + KCl (45) + Na$_3$AlF$_6$ (10)	γ-Al$_2$O$_3$	1.26
950	NaCl (42.5) + KCl (42.5) + Na$_3$AlF$_6$ (15)	γ-Al$_2$O$_3$	1.86
950	NaCl (40) + KCl (40) + Na$_3$AlF$_6$ (20)	γ-Al$_2$O$_3$	2.32
850	NaCl (45) + KCl (45) + AlF$_3$ (10)	γ-Al$_2$O$_3$	2.00
850	NaCl (45) + KCl (45) + KF (10)	γ-Al$_2$O$_3$	1.30
850	NaCl (45) + KCl (45) + NaF (10)	γ-Al$_2$O$_3$	0.07
850	MgCl$_2$ (50) + NaCl (30) + KCl (20)	γ-Al$_2$O$_3$	0.08
750	NaCl (45) + KCl (45) + AlCl$_3$ (10)	γ-Al$_2$O$_3$	1.40
750	NaCl (45) + KCl (45) + AlCl$_3$ (10)	γ-Al$_2$O$_3$	1.35

against the NaCl–KCl–Na$_3$AlF$_6$ melt without alumina addition. With 10 wt% of cryolite (Na$_3$AlF$_6$) addition, the solubility of Al$_2$O$_3$ was increased from 0 to 1.20 wt% at 850 °C. Increasing temperature from 750 to 850 °C has resulted in a slightly increased solubility of Al$_2$O$_3$ from 1.10 to 1.20 wt%, possibly due to the high fusion enthalpy of alumina. The results are shown in Figures 24.1 and 24.2. It is clear that addition of cryolite from 10 to 20 wt% has more significant effect on the solubility than the temperature. If comparing to the solubility values of aluminium oxide (12 wt% at 1000 °C [3]) in the literature, the solubility of Al$_2$O$_3$ in 10 wt% cryolite salt flux in a NaCl–KCl–Na$_3$AlF$_6$ system of 1.1–1.3 wt% in the temperature range of 750 to 950 °C is reasonable.

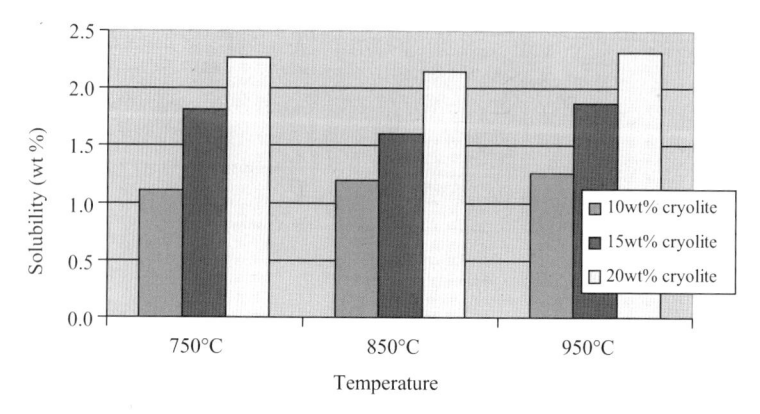

FIGURE 24.1 The measured solubility of Al$_2$O$_3$ in the NaCl–KCl–Na$_3$AlF$_6$ salt system.

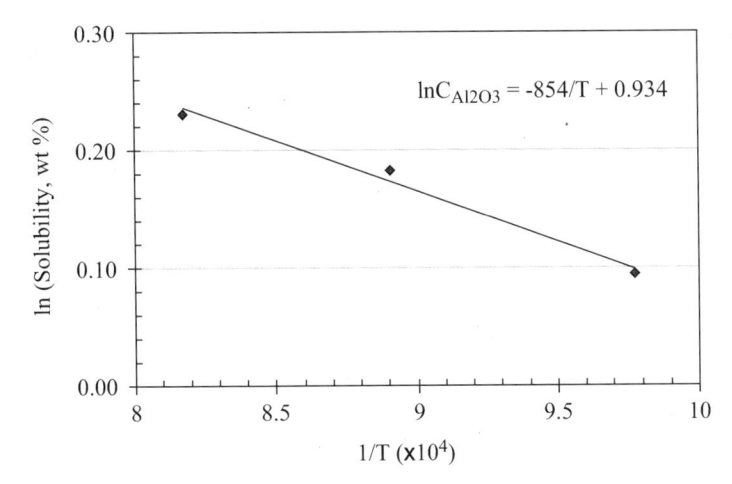

FIGURE 24.2 The relationship between Al_2O_3 solubility and temperature in 45 wt% NaCl–45 wt% KCl–10 wt% Na_3AlF_6 salt system.

According to the XRD patterns of the solidified melt with dissolved Al_2O_3, it appears that a new compound was formed. The newly formed compound is most likely K_2NaAlF_6. However, it may also be $K_2NaAlF_6 \cdot Na_2O$, which is comparable to the compound formed in the Al_2O_3–Na_3AlF_6 binary system. Due to lack of XRD database information on $K_2NaAlF_6 \cdot Na_2O$, this cannot be confirmed at the present stage.

In aluminium recycling, to recover the aluminium metal from the scrap, cryolite in the salt flux plays an important role in lowering the interfacial energy between oxide layers and the aluminium melt: the solubility of oxide in the molten slag is also important. Higher solubility of oxide in the salt melt can lead to higher metal recovery. As an example, Figure 24.3 shows the effect of cryolite on metal coalescence during melting experiments with margarine foils. The samples of margarine foils were melted at 800 °C for 2 hours under an inert atmosphere, with the added salt flux of 70 wt% NaCl–30 wt% KCl and additional 10, 15, and 20 wt% cryolite. Due to the particular surface conditions of the margarine foils, the agglomeration of the metal

FIGURE 24.3 Effect of cryolite on the coalescence of the metal droplets in the melting of aluminium margarine foils at 800 °C [6].

droplets was difficult. From the present experimental study, it is obvious that higher concentrations of cryolite in the salt flux give better coalescence of the metal droplets. An increase in the cryolite content of the salt flux means a steady increase in the size of metal beads. The addition of cryolite promotes the oxide film removal from the margarine foils, and increases the solubility of oxide in the salt flux, as proved from the present experimental measurement. However, one must also remember that cryolite addition will at the same time increase the density of the salt flux, which reduces the density difference between the salt and the metal and hence hinders the metal separation from the salt and the metal phase settling. Therefore cryolite addition should be kept to the minimum necessary for adequate performance of the process.

24.3.3 The Effect of Additives

The effect of various additives on the solubility of Al_2O_3 was investigated. Figure 24.4 shows the experimental results. Comparing various additives at a 10 wt % level, AlF_3 gives the highest solubility (2.0 wt %) for Al_2O_3, followed by KF (1.3 wt %), Na_3AlF_6 (1.2 wt %), and NaF (0.07 wt %) at 850 °C. However, with 50 g salt mixture, 10 wt % additives are equivalent to 0.024 mole Na_3AlF_6, 0.060 mole AlF_3, 0.086 mole KF, and 0.119 mole NaF, respectively. So far, the mechanism of alumina dissolution in chloride-based salt systems is not clear, and further research is required to understand the ionic structure of the system (e.g., with Raman spectra).

In general, no stirring was applied in the present experimental research. Due to the added fine Al_2O_3 powder in the experiments, oversaturation or entrapment problems may occur. However, a uniform concentration was confirmed using three samples taken at different locations in the melted salt flux and the repeated AAS analysis. The effect of salt vapourisation on the solubility was considered small, and thus ignored.

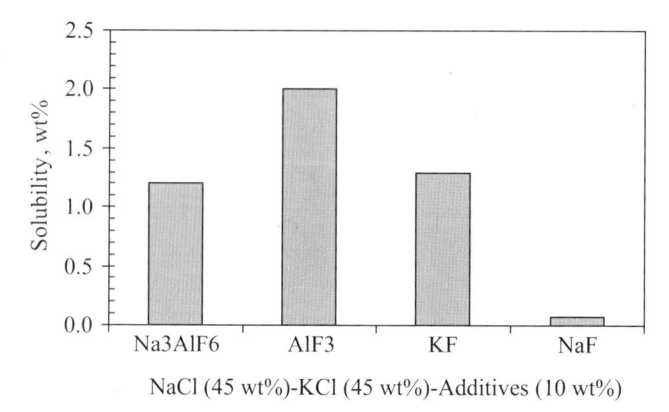

FIGURE 24.4 Effect of additives on the solubility of Al_2O_3 in a NaCl–KCl based salt system at 850 °C.

24.4 CONCLUSION

According to the experimental results from the solubility measurements, Al_2O_3 does not dissolve in the NaCl–KCl system, even at 850 °C. Cryolite addition and temperature increases increase the solubility of Al_2O_3 in the chloride melt. The influence of temperature is less significant than the cryolite addition. The solubility of Al_2O_3 in the NaCl–KCl–Na_3AlF_6 system increases from 1.2 to 2.3 wt %, when the Na_3AlF_6 increases from 10 to 20 wt % in the chloride melt. For the effect of temperature on the NaCl–KCl–Na_3AlF_6 system containing 10 wt % of cryolite, when the temperature increases from 750 to 950 °C, the solubility of Al_2O_3 increases from 1.1 to 1.26 wt %. Comparing various additives at the 10 wt % level, AlF_3 gives the highest solubility of Al_2O_3 (2.0 wt %), followed by KF (1.3 wt %), Na_3AlF_6 (1.2 wt %), and NaF (0.07 wt %) at 850 °C.

REFERENCES

1 Castrillejo, Y., Bermejo, M. R., Barrado, E., Martinez, A. M., and Arocas P. D., Solubilisation of rare earth oxides in the eutectic LiCl–KCl mixture at 450 °C and in the equimolar $CaCl_2$–NaCl melt at 550 °C, *J. Electroanalytical Chem.* **545**, 141–157 (2003).

2 Zhang, Y., Wu, X., and Rapp, R. A., Solubility of alumina in cryolite melts: measurements and modelling at 1300 K, *Metall. Mater. Trans. B*, **34B**, 235–242 (2002).

3 Skybakmoen, E., Solheim, A., and Sterten A., Alumina solubility in molten salt of interest for aluminum electrolysis and related phase diagram data, *Metall. Mater. Trans. B* **28B**, 81–86 (1997).

4 Fenerty, A. and Hollingshead, E. A., Liquidus curves for aluminum cell electrolyte, *J. Electrochem. Soc.* **107**, 993–997 (1960).

5 Foster, P. A. Jr., Phase diagram of a portion of the system Na_3AlF_6–AlF_3–Al_2O_3, *J. Am. Ceram. Soc.* **58**, 288–291 (1975).

6 Xiao, Y., Reuter, M. A., and Boin, U., Aluminum recycling and environmental issues of slag treatment, *J. Environ. Sci. Health A: Toxic/Hazardous Substance Environ. Eng.* **40** (10), 1861–1875 (2005).

25 Molten Salt Synthesis of Ceramic Materials

SHAOWEI ZHANG

Department of Engineering Materials, University of Sheffield, Sheffield, United Kingdom

D. D. JAYASEELAN, ZUSHU LI, and WILLIAM EDWARD LEE

Department of Materials, Imperial College, London, United Kingdom

Abstract

Two different reaction mechanisms, template formation and dissolution–precipitation, are involved in the molten salt synthesis (MSS) process. Based on these two mechanisms, MSS of $CaZrO_3$ powder, $MgAl_2O_4$ platelet, and $LaAlO_3$ powder have been carried out. The synthesis temperature of $CaZrO_3$ is reduced to 1050 °C by using calcium chloride ($CaCl_2$), sodium carbonate (Na_2CO_3), and zirconia (ZrO_2) powders as starting materials. Based on the fact that Al_2O_3 has much lower solubility than MgO in molten K_2SO_4, $MgAl_2O_4$ platelets are readily prepared by treating Al_2O_3 platelets (template particles) with MgO at 1150 °C for 3 h in the salt. MSS of $LaAlO_3$ in KF–KCl eutectic salts is dominated by the dissolution–precipitation mechanism: its synthesis temperature is reduced from about 1600 °C (used by CMOS) to as low as 630 °C.

25.1 INTRODUCTION

In ceramics engineering, preparation and use of high quality raw materials is key for manufacturing components with guaranteed properties and performance. A number of different oxide raw materials, either single (unitary) or complex oxides, are used to fabricate ceramics. The complex oxides used are conventionally synthesised by a high temperature solid–solid reaction of appropriate precursor compounds (conventional mixed oxide synthesis, CMOS). Since the reactions are generally controlled by a slow

---

Molten Salts and Ionic Liquids: Never the Twain? Edited by Marcelle Gaune-Escard and Kenneth R. Seddon
Copyright © 2010 John Wiley & Sons, Inc.

diffusion mechanism, highly reactive precursor powders, high temperatures, and long times have to be used to complete the reaction. Moreover, the reaction product is often strongly agglomerated, requiring repeated crushing and grinding to achieve the desired size before fabrication into ceramic articles. Uncontrollable quantities of undesirable impurities from the grinding media may be introduced during this process, which, in many cases, will affect the properties of the final product [1]. Another drawback of CMOS is that the synthesised powders are often not fully reacted to yield uniform stoicheiometry at the nano or even micro level. Furthermore, the milled powders have low reactivity, so higher firing temperature and longer times are needed.

Electrofusion processes are sometimes used instead. Reactions via this route are more complete than by CMOS and large crystals, when desired, can be obtained [1]. However, this process requires even higher temperatures (above the mix melting point) and so is highly energy intensive and thus expensive. Furthermore, the solidified aggregates also need to be crushed and ground to fine powders before being used as raw material powders, and the powder surface reactivity is even lower than that of the powder prepared by CMOS.

To prepare high quality (uniform and reactive) complex oxide powders at low temperature, several *wet chemical synthesis* methods have been developed, including sol-gel routes [1, 2] and hydrothermal processes [3]. Complex oxide powders prepared by these processes are pure, homogeneous, and highly reactive, conferring excellent properties on the final products. Unfortunately, these processes still suffer from several drawbacks. Sol-gel processes often require expensive (and environmentally unfriendly) organic precursors and solvents (e.g., metal alkoxides and alcohol), and so are applicable only to niche cases and for small scale products (e.g., thin films or coatings or nanoscale powders). Hydrothermal synthesis sometimes uses similar precursors and requires a high pressure autoclave and often a long reaction time.

Besides wet chemical synthesis techniques, a low temperature technique, *molten salt synthesis* (MSS), is beginning to attract interest, although so far it has been used mainly to synthesise relatively low melting oxide powders [4, 5]. To illustrate the feasibility of MSS for synthesis of high melting complex oxides, investigations of MSS of $CaZrO_3$, $MgAl_2O_4$, and $LaAlO_3$ powders, which are used extensively in the ceramics/refractories industries, are described in this chapter.

25.2 EXPERIMENTAL

25.2.1 Batch Mixing and Firing

CaZrO₃ Powder Aldrich ZrO_2 (<5 μm, 99% pure), Na_2CO_3 (<1 mm, ≥99.5% pure), and $CaCl_2$ were mixed and ground in molar ratio of $ZrO_2:CaCl_2:Na_2CO_3 = 1.0:1.0:1.2$ using an agate mortar. The powder mixture (20 g) was placed in an alumina crucible covered with an alumina lid, heated to a given temperature between 600 and 1050 °C and held for 3–5 h.

MgAl₂O₄ Platelet An equimolar composition of MgO (Cerac, Grade-M1016, purity 99.14%) and α-Al₂O₃ platelets (prepared from partially decomposed Al(SO₄)₃ using Na₂SO₄ flux [6]) was homogeneously mixed in propanone, using an agate mortar, without disrupting the morphology of α-Al₂O₃ platelets and dried. The dried powder was later mixed with K₂SO₄ (Aldrich, >99%) in the weight ratio of 3:1. The powder mixture was placed in an alumina crucible covered with an alumina lid, heated to 1150 °C and held for 3 h.

LaAlO₃ Powder La₂O₃ powder (Aldrich, ~1 μm, ≥99.9% pure), low-soda calcined Bayer-derived Al₂O₃ powder (Almatis, D50 = 0.80 μm, 99.8% pure), ACS reagent KCl (Aldrich, ≥99.0% pure), and reagent grade KF (Aldrich, 98% pure) were used as starting materials. Equimolar La₂O₃ and Al₂O₃ powders were mixed in an agate mortar with KF–KCl eutectic salt. The molar ratio of KF to KCl in the eutectic salt was 46.7:53.3 and the weight ratio of (KF + KCl) to (La₂O₃ + Al₂O₃) was 3:1. The samples were heated in an alumina crucible to 630, 700, and 800 °C and held at the temperature for 3 h.

25.2.2 Leach of Salt

After furnace-cooling to room temperature, the reacted mass was washed for 2 h in hot distilled water, followed by filtration to remove the salts. This process was repeated five times. The resulting oxide powder was oven-dried prior to further characterisation.

25.2.3 Characterisation

Phases and extents of reaction in the powder sample were identified by powder X-ray diffraction (XRD) analysis (Siemens D500 Reflection Diffractometer). XRD patterns were recorded at 30 mA and 40 kV, using Ni-filtered Cu Kα radiation. The scan rate was 1° (2θ) per minute. Microstructural morphologies of the as-received raw materials and the synthesised powders were observed using SEM (JEOL6400, Japan) and in some cases a high resolution FEI Sirion FEG-SEM. EDS was also used to assist the phase identification.

25.3 RESULTS AND DISCUSSION

25.3.1 MSS of CaZrO₃ Powder

Figure 25.1 shows XRD from samples heated for 3 h at different temperatures. After heating for 3 h at 600–650 °C, CaCO₃ (formed from the reaction between CaCl₂ and Na₂CO₃) and *m*-ZrO₂ were identified as the main phases, but CaZrO₃ was not detected. CaZrO₃ peaks began to appear at 700 °C. With increasing temperature from 700 to 1000 °C, CaZrO₃ peak height increased whereas *m*-ZrO₂ and CaCO₃ peak heights decreased significantly. Small amounts of CaO-stabilised ZrO₂ (CSZ) were

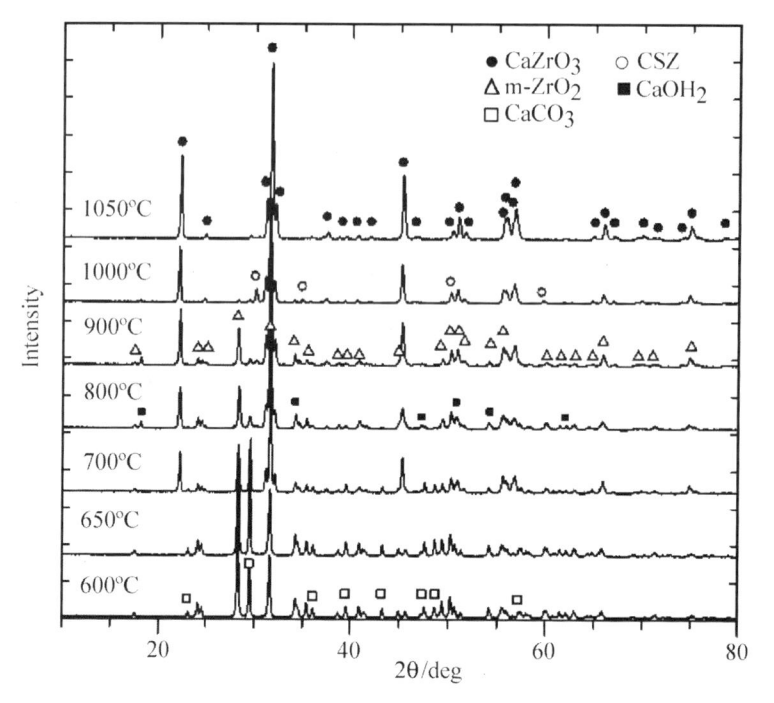

FIGURE 25.1 XRD of water-washed powders showing the effect of a 3 h hold at the heating temperature on formation of CaZrO$_3$.

detected after heating for 3 h at 900–1000 °C. On further increasing the temperature to 1050 °C, other phases such as m-ZrO$_2$ and CSZ disappeared, and only CaZrO$_3$ along with small amounts of CaCO$_3$ were detected.

Figure 25.2 further gives XRD from samples heated at 1050 °C for 3 and 5 h. On increasing reaction time from 3 to 5 h, small amounts of CaCO$_3$ found in the samples heated for 3 h at 1050 °C disappeared, and only single-phase CaZrO$_3$ was detected. This synthesis temperature is at least 250 °C lower than that required by CMOS [7], and close to that used by wet chemical synthesis routes, which is attributable to the more homogeneous mixing and more rapid diffusion of species such as Ca^{2+} and O^{2-} in the MSS liquid–solid system than in the CMOS solid–solid system.

The morphologies of the as-received ZrO$_2$ and CaZrO$_3$ powders synthesised for 5 h at 1050 °C are illustrated in Figure 25.3. The shape of CaZrO$_3$ grains (rounded/angular) was similar to that of the ZrO$_2$ powder, although their size (0.5–1.0 μm) was slightly greater than the ZrO$_2$ (0.2–0.5 μm). The similarity can be explained as follows. As reported previously [8], the reactant solubilities in the molten salt play an important role in MSS, because they affect not only the reaction rate but also the morphologies of the synthesised grains. If both of the reactants are soluble in the molten salt, then the product phase will readily be synthesised via precipitation

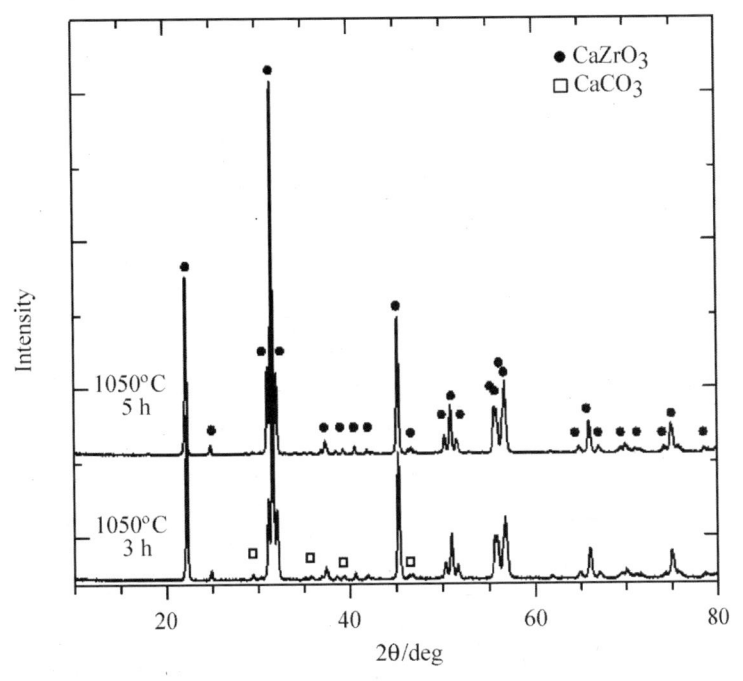

FIGURE 25.2 XRD of water-washed powders after heating at 1050 °C for various times showing effect of heating times on CaZrO$_3$ formation.

from the salt containing the dissolved reactants (*dissolution–precipitation* mechanism). In this case, the morphologies of the product phases generally will be different from those of the reactants. On the other hand, if one reactant is much more soluble than another, then the more soluble reactant will dissolve into the salt first and then diffuse onto surfaces of the less soluble reactant and react in situ to form the product phase. In this case, the morphology of the synthesised phase will, to a large extent, retain that of the less soluble reactant (*template formation* mechanism). According to Zhu et al. [9–12], both CaCO$_3$ and CaO (formed from the decomposition of CaCO$_3$) are soluble in a chloride molten salt. Their solubilities in a NaCl-based salt at 700–1000 °C are on the order of 10^{-3} (molar fraction), which is about 1000 times higher than that of ZrO$_2$ (on the order of 10^{-6}) [10–12]. Therefore, during the MSS process, CaCO$_3$/CaO would dissolve more in the NaCl–Na$_2$CO$_3$ molten salt and react with ZrO$_2$ "templates" to form in situ CaZrO$_3$.

25.3.2 MSS of MgAl$_2$O$_4$ Platelets

Figure 25.4 shows the SEM of the as-received α-Al$_2$O$_3$ platelets (template) and the MgAl$_2$O$_4$ platelets obtained by MSS in molten K$_2$SO$_4$. MgAl$_2$O$_4$ platelets with an edge length of ~5 μm and thickness of ~1 μm were observed. They retained

FIGURE 25.3 SEM of (a) as-received ZrO$_2$ powder and (b) CaZrO$_3$ particles synthesised after 5 h at 1050 °C.

the typical hexagonal plate morphology, indicating that MgO is the fast dissolving component of the constituent oxides in molten K$_2$SO$_4$ and that the formation of MgAl$_2$O$_4$ initiates on the Al$_2$O$_3$ platelet surfaces, that is, the template formation mechanism already discussed (vide supra) has dominated the MSS in this case as well.

25.3.3 MSS of LaAlO$_3$ Powder

XRD (Fig. 25.5) of the powders from samples heated for 3 h at 630, 700 and 800 °C, respectively, revealed that only LaAlO$_3$ with a rhombohedral perovskite crystal structure, and no other phases were identified in the samples. The lowest synthesis temperature (630 °C) is about 1000 °C lower than by CMOS, and close to or lower than those used by most wet chemical techniques.

FIGURE 25.4 SEM of (a) the as-received Al_2O_3 platelet and (b) the $MgAl_2O_4$ platelet synthesised in molten K_2SO_4 at 1150 °C for 3 h.

Figure 25.6 further shows SEM of the as-received raw materials and the $LaAlO_3$ particles synthesised at 630 °C for 3 h. Although spheroidal La_2O_3 and Al_2O_3 starting powders were used (Fig. 25.6a,b), the synthesised $LaAlO_3$ particles showed well-crystallised euhedral shapes (Fig. 25.6c), indicating that the template formation mechanism involved in the MSS processes of $CaZrO_3$ and $MgAl_2O_4$ (vide supra) is not significant: in other words, the dissolution–reaction–precipitation mechanism is dominant in this case. As reported elsewhere [13–15], both La_2O_3 and Al_2O_3 are highly soluble in molten potassium fluoride/chloride salts. The La_2O_3 and Al_2O_3 dissolved in the molten salt are mixed homogeneously at the atomic level, and diffusion is more rapid than in the solid state. Consequently, the dissolved La_2O_3 and Al_2O_3 react rapidly to form $LaAlO_3$ in the molten salt.

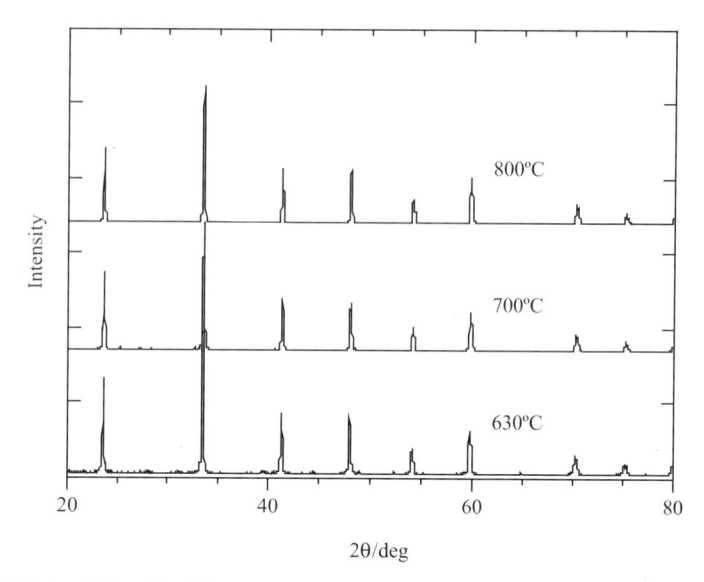

FIGURE 25.5 XRD of LaAlO$_3$ powders synthesised at various temperatures for 3 h, showing that only LaAlO$_3$ was identified.

Once the molten salt is oversaturated with LaAlO$_3$, LaAlO$_3$ crystals start to precipitate from the salt. The precipitation of LaAlO$_3$ from the oversaturated salt leads to further dissolution and reaction of La$_2$O$_3$ and Al$_2$O$_3$, and then more precipitation of LaAlO$_3$. This cycle is repeated until all of the starting La$_2$O$_3$ and Al$_2$O$_3$ are used up in forming LaAlO$_3$.

25.4 CONCLUSION

1. Depending on the reactant solubilities in the molten salt, two main reaction mechanisms are involved in MSS. If both of the reactants are soluble in the molten salt, then the dissolution–precipitation mechanism will be dominant. On the other hand, if one reactant is much more soluble than another, then the template formation mechanism will be dominant.
2. Although MSS of CaZrO$_3$ is dominated by the template formation mechanism, it can be synthesised at 1050 °C using calcium chloride (CaCl$_2$), sodium carbonate (Na$_2$CO$_3$), and zirconia (ZrO$_2$) powders as starting materials. This synthesis temperature is at least 250 °C lower than that used by CMOS, and close to that used by wet chemical synthesis.
3. Based on the template formation mechanism, MgAl$_2$O$_4$ platelets can readily be prepared by treating Al$_2$O$_3$ platelets (template) with MgO in a molten K$_2$SO$_4$ salt at 1150 °C for 3 h.
4. MSS of LaAlO$_3$ in KF–KCl eutectic salts is dominated by the dissolution–precipitation mechanism. Its synthesis temperature is reduced to as low as

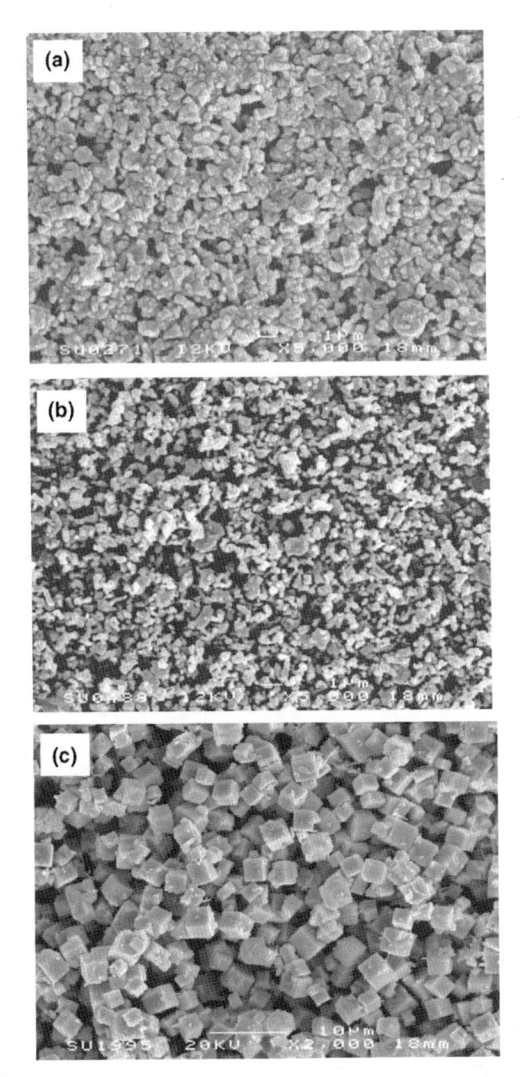

FIGURE 25.6 SEM of as-received raw (a) La_2O_3 and (b) Al_2O_3 powders and (c) $LaAlO_3$ particles synthesised for 3 h at 630 °C.

630 °C, which is about 1000 °C lower than by CMOS, and close to or lower than those used by most wet chemical techniques.

REFERENCES

1. Lee, W. E. and Rainforth, M., *Ceramic Microstructures: Property Control by Processing*, Chapman and Hall, London, 1994.

2. Brinker, C. J. and Scherer, G. W., *Sol-Gel Science: The Physics and Chemistry of Sol-Gel Processing*, Academic Press, San Diego, CA, 1990.
3. Wang, D., Yu, R., Feng, S., Zheng, W., Takei, T., Kumada, N., and Kinomura, N., Hydrothermal synthesis of perovskite-type solid solution of $(1-x)BaTiO_3 \cdot xLa_{2/3}TiO_3$, *Solid State Ionics* **151**, 329–33 (2002).
4. Battisha, I. K., Speghini, A., Polizzi, S., Agnoli, F., and Bettinelli, M., Molten chloride synthesis, structural characterisation and luminescence spectroscopy of ultrafine Eu^{3+}-doped $BaTiO_3$ and $SrTiO_3$, *Mater. Lett.* **57**, 183–187 (2002).
5. Wang, X., Gao, L., Zhou, F., Zhang, Z., Tang, M. Ji C., Shen, T., and Zheng, H., Large-scale synthesis of α-$LiFeO_2$ nanorods by low-temperature molten salt synthesis (MSS) method, *J. Crystal Growth* **265** (1–2), 220–223 (2004).
6. Hashimoto, S. and Yamaguchi, A., Synthesis of α-Al_2O_3 platelets using sodium sulfate flux, *J. Mater. Res.* **14** (12), 4667–4672 (1999).
7. Nadler, M. R. and Fitzsimmons, E. S., Preparation and properties of calcium zirconate, *J. Am. Ceram. Soc.* **38**, 214–217 (1955).
8. Zhang, S., Jayaseelan, D. D., Bhattacharya, G., and Lee, W. E., Molten salt synthesis of magnesium aluminate ($MgAl_2O_4$) spinel powder, *J. Am. Ceram. Soc.* **89**, 1724–1726 (2006).
9. Roth, R. S., Clevinger, M. A., and McKenna, D., *Phase Diagrams for Ceramists*, G. Smith (Ed.), American Ceramic Society, Columbus, OH, 1984.
10. Boyarchuk, T. P., Khailova, E. G., and Cherginets, V. L., Potentiometric measurements in molten chlorides: solubilities of metal oxides in the molten eutectic mixture CsCl–KCl–NaCl at 600°C, *Electrochim. Acta* **38** (10), 1481–1485 (1993).
11. Cherginets, V. L., Deineka, T. G., Demirskaya, O. V., and Rebrova, T. P., Potentiometric investigation of oxide solubilities in molten KCl–NaCl eutectic. The effect of surface area of solid particles on the solubilities, *J. Electroanal. Chem.* **531**, 171–178 (2002).
12. Cherginets, V. L. and Khailova, E. G., On the solubility of bivalent metal oxides in molten alkaline chlorides, *Electrochim. Acta* **39** (6), 823–829 (1994).
13. Balaraju, J. N., Ananth, V., and Sen, U., Studies on low temperature Al electrolysis using composite anodes in NaF–KCl bath electrolyte, *J. Electrochem. Soc.* **142** (2), 439–444 (1995).
14. Belov, S. F., Gladneva, A. F., Matveev, V. A., and Igumnov, M. S., Solubility of lanthanum and neodymium oxides in molten fluorides, *Izv. Akad. Nauk SSSR Neorg. Mater.* **8** (5) 966–967 (1972).
15. Dewing, E. W., The chemistry of the alumina reduction cell, *Can. Metall. Q.* **30** (3), 153–161 (1991).

26 Fuel Cell and Electrolysis Studies with Dual Phase Proton and Oxide Ion Conduction

BIN ZHU

Department of Energy Technology, Royal Institute of Technology (KTH), Stockholm, Sweden and Institute of Materials and Technology, Dalian Maritime University, Dalian, People's Republic of China

S. LI, X. L. SUN, and J. C. SUN

Institute of Materials and Technology, Dalian Maritime University, Dalian, People's Republic of China

Abstract

Ceria-based dual-phase composite electrolytes have been investigated based on fuel cell and electrolysis studies. The results showed that both proton conduction and oxygen ion conduction exist in the ceria-based composite electrolytes, resulting in high current outputs in both fuel cell and electrolysis. Corresponding to high current output in the electrolysis cell, a high hydrogen production path can be expected.

Dual or hybrid proton and oxide ion conduction is based on a consideration that both proton (H^+) and oxide ion (O^{2-}) are the fuel cell source ions. Proton conduction is important for low temperature solid oxide fuel cells (LTSOFCs), because it can be activated easier than oxide ions in the low temperature (300–600 °C) region, thus making significant conducting and electrical contributions. The composite approach can combine or integrate multi-ion functions, typically dual H^+ and O^{2-} conduction, to enhance the material conductivity and device performance.

26.1 INTRODUCTION

Extensive studies have contributed to ion-doped ceria forms (e.g., Gd^{3+}- or Sm^{3+}-doped CeO_2 (GDC or SDC)) that are regard as the most promising electrolyte

---

Molten Salts and Ionic Liquids: Never the Twain? Edited by Marcelle Gaune-Escard and Kenneth R. Seddon
Copyright © 2010 John Wiley & Sons, Inc.

candidates for intermediate temperature (IT: 600–800 °C) solid oxide fuel cells (SOFCs) [1–4]. The ion-doped ceria solids are single-phase materials, and also single ion (O^{2-}) conductors. However, there is a problem with ceria material instability for electronic conduction in fuel cell applications. In order to improve the single-phase ceria electrolytes and develop intermediate and low temperature (300–800 °C) solid oxide fuel cell (IT/LTSOFC) applications, the ceria-based dual-phase composite electrolytes have been investigated widely, including various ion-doped ceria, such as SDC (samarium-doped ceria), GDC (gadolinium-doped ceria), YDC (yttrium-doped ceria), and CDC (caesium-doped ceria) and composites with different salts and hydrates, such as chlorides [5–7], fluorides [8], carbonates [8–12], sulfates [13], and sodium or potassium hydrates [8, 14]. Among these ceria composites, the SDC (or GDC and CDC) carbonate composites are the most commonly used materials, exhibiting the best performances in many LTSOFC applications [9–12, 15]. The two-phase ceria composite electrolytes, specifically the ceria–carbonate ($Li_2[CO_3]$–$Na_2[CO_3]$) containing one molten phase, realised superionic conductivity (0.1 S·cm^{-1} at 600 °C). This solid-like, or semi-solid and soft-solid, composite electrolyte with high ion conduction based on the composite containing one molten phase was discovered in 1993 [16].

Dual- or multiphase composite electrolytes can also merge or integrate various kinds of ion conduction from two material phases into one composite bulk, forming the hybrid or dual-ion conductors (e.g., for proton and oxide ion conduction), with new functions. Our innovative development of dual H^+/O^{2-} conductors in a two-phase composite to create superfunctional materials for advanced LTSOFCs is very different from other SOFC developments using the ion-doped single-phase and single-ion-conducting oxides.

Ceria-based composite materials (usually with two/dual phases) have demonstrated major advantages in fuel cell applications over single-phase electrolytes. Proton conduction is important for LTSOFCs because it can be thermally activated at low temperatures. The dual proton and oxide ion conduction creates new advantages for LTSOFCs.

26.2 EXPERIMENTAL

26.2.1 Ceria-Based Two-Phase Composite Preparations

Typical SDC–carbonate composite electrolytes were prepared in two steps. The first step is the SDC preparation. The SDC was prepared by the solution route with coprecipitation. Starting chemicals used were as follows: cerium(III) nitrate hexahydrate (supplied by Sigma-Aldrich) and samarium(III) nitrate hexahydrate (Aldrich). These chemicals were prepared as 1 M aqueous solutions. Then the samarium(III) nitrate solution was mixed with the cerium(III) nitrate solution, according to the desired molar ratios. An appropriate amount of aqueous oxalic acid solution (2 M) was added to prepare the SCO precursor in the oxalate state. The precipitate was rinsed several times in deionised water, followed by several ethanol washings, in order to remove water from the particle surfaces. The obtained precipitates were dried in an oven at 100 °C overnight and then ground in a mortar. The resulting powder was sintered at 700 °C for 2 h to obtain the SDC.

For the second step, the SDC–carbonate composites were prepared by (1) mixing the SDC and (Li–Na)$_2$[CO$_3$] (Li: Na = 2:1 molar ratio) (represented as LNCO) powders in a 80:20 wt % ratio, (2) grinding the mixtures, which were then sintered at 650 °C for 30 min for SDC–LNCO, and at 750 °C for 30 min for SDC–CBCO, and (3) finally grinding to yield the SDC–carbonate composite electrolytes.

26.2.2 Fuel Cell Studies

The characteristics of a fuel cell are often determined by an *I–V* diagram showing the cell performance in terms of voltage versus current. The *I–V* plot shows the fuel cell voltage as a function of current. The current and voltage are measured under various loads. The measurements give information about the electrolyte resistance, the reaction kinetics at the electrodes, and gas diffusion limitations at the electrodes. In this work, the main experimental part of the *I–V* measurements is made in order to investigate the performance of the prepared fuel cell PEN (i.e., the anode/electrolyte/cathode assembly). We used an electrolyte-supported cell using Pt paste (Leitplatin 308A, Hanau) electrodes, where the electrolyte was about 1 mm thick. The fuel cell PEN assemblies were sandwiched between two gas-distributing rings with an inner diameter of 0.95 mm and an effective gas area of 0.71 cm^2. The rings were made of stainless steel and painted with silver paste (Leitsilber 200, Hanau), which provides better contact and sealing between the electrodes and the gas-distributing rings. The fuel (hydrogen) and the oxidant (air) were supplied via tubes to the anode and cathode chambers, respectively. A variable load unit was used to measure the current and voltage. The fuel cell was inserted into a cylinder furnace and the fuel cell temperature was measured by a thermocouple. After introducing fuel at the set temperature, the OCV increased to a steady value. The current–voltage characteristic was obtained by drawing current at incremental steps and measuring the cell voltage at each step at steady state.

26.2.3 Electrolysis

The electrolysis experiment was performed using the fuel cell setup described above, but operated in a reverse process mode. The electrolysis cell was connected to a constant voltage supply and a multimeter was used to measure the current through the circuit. The applied voltage was varied between 0.5 and 3.0 V, and the current–voltage characteristics were obtained by drawing the voltage versus current in 0.5 V steps. When a more exact determination was needed, for example, at the transition point of the electrolysis, smaller voltage steps (e.g., 0.2 V) were used. The current was measured at each step at steady state.

Water (steam) was supplied to one chamber, either to the anode or to the cathode, through an air flow saturated by means of a bubbler that contained boiling, or alternatively room temperature, water. The other chamber (cathode or anode) was supplied by nitrogen gas in order to prevent the water (steam) from the air entering the second chamber.

26.3 RESULTS AND DISCUSSION

26.3.1 Structural Characteristics

The SDC–carbonate composites are a typical ceria composite containing either a solid Ba[CO$_3$]–Ca[CO$_3$] (CBCO) or molten Li$_2$[CO$_3$]–Na$_2$[CO$_3$] (LNCO) carbonate phase in the measured temperatures for fuel cell or electrolysis. Figure 26.1 shows the XRD results obtained from the SDC–NLCO composites. The XRD results obtained from the SDC–LNCO composites show only the SDC phase—the molten carbonate is invisible from the XRD pattern. The EDX analysis, however, suggests that the LNCO exists in the SDC–LNCO composite, where Na is clearly identified; see Figure 26.2 The XRD results demonstrated there was neither chemical reaction nor new compound between the SDC and carbonate phases. The carbonate component was amorphous and highly distributed among the SDC phase.

Differential thermal analysis (DTA) was carried out on the carbonate and SDC–LNCO composite samples (see Fig. 26.3). A mixture of pure Li$_2$[CO$_3$] and Na$_2$[CO$_3$] showed a melting point at 515.7 °C, and no DTA peak was detected for the pure SDC from room temperature to 700 °C (Fig. 26.3a). In contrast, the SDC–carbonate composites, heat-treated at 500, 680, and 800 °C, or without heat treatment, respectively, showed significantly different DTA curves, usually two endothermic peaks appearing, where one appears at 507 °C, 10 °C below the Li$_2$[CO$_3$] and Na$_2$[CO$_3$] melting point; and the other occurs at a higher temperature, up to 560–580 °C (Fig. 26.3b).

The two thermal effects (endothermic peaks) cannot be interpreted as a simple melting effect caused by the carbonate, since the XRD analysis of the SDC–LNCO composite showed that there were only two phases: the SDC and carbonate

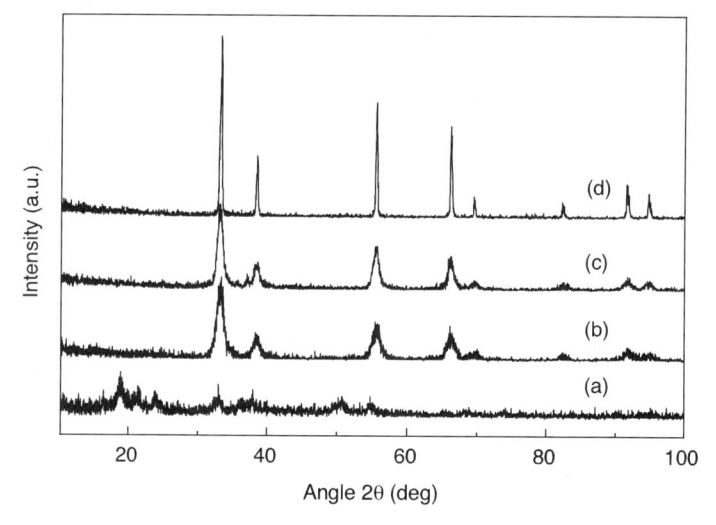

FIGURE 26.1 XRD patents of the SDC–carbonate composites: (a) containing solid carbonate, SDC–CBCO, and (b) containing molten carbonate, SDC–NLCO.

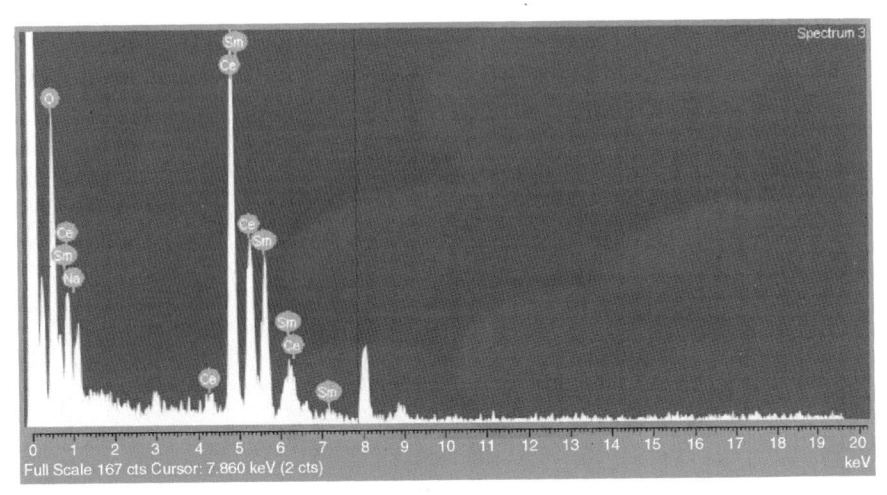

FIGURE 26.2 EDX analysis on the SDC–NLCO sample.

phases coexisting without new compounds being formed. To explain the two endothermic peaks, an interfacial structure or boundary phase between the SDC and carbonate phases may be suggested, which could be responsible for the material properties (e.g., superionic conduction as the electrolytes for fuel cell applications).

A certain amount of the molten carbonate highly distributed and incorporated in the interfacial regions of the ceria oxide grains/phases can form composites with a controlling level of the microstructure. In this case, a certain amount of the molten carbonate can cause extremely high conductivity due to interfacial effects, but cannot weaken the mechanical strength. So the system can maintain a solid-like state. This new material concept provides a mechanism to achieve high ionic conductivity, $>10^{-1}$ S·cm^{-1}, in the low temperature region. The chemical stability and strong mechanical strength of the oxide benefit the system at the same time. They can also provide the advantages of less corrosion compared with the pure molten salt phase, and higher conductivity than the solid state. Therefore the SDC–LNCO composite represents some kind of intermediate phase between solid and molten (liquid) states, which can combine the advantages from the SOFCs and MCFCs (molten carbon fuel cells).

26.3.2 Fuel Cell Studies

For a H_2/O_2 fuel cell with an O^{2-} ion conducting electrolyte, oxygen is reduced at the cathode side to form oxide ions, which are transported through the electrolyte to react with hydrogen at the anode side to form water. In contrast, if a proton conducting electrolyte is used, hydrogen will be oxidised at the anode to form protons that are transported through the electrolyte to react with O_2 to produce water at the cathode.

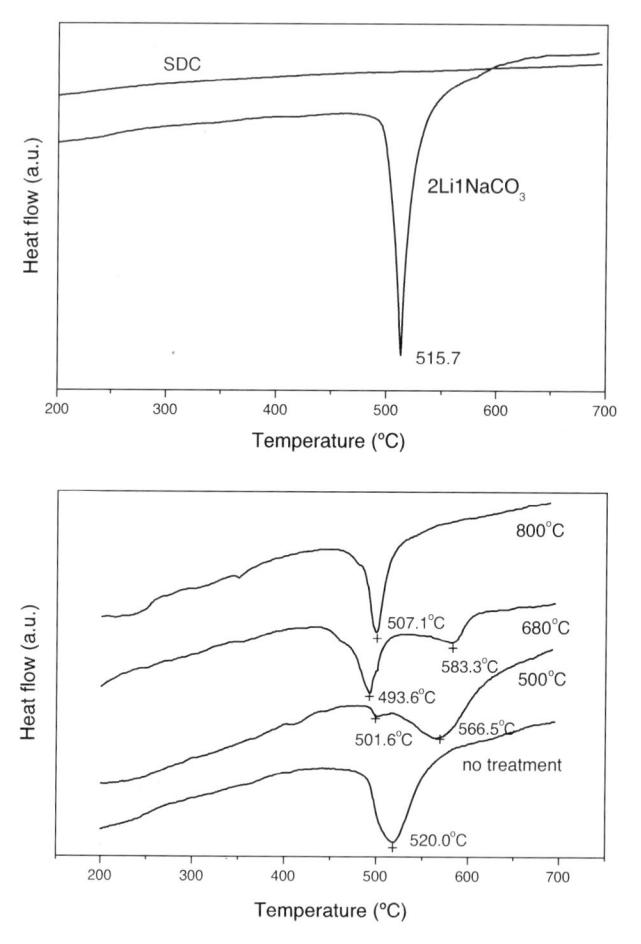

FIGURE 26.3 DTA effect profiles on (a) the LiNaCO₃ (Li:Na in 2:1 weight ratio) and (b) the SDC–20 wt % LiNaCO₃ without and with heat treatment at 500, 680, and 800 °C, respectively.

In a dual H^+/O^{2-} electrolyte, there are two ionic transporting processes, H^+ and O^{2-}, and the following cell reactions take place:

$$\text{at the anode,} \quad H_2(a) \Leftrightarrow 2H^+ + 2e \quad (26.1)$$

$$H_2(a) + O^{2-} \Leftrightarrow H_2O(a) + 2e \quad (26.2)$$

$$\text{at the cathode,} \quad \tfrac{1}{2}O_2(c) + 2H^+ + 2e \Leftrightarrow H_2O(c) \quad (26.3)$$

$$\tfrac{1}{2}O_2 + 2e \Leftrightarrow O^{2-} \quad (26.4)$$

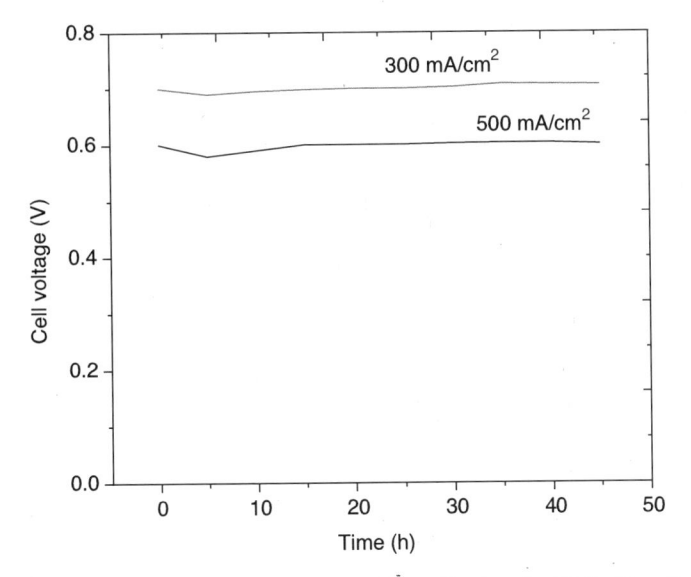

FIGURE 26.4 Discharge performances for the fuel cell using the SDC–20 wt % carbonate electrolyte at 550 °C.

Overall,

$$(26.1) \text{ and } (26.3) \text{ give} \quad H_2(a) + \tfrac{1}{2}O_2(c) \Leftrightarrow H_2O(c) \quad (26.5)$$

$$(26.2) \text{ and } (26.4) \text{ give} \quad H_2(a) + \tfrac{1}{2}O_2(c) \Leftrightarrow H_2(a) \quad (26.6)$$

Hence water is produced at both the anode and the cathode.

When the fuel cell is discharged under a constant load, the discharge curve shown in Figure 26.4 displays a constant output at a current density of 500 mA·cm^{-2}. During the cell operation (discharge), water formation was tested for at both the anode and cathode.

26.3.3 Electrolysis Studies

During electrolysis (i.e., the reverse operational mode of the fuel cell), the cell reactions are dependent on which electrode is supplied with water. Due to the changed polarities compared to the fuel cell, the anode and the cathode defined in the electrolysis cell are reversed to those of the fuel cell operating mode. In the proton conduction case, water is supplied to the anode, protons are transported through the electrolyte from the anode to the cathode, and hydrogen is produced at the cathode (Fig. 26.5a), while in the O^{2-} conduction case, water is supplied to the cathode, oxide ions are transported through the electrolyte in the opposite direction compared to that of protons (i.e., from the cathode to the anode), thus producing the hydrogen again at

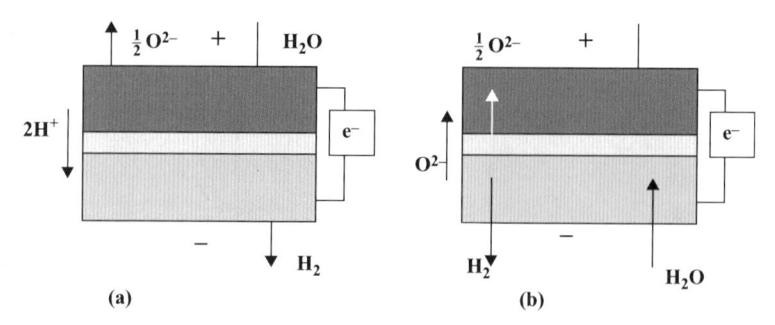

FIGURE 26.5 Electrolysis operational mode: cell reactions taking place (a) with water supplied at the positive electrode, in the H^+ conducting mode, and (b) with water supplied at the negative electrode, in the O^{2-} conducting mode.

the cathode (Fig. 26.5b). These two kinds of electrolysis cell reactions are described below. In the case of water supplied at the cathode for an oxide ion conductor, the reactions are:

$$\text{at the anode,} \qquad O^{2-} \rightarrow \tfrac{1}{2}O_2 + 2e^- \qquad (26.7)$$

$$\text{at the cathode,} \qquad H_2O + 2e^- \rightarrow O^{2-} + H_2 \qquad (26.8)$$

When water is supplied at the anode for a proton conductor, the following reactions take place:

$$\text{at the anode,} \qquad H_2O \rightarrow \tfrac{1}{2}O_2 + 2H^+ + 2e^- \qquad (26.9)$$

$$\text{at the cathode,} \qquad 2H^+ + 2e^- \rightarrow H_2 \qquad (26.10)$$

The typical current–voltage curve of the electrolysis, shown in Figure 26.6, appears when the applied voltage is gradually increased. The current is small until a certain voltage—the decomposition voltage—is reached, and thereafter the current increases more rapidly. The small current is caused by the diffusion of ions away from the electrodes, and a certain amount of external electric energy is needed to counterbalance this diffusion and to keep the electrodes polarised [17].

Figure 26.7 shows electrolysis curves obtained for the electrolyte-supported cell with platinum electrodes, where a clear electrolysis voltage transition, above 1.0 V for H^+ or 1.75 V for O^{2-}, is observed.. Water-saturated air at room temperature supplied to the electrolysis cell shows a lower current density than that of water-saturated hot air. This could be due to the fact that low temperature gives rise to a low water pressure, and consequently a low concentration of charge carriers. Finally, the cells showed significant current outputs, both when water was supplied to the anode or to the cathode, which implies the two charge carriers, H^+ and O^{2-}, are present.

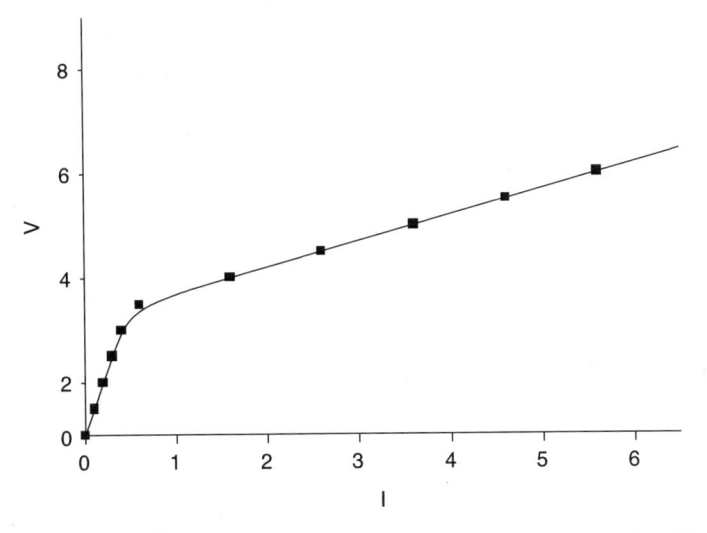

FIGURE 26.6 A typical current–voltage curve for an electrolysis cell.

26.3.4 Dual or Hybrid H$^+$ and O^{2-} Conduction

In general, proton conduction exists commonly in many types of salts, and it may often arise from transport in the two-phase interfaces of the composites. In the SDC–carbonate composites, the ceria is the host O^{2-} conducting phase, while H$^+$ conduction may exist in the second (guest) salt phase, and in the two-phase interfaces [8, 18]. Proton and oxide ion conduction takes place in the SDC and

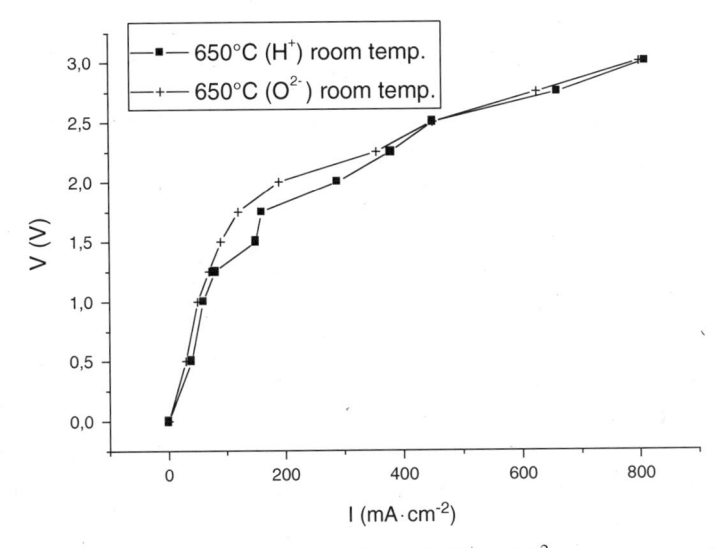

FIGURE 26.7 Performance for the electrolysis cell in H$^+$ and O^{2-} conduction modes. Water saturated by air was supplied during the process.

carbonate phases, or in the two-phase interfaces based on different mechanisms. The O^{2-} conduction in ceria phase is determined by the oxide ion vacancies (mechanism) in the ceria lattice; H^+ conduction may occur in the carbonate through a temporal bonding mechanism, $H^+ - [CO_3]^{2-}$ ($H[CO_3]^-$) [9]. Both H^+ and O^{2-} conduction can be integrated in one composite to enhance the system conductivity based on the composite effect; this effect concerns the ion conduction at interfaces, or interfacial conduction.

On one hand, it is highly possible for oxide ions/atoms in the SDC particle/phase surfaces to meet and capture protons. This acts as a basic driving force by oxide atom/ion on protons moving from one site to another in the interfacial range. On the other hand, because of the high temperature and the high mobility of the oxide ions in SDC, the captured protons cannot be stable with oxide atoms/ions. There is a high tendency to release them. While at the same time, the neighbouring oxide atoms/ions are free from protons and can accept them. Thus there is another driving force to push protons. Protons can only temporarily stay with oxide atoms/ions, and then have to jump to another site. This process of proton capture and release by oxide atoms/ions results in an interfacial conducting chain/mechanism. Correspondingly, this dynamic process/mechanism can also further promote oxide ion mobility and vice versa. When both H^+ and O^{2-} conducting phases form a unified bulk, the results are good to excellent. This can never be the case when each phase is independent. This is the advanced function and mechanism of these novel hybrid/dual composite conductors.

The combination of the H^+ and O^{2-} conduction can, on the one hand, enhance the transportation of ion concentration/flow, resulting in higher fuel cell current/power outputs; on the other hand, it also improves the electrode dynamics/kinetics and processes to create excellent fuel cell performance. These phenomena are significantly different from other fuel cells with only one kind of ion function for the processes. Recent theoretical studies [19] have shown that the SOFC-H^+ operation mode is more preferential than the SOFC-O^{2-} mode because the former has a higher efficiency, especially for direct operation with fuels such as methanol. Therefore the combination of H^+/O^{2-} contributions creates another excellent advance for LTSOFC technology.

26.4 CONCLUSION

The innovative dual-phase SDC–carbonate composites/materials possess many advantages, overcoming some of the drawbacks of the single-phase materials. This work demonstrates a new promising technology to produce hydrogen with the ceria-based dual-phase composite electrolyte fuel cell technology. The device can operate either in the fuel cell mode (to produce electricity from hydrogen and oxygen) or in the electrolysis mode (to produce hydrogen and oxygen from electricity and steam).

Our dual H^+/O^{2-} conductors are based on proton and oxide ion two-phase conduction, in addition to the interfacial mechanism. It can greatly enhance the electrolyte material conductivity, resulting in high current outputs, and exhibits an enormous potential to develop advanced LTSOFC technology for practical application and commercialisation.

ACKNOWLEDGEMENTS

This work is supported by Carl Tryggers Stiftelse for Vetenskap Forskning (CTS).

REFERENCES

1. Steele, B. C. H. and Heinzel, A., Materials for fuel-cell technologies, *Nature* **414**, 345 (2001).
2. Huijsmans, J. P. P., van Berkel, F. P. F., and Christie, G. M. J., *J. Power Sources* **71**, 107 (1998).
3. Maric, R., Ohara, S., Fukui, T., Yoshida, H., Nishimura, M., Inagaki, T., and Miura, K., *J. Electrochem. Soc.* **146**, 2006–2010 (1999).
4. Shao, Z. P. and Haile, S. M., *Nature* **431**, 170 (2004).
5. Zhu, B. and Mellander, B.-E., Performance of intermediate temperature SOFCs with composite electrolytes, in *Solid Oxide Fuel Cells VI*, S. C., Singhal, and M. Dokiya (Eds), The Electrochemical Society, Pennington, NJ, 1999, pp. 244–253.
6. Fu, Q. X., Peng, D. K., Meng, G. Y., and Zhu, B., *Mater. Lett.* **53**, 186 (2002).
7. Fu, Q. X., Peng, D. K., Meng, G. Y., and Zhu, B., *J. Power Sources* **104**, 73 (2002).
8. Zhu, B., *J. Power Sources* **114**, 1 (2003).
9. Zhu, B., Liu, X. R., Zhou, P., Yang, X. T., Zhu, Z. G., and Zhu, W., *Electrochem. Commun.* **3**, 566 (2001).
10. Zhu, B., Liu, X. R., Zhu, Z. G., Zhu, W., and Zhou, S. F., *J. Mater. Sci. Lett.* **20**, 591 (2001).
11. Zhu, B., Yang, X. T., Xu, J., Zhu, Z. G., Ji, S. J., Sun, M. T., and Sun, J. C., *J. Power Sources* **118**, 47 (2003).
12. Zhu, B., Liu, X. R., Sun, M. T., Ji, S. J., and Sun, J. C., *Solid State Sci.* **5**, 1127 (2003).
13. Liu, X. R., Zhu, B., Xu, J., Sun, J. C., and Mao, Z. Q., *Key Eng. Mater.* **425**, 280–283 (2004).
14. Hu, J. D., Tosto, S., Guo, Z. X., and Wang, Y. F., *J. Power Sources* **154**, 106 (2006).
15. Huang, J. B., Yang, L. Z., Gao, R. F., Mao, Z. Q., and Wang, C., *Electrochem. Commun.* **8**, 785 (2006).
16. Zhu, B. and Mellander, B.-E., *Solid State Phenomena* **19**, 39–40 (1994).
17. Colomban, P., *Proton Conductors: Solids, Membranes and Gels–Materials and Devices*, Cambridge University Press, Cambridge, UK, 1992.
18. Zhu, B., Proton and oxygen ion-mixed-conducting composites and fuel cells, *Solid State Ionics* **145**, 371 (2001).
19. Assabumrungrat, S., Sangtongkitcharoen, W., Laosiripojana, N., Arpornwichanop, A., Charojrochkul, S., and Praserthdam, P., *J. Power Sources* **148**, 18 (2005).

INDEX

Accelerator-driven systems (ADSs), spent nuclear fuel management, research background, 265–266
Actinide separation. *See also* Minor actinides (MAs), spent nuclear fuel management
  pyrometallurgical processes, 64–65
  spent nuclear fuel management
    chemical thermodynamic techniques, 267–276
    research background, 265–266
Additive effects, sodium chloride/potassium chloride molten salts, aluminum solubility in, 394
Aggregate behaviour
  hydrophilic ionic liquids in aqueous solutions
    basic principles, 38–39
    head type, 45–46
  mass spectrometric studies
    Cook's kinetic method, hydrogen bond strengths, 55–59
    electrospray ionisation mass spectra results, 51–52
    experimental protocols, 51
    research overview, 50–51
    tandem electrospray ionisation mass spectra-mass spectra and breakdown graphs, 52–55
Air-sensitive melts, micro-Raman spectroscopy, 322–323
Air-water mixture, microstructured molybdenum reactor corrosion resistance
  molybdenum disilicide boronisation, 210–214
  siliciding molybdenum, 206–207
Alkali chlorides, microstructured molybdenum reactor corrosion resistance, hafnium/hafnium diboride coating electrodeposition, 197
Alkali halides, liquid electrolytes, pressure effects, 187–188

Alkali metal iodides
  melting compounds (M_3LnX_6)
    heat capacity, 347–350
    phase transition enthalpies, 345–347
    thermodynamic analysis, sample preparation, 342–343
Alkali metal peroxide
  catalyst enhanced molten salt oxidation, 156–159
  fission product elements, 162
  enhanced molten salt oxidation, 154–155
Alkali nitrates, liquid electrolyte pressure effects, 187–188
Alkoxyphenylphosphine, BASF BASIL preparation of, 189
1-Alkyl-3-methylimidazolium chlorides ([$C_n$mim] Cl)
  aqueous solution-air interface, 41
  chemical structure, 39–40
  interfacial tension measurements, 41–46
Alkyltrimethylammonium/alkylammonium chloride families, cationic surfactants, 44–46
Aluminum/aluminum alloys
  cryolite-alumina melts
    potassium cryolites, solubility
      electrical conductivity, 76–78, 80–81
      electrolyte preparation, 76
      metal solubility determination, 79–80, 82–83
      potentiometric titration analysis, 78–79, 81–82
      temperature-dependent properties, 75–76
    thermodynamic behaviour, 133–140
  electrolysis, metallic inert anodes
    experimental protocols, 125–126
    research background, 123–124
    results, 126–131
  metal/semiconductor electrodeposition, 90–91
  spent nuclear fuel management, cathodes, electrowinning of TRU elements, 275–276

Molten Salts and Ionic Liquids: Never the Twain? Edited by Marcelle Gaune-Escard and Kenneth R. Seddon
Copyright © 2010 John Wiley & Sons, Inc.

Aluminum(III) chloride molten salts
  applications, 282
  ionic liquids, pericyclic reactions, 291–292
  organic reactions, 283–284
Aluminum oxide, sodium chloride/potassium chloride molten salts, solubility in
  additives, effect of, 394
  cryolite addition, 391–394
  measurement techniques, 391
  raw materials, 390
  research background, 389–390
  thermal physical properties, 391
Ambient temperature, organic cation halides, spectroscopic analysis, 12–15
Americium, spent nuclear fuel management electrowinning
  current density limits, 275–276
  with liquid cadmium cathode, 272–276
  lithium chloride-potassium chloride eutectic melt, 268–272
Ammonium salts
  liquid electrolytes, pressure effects, 188
  as protic electrolytes, 18–20
Anharmonicity, Raman spectra, high-temperature melts, temperature-based frequency shifts, 308
Anions
  hydrophilic ionic liquids in aqueous solutions, 44–46
  ionic liquids, ambient temperature, 10–12
  water-ionic liquid systems thermodynamics, COSMO-RS technique
    liquid-liquid equilibria modeling, 111–113
    vapour-liquid equilibria modeling, cation alkyl chain length, 114–116
Anisotropic Raman scattered intensities, Raman spectra, high-temperature melts, 305–307
Anodic discharge, plasma-induced molten salt electrolysis, fine particle formation, 175–179
  carbon nitride particles, 177–179
  metal oxide particles, 176–177
  metal sulfide particles, 177
Anthracene, photochemistry, 294–295
Antimony(III), molten salts and, 282
  organic reactions, 283–284
Anti-Stokes Raman spectrum, high-temperature melts, 303
  blackbody radiation, 311
  container-free melts, macro-Raman spectroscopy, 326
  selection rule, 304–305
Aqueous solutions
  fission products, catalyst enhanced molten salt oxidation, 162–163

hydrophilic ionic liquids
  basic principles, 38–39
  chemical compounds, 39–40
  experimental results, 41–46
  fluorescence spectroscopy, 41
  interfacial tension measurements, 40
  proton NMR, 41
ionic liquid/molten salt physicochemistry, electrical conductivity, 245–246
selenium metal/semiconductor electrodeposition, 95–97
Arrhenius plots
  gelled polymer electrolyte systems, 29–30
  conductivity plots, 31–35
  ionic liquids, 6–7
  strong and fragile liquids, 7–10
  liquid-based electrolyte solutions, 30–31
Aryl alkyl ethers, pyridinium chloride reaction, 285
As-received raw materials, lanthanum aluminate, molten salt synthesis, 221–222
Atomic absorption spectrophotometry (AAS), sodium chloride/potassium chloride molten salts, aluminum solubility in, 391

B88-VWN/DNP, water-ionic liquid systems, COSMO-RS thermodynamic analysis, 106–107
Backscattering geometry, in situ Raman spectroscopy, electrode surfaces, high-temperature melts, 333–336
Bandshapes, Raman spectra, high-temperature melts, 307–308
Barium
  catalyst enhanced molten salt oxidation, 162
  phosphate precipitation, 164
Basis set parameterisation, water-ionic liquid systems, COSMO-RS thermodynamic analysis, 105–107
Batch mixing and firing, molten salt synthesis, ceramic materials, 398–399
Baylis-Hillman reaction, ionic liquids, carbon-carbon bond formation, 288–291
Bench scale experiments, plasma-induced molten salt electrolysis, fine particle formation, 173–175
Biopreservation, ionic liquids, 20
Biphasic ionic liquid-ionic liquid systems
  chemicals in, 148
  molar balance, 149
  partition coefficients, NMR determination, 148–149
  phase compositions
    basic characteristics, 143–148
    NMR determination, 148

spectroscopic measurements, 148
upper solution temperature determination, 149
Bismuth cathodes, spent nuclear fuel management, electrowinning of TRU elements, 275–276
Bismuth trichloride ($BiCl_3$), liquid electrolytes, pressure effects, 188
Blackbody radiation
near-infrared Raman spectroscopy, 332
Raman spectroscopy, high-temperature melts, 310–311
Blend methods, ionic liquid-polymer composite electrolyte formation, 371
film formation, 377–379
Block copolymers, gelled polymer electrolyte systems, 28–30
Boltzmann constant, Raman spectra, high-temperature melts, 307
blackbody radiation, 311
Boron compounds, microstructured molybdenum reactor corrosion resistance, electrochemical boronising, 194–196
Boronising, microstructured molybdenum reactor corrosion resistance
molybdenum boronising, 197–198, 202–205
molybdenum disilicide, 198, 207–214
research background, 194–196
Bose factor, Raman spectra, high-temperature melts, 306–307
BP/SVP//AM1 calculations, water-ionic liquid systems, COSMO-RS thermodynamic analysis, 106–107
BP/TZVP calculations, water-ionic liquid systems, COSMO-RS thermodynamic analysis, 106–107
liquid-liquid equilibria modeling, 109–113
vapour-liquid equilibria modeling, anion identity, 116–118
Breakdown graphs, tandem electrospray ionisation mass spectrometry, ionic liquid aggregate behaviour, 52–55
1-Butyl-3-methylimidazolium cations, fluidity data, 11–12
1-Butyl-3-methylimidazolium tetrafluoroborate [$C_n$mim][BF_4]
aggregate behaviour
Cook's kinetic method, hydrogen bond strengths, 55–59
electrospray ionisation mass spectra results, 51–52
tandem electrospray ionisation mass spectrometry, 52–55
in aqueous solutions
aggregate formation, 38–39

critical micelle concentration, 41–46
ionic liquids
Mukaiyama aldol condensation, 289
protection-deprotection, 287
Wittig reaction, 289
1-Butyl-1-methylpyrrolidinium bis(trifluoroethylsulfonyl)amide ([C_4mpyrr][NTf_2])
electrolyte solutions, 27–28
metal/semiconductor electrodeposition
experimental protocol, 86–87
selenium electrodeposition, 95–97
tantalum electrodeposition, 88–90

Cadmium liquid metal cathodes, spent nuclear fuel management plutonium/americium, liquid cadmium cathode, 272–276
pyroprocessing technology, 266–267
Caesium
in aqueous solutions, 162–163
catalyst enhanced molten salt oxidation, 162
melting compounds (M_3LnX_6), phase transition enthalpies, 345–347
phosphate precipitation, 163–164
Calcium compounds
cryolite-alumina melts, phosphorus/sodium thermodynamics, 137–140
molten salt synthesis, ceramic materials, 398–401
Calcium-potassium-nitrate (CKN) system, glass-forming liquids, 5–7
Capillary cells, potassium cryolites, alumina solubility analysis, 77–78
Capillary viscometry, ionic liquid/molten salt physicochemistry, 231–232
Carbocation formation, aluminum(III)/antimony (III) molten salts, 284
Carbon anodes, aluminum electrolysis, 123–131
Carbonate melts. See Molten carbonates
Carbon atoms, 1-alkyl-3-methylimidazolium chlorides ([$C_n$mim]Cl), CMC values, 43–46
Carbon-carbon bond formation, ionic liquids, 287–291
Carbon dioxide, aluminum electrolysis, metal inert anodes, 123–131
Carbon dioxide laser hearth melting
container-free melts, macro-Raman spectroscopy, 327–328
Raman spectra, high-temperature melts, purification, 313–314
Carbon nitride particles, plasma-induced molten salt electrolysis, fine particle formation, 177–179

Carbon NMR spectroscopy (^{13}C NMR)
  biphasic ionic liquid-ionic liquid systems, 148
  hydrophilic ionic liquids in aqueous solutions, 40
Carbon tetrafluoride, aluminum electrolysis, metal inert anodes, 123–131
Catalyst enhanced molten salt oxidation (CEMSO), spent fuel reprocessing
  ceramic UO$_2$, 156
  component abundance, 162
  electron transfer, 156
  experimental protocols, 153–155
  fission product precipitation, 162–164
  fission product separation, 161–162
  mathematical modelling, 160–161
  molten chlorides, phosphate precipitation, 163–164
  phosphate precipitation, 163
  plutonium data, 160
  reprocessing cycle, 160
  research background, 151–153
  uranate(VI) solubility, molten carbonates, 156–159
  water treatment, 162–163
Catalytic coatings, microstructured molybdenum reactor corrosion resistance, 196
Cathode characteristics, samarium-aluminum alloy formation, lithium chloride-potassium chloride eutectic, 65–66
Cathodic discharge electrolysis, plasma-induced molten salt electrolysis, fine particle formation, 170–175
Cation alkyl chain length, water-ionic liquid systems, COSMO-RS thermodynamic analysis
  liquid-liquid equilibria modeling, 111–112
  vapour-liquid equilibria modeling, 113–116
Cation-anion pairs, ionic liquid aggregate behaviour
  Cook's kinetic method hydrogen bond strengths, 57–59
  organic reactions, 286
Cation methyl inclusion, water-ionic liquid systems thermodynamics, COSMO-RS technique, liquid-liquid equilibria modeling, 113
CEMSO. *See* Catalyst enhanced molten salt oxidation (CEMSO)
Ceramic inert anodes, aluminum electrolysis, 124
Ceramic materials, molten salt synthesis
  batch mixing and firing, 398–399
  calcium-zirconia powders, 398–401
  characterisation, 399
  lanthanide aluminum powder, 399, 402–404
  leaching, 399
  magnesium-aluminum oxide platelets, 399, 401–402
  research background, 397–398
Ceramic uranium oxide (UO$_2$), catalyst enhanced molten salt oxidation, 156
Ceria-based dual-phase composite electrolytes, fuel cell and electrolysis studies
  electrolysis, 409–415
  hybrid hydrogen ion/oxygen ion conduction, 415–416
  research background, 407–408
  structural characteristics, 410–415
  two-phase composites, 408–409
Cermet anodes, aluminum electrolysis, 124
Charge coupled device (CCD), Raman spectroscopy, high-temperature melts, 309–314
  macro-Raman spectroscopy, 314–318
Charge-discharge capacities, ionic liquid-polymer composite electrolyte, lithium ion batteries, 382–383
Chloride ions, tantalum electrodeposition, 89–90
Chloroaluminate(III) ionic liquids, metal/semiconductor electrodeposition, aluminum/aluminum alloys, 90–91
Chlorofluorocarbons, catalyst enhanced molten salt oxidation, mathematical modelling, 160–161
Claisen rearrangement, ionic liquids, 292
Cluster ions, ionic liquid aggregate behaviour, Cook's kinetic method, 55–59
Coating analysis, microstructured molybdenum reactor corrosion resistance, molten salt coating electrodeposition, 198–199
Cobalt cations
  liquid electrolyte structure, 184
  spectroscopic analysis, 14–15
Collision energy, ionic liquid aggregate behaviour, tandem electrospray ionisation mass spectrometry, 52–55
Collision-induced dissociation, ionic liquid aggregate behaviour, tandem electrospray ionisation mass spectrometry, 54–55
Combustion testing, ionic liquid-polymer composite electrolyte, 380–383
Competitive dissociations, ionic liquid aggregate behaviour, Cook's kinetic method, 55–59
Concentric cylinder geometry, ionic liquid/molten salt thermal conductivity, steady-state technique, 242–244
Condensation reactions, ionic liquids, carbon-carbon bond formation, 288
Conductivity, liquid electrolytes, 185–187
Conductor-like Screening Model for Real Solvents (COSMO-RS) predictive model, water-ionic liquid systems thermodynamics

conformers influence, 107–109
liquid-liquid equilibria modeling, 109–113
  anion identity, 113
  cation alkyl chain length, 111–112
  cation methyl inclusion, 112
  LLE/VLE experimental database, 105
  parameterisation calculations, 105–107
  phase equilibria prediction, 103–104, 118
  research background, 102–103
  vapour-liquid equilibria modeling, 113–116
  anion identity, 115–116
  cation alkyl chain length, 113–116
  temperature dependence, 116–118
Conformer molecules, water-ionic liquid systems thermodynamics, COSMO-RS model, 107–109
Constant potential experiment, in situ Raman spectroscopy, electrode surfaces, high-temperature melts, 333–334
Container equipment, macro-Raman spectroscopy, high-temperature melts, 315–317
Container-free melts
  Raman spectroscopy, 323–328
    levitation method, 324–326
    pyrometry, 326
    self-support method, 327–328
    temperature estimation, 326
  ultraviolet Raman spectroscopy, 329–330
Continuous-wave laser experiments, Raman spectroscopy, high-temperature melts, 309–314
Conventional mixed oxide synthesis (CMOS)
  ceramic materials, research background, 397–398
  lanthanum aluminate, 220
Cook's kinetic method, ionic liquid aggregate behaviour, 55–59
Cooling rates, melting compounds (M_3LnX_6), phase transition enthalpies, 346–347
Corrosive melts, micro-Raman spectroscopy, 322–323
COSMO-RS. See Conductor-like Screening Model for Real Solvents (COSMO-RS) predictive model
COSMOtherm program, water-ionic liquid systems thermodynamics, COSMO-RS model, 104
Counterion binding, hydrophilic ionic liquids, aqueous solutions, 42–46
Critical micelle concentration (CMC)
  1-alkyl-3-methylimidazolium chlorides ([$C_n$mim]Cl), interfacial tension measurements, 41–46
  hydrophilic ionic liquids in aqueous solutions, basic principles, 38–39

Cross-validation techniques, molten salts thermodynamic properties, partial least squares calculation, 357–363
Crucible melting, Raman spectra, high-temperature melts, purification, 313–314
Cryolite-alumina melts
  aluminum electrolysis, metal inert anodes, 123–131
  metal/semiconductor electrodeposition, aluminum/aluminum alloys, 90–91
  phosphorus and sulfur thermodynamics
    experimental protocols, 133–134
    research background, 133
    results and discussion, 134–140
  potassium cryolites, alumina solubility
    electrical conductivity, 76–78, 80–81
    electrolyte preparation, 76
    metal solubility determination, 79–80, 82–83
    potentiometric titration analysis, 78–79, 81–82
    temperature-dependent properties, 75–76
  sodium chloride/potassium chloride molten salts, aluminum solubility in, 391–394
Crystal field stabilisation energies (CFSEs), liquid electrolyte structure, 184
Current density, spent nuclear fuel management, transuranium electrowinning and limits on, 273–276
Cyclic oxidation tests, microstructured molybdenum reactor corrosion resistance, molybdenum disilicide boronisation, 210–214
Cyclic voltammetry
  samarium-aluminum alloy formation, lithium chloride-potassium chloride eutectic, 66–68
  selenium metal/semiconductor electrodeposition, 95–97
  silicon metal/semiconductor electrodeposition, 91–94
  tantalum electrodeposition, 88–90
Cycloaddition reactions, ionic liquids, 291–292

DABCO (1,4-diazabicyclo[2.2.2]octane), ionic liquids, Baylis-Hillman reaction, 289–290
Dark coloured melts
  macro-Raman spectroscopy, 317–318
  near-infrared Raman spectroscopy, 332
Debye-Hückel-Onsager equation, liquid electrolyte conductivity, 186–187
Demal scale, ionic liquid/molten salt physicochemistry, electrical conductivity, 245–246
Density functional theory (DFT)
  ionic liquid cation-anion ion pairs, 58–59
  water-ionic liquid systems, COSMO-RS thermodynamic analysis, liquid-liquid equilibria modeling, 109–113

Density measurements
  ionic liquid/molten salt physicochemistry, 234
  liquid electrolytes, 185
    pressure effects, 187–188
Depolarisation ratio, Raman spectra, high-temperature melts, 307
Desilylation, ionic liquid protection-deprotection, 287
1,3-Dialkylimidazolium-based ionic liquids, cation-anion ion pairs, 58–59
Diamond anvil high pressure cell, organic cation halides, 12–15
1,1-Didodecylpiperidinium bromide, [$C_{12}C_{12}$pip]Br, chemical structure, 39–40
Dielectric constant, ionic liquid/molten salt physicochemistry, electrical permittivity, 247–250
Diels-Alder reactions, ionic liquids, 291–292
Diethyl carbonate (DEC), electrolyte solutions, 27–28
Differential thermal analysis (DTA), fuel cell and electrolysis studies, samarium-doped ceria composite, 410–415
Diffusion coefficients
  ionic liquid/molten salt physicochemistry, 250–257
  spent nuclear fuel management, plutonium electrowinning, current density limits, 274–276
Dimethyl carbonate (DMC), electrolyte solutions, 27–28
Diode-pumped solid state (DPSS) lasers, melt spectra, near-infrared Raman spectroscopy, 330–332
Direct current (DC) polarisation, gelled polymer electrolyte systems, 29–30
Dissociation rates, ionic liquid aggregate behaviour
  Cook's kinetic method hydrogen bond strengths, 55–59
  tandem electrospray ionisation mass spectrometry, 52–55
Dissolution-precipitation mechanism
  lanthanum aluminate, molten salt synthesis, 226
  molten salt synthesis, ceramic materials, 401
Diuranate precipitation, catalyst enhanced molten salt oxidation
  ceramic uranium oxide, 156
  solubility in molten carbonates, 156–159
Droplet fission, ionic liquids, electrospray ionisation mass spectrometry, 50–51
Dual-phase proton and oxide ion conduction, fuel cell and electrolysis studies
  ceria-based two-phase composites, 408–409
  electrolysis, 409–415

hybrid hydrogen ion/oxygen ion conduction, 415–416
  research background, 407–408
  structural characteristics, 409–415
Dufour effect, ionic liquid/molten salt physicochemistry, 250–257
Dynamic parameters, Raman spectra, high-temperature melts, 307–308

Electrical conductivity
  ionic liquid/molten salt physicochemistry, 234, 243, 245–246
  potassium cryolites, alumina solubility analysis
    experimental cell design, 76–78
    melt data, 80
    temperature dependence, 80–82
Electrical permittivity, ionic liquid/molten salt physicochemistry, 234, 246–250
Electrochemistry, liquid electrolytes, 184–185
Electrodeposition
  metals and semiconductors, ionic liquid solvents
    aluminum electrodeposition, 90–91
    experimental protocol, 86–87
    research background, 86
    selenium electrodeposition, 94–97
    silicon electrodeposition, 91–94
    tanatlum electrodeposition, 87–90
  microstructured molybdenum reactors, corrosion resistance, coatings in molten salts
    boronisation, 197–198, 202–205
    molybdenum disilicide, 198, 207–214
    coating characterisation, 198–199
    hafnium/hafnium diboride coatings, 197, 199–202
    research background, 194–196
    siliciding of, 198, 205–207
    substrate materials, 196
Electrode surfaces, high-temperature melts, in situ Raman spectroscopy, 333–336
  constant potential, 333–334
  modulation, 334–336
Electrofusion process, molten salt synthesis, ceramic materials, 398
Electrolysis
  aluminum/aluminum alloys, metallic inert anodes, 123–131
  dual-phase electrolyte composites, proton and oxide ion conduction
    ceria-based two-phase composites, 408–409
    hybrid hydrogen ion/oxygen ion conduction, 415–416
    research background, 407–408
    structural characteristics, 409–415

INDEX 425

plasma-induced molten salt electrolysis, fine particle formation
  anodic discharge, 175–179
  basic principles, 170
  bench scale experiments, 173–175
  carbon nitride particles, 177–179
  cathodic discharge, 170–175
  metal oxide particles, 176–177
  metal sulfide particles, 177
  research background, 169–170
  samarium-aluminum alloy formation, lithium chloride-potassium chloride eutectic, 69–71

Electrolytes
  ionic liquid-polymer electrolyte, lithium ion polymer battery
    combustion testing, 380–382
    conjugated polymer composite electrolyte film, 368–371
    film formation and structure, 377–379
    future research issues, 385–386
    gel electrolyte, film formation method, 371–372
    host polymer selection, 372
    ICE electrolyte application, 377–383
    ICE electrolyte formulation, 376–377
    ionic liquid formulation, 372–376
    lithium gel polymer rechargeable battery applications, 382–383
    microphase separation, TEM analysis, 379–380
    Pionics LIP cell characteristics, 384–385
    PLB cell performance, 383–384
    research background, 368
    simple blend and blend method with graft polymerisation, 371–377
  liquid electrolytes, 182
    applications, 188–189
    conductivity, 185–187
    density, 185
    electrochemistry, 184–185
    formation, 183
    ionic systems, 188
    pressure effects, 187–188
    solids and vapours, 183
    spectroscopy and structure, 184
    theories, 183
  preparation, potassium cryolites, alumina solubility analysis, 76

Electromotive force (EMF)
  potassium cryolites, alumina solubility analysis, potentiometric titration, 79
  samarium-aluminum alloy formation, lithium chloride-potassium chloride eutectic, 66–68

Electron transfer, catalyst enhanced molten salt oxidation, 156
Electrospray ionisation mass spectrometry (ESI-MS)
  basic principles, 50–51
  ionic liquid aggregates, 51–52
Electrowinning process, spent nuclear fuel management, plutonium/americium, liquid cadmium cathode, 272–276
Elution concentration profile, ionic liquid/molten salt physicochemistry, 254–257
Energy-dispersive X-ray spectroscopy
  ionic liquid-polymer composite electrolyte formation, 378–379
  microstructured molybdenum reactor corrosion resistance
    molten salt coating electrodeposition analysis, 198–199
    molybdenum disilicide boronisation, 212–214
Energy level diagram, protic ionic liquids, 15–17
Enhanced molten salt oxidation, spent fuel reprocessing, 153–155
Enthalpy
  catalyst enhanced molten salt oxidation, mathematical modelling, 160–161
  LnI_3-MI binary systems, phase transitions, 344–347
  samarium-aluminum alloy formation, lithium chloride-potassium chloride eutectic, 71–73
  spent nuclear fuel management, plutonium electrowinning, current density limits, 273–276
Entropy
  samarium-aluminum alloy formation, lithium chloride-potassium chloride eutectic, 71–73
  spent nuclear fuel management, plutonium electrowinning, current density limits, 273–276
ESI. *See* Electrospray ionisation mass spectrometry
ESI-MS-MS. *See* Tandem electrospray ionisation mass spectrometry
1-Ethyl-3-methylimidazolium bis(trifluoromethylsulfonyl) amide ([C_2mim][NTf_2]), metal/semiconductor electrodeposition
  aluminum/aluminum alloys, 90–91
  experimental protocol, 86–87
1-Ethyl-3-methylimidazolium hexafluorosilicate, silicon metal/semiconductor electrodeposition, 91–94
Ethylammonium nitrat, ionic liquids, pericyclic reactions, 291–292
Ethylammonium nitrate, spectroscopic analysis, 17

Ethylene carbonate (EC), electrolyte solutions, 27–28
Ethyl ethanoate, hydrophilic ionic liquids in aqueous solutions, 40
Ethylmethyl carbonate (EMC), electrolyte solutions, 27–28
Eutectics
  lanthanum aluminate, molten salt synthesis, 224
  lithium chloride-potassium chloride, samarium-aluminum alloy formation
    cyclic voltammetry study, 66–68
    electrolysis, 69–71
    experimental protocol, 65–66
    goals, 64–65
    open-circuit chronopotentiometry study, 68–69
    thermodynamic properties, 71–73
  lithium fluoride-sodium fluoride-potassium fluoride eutectic
    carbonate melts, 155
    tantalum electrodeposition, 88–90
  molten salt reactions and, 281–282
  in situ Raman spectroscopy, electrode surfaces, high-temperature melts, 333–334
  spent nuclear fuel management
    pyroprocessing technology, 267
    transuranium elements, dichloride/trichloride relative stability, 268–272

Faraday constant, microstructured molybdenum reactor corrosion resistance, boronising molybdenum, 203–205
Fast reactors (FRs), spent nuclear fuel management
  pyroprocessing technology, 266–267
  research background, 265–266
Fick's second law of diffusion, ionic liquid/molten salt physicochemistry, 250–257
Film formation process, ionic liquid-polymer composite electrolyte formation, 371–372
  combustion testing, 380–383
  structure and application, 377–379
Fine particle formation, plasma-induced molten salt electrolysis
  anodic discharge, 175–179
  basic principles, 170
  bench scale experiments, 173–175
  carbon nitride particles, 177–179
  cathodic discharge, 170–175
  metal oxide particles, 176–177
  metal sulfide particles, 177
  research background, 169–170
First-order Raman effect, high-temperature melt spectroscopy, 304–305, 307

Fission products
  catalyst enhanced molten salt oxidation
    molten chlorides, phosphate precipitation, 163–164
    phosphate precipitation, 163
    precipitation, 162–164
    separation, 161–162
    water treatment, 162–163
  partitioning and transmutation strategy, 64
Fluorescence spectroscopy
  hydrophilic ionic liquids in aqueous solutions, 41
  solvatochromic probe, 42–46
  liquid electrolytes, 184
Fluorinated anion salts
  ionic liquids, ambient temperature, 11–12
  macro-Raman spectroscopy, high-temperature melts, 315–316
Fourier transform (FT) Raman spectroscopy, high-temperature melts, 331–332
Fragmentation dynamics, ionic liquids
  Cook's kinetic method hydrogen bond strengths, 56–59
  tandem electrospray ionisation mass spectrometry, 52–55
Frequency shifts, Raman spectra, high-temperature melts, anharmonicty, 308
Friedel-Crafts reaction, aluminum(III) chloride molten salts, 283–284
Frozen samples analysis technique, potassium cryolites, alumina solubility analysis, metal solubility, 79–80
Fuel cell electrolytes
  dual-phase composites, proton and oxide ion conduction
    ceria-based two-phase composites, 408–409
    electrolysis, 409–415
    hybrid hydrogen ion/oxygen ion conduction, 415–416
    research background, 407–408
    structural characteristics, 409–415
  inorganic compounds, 18–20
  protic ionic liquids, 17–18
Furnace equipment, macro-Raman spectroscopy, high-temperature melts, 315
Fused silica containers, macro-Raman spectroscopy, high-temperature melts, 315–318

Gallium cathodes, spent nuclear fuel management, electrowinning of TRU elements, 275–276
Galvanostatic electrolyses, samarium-aluminum alloy formation, lithium chloride-potassium chloride eutectic, 70–71
Gaseous elemental phosphorus, cryolite-alumina melts, phosphorus/sodium thermodynamics, 137–140

Gas-phase salts, ionic systems, 188
Gelled polymer electrolyte systems
  composition, 28–30
  conductivity analysis, 31–35
  ionic liquid-polymer composite electrolyte
    combustion testing, 380–382
    conjugated polymer composite electrolyte film, 368–371
    film formation and structure, 377–379
    future research issues, 385–386
    gel electrolyte, film formation method, 371–372
    host polymer selection, 372
    ICE electrolyte application, 377–383
    ICE electrolyte formulation, 376–377
    ionic liquid formulation, 372–376
    lithium gel polymer rechargeable battery applications, 382–383
    microphase separation, TEM analysis, 379–380
    Pionics LIP cell characteristics, 384–385
    PLB cell performance, 383–384
    research background, 368
    simple blend and blend method with graft polymerisation, 371–377
  lithium and lithium-ion batteries, 26–27
Gibbs free energy
  cryolite-alumina melts, phosphorus/sodium thermodynamics, 134–140
  molten salts thermodynamic properties, material informatics, 358–363
  samarium-aluminum alloy formation, lithium chloride-potassium chloride eutectic, 71–73
  water-ionic liquid systems, COSMO-RS thermodynamic analysis, 102–103
Gibbs-Helmholtz equation, samarium-aluminum alloy formation, lithium chloride-potassium chloride eutectic, 72–73
Glass-former phenomenology, ionic liquids, 4–7
Glass transition temperature, ionic liquids, 5–7
  strong and fragile liquids, 7–10
Gold surfaces, metal/semiconductor electrodeposition, aluminum/aluminum alloys, 91–92
Götze mode-coupling theory, glass-forming liquids, 5–7
Graft polymerisation
  ionic liquid-polymer composite electrolyte, polymerised ionic liquids, 373–377
  ionic liquid-polymer composite electrolyte formation, 371–372
  film formation, 377–379
Graphite crucible counter electrode, aluminum electrolysis, 125–131

Grignard reagent, ionic liquids, carbon-carbon bond formation, 287–288
Grüneisen parameter, Raman spectra, high-temperature melts, anharmonicty temperature-based frequency shifts, 308
Gurney proton energy level diagram, protic ionic liquids, 15–17

Hafnium/hafnium diboride coatings, microstructured molybdenum reactor corrosion resistance
  coating electrodeposition, 197, 199–202
  electrochemical synthesis, 199–202
  impurities, 199–202
  research background, 194–196
Halides. See also Rare earth halides; specific halides, e.g. Alkali halides
  complex anions, 10–12
  hydrophilic ionic liquids in aqueous solutions, 39–40
  organic cation halides, spectroscopic analysis, 12–15
Hall-Heroult process
  metal/semiconductor electrodeposition, aluminum/aluminum alloys, 90–91
  sodium chloride/potassium chloride molten salts, aluminum solubility in, 389–390
Heat capacity
  ionic liquid/molten salt physicochemistry, 234
  LnI$_3$-MI binary systems, step measurement method, 344
  melting compounds (M$_3$LnX$_6$), 347–350
Heating temperatures, lanthanum aluminate, molten salt synthesis, 221–222
  salt/oxide ratios, 226
Henry's law, molten salt oxidation, 153
Hex-1-ene, biphasic ionic liquid-ionic liquid systems, 146–148
Highly oriented pyrolytic graphite (HOPG), metal/semiconductor electrodeposition
  experimental protocol, 87
  silicon, 91–94
High-temperature ionic liquids (HTILs)
  physicochemical measurement, 230
  electrical conductivity, 245–246
  viscosity measurements, 237–240
Hildebrand solubility parameter, ionic liquids, 285–286
Homogeneity, Raman spectra, high-temperature melts, 308
Host polymer selection, ionic liquid-polymer composite electrolyte, 372

INDEX 427

Hydrogen bond strengths
  ionic liquid aggregate behaviour, Cook's kinetic method, 55–59
  ionic liquid/molten salt physicochemistry, electrical permittivity, 248–250
  water-ionic liquid systems, COSMO-RS thermodynamic analysis, liquid-liquid equilibria modeling, 110–113
Hydrogen ion conduction, dual-phase composite electrolytes, 411–416
Hydrophilic ionic liquids, aqueous solutions
  basic principles, 38–39
  chemical compounds, 39–40
  experimental results, 41–46
  fluorescence spectroscopy, 41
  interfacial tension measurements, 40
  proton NMR, 41
Hydrophobicity, water-ionic liquid systems thermodynamics, COSMO-RS technique
  liquid-liquid equilibria modeling, cation alkyl chain length, 112
  vapour-liquid equilibria modeling, cation alkyl chain length, 116
Hyperpolarisability tensor, Raman spectra, high-temperature melts, 303–305

ICE. See Ionic liquid-polymer composite electrolyte (ICE)
Imidazolium-based ionic liquids
  mass spectrometric studies, aggregate behaviour
    Cook's kinetic method, hydrogen bond strengths, 55–59
    electrospray ionisation mass spectra results, 51–52
    experimental protocols, 51
    research overview, 50–51
    tandem electrospray ionisation mass spectra-mass spectra and breakdown graphs, 52–55
  organic reactions, 286
  water-ionic liquid systems, COSMO-RS thermodynamic analysis
    experimental database, 105
    liquid-liquid equilibria modeling, 109–113
    vapour-liquid equilibria modeling, 113–116
Immiscible phases, biphasic ionic liquid-ionic liquid systems, 143–148
Impurity levels, lanthanum aluminate, molten salt synthesis, 222–224
Infrared absorption spectroscopy, liquid electrolytes, 184
Inhomogeneous broadening, Raman spectra, high-temperature melts, 307–308
Inorganic fuel cell electrolytes, 18–20

Inorganic salts, ionic systems, 188
In situ Raman spectroscopy, high-temperature melts, electrode surfaces, 333–336
  constant potential, 333–334
  modulation, 334–336
Interdiffusion coefficient, ionic liquid/molten salt physicochemistry, 250–257
Interfacial tension (IFT) measurements
  1-alkyl-3-methylimidazolium chlorides ([$C_n$mim]Cl), 41–46
  hydrophilic ionic liquids in aqueous solutions, 40
Intradiffusion coefficient, ionic liquid/molten salt physicochemistry, 250–257
Iodines. See Alkali metal iodides; Lanthanide triiodides (LaI_3 and NdI_3)
Ion-dependent liquid electrolyte applications, 189
  dual-phase composite electrolytes, fuel cell and electrolysis studies, 408
Ionic conductivity, ionic liquid-polymer composite electrolyte, 372–376
Ionic liquid-polymer composite electrolyte (ICE), lithium ion polymer battery
  combustion testing, 380–382
  conjugated polymer composite electrolyte film, 368–371
  film formation and structure, 377–379
  future research issues, 385–386
  gel electrolyte, film formation method, 371–372
  host polymer selection, 372
  ICE electrolyte application, 377–383
  ICE electrolyte formulation, 376–377
  ionic liquid formulation, 372–376
  lithium gel polymer rechargeable battery applications, 382–383
  microphase separation, TEM analysis, 379–380
  Pionics LIP cell characteristics, 384–385
  PLB cell performance, 383–384
  research background, 368
  simple blend and blend method with graft polymerisation, 371–377
Ionic liquids. See also High-temperature ionic liquids (HTILs); Low-temperature ionic liquids (LTILs)
  biopreservation, 20
  biphasic systems, see Biphasic ionic liquid-ionic liquid systems
  cationic halids, spectroscopic aspects, 12–15
  complex anions, ambient temperature development, 10–12
  fuel cell electrolytes
    inorganic fuel cells, 18–20
    protic ionic liquids, 17–18
  historical perspectives, 2–4

ionic liquid-polymer composite electrolyte, formulation selection, 372–376
liquid electrolytes, basic properties, 182, 188
nucleophilic aromatic substitution, 282–283
physicochemical measurement
  density, 234
  diffusion coefficients, 250–257
  electrical conductivity, 240–246
  electrical permittivity, 246–250
  melting points, 233–234
  principal characteristics, 230–233
  purity characteristics, 234–236
  research background, 230
  thermal conductivity, 234, 240–244
  viscosity, 237–240
protic ionic liquids
  fuel cell electrolytes, 17–18
  physical properties, 15–17
room temperature characterisation, 285–286
  carbon-carbon bond formation, 287–291
  pericyclic reactions, 291–292
  photochemistry, 294–295
  protection-deprotection, 287
  reductions and oxidations, 292–294
spectroscopic analysis
  cationic liquids, 12–15
  overview, 15
strong and fragile compounds, 7–10
viscosity and glass-former phenomenology, 4–7
Ionic surfactants, in aqueous solutions, 38–39
Iron/iron alloys, metallic inert anodes, aluminum electrolysis, 123–131
Isobaric activation energy, liquid electrolytes, pressure effects, 187–188
Isochoric activation, liquid electrolytes, pressure effects, 187–188
Isomerisation, molten salts, 284
Isotropic Raman scattered intensities, Raman spectra, high-temperature melts, 305–307

Kanthal-resistant muffle furnace, aluminum electrolysis, 125–131
Kapustinskii formulations, spent nuclear fuel management, transuranium elements, dichloride/trichloride stability, 269–272
Karl-Fischer analysis
  hydrophilic ionic liquids in aqueous solutions, 40
  metal/semiconductor electrodeposition, experimental protocol, 86–87
Kinetic hindrance, melting compounds (M_3LnX_6), phase transition enthalpies, 346–347
Knudsen effusion mass spectrometry (KEMS), lanthanide binary systems (LnI_3-MI), 350

Lanthanides
  catalyst enhanced molten salt oxidation, 162
  LnI_3-MI binary systems
    heat capacity, 347–350
    Knudsen effusion mass spectrometry (KEMS), 351
    measurements, 344
    phase transition enthalpies, 344–347
    research background, 342
    sample preparation, 342–343
  molten salts thermodynamic properties, partial least squares calculation, 357–363
  molten salt synthesis, ceramic materials, 399, 402–404
  phosphate precipitation, 163–164
Lanthanide triiodides (LaI_3 and NdI_3), thermodynamic analysis, sample preparation, 342–343
Lanthanum aluminate ($LaAlO_3$), molten salt synthesis
  as-received raw materials, 221
  experimental protocols, 220–221
  heating temperature effects, 221–222, 226
  impurity levels, 222–224
  mechanisms, 224–226
  research background, 219–220
  salt/oxide ratios, 222
Laser technology, melt spectra
  container-free melts, macro-Raman spectroscopy, 324–326
  macro-Raman spectroscopy, 315
  near-infrared Raman spectroscopy, 330–332
  Raman experimental protocol, 309–314
  research background, 301–302
  in situ Raman spectroscopy, 333–336
  ultraviolet Raman spectroscopy, 328–330
Latent variables (LVs), molten salts thermodynamic properties, material informatics, 357–363
Lattice structure
  ionic liquids, 182
  liquid electrolytes, solid/vapour relationship, 183
Leaching of salts, molten salt synthesis, ceramic materials, 399
Leave-one-out (LOO) cross-validation, molten salts thermodynamic properties, partial least squares calculation, 359–363
Levitation device, container-free melts, macro-Raman spectroscopy, 324–326
Lewis acids
  liquid electrolytes, 184
  organic cation halides, 12–15
Light scattering, Raman spectroscopy, high-temperature melts, 302–303
Light water reactors (LWRs), spent nuclear fuel management, molten salt physics and chemistry, 265–266

Linewidths, Raman spectra, high-temperature melts, 307–308
Liquid-based electrolyte solutions
 composition, 27–28
 linear plot conductivity, 30–31
 lithium and lithium-ion batteries, 26–27
Liquid cadmium cathode (LCC), spent nuclear fuel management, pyroprocessing technology, 267
Liquid cadmium-lithium cathodes, spent nuclear fuel management, plutonium/americium electrowinning, 272–276
Liquid electrolytes, 182
 applications, 188–189
 conductivity, 185–187
 density, 185
 electrochemistry, 184–185
 formation, 183
 ionic systems, 188
 pressure effects, 187–188
 solids and vapours, 183
 spectroscopy and structure, 184
 theories, 183
Liquid-liquid equilibria (LLE) modeling, water-ionic liquid systems, COSMO-RS thermodynamic analysis, 109–113
 anion identity, 113
 cation alkyl chain length, 111–112
 cation methyl inclusion, 112
 conformer influence, 107–109
 experimental data base, 105
 parameterisation, 105–107
 research background, 102–103
Lithium and lithium alloys
 gelled polymer electrolyte lithium batteries, 34–35
  ionic liquid-polymer composite electrolyte
   combustion testing, 380–382
   conjugated polymer composite electrolyte film, 368–371
   film formation and structure, 377–379
   future research issues, 385–386
   gel electrolyte, film formation method, 371–372
   host polymer selection, 372
   ICE electrolyte application, 377–383
   ICE electrolyte formulation, 376–377
   ionic liquid formulation, 372–376
   lithium gel polymer rechargeable battery applications, 382–383
   microphase separation, TEM analysis, 379–380
   Pionics LIP cell characteristics, 384–385
   PLB cell performance, 383–384
   research background, 368

   simple blend and blend method with graft polymerisation, 371–377
  organic electrolyte solutions, 26–27
  spent nuclear fuel management, plutonium/americium electrowinning, 272–276
Lithium chloride-potassium chloride eutectic samarium-aluminum alloy formation
 cyclic voltammetry study, 66–68
 electrolysis, 69–71
 experimental protocol, 65–66
 goals, 64–65
 open-circuit chronopotentiometry study, 68–69
 thermodynamic properties, 71–73
 spent nuclear fuel management, transuranium elements, dichloride/trichloride stability, 268–272
Lithium fluoride-sodium fluoride-potassium fluoride eutectic, tantalum electrodeposition, 88–90
Lithium nitrates, viscosity measurement, 230–232
Lithium tetrahydroaluminate(III), ionic liquids, reduction reactions, 293–294
Low-temperature ionic liquids (LTILs), physico-chemical measurement
 density, 234
 diffusion coefficients, 250–257
 electrical conductivity, 240–246
 electrical permittivity, 246–250
 melting points, 233–234
 principal characteristics, 230–233
 purity characteristics, 234–236
 research background, 230
 thermal conductivity, 234, 240–244
 viscosity, 237–240
Low-temperature solid oxide fuel cells (LTSOFC), dual-phase composite electrolytes, fuel cell and electrolysis studies, 408

Macro-Raman spectroscopy
 container-free melts, levitation method, 324–326
 molten salts, 314–318
  container design, 315–317
  corrosive and air-sensitive melts, 322–323
  dark coloured melts, 317–318
  furnace design, 315
  laser sources, 315
Madelung energy, liquid electrolytes, solid/vapour relationship, 183
Magnesium-aluminum oxide platelets, molten salt synthesis, ceramic materials, 399, 401–402
Mass spectrometric studies, ionic liquid aggregate behaviour

Cook's kinetic method, hydrogen bond strengths, 55–59
electrospray ionisation mass spectra results, 51–52
experimental protocols, 51
research overview, 50–51
tandem electrospray ionisation mass spectra-mass spectra and breakdown graphs, 52–55
Materials informatics, molten salts chemistry mathematical modelling, 356–357
partial least squares mathematical foundation, 363–365
research background, 356
thermodynamic property prediction, 357–363
Mathematical modelling
catalyst enhanced molten salt oxidation, 160–161
molten salts chemistry, material informatics, 356–357
Melting points
ionic liquid/molten salt physicochemistry, 233–234
ionic liquids, strong and fragile liquids, 8–10
molten salts, 281–282
Mercury halides, liquid electrolytes, pressure effects, 188
Metal block furnace, macro-Raman spectroscopy, high-temperature melts, 315–316
Metal-catalysed reactions, liquid organic salts, 189
Metal electrodepositions, ionic liquid solvents
aluminum electrodeposition, 90–91
experimental protocol, 86–87
research background, 86
selenium electrodeposition, 94–97
silicon electrodeposition, 91–94
tantalum electrodeposition, 87–90
Metal halides, mole fraction, 11–12
Metallic inert anodes, aluminum electrolysis, 123–131
Metallographic optical microscopy, aluminum electrolysis, 127–131
Metal oxide particles
plasma-induced molten salt electrolysis, fine particle formation, 176–177
Raman spectra, high-temperature melts, purification, 313–314
Metal solubility determination, potassium cryolites, alumina solubility analysis, 79–80
tin-electrolyte systems, 82–83
Metal sulfide particles, plasma-induced molten salt electrolysis, fine particle formation, 177
Methyl-trioctylammonium bis(trifluoromethyl-sulfonyl)amide ([N$_{1888}$][NTf$_2$]), biphasic ionic liquid-ionic liquid systems, 144–148

Microphase separated structure, ionic liquid-polymer composite electrolyte formation, 378–379
transmission electron spectroscopic analysis, 379–380
Micro-Raman spectroscopy, high-temperature melts, 318–323
corrosive and air-sensitive melts, 322–323
intermediate temperatures, 320–322
wire-loop technique, 322
Microstructured molybdenum reactors, corrosion resistance, electrodeposition of coatings
boronisation, 197–198, 202–205
molybdenum disilicide, 198, 207–214
coating characterisation, 198–199
hafnium/hafnium diboride coatings, 197, 199–202
research background, 194–196
siliciding of, 198, 205–207
substrate materials, 196
Microwave dielectric spectroscopy, ionic liquid/molten salt physicochemistry, electrical permittivity, 248–250
Minimum energy conformations, water-ionic liquid systems thermodynamics, COSMO-RS model, 107–109
Minor actinides (MAs), spent nuclear fuel management, 265–266
Miscibility gap, biphasic ionic liquid-ionic liquid systems, 143–148
Modulation experiments, in situ Raman spectroscopy, electrode surfaces, high-temperature melts, 334–336
Molar balance
biphasic ionic liquid-ionic liquid systems, 149
liquid electrolytes
conductivity, 186–187
pressure effects, 187–188
LnI$_3$-MI binary systems, phase transition enthalpies, 345–347
phosphate precipitation, molten chlorides, 163–164
Molecular orientation/reorientation, Raman spectra, high-temperature melts, 305–307
Molten carbonates
catalyst enhanced molten salt oxidation, 155–162
fission product separation, 161–162
uranate solubility, 156–159
fuel cell and electrolysis studies, samarium-doped ceria composite, 410–415
molten salt oxidation, 154
phosphate precipitation, molten chlorides, 163–164

Molten salts
  catalyst enhanced molten salt oxidation, spent fuel reprocessing
    ceramic UO_2, 156
    component abundance, 162
    electron transfer, 156
    experimental protocols, 153–155
    fission product precipitation, 162–164
    fission product separation, 161–162
    mathematical modelling, 160–161
    molten chlorides, phosphate precipitation, 163–164
    phosphate precipitation, 163
    plutonium data, 160
    reprocessing cycle, 160
    research background, 151–153
    uranate(VI) solubility, molten carbonates, 156–159
    water treatment, 162–163
  ceramic materials synthesis
    batch mixing and firing, 398–399
    calcium-zirconia powders, 398–401
    characterisation, 399
    lanthanide aluminum powder, 399, 402–404
    leaching, 399
    magnesium-aluminum oxide platelets, 399, 401–402
    research background, 397–398
  coatings, electrodeposition, microstructured molybdenum reactor corrosion resistance
    boronisation, 197–198, 202–205
      molybdenum disilicide, 198, 207–214
    coating characterisation, 198–199
    hafnium/hafnium diboride coatings, 197, 199–202
    research background, 194–196
    siliciding of, 198, 205–207
    substrate materials, 196
  high-temperature compounds
    aluminum(III) chloride salts, 283–284
    isomerisation, 284
    melting points, 281–282
    nucleophilic aromatic substitution, 282–283
    pyridinium chloride, 285
    Raman spectroscopy
      anharmonicity and temperature frequency shifts, 308
      band linewidths and bandshapes, 307–308
      blackbody radiation, 310–311
      container-free melts, 323–328
        levitation method, 324–326
        pyrometery, 326
        self-support method, 327–328
        temperature estimation, 326
      experimental protocols, 308–309
      light scattering, 302–303
      macro-Raman spectroscopy, 314–318
        container design, 315–317
        dark coloured melts, 317–318
        furnace design, 315
        laser sources, 315
      materials purity and preparation, 311–314
      micro-Raman spectroscopy, 318–323
        corrosive and air-sensitive melts, 322–323
        intermediate temperatures, 320–322
        wire-loop technique, 322
      near-infrared spectroscopy, 330–332
      research backgrounds, 301–302
      selection rules and band intensities, 303–305
      in situ spectra, electrode surfaces, 333–336
        constant potential, 333–334
        modulation, 334–336
      spectral intensity, 305–307
      ultraviolet spectroscopy, 328–330
    research background, 279–281
  historical perspective, 3–4
  lanthanum aluminate synthesis
    as-received raw materials, 221
    experimental protocols, 220–221
    heating temperature effects, 221–222, 226
    impurity levels, 222–224
    mechanisms, 224–226
    research background, 219–220
    salt/oxide ratios, 222
  liquid electrolytes, 182
    solids and vapours, 183
    spectroscopy and structure, 184
  materials informatics
    mathematical modelling, 356–357
    partial least squares mathematical foundation, 363–365
    research background, 356
    thermodynamic property prediction, 357–363
  oxidation, historical perspective, 152–153
  photochemistry, 294–295
  physicochemical measurement
    density, 234
    diffusion coefficients, 250–257
    electrical conductivity, 240–246
    electrical permittivity, 246–250
    melting points, 233–234
    principal characteristics, 230–233
    purity characteristics, 234–236
    research background, 230

thermal conductivity, 234, 240–244
viscosity, 237–240
plasma-induced electrolysis, fine particle formation
anodic discharge, 175–179
basic principles, 170
bench scale experiments, 173–175
carbon nitride particles, 177–179
cathodic discharge, 170–175
metal oxide particles, 176–177
metal sulfide particles, 177
research background, 169–170
sodium chloride/potassium chloride systems, aluminum oxide solubility
additives, effect of, 394
cryolite addition, 391–394
measurement techniques, 391
raw materials, 390
research background, 389–390
thermal physical properties, 391
spent nuclear fuel management
pryoprocess research, 266–267
research background, 265–266
transuranium elements electrorefining, chemical thermodynamics, 267–276
dichloride/trichloride relative stability, 268–272
plutonium/americium electrowinning, liquid cadmium cathodes, 272–276
Molybdenum
anodes, plasma-induced molten salt electrolysis, fine particle formation, 177
corrosion resistance, electrodeposition of coatings
boronisation, 197–198, 202–205
molybdenum disilicide, 198, 207–214
coating characterisation, 198–199
hafnium/hafnium diboride coatings, 197, 199–202
research background, 194–196
siliciding of, 198, 205–207
substrate materials, 196
Molybdenum diboride, microstructured molybdenum reactor corrosion resistance, 212–214
Molybdenum disilicide, microstructured molybdenum reactor corrosion resistance
boronising, 207–214
research background, 195–196
Mukaiyama aldol condensation, ionic liquids, carbon-carbon bond formation, 288–291
Mutual diffusivity, ionic liquid/molten salt physicochemistry, 254–257

Near-infrared (IR) Raman spectroscopy, high-temperature melts, 330–332

Neutron diffraction, liquid electrolytes, 184
Nickel/nickel alloys
aluminum electrolysis, 127–131
liquid electrolyte structure, 184
plasma-induced molten salt electrolysis, fine particle formation, titanium bench scale experiments, 174–175
pyridinium chloride concentrations, 12–13
Nitrate
catalyst enhanced molten salt oxidation, 155–162
molten salt oxidation, 154–155
Nitrite
catalyst enhanced molten salt oxidation, 155–162
molten salt oxidation, 154–155
Nitrogen cations, hydrophilic ionic liquids in aqueous solutions, 45–46
Noble metals. *See also* Rare earth halides
aluminum electrolysis, 127–131
catalyst enhanced molten salt oxidation, fission product separation, 162
Noncurrent transfer, microstructured molybdenum reactor corrosion resistance, molybdenum disilicide, boronisation, 208–214
Noniterative partial least squares (NIPALS), molten salts thermodynamic properties, partial least squares calculation, 364–365
Non-Linear Least-Square (NLLSQ) fit software, gelled polymer electrolyte systems, 29–30
Nonlinear relaxation, organic cation halides, spectroscopic analysis, 14–15
Nonpolar organic solvents, biphasic ionic liquid-ionic liquid systems, 146–148
Nonrandom two-liquid (NRTL) model, water-ionic liquid systems, COSMO-RS thermodynamic analysis, 102–103
Nuclear magnetic resonance (NMR) spectroscopy. *See also* Carbon NMR spectroscopy (^{13}C NMR); Proton NMR spectroscopy (^{1}H NMR)
biphasic ionic liquid-ionic liquid systems
partition coefficient determination, 148–149
phase composition determination, 148
Nucleophilic aromatic substitution, molten salts, 282–283

OCP. *See* Open-circuit chronopotentiometry
Oct-1-ene, biphasic ionic liquid-ionic liquid systems, 146–148
Olefins, biphasic ionic liquid-ionic liquid systems, 145–148
Open-circuit chronopotentiometry (OCP), samarium-aluminum alloy formation, lithium chloride-potassium chloride eutectic, 68–69

Ordinary least squares (OLS), molten salts thermodynamic properties, partial least squares calculation, 363–365
Organic cation halides, spectroscopic analysis, 12–15
Organic electrolyte solutions
 gelled polymer systems, 28–30
 conductivity plots, 31–35
 liquid-based solutions, 27–28
 conductivity plots, 30–31
 lithium and lithium-ion batteries, 26–27
Organic salts. *See also* Molten salts
 applications, 189
 ionic systems, 188
Organic solvents
 ionic liquid-polymer composite electrolyte, lithium ion batteries, combustion testing, 381–383
 metal/semiconductor electrodeposition, aluminum/aluminum alloys, 90–91
Orthogonality constraint, molten salts chemistry, material informatics, mathematical modelling, 357
Outokumpu program (HSC Chemistry for Windows), cryolite-alumina melts, phosphorus thermodynamics, 134–140
Oxidation kinetics
 ionic liquids, organic reactions, 292–294
 microstructured molybdenum reactor corrosion resistance
 boronising molybdenum, 204–205
 molten salt coating electrodeposition, 196
 molybdenum disilicide boronisation, 209–214
 siliciding molybdenum, 206–207
Oxygen ion conduction, dual-phase composite electrolytes, 411–416

Palladium-catalysed reactions, ionic liquids, 290–291
Parameterisation, water-ionic liquid systems, COSMO-RS thermodynamic analysis, 105–107
Partial least squares (PLS) analysis, molten salts chemistry, material informatics
 mathematical modelling, 356–357
 partial least squares mathematical foundation, 363–365
 research background, 356
 thermodynamic property prediction, 357–363
Particle size
 lanthanum aluminate, molten salt synthesis, salt/oxide ratios, 226

plasma-induced molten salt electrolysis, fine particle formation, 172–175
Partition coefficients, biphasic ionic liquid-ionic liquid systems, NMR determination, 148–149
Partitioning and transmutation (P&T) strategy, radionuclide management
 cyclic voltammetry study, 66–68
 electrolysis, 69–71
 experimental protocol, 65–66
 goals, 64–65
 open-circuit chronopotentiometry study, 68–69
 thermodynamic properties, 71–73
PelLicle® lithium ion battery, 381–383
"Pendant drop" technique, container-free melts, macro-Raman spectroscopy, 327–328
1,1,1,3,3-Pentafluoropropane, ionic liquid/molten salt physicochemistry, electrical permittivity, 248–250
Pericyclic reactions, ionic liquids, 291–292
Peroxide (Na_2O_2)
 catalyst enhanced molten salt oxidation, 155–162
 molten salt oxidation, 152–155
Peroxycarbonate (Na_2CO_4), molten salt oxidation, 152–155
Peroxydicarbonate ($Na_2C_2O_6$), molten salt oxidation, 152–153
Pesting, microstructured molybdenum reactor corrosion resistance
 molten salt coating electrodeposition, 195–196
 siliciding molybdenum, 206–207
Phase array, microstructured molybdenum reactor corrosion resistance, molybdenum disilicide boronisation, 209–214
Phase discrimination, biphasic ionic liquid-ionic liquid systems, 144–148
Phase equilibria prediction, water-ionic liquid systems thermodynamics, COSMO-RS technique, 103–104, 118
 liquid-liquid equilibria modeling, cation alkyl chain length, 111–112
 vapour-liquid equilibria modeling, 113–116
Phase stability diagrams, cryolite-alumina melts, phosphorus/sodium thermodynamics, 135–140
Phase transition enthalpy, LnI_3-MI binary systems, 344–347
Phosphate precipitation, catalyst enhanced molten salt oxidation, 163
 molten chlorides, 163–164
Phosphoric acid electrolyte, spectroscopic analysis, 18–20
Phosphorus, cryolite-alumina melts, thermodynamic behaviour, 133–140
Photochemistry, ionic liquids, 294–295

INDEX **435**

Photomultiplier tubes (PMT), Raman spectroscopy,
  high-temperature melts, 309–314
  macro-Raman spectroscopy, 314–318
  in situ Raman spectroscopy, electrode surfaces,
    334–336
Photon-correlation spectroscopy, ionic liquid/
  molten salt thermal conductivity, 241–244
Photovoltaic technology, selenium metal/semiconductor electrodeposition, 95–97
Pionics' lithium battery (PLB)
  cell performance, 383–384
  combustion testing, 380–383
  future research issues, 385–386
  summary of characteristics, 384–385
Planck's constant, Raman spectra,
  high-temperature melts, 307
Plasma-induced molten salt electrolysis, fine
  particle formation
  anodic discharge, 175–179
  basic principles, 170
  bench scale experiments, 173–175
  carbon nitride particles, 177–179
  cathodic discharge, 170–175
  metal oxide particles, 176–177
  metal sulfide particles, 177
  research background, 169–170
Platinum-platinum-rhodium thermocouple design
  container-free melts, macro-Raman spectroscopy, 327–328
  potassium cryolites, alumina solubility
    analysis, electrical conductivity
    experiments, 77–78
Plutonium, spent nuclear fuel management
  electrowinning
  current density limits, 273–276
  liquid cadmium cathode, 272–276
  transuranium elements, dichloride/trichloride
    stability, 269–272
Plutonium oxide, catalyst enhanced molten salt
  oxidation, 160
  mathematical modelling, 160–161
  reprocessing cycle, 160
Polarisability tensor, Raman spectra, high-
  temperature melts, 303–305
  symmetric/asymmetric parts, 305–307
Polarisation technique, ionic liquid/molten salt
  thermal conductivity, 242–244
Polar liquids, liquid electrolytes, pressure effects,
  188
Polymerised ionic liquids (PILs), ionic liquid-
  polymer composite electrolyte, 372–377
"Polymer-like" G^E model, water-ionic liquid
  systems, COSMO-RS thermodynamic
  analysis, 102–103

Poly(tetrafluoroethylene) (PTFE), ionic liquid-
  polymer composite electrolyte, gelled
  polymer electrolyte lithium batteries, research
  background, 368
Poly(vinylidene fluoride) (PVdF), ionic liquid-
  polymer composite electrolyte
  composite materials technology, 369–371
  film formation process, 371–372
  gelled polymer electrolyte lithium batteries,
    research background, 368
  polymerised ionic liquid formulation, 372–377
  tensile strength properties, 372
Potassium chlorate(VII), enhanced molten salt
  oxidation, 155
Potassium chloride (KCl)
  ionic liquid/molten salt physicochemistry,
    electrical conductivity, 245–246
  ionic liquids and, 182
Potassium cryolites, alumina solubility
  electrical conductivity, 76–78, 80–81
  electrolyte preparation, 76
  metal solubility determination, 79–80, 82–83
  potentiometric titration analysis, 78–79, 81–82
  temperature-dependent properties, 75–76
Potassium melting compounds (M_3LnX_6)
  heat capacity, 348–350
  phase transition enthalpies, 345–347
Potassium nitrate, viscosity measurement, 230–232
Potentiometric titration technique, potassium
  cryolites, alumina solubility analysis
  cell design and characteristics, 78–79
  temperature dependence, 81–82
Pressure effects, liquid electrolytes, 187–188
Primary measurement techniques, ionic liquid/
  molten salt physicochemistry, viscosity measurements, 237–238
Principal component analysis (PCA), molten salts
  thermodynamic properties, 357–363
Protection-deprotection, ionic liquids, 287
Protic ionic liquids
  fuel cell electrolytes, 17–18
  dual-phase composite electrolytes, 411–416
  physical properties, 15–17
Proton affinity, ionic liquid aggregate behaviour,
  Cook's kinetic method hydrogen bond
  strengths, 57–59
Proton NMR spectroscopy (^{1}H NMR)
  biphasic ionic liquid-ionic liquid systems, 144–148
  molar balance determination, 149
  partition coefficient determination, 148–149
  phase composition determination, 148
  hydrophilic ionic liquids in aqueous solutions, 40
  CMC values, 41–46
  interfacial tension measurements, 41

Proton NMR spectroscopy (*Continued*)
  ionic liquids, 15
  liquid electrolytes, 184
Pseudo-binary approach, water-ionic liquid systems thermodynamics, COSMO-RS model, 104
Pseudo-Madelung energy, ionic liquids, 182
Purity standards
  ionic liquid/molten salt physicochemistry, 234
  Raman spectra, high-temperature melts, 311–314
PVdF-HFP, gelled polymer electrolyte systems, 28–30
Pyrenes, hydrophilic ionic liquids, aqueous solutions, CMC values, 42–46
Pyridinium cations, organic reactions, 286
Pyridinium chloride, in molten salts, 285
Pyrolytic boron nitride (PBN) cells, potassium cryolites, alumina solubility analysis, electrical conductivity experiments, 77–78
Pyrometric equipment, container-free melts, macro-Raman spectroscopy, 326
Pyroprocessing technology
  radionuclide management, 64–65
  spent nuclear fuel management, 266–267

Quantum chemistry, water-ionic liquid systems thermodynamics, COSMO-RS model, 103–104
  basis set parameterisation, 107
Quantum mechanics, Raman spectra, high-temperature melts, 304–305
Quasi-harmonic model, Raman spectra, high-temperature melts, anharmonicty temperature-based frequency shifts, 308

Radial distribution functions (RDFs), liquid electrolytes, 184
Radionuclides, partitioning and transmutation strategy
  cyclic voltammetry study, 66–68
  electrolysis, 69–71
  experimental protocol, 65–66
  goals, 64–65
  open-circuit chronopotentiometry study, 68–69
  thermodynamic properties, 71–73
Raman spectroscopy
  liquid electrolytes, 184
  molten salts
    anharmonicity and temperature frequency shifts, 308
    band linewidths and bandshapes, 307–308
    blackbody radiation, 310–311
    container-free melts, 323–328
      levitation method, 324–326
      pyrometry, 326
      self-support method, 327–328
      temperature estimation, 326
    experimental protocols, 308–309
    light scattering, 302–303
    macro-Raman spectroscopy, 314–318
      container design, 315–317
      dark coloured melts, 317–318
      furnace design, 315
      laser sources, 315
    materials purity and preparation, 311–314
    micro-Raman spectroscopy, 318–323
      corrosive and air-sensitive melts, 322–323
      intermediate temperatures, 320–322
      wire-loop technique, 322
    near-infrared spectroscopy, 330–332
    research backgrounds, 301–302
    selection rules and band intensities, 303–305
    in situ spectra, electrode surfaces, 333–336
      constant potential, 333–334
      modulation, 334–336
    spectral intensity, 305–307
    ultraviolet spectroscopy, 328–330
Raoult's law, water-ionic liquid systems thermodynamics, COSMO-RS technique, vapour-liquid equilibria modeling, 113, 116
Rare earth halides. *See also* Noble metals
  LnI$_3$-MI binary systems
    heat capacity, 347–350
    Knudsen effusion mass spectrometry (KEMS), 351
    measurements, 344
    phase transition enthalpies, 344–347
    research background, 342
    sample preparation, 342–343
  molten salts thermodynamic properties, partial least squares calculation, 358–363
  Raman spectra, high-temperature melts, purification, 312–314
  spent nuclear fuel management, chemical thermodynamic techniques, 268–276
Rayleigh scattering, ionic liquid/molten salt thermal conductivity, 241–244
"Rayleigh" wing, Raman spectroscopy, high-temperature melts, blackbody radiation, 310–311
Reduced representation, Raman spectra, high-temperature melts, 307
Reduction reactions, ionic liquids, 292–294
Reprocessing cycle, catalyst enhanced molten salt oxidation, 160

RE/TRU atomic ratio, spent nuclear fuel management, chemical thermodynamic techniques, 268–276
Room-temperature electroplated aluminum (REAL), metal/semiconductor electrodeposition, aluminum/aluminum alloys, 90–91
Root-mean-square error of cross-validation (RMSECV), molten salts thermodynamic properties, partial least squares calculation, 357–363
Rubidium melting compounds (M_3LnX_6)
  heat capacity, 348–350
  phase transition enthalpies, 345–347

Salt/oxide ratios, lanthanum aluminate, molten salt synthesis, 222
  heating temperatures and, 226
Samarium-aluminum alloy formation, lithium chloride-potassium chloride eutectic
  cyclic voltammetry study, 66–68
  electrolysis, 69–71
  experimental protocol, 65–66
  goals, 64–65
  open-circuit chronopotentiometry study, 68–69
  thermodynamic properties, 71–73
Samarium-doped ceria (SDC), dual-phase composite electrolytes
  fuel cell and electrolysis studies, 408–409
  hydrogen/oxygen ion conduction, 415–415
  structural characteristics, 410–415
Scanning electron microscopy (SEM)
  lanthanum aluminate, molten salt synthesis, salt/oxide ratios, 222
  metal/semiconductor electrodeposition
    aluminum/aluminum alloys, 91–92
    selenium, 96–97
    silicon, 92–94
  microstructured molybdenum reactor corrosion resistance
    boronising molybdenum, 202–205
    hafnium/hafnium diboride coatings, 199–202
    siliciding molybdenum, 198, 205–207
  molten salt synthesis, ceramic materials, 401–402
  plasma-induced molten salt electrolysis, fine particle formation, 171–175
Scanning tunneling microscopy (STM), metal/semiconductor electrodeposition, silicon, 93–94
Scholl reaction, aluminum(III) chloride molten salts, 283–284
Second-order Raman effect, high-temperature melt spectroscopy, 304–305
  reduced representation, 307

Selection rule, Raman spectroscopy, high-temperature melts, 304–305
Selenium, metal/semiconductor electrodeposition, 95–97
Self-aggregation, 1-alkyl-3-methylimidazolium chlorides ($[C_n mim]Cl$), 41–46
Self-diffusion coefficient, ionic liquid/molten salt physicochemistry, 250–257
Self-support techniques, container-free melts, macro-Raman spectroscopy, 327–328
SEM-EDX analysis, samarium-aluminum alloy formation, lithium chloride-potassium chloride eutectic, 69–71
Semiconductor electrodeposition, ionic liquid solvents
  aluminum electrodeposition, 90–91
  experimental protocol, 86–87
  research background, 86
  selenium electrodeposition, 94–97
  silicon electrodeposition, 91–94
  tantalum electrodeposition, 87–90
Siemens galvano-aluminum (SIGAL) processes, metal/semiconductor electrodeposition, aluminum/aluminum alloys, 90–91
Signal saturation, Raman spectra, high-temperature melts, 312–314
Siliciding treatments, microstructured molybdenum reactor corrosion resistance, molybdenum, 198, 205–207
Silicon
  metal/semiconductor electrodeposition, 91–94
  microstructured molybdenum reactor corrosion resistance, 207–214
Silver chloride/silver reference electrodes, samarium-aluminum alloy formation, lithium chloride-potassium chloride eutectic
  electrolysis, 69–71
  thermodynamic analysis, 71–73
Silver ion-catalysed isomerisation, molten salts, 284
SIMPLS algorithm, molten salts thermodynamic properties, partial least squares calculation, 358–365
Sodium, cryolite-alumina melts, thermodynamic behaviour, 133–140
Sodium chloride/potassium chloride molten salts, aluminum oxide solubility
  additives, effect of, 394
  cryolite addition, 391–394
  measurement techniques, 391
  raw materials, 390
  research background, 389–390
  thermal physical properties, 391
Sodium sulfate, molten salt oxidation, 153

Solid oxide fuel cells (SOFC), dual-phase composite electrolytes, fuel cell and electrolysis studies, 408
Solids
  ionic liquid-polymer composite electrolyte, 373–377
  liquid electrolytes and, 183
Solid state reactions, melting compounds (M_3LnX_6), phase transition enthalpies, 346–347
Solubility parameters, ionic liquids, 285–286
Solvatophobic interactions, hydrophilic ionic liquids, 38–39
  fluorescence spectroscopic measurements, 42–46
Solvent polarity scale (I3/I1), hydrophilic ionic liquids in aqueous solutions, 41
  fluorescence measurements, 42–46
Sonogashira reaction, ionic liquids, 290–291
Specific conductivity, liquid electrolytes, 186–187
Spectroscopic analysis, liquid electrolytes, 184
Spent nuclear fuel management
  catalyst enhanced molten salt oxidation
    ceramic UO_2, 156
    component abundance, 162
    electron transfer, 156
    experimental protocols, 153–155
    fission product precipitation, 162–164
    fission product separation, 161–162
    mathematical modelling, 160–161
    molten chlorides, phosphate precipitation, 163–164
    phosphate precipitation, 163
    plutonium data, 160
    reprocessing cycle, 160
    research background, 151–153
    uranate(VI) solubility, molten carbonates, 156–159
    water treatment, 162–163
  molten salt physics and chemistry
    pryprocess research, 266–267
    research background, 265–266
    transuranium elements electrorefining, chemical thermodynamics, 267–276
      dichloride/trichloride relative stability, 268–272
      plutonium/americium electrowinning, liquid cadmium cathodes, 272–276
Steady-state technique, ionic liquid/molten salt thermal conductivity, 242–244
Steel alloy anodes, aluminum electrolysis, 126–131
Steric hindrance, hydrophilic ionic liquids in aqueous solutions, 45–46

Stern layer, hydrophilic ionic liquids in aqueous solutions, 45–46
Stilbene compounds, photochemistry, 295
Stokes's law
  gelled polymer electrolyte systems, 32–35
  Raman spectra, high-temperature melts, 303
    blackbody radiation, 311
    container-free melts, macro-Raman spectroscopy, 326
    reduced representation, 307
    selection rule, 304–305
    ultraviolet Raman spectroscopy, 328–330
Strontium
  catalyst enhanced molten salt oxidation, 162
  phosphate precipitation, 164
"Structural disorder," melting compounds (M_3LnX_6), heat capacity, 350
Substrate materials, microstructured molybdenum reactor corrosion resistance, molten salt coating electrodeposition, 195–196
Supercritical temperatures, liquid electrolytes, pressure effects, 187–188
Superionic liquids, historical perspective, 3
Superoxide (NaO_2)
  catalyst enhanced molten salt oxidation, 155–162
  mathematical modelling, 160–161
  molten salt oxidation, 152–1535
Superprotonic electrolytes
  historical perspective, 3
  spectroscopic analysis, 18–20
Surface tension, ionic liquid/molten salt physicochemistry, 234
Suzuki reaction, ionic liquids, 290–291
Synthesis mechanism, lanthanum aluminate, molten salt synthesis, 224–226

Tafel plot, inorganic fuel cell electrolytes, 18–20
Tandem electrospray ionisation mass spectrometry (ESI-MS-MS)
  aggregate behaviour, ionic liquids
    breakdown graphs, 52–55
    Cook's kinetic method hydrogen bond strengths, 57–59
    basic principles, 50–51
Tantalum
  metal/semiconductor electrodeposition, 87–90
  plasma-induced molten salt electrolysis, fine particle formation, 171–175
Taylor diffusion technique, ionic liquid/molten salt physicochemistry, 240–257
Temperature effects
  gelled polymer electrolyte systems, 32–35

lanthanum aluminate, molten salt synthesis,
heating temperature effects, 221–222, 226
liquid electrolyte conductivity, 187
LnI$_3$-MI binary systems, 344
metals and semiconductor electrodeposition,
ionic liquid solvents
aluminum electrodeposition, 90–91
experimental protocol, 86–87
research background, 86
selenium electrodeposition, 94–97
silicon electrodeposition, 91–94
tantatlum electrodeposition, 87–90
microstructured molybdenum reactor corrosion
resistance, molybdenum disilicide boronisation, 209–214
potassium cryolites, alumina solubility analysis
electrical conductivity, 80–82
metal solubility, 82–83
potentiometric titration, 81–82
Raman spectra, high-temperature melts
anharmonicity and frequency shifts, 308
container-free melts, macro-Raman spectroscopy, 326
micro-Raman spectroscopy, intermediate
temperatures, 320–322
wire-loop technique, temperatures to 2000 K, 322
sodium chloride/potassium chloride molten
salts, aluminum solubility in, 392–394
water-ionic liquid systems thermodynamics,
COSMO-RS technique, vapour-liquid
equilibria modeling, 116–118
Temperature gradients, ionic liquid/molten salt
thermal conductivity, 240–244
Template formation, molten salt synthesis, ceramic
materials, 401
Tensile strength, ionic liquid-polymer composite
electrolyte
ionic liquid formulation, 372–376
poly(vinylidene fluoride) properties, 369–371
Tetraalkylammonium, organic reactions, 286
Tetraalkylphosphonium, organic reactions, 286
Tetrachloroferrate(III) salt, complex anions, 10–12
Tetrahydrofuran, biphasic ionic liquid-ionic liquid
systems, 145–148
Tetrapropylammonium perruthenate, ionic liquids,
oxidation reactions, 293–294
Thermal conductivity, ionic liquid/molten salt
physicochemistry, 234, 240–244
Thermal population, Raman spectra, high-temperature melts, 306–307
Thermodynamic analysis
cryolite-alumina melts, phosphorus and sulfur
behaviour, 133–140

LnI$_3$-MI binary systems
heat capacity, 347–350
Knudsen effusion mass spectrometry
(KEMS), 351
measurements, 344
phase transition enthalpies, 344–347
research background, 342
sample preparation, 342–343
molten salts chemistry, material informatics,
357–363
samarium-aluminum alloy formation, lithium
chloride-potassium chloride eutectic,
71–73
spent nuclear fuel management, transuranium
elements electrorefining, 267–276
water-ionic liquid systems, COSMO-RS
conformers influence, 107–109
liquid-liquid equilibria modeling, 109–113
anion identity, 113
cation alkyl chain length, 111–112
cation methyl inclusion, 112
LLE/VLE experimental database, 105
parameterisation calculations, 105–107
phase equilibria prediction, 103–104, 118
research background, 102–103
vapour-liquid equilibria modeling, 113–116
anion identity, 115–116
cation alkyl chain length, 113–116
temperature dependence, 116–118
Titanium, plasma-induced molten salt electrolysis,
fine particle formation, 171–175
anodic discharge, 176–179
bench scale experiments, 173–175
Toluene, biphasic ionic liquid-ionic liquid systems,
146–148
Transient hot-wire technique, ionic liquid/molten
salt thermal conductivity, 241–244
Transmission electron microscopy (TEM), ionic
liquid-polymer composite electrolyte
formation, 379–380
Transport numbers, gelled polymer electrolyte
systems, 32–35
Transuranium (TRU) elements, spent nuclear fuel
management
chemical thermodynamic techniques, 267–276
dichloride/trichloride relative stability, 268–272
pyroprocessing technology, 266–267
research background, 265–266
Trialkylsulfonium cations, organic reactions,
286
Tributyl phosphate (TBP), spent nuclear fuel
management, 265–266
Triflates, ambient temperature water-stable ionic
liquids, 12

Triphasic ionic liquid systems, nonpolar organic solvents, 146–148
Triple-ζ valence polarised large basis set, water-ionic liquid systems, COSMO-RS thermodynamic analysis, liquid-liquid equilibria modeling, 109–113
Tungsten compounds, samarium-aluminum alloy formation, lithium chloride-potassium chloride eutectic, cyclic voltammetry study, 67–68
Tyndall scattering, Raman spectra, high-temperature melts, 312

Ultralarge-scale integrated (ULSI) circuits, tantalum electrodeposition, 87–90
Ultraviolet Raman spectroscopy, high-temperature melts, 328–330
Underpotential deposition (UPD), metal/semiconductor electrodeposition, aluminum/aluminum alloys, 91–92
UNIFAC equation, water-ionic liquid systems, COSMO-RS thermodynamic analysis, 102–103
UNIQUAC equation, water-ionic liquid systems, COSMO-RS thermodynamic analysis, 102–103
Upper solution temperature, biphasic ionic liquid-ionic liquid systems, 145–148
determination of, 149
Uranates
catalyst enhanced molten salt oxidation
fission product separation, 162
molten carbonates, solubility in, 156–159
enhanced molten salt oxidation, 154–155
Uranium oxide (UO_2)
catalyst enhanced molten salt oxidation
complete oxidation conditions, 156
mathematical modelling, 160–161
reprocessing cycle, 160
enhanced molten salt oxidation, 154–155

v^4 Law, near-infrared Raman spectroscopy, 332
Vacuum capacitance, ionic liquid/molten salt physicochemistry, electrical permittivity, 247–250
van der Waals interactions, fluorinated anion salts, 11–12
Vapour-liquid equilibria (VLE) modeling, water-ionic liquid systems, COSMO-RS thermodynamic analysis
anion identity, 115–116
cation alkyl chain length, 113–116
conformer influence, 107–109
experimental data base, 105
parameterisation, 105–107

research background, 102–103
temperature dependence, 116–118
Vapours, liquid electrolytes and, 183
Vibrational relaxation
near-infrared Raman spectroscopy, 332
Raman spectra, high-temperature melts, 308
Virtual combinatorial experiments, molten salts chemistry, material informatics, 356
Viscosity
ionic liquid/molten salt physicochemistry, 234–235, 237–240
ionic liquids, 4–7
liquid-based electrolyte solutions, conductivity plots, 31
Vogel-Tammann-Fulcher (VTF) equation
gelled polymer electrolyte systems, 31–35
glass-forming liquids, 5–7
liquid-based electrolyte solutions, conductivity plots, 31
liquid electrolyte conductivity, 187

Walden rule
gelled polymer electrolyte systems, 33–35
molten salts, 3
superprotonic electrolytes, 18–20
Water-ionic liquid systems, COSMO-RS thermodynamic analysis
conformers influence, 107–109
liquid-liquid equilibria modeling, 109–113
anion identity, 113
cation alkyl chain length, 111–112
cation methyl inclusion, 112
LLE/VLE experimental database, 105
parameterisation calculations, 105–107
phase equilibria prediction, 103–104, 118
research background, 102–103
vapour-liquid equilibria modeling, 113–116
anion identity, 115–116
cation alkyl chain length, 113–116
temperature dependence, 116–118
Water treatment, fission products, catalyst enhanced molten salt oxidation, 162–163
Wavenumber calculations, Raman spectroscopy, high-temperature melts, 303
Wet chemical synthesis, molten salt synthesis, ceramic materials, 398
Wilson equation, water-ionic liquid systems, COSMO-RS thermodynamic analysis, 102–103
Wire-loop technique, Raman spectra, high-temperature melts, 322
Wittig reaction, ionic liquids, carbon-carbon bond formation, 288–291